Molecular Approaches for the Detection of Fungal Phytopathogens

This book discusses recent [illegible] [illegible] molecular diagnostics of fungal phytopathogens. The techniques [illegible] [illegible] help in [illegible] the importance of early detection of fungal pathogens [illegible] [illegible] stages of crop production. *Molecular Approaches for* [illegible] [illegible] the fungal identification and isolation from different [illegible] [illegible] molecular approaches. These include variations of polymerase [illegible] [illegible] (genus and species) and gene-specific primers, RNA [illegible] [illegible] microarrays, serial analysis of gene expression, fluorescence [illegible] [illegible] amplification, macroarrays, PCR-denaturing gradient gel [illegible] magnetic capture [illegible] PCR, nanotechnology, and immunoassays provide [illegible] [illegible] traditional methods that overcome challenges [illegible] [illegible] control strategies.

Features:

- Explores various diagnostic [illegible] phytopathogens.
- Discusses early detection [illegible] fungal spores [illegible] are no visible symptoms.
- Provides a balanced [illegible] compares the traditional methods of detection with the current molecular techniques.
- Describes details of the [illegible] technique mentioned.
- Covers point-of-care [illegible] [illegible] for molecular testing in plant health.

Molecular Approaches for the Detection of Fungal Phytopathogens

Edited by

Deepa Nagaraju

Sreenivasa Marikunte Yanjarappa

Charith Raj Adkar Purushothama

CRC Press
Taylor & Francis Group
Boca Raton London New York

CRC Press is an imprint of the
Taylor & Francis Group, an **informa** business

Designed cover image: ShutterStock Images

First edition published 2025
by CRC Press
2385 NW Executive Center Drive, Suite 320, Boca Raton FL 33431

and by CRC Press
4 Park Square, Milton Park, Abingdon, Oxon, OX14 4RN

CRC Press is an imprint of Taylor & Francis Group, LLC

ISBN: 978-1-032-50395-0 (hbk)
ISBN: 978-1-032-50796-5 (pbk)
ISBN: 978-1-003-39970-4 (ebk)

DOI: 10.1201/9781003399704

Typeset in Times
by SPi Technologies India Pvt Ltd (Straive)

Contents

Preface

Molecular Approaches for the Detection of Fungal Phytopathogens mainly deals with recent updates and molecular advancements in the field of early detection of fungal phytopathogens. This book aids students and researchers in helping farmers in both post- and pre-harvest stages of crop production fight phytopathogens and their toxins from affecting human and animal health. Phytopathogen growth occurs because of poor agricultural practices that increase the risk of toxin production in crops entering the food chain. Molecular methods may even be used when no visible symptoms are observed or with low concentrations of fungal spores; the methods may be applied to different plant parts, which is not possible through conventional methods.

The book describes the various molecular-based techniques available for the detection of fungal phytopathogens to meet the industry's need for quick results like PCR and its types but even when phytopathogen attack is not minimized. Detection of such phytopathogenic fungi using conventional methods is laborious and time consuming. Current advanced molecular methods like variants of polymerase chain reaction—such as n-PCR, m-PCR, Q-PCR, restriction fragment length polymorphism (RFLP), randomly amplified polymorphic DNA (RAPD), microarray involving gene or species-specific primers, DNA barcoding, RNA-sequence-based next-generation sequence, serial analysis of gene expression (SAGE), fluorescence in-situ hybridization (FISH), loop-mediated isothermal amplification (LAMP), macroarray, PCR-DGGE, MCH-PCR, in-situ hybridization, aptamers, serological methods, nanotechnology and DNA barcoding—provide better approaches than traditional methods. To overcome the challenges, this book explains novel disease control strategies such as early detection of the pathogen by on-site testing for phytopathogens, and the institution of certain policies and regulations for sustainable agriculture and agri-food industries, which are also more labour intensive but less time consuming.

Deepa Nagaraju, Sreenivasa Marikunte Yanjarappa, and Charith Raj Adkar Purushothama

Editors

Deepa Nagaraju is a research associate for the Indian Council of Medical Research's (ICMR) funded project, "Potential Antifungals from PGPR Bacteria against Fumonisin" in the Department of Studies in Microbiology, University of Mysore, Mysuru, India. Her research interests include molecular mycology and myco-toxicology and its prevention. She earned her doctoral degree from the University of Mysore in 2018 with her thesis, "Development of Nested and Multiplex PCR for the Early Detection of Fumonisin-Producing *Fusarium verticillioides* Associated with Cereals". She worked as a project assistant for 3 years on the DST-Fastrack-funded project, as well as a senior research fellow for ICMR funding for 3 years. She served on the faculty for postgraduate students for 4 years in the Department of Studies in Microbiology, University of Mysore. She has authored 31 scientific papers, including five book chapters and a book.

Sreenivasa Marikunte Yanjarappa is a professor of Microbiology and has more than 20 years of teaching experience at the Department of Studies in Microbiology, University of Mysore, Mysuru, India. His research interests include mycology, myco-toxicology, probiotics and postbiotics. In recognition of his research output, he was awarded the Raman post-doctoral fellow position to carry out his post-doctoral studies at the Department of Plant Pathology, University of Arkansas, USA. In recognition of the Indo-US Research collaboration, he was awarded the International Faculty Development award by the Kennesaw State University, Kennesaw, USA. He was the Visiting Research Scientist at the University of Ottawa, Canada under the Shastri Indo-Canadian Institute, India. He has successfully completed seven research projects in the field of microbiology funded by national and international agencies. He has an Indian patent and has published more than 100 scientific papers, book chapters and editorial articles with an H-index of 30, i10 index of 55 and more than 2200 citations. He is a member of the editorial review boards of several international journals. As an editor, he has also published two books. He has been invited as the keynote speaker and expert in various plant pathology, microbiology, mycology and life sciences-based conferences and workshops.

Charith Raj Adkar Purushothama is an adjunct professor in the Department of Biochemistry and Functional Genomics, Université de Sherbrooke, Canada and a biotechnology entrepreneur. He earned his master's in microbiology from the University of Mysore, Mysuru, India and his doctorate from the School of Plant Biology and Crop Production, University of Milano, Italy. He did his post-doctoral training in the laboratory of Prof. Teruo Sano at Hirosaki University, Japan, and in the laboratory of Prof. Jean-Pierre Perreault at Université de Sherbrooke, Canada. During his post-doctoral training, he worked on plant-pathogen interactions and management strategies. He received several research scholarships and grants at national and international levels. He has published over 60 original research articles, two review articles and several book chapters, and presented in more than 30 international scientific conferences. In 2017, one of his research publications on viroid received the "Paper of Excellence" award from The Phytopathological Society of Japan.

Contributors

G. Adithi
Applied Mycology Lab, Department of Studies in Microbiology
University of Mysore
Mysuru, India

K. Ajithkumar
Main Agricultural Research Station
University of Agricultural Sciences
Bangalore, India

G. Balaji
Lactalis India
Chennai, India

Thippeswamy Basaiah
Department of P.G. Studies and Research in Microbiology, Bioscience Complex
Kuvempu University
Shimoga Dist, India

Anuja Bejoy
Food Safety and Standards Authority of India
National Food Laboratory
New Delhi, India

P. V. Bhavani
Pentavalent Bio Sciences Private Limited
Bengaluru, India

Manjunatha Dadayya
Department of P.G. Studies and Research in Microbiology
Bioscience Complex Kuvempu University
Shimoga Dist, India

Vignesh Dhandapani
Department of Chemical and Biotechnological Engineering
Université de Sherbrooke
Sherbrooke, Canada

S. Dhanush
Department of Microbiology
Yuvaraja's College, University of Mysore
Mysuru, India

H. V. Girish
Department of Studies in Botany
University of Mysore
Mysuru, India

Sandra V. Gomez-Gutierrez
Department of Botany and Plant Pathology
Purdue University
West Lafayette, IN, USA

Stephen B. Goodwin
Crop Production and Pest Control Research Unit
USDA–Agricultural Research Service
West Lafayette, IN, USA

Sanu Jacob
Food Safety and Standards Authority of India
National Food Laboratory
Chennai, India

V. Jaiganesh
Rice Research Station (RRS)
Tamil Nadu Agricultural University (TNAU)
Ambasamudram, Tirunelveli, India

K. Joseph
Defence Food Research Laboratory
Defence Research & Development Organisation (DRDO)
New Delhi, India

J. Joy
Department of Studies in Microbiology
University of Mysore
Mysuru, India

S. Mahadevakumar
Botanical Survey of India, Andaman and Nicobar Regional Centre
Sri Vijaya Puram, A&N Islands, India

M. Mahesh
Department of Studies in Botany
University of Mysore
Mysuru, India

P. K. Maheshwar
Department of Microbiology, Yuvaraja's College
University of Mysore
Mysuru, India

Deepa Nagaraju
Molecular Mycotoxicology Laboratory, Department of Studies in Microbiology
University of Mysore
Mysuru, India

Subhaprada Nishtala
Association of Food Scientists and Technologists
India

Charith Raj Adkar Purushothama
RNA Group, Department of Biochemistry and Functional Genomics, Faculty of Medicine and Health Sciences
Université de Sherbrooke
Sherbrooke, Canada

M. P. Raghavendra
Department of Microbiology
Government Science College (Autonomous)
Hassan, Karnataka

R. Rakesh
Defence Food Research Laboratory
Defence Research & Development Organisation (DRDO)
New Delhi, India

M. Renuka
Main Agricultural Research Station
University of Agricultural Sciences
Bangalore, India

S. S. Hari Sankar
Food Safety and Standards Authority of India
Tuticorin, India

Madanraj Appiya Santharam
Department of Immunology and Cell Biology, Faculty of Medicine and Health Sciences
Université de Sherbrooke
Sherbrooke, Canada

Teruo Sano
Hirosaki University
Hirosaki, Japan

M. Sarvajith
WDRC, Biological and Environmental Sciences and Engineering Division
King Abdullah University of Science and Technology
Kingdom of Saudi Arabia

A. S. Savitha
Department of Plant Pathology, Agricultural College
University of Agricultural Sciences
Bangalore, India

B. Shruthi
Applied Mycology Lab, Department of Studies in Microbiology
University of Mysore
Mysuru, India

R. Shylaja
Defence Food Research Laboratory
Defence Research & Development Organisation (DRDO)
New Delhi, India

K. R. Sridhar
Department of Biosciences
Mangalore University
Mangaluru, India

S. Sundaramoorthy
Horticultural College and Research Institute (HC&RI)
Tamil Nadu Agricultural University (TNAU)
Coimbatore, India

Suriya Tamilselvan-Nattar-Amutha
Leibniz Institute of Plant Genetics and Crop Plant Research (IPK) OT Gatersleben
Gatersleben, Germany

Sowmya Hirakannavar Veeranna
Department of P.G. Studies and Research in Microbiology, Bioscience Complex, Kuvempu University
Shimoga Dist, India

S. Wily
Department of Botany and Plant Pathology
Purdue University
West Lafayette, USA

Sreenivasa Marikunte Yanjarappa
Molecular Mycotoxicology Laboratory, Department of Studies in Microbiology
University of Mysore
Mysuru, India

Introduction

Deepa Nagaraju and Sreenivasa Marikunte Yanjarappa
University of Mysore, India
Charith Raj Adkar Purushothama
Université de Sherbrooke, Canada

Phytopathogenic fungi represent toxigenic fungi with confusing and unstable taxonomic history. These fungi, namely, *Aspergillus*, *Fusarium*, *Penicillium*, *Rhizopus*, *Magnotophora*, *Botrytis*, *Macrophomina*, *Alternaria*, *Claviceps*, *Trichoderma*, have been identified on the basis of morphological characteristics, which is often difficult to identify and requires highly skilled personnel (Nelson et al., 1983). This has led to inaccurate identification of fungal species, leading to assigning of different species to the same fungus. Identification of fungal species based on cultural and morphological features is often complicated and confusing, even for expert taxonomists (Yoder and Christianson, 1997). The result of such erroneous identification has given rise to rampant and inconsistent misapplication of species names to toxigenic isolates.

Several analytical methods are available today to assess the degree of contamination of cereals with phytopathogenic fungi, such as thin layer chromatography (TLC), liquid chromatography (LC), gas chromatography (GC), and immunological and molecular methods (Musser and Plattner, 1997). Methods suitable for the detection and quantification of mycotoxins are usually included by primary extraction and clean-up steps prior to determination of the sample. The method of choice depends on available equipment and analytical requirements such as sensitivity and time required for analysis. Recent advances in analytical instrumentation have resulted in the introduction of liquid chromatography–mass spectrometers (LC-MS), which are more sensitive and specific for the detection and quantification of mycotoxins (Shephard, 1998). High sensitivity can be achieved, although disadvantages include high running and maintenance costs. Conventional methods comprise culturing and identification from whole or processed grains on complex or synthetic media (Nirenberg, 1981) or on selective media (Bragulat et al., 2004). The food industry often demands answers in minutes, an impossible request for traditional mycological techniques. It is often important to determine whether a certain fungus might have been growing in a raw material before it was killed by processing. However, these traditional methods are relatively simple and inexpensive with regard to materials, but they can be labour intensive and results may not be obtained for weeks. As an alternative analytical approach to fungal detection assays rely on molecular tools application.

The application of molecular methods for the diagnosis of phytopathogens has been bifurcated into different chapters. The most important methods used for the detection of phytopathogenic fungi are polymerase chain reaction (PCR) and its variants such as nested PCR, multiplex PCR, q-PCR, RT-PCR, next-generation PCR, conventional PCR, end-point PCR and bio-PCR, PCR coupled methods like MCH-PCR, PCR-ELISA, in-situ PCR, PCR-DGGE, PCR-based methods like restriction fragment length polymorphism (RFLP), amplified length polymorphism (AFLP), RAPD, and some PCR post amplification methods such as microarray, macroarray, LAMP and advance PCR methods like FISH, nucleic acid sequence-based amplification (NASBA), SAGE, RNAi, Northern blot and DNA/RNA probe, which are in practice to reduce the spread of disease and minimizing the yield and crop loss (Deepa and Sreenivasa, 2019a).

Mycotoxigenic fungal species are ubiquitous in distribution, infecting a wide range of crop plants and resulting in significant post-harvest losses (Desjardins et al., 2000, Nagaraja et al., 2016). In cereals, fungal species are known to cause root, stem and ear rots. Depending on geographical location and environmental conditions, different phytopathogenic fungi may be associated with different diseases of major crops (Sreenivasa et al., 2011a; Sreenivasa et al., 2011b). Many fungal species are known to produce different mycotoxins of which fumonisins are important. The consumption of mycotoxin-contaminated food and feed products poses a serious risk to human and animal health, as these mycotoxins are carcinogenic and can potentially impair the immune system (Deepa and Sreenivasa, 2019a, 2019b). Recent outbreaks of diseases caused by these fungi pose a great problem to the agricultural industry and are potentially threatening the global food supply (McMullen et al., 1997; Clear et al., 2000).

PCR technique offers much faster alternatives to standard identification procedures since fungal DNA can be extracted from food samples without incubation. Killed fungi can also be detected, which might be an additional advantage. The technique offers effective detection of fungal genera or species if a suitable set of primers are used. Some common amplification targets have been 5.8s, 16s, 18s and 28s rDNA (Sreenivasa et al., 2008; Deepa et al., 2016), the internal transcribed spacer (ITS) region of the ribosomal genes (Abd-elsalam et al., 2003; Dass et al., 2007), the larger intergenic spacer (IGS) region (Gonzalez-Jaen et al., 2004; Jurado et al., 2005; Mule et al., 2005) and mitochondrial DNA (Smith et al., 1996), as well as genes coding for specific proteins (Niessen and Vogel, 1998). The latest development in the field of molecular biology has resulted in the introduction of real-time PCR, a new technology to detect and study phytopathogenic fungal species (Schena et al., 2004; Pasquali et al., 2004; Lopez-Errasquin et al., 2007). Real-time PCR is more sensitive and specific for the detection and quantification of toxigenic fungi (Ginzinger, 2002; Schaad and Frederick, 2002). The real-time PCR could detect as little as 300fg of genomic DNA. It can also identify symptomless infections with higher sensitivity than conventional PCR (Schena et al., 2004; Janardhana et al., 2009).

In developing a PCR-detection protocol, it is essential that the parameters used in the PCR are optimal (Arnheim and Erlich, 1992). PCR parameters such as temperature conditions, reaction mixtures and primers have to be optimized prior to verifying the specificity and determining the sensitivity of the PCR amplification. Some common amplification targets have been 5.8s, 16s, 18s and 28s rDNA (Bruns et al., 1991; Yao et al., 1992; Fletcher et al., 1998; Kappe et al., 1998); the ITS region of the ribosomal genes (O'Donnell and Cigelnik, 1997; Pedersen et al., 1997; Grim and Geisen, 1998; Abd-elsalam et al., 2003) and the IGS region (Gonzalez-Jaen et al., 2004; Jurado et al., 2005; Mule et al., 2005). Bluhm et al. (2004) used *fum1* gene (fumonisin biosynthesis gene) to discriminate fumonisin-producing species from non-fumonisin-producing *Fusarium* species. The ITS, IGS regions and *fum1* regions have been used successfully to generate genus, species and toxin-specific primers capable of differentiating closely related *Fusarium* species.

Recently molecular detection methods have provided an alternative to microbiological identification of phytopathogenic fungi. Amoah et al. (1995) reported the use of RFLPs and RAPDs of ribosomal DNA to study the genetic variability among *F. verticillioides* isolates. Schlacht et al. (1997) analysed the DNA of members of *Fusarium* species belonging to mating populations A, D and E, the teleomorphs of *F. verticillioides*, *F. proliferatum* and *F. subglutinans*, respectively. RFLP studies have been conducted by Gonzalez-Jaen (2004) to reveal the intraspecific variability in the IGS region within *F. verticillioides*. Moeller et al. (1999) developed specific primers for *F. verticillioides* and *F. subglutinans* based on sequences of RAPD fragments. On the other hand, PCR has widespread use in plant science diagnostics, and a number of PCR-based assays have been developed for use with toxigenic phytopathogens (Deepa et al., 2018).

The current method for identification of the phytopathogens mostly relies on micro-morphological characteristics such as presence or absence of microconidia, shape and size of macroconidia, colony morphology, pigmentation and growth rates (Nelson et al., 1983). These morphological traits could be highly variable depending on the media and cultural conditions. In addition, degeneration of the cultures and production of mutants may further add to the problems in fungal identification and diagnosis. The result of such erroneous identification has given rise to rampant and inconsistent misapplication of species names to toxigenic isolates. However, an accurate identification of the isolates is also necessary to predict whether the isolate is able to produce a particular toxin or not, since different species or populations produce different toxins and secondary metabolites (Huss et al., 1996). Host preferences are also associated to a certain extent with different species or populations (Desjardins and Plattener, 2003; Patino et al., 2006). In view of the above, there is demand for a simple, rapid and reliable method for the rapid identification and differentiation of fungal species that does not use morphological characters.

To avoid these cumbersome methods, faster diagnostic methods such as polymerase chain reaction (PCR) protocols have emerged as useful and rapid techniques. A number of PCR-based assays have been developed by several researchers throughout the world for the detection and differentiation of toxigenic fungal species (Patino et al., 2004; Jurado et al., 2005; Patino et al., 2006). The introduction of PCR has found widespread use and has become one of the important tools in the field of plant pathology. Diagnostic methods based on the PCR are rapid, as there is no need to culture organisms prior to their identification. They are specific, since identification of species is made on the basis of genotypic differences, and are highly sensitive, detecting the target DNA molecules in complex mixtures, even when the mycelia are no longer viable (Patino et al., 2004; Deepa and Sreenivasa, 2019b).

Our edited book *Molecular Approaches for the Detection of Fungal Phytopathogens* consists of 17 chapters discussing various molecular methods for early detection of fungal phytopathogens. We have hereby covered all the major pathogens, their history, importance, methodology on identifying fungal pathogens, molecular approaches such as PCR and its variants, LAMP, microarray, macroarray, lateral flow assay, DNA barcoding, sequencing methods, *in-situ* hybridization, serological methods, nanotechnology-based assays, aptamer assays, on-site testing for fungal phytopathogens, certain policies and regulations required for sustainable agriculture and agri-food industry. We have also discussed the further direction and on diagnosing the fungal phytopathogens. Our contributors have majorly discussed all the recent advanced strategies: mechanisms, advantages, disadvantages, and applications for the early detection of fungal phytopathogens. Further aspects of reading and study questions pertaining to each methodology have been discussed in each of the chapters.

REFERENCES

Abd-Elsalam, K.A., Aly, N.I., Abdel-Satar, A.M., Khalil, S.M., Verreet, A.J. 2003. PCR identification of *Fusarium* genus based on nuclear ribosomal-DNA sequence data. *African Journal of Biotechnology*. 2: 82–85.

Amoah, B.K., Rezanoor, N., Nicholson, P., MacDonald, M.V. 1995. Variation in the *Fusarium* section *Liseola*: Pathogenicity and genetic studies of isolates *Fusarium moniliforme* Sheldon from different hosts in Ghana. *Plant Pathology*. 44: 563–572.

Arnheim, N., Erlich, H. 1992. Polymerase chain reaction strategy. *Annual Review of Biochemistry*. 61: 131–156.

Bluhm, B.H., Flaherty, J.E., Cousin, M.A., Woloshuk, C.P. 2004. Multiplex polymerase chain reaction assay for the differential detection of trichothecene- and fumonisin-producing species of *Fusarium* in cornmeal. *Journal of Food Protection*. 65: 1955–1961.

Bragulat, M.R., Martinez, E., Castella, G., Cabanes, F.J. 2004. Selective efficacy of culture media recommended for isolation and enumeration of *Fusarium* species. *Journal of Food Protection*. 67: 207–211.

Bruns, T.D., White, T.J., Talyor, J.W. 1991. Fungal molecular systematics. *Annual Review of Ecological System*. 22: 525–564.

Clear, R.M., Patrick, S.K., Gaba, D. 2000. Prevalence of fungi and fusariotoxins on barley seed from western Canada, 1995 to 1997. *Canadian Journal of Plant Pathology*. 22: 44–50.

Dass, R.S., Sreenivasa, M.Y., Janardhana, G.R. 2007. High incidence of *Fusarium verticillioides* in animal and poultry feed mixtures produced in Karnataka, India. *Plant Pathology Journal*. 6(2), 174–178.

Deepa, N., Nagaraja, H., Sreenivasa, M.Y. 2016. Prevalence of fumonisin producing Fusarium verticillioides associated with cereals grown in Karnataka (India). *Food Science Human Wellness*. 5, 156–162.

Deepa, N., Rakesh, S., Sreenivasa, M.Y. 2018. Morphological, pathological and mycotoxicological variations among *Fusarium verticillioides* isolated from cereals. *3 biotech*. 8: 105.

Deepa, N., Sreenivasa, M.Y. 2019a. Molecular methods and key genes targeted for the detection of fumonisin producing *Fusarium verticillioides* - An updated review. *Food Bioscience*. 32, 100473.

Deepa, N., Sreenivasa, M.Y. 2019b. Biocontrol Strategies for Effective Management of Phytopathogenic Fungi Associated with Cereals, In: Singh J.S., Singh D.P. (eds) *New and Future Developments in Microbial Biotechnology and Bioengineering*, Elsevier, pp, 177–189.

Desjardins, A.E., Manandhar, G.G., Plattner, R.D., Maragos, C.M., Shrestha, K., McCormick, S.P. 2000. Occurrence of *Fusarium* species and mycotoxins in Nepalese maize and wheat and the effect of traditional processing methods on mycotoxin levels. *Journal of Agricultural Food Chemistry*. 48: 1377–1383.

Desjardins, A.E., Plattener, R.D. 2003. Fumonisin B_1-nonproducing strains of *Fusarium verticillioides* cause maize (*Zea mays*) ear infection and ear rot. *Journal of Agricultural Food Chemistry*. 48: 5773–5780.

Fletcher, H.A., Barton, R.C., Verweij, P.E., Evans, E.G.V. 1998. Detection of *Aspergillus fumigatus* PCR products by a microtitre plate-based DNA hybridization assay. *Journal of Clinical Pathology*. 51: 617–620.

Ginzinger, D.G. 2002. Gene quantification using real-time quantitative PCR: An emerging technology hits the mainstream. *Experimental Hematology*. 30: 503–512.

Gonzalez-Jaen, M.T., Mirete, S., Patino, B., Lopez-Errasquın, E., Vazquez, C. 2004. Genetic markers for the analysis of variability and for production of specific diagnostic sequences in fumonisin-producing strains of *Fusarium verticillioides*. *European Journal of Plant Pathology*. 110: 525–532.

Grim, C., Geisen, R. 1998. A PCR-ELISA for the detection of potential fumonisin-producing *Fusarium* species. *Letters in Applied Microbiology*. 26: 456–462.

Huss, M.J., Campbell, C.H.L., Jennings, D.B., Leslie, J.F. 1996. Isozyme variation among biological species in the *Gibberella fujikuroi* species complex (*Fusarium* section *Liseola*). *Applied Environmental Microbiology*. 62: 3750–3756.

Janardhana, G. R., Dass, R.S., Sreenivasa, M.Y. 2009. Fumonisins – A new class of Fusarial toxins. In *Frontiers in Fungal Ecology, Diversity and Metabolites*. Ed: K. A. Sridhar et al. New Delhi: I. K. International Publishing House Pvt. Ltd.

Jurado, M., Vazquez, C., Patino, B., Gonzalez-Jaen, M.T. 2005. PCR detection assays for the trichothecene-producing species *Fusarium graminearum, Fusarium culmorum, Fusarium poae, Fusarium equiseti* and *Fusarium sporotrichioides*. *Systemic Applied Microbiology*. 28: 562–568.

Kappe, R., Okeke, C.N., Fauser, C., Maiwald, M., Sonntag, H.G. 1998. Molecular probes for the detection of pathogenic fungi in the presence of human tissue. *Journal of Medical Microbiology*. 47: 811–820.

Lopez-Errasquin, E., Vazquez, C., Jimenez, M., Gonzalez-Jaen, M.T. 2007. Real-Time RT-PCR assay to quantify the expression of fum1 and fum19 genes from the Fumonisin-producing *Fusarium verticillioides*. *Journal of Microbiology Methodology*. 68: 312–317.

McMullen, M., Jones, R., Gallenberg, D. 1997. Scab of wheat and barley: A re-emerging disease of devastating impact. *Plant Diseases*. 81: 1340–1348.

Moeller, E.M., Chelkowski, J., Geiger, H.H. 1999. Species-specific PCR assays for the fungal pathogens *Fusarium moniliforme* and *Fusarium subglutinans* and their application to diagnose maize ear rot. *Journal of Phytopathology*. 147: 497–508.

Mule, G., Gonzalez-Jaen, M.T., Hornok, L., Nicholson, P., Waalwijk, C. 2005. Advances in molecular diagnosis of toxigenic *Fusarium species*: A review. *Food Additives Contaminants*. 22: 316–323.

Musser, S.M., Plattner, R.D. 1997. Fumonisin composition in cultures of *Fusarium moniliforme, Fusarium proliferatum, and Fusarium nygamai*. *Food Additives Contaminants*. 45: 1169–1173.

Nagaraja, H., Chennappa, G., Rao, P. C. K., Mahadevprasad, G., Sreenivasa, M.Y. 2016. Diversity of toxic and phytopathogenic *Fusarium* species occurring on cereals grown in Karnataka state, India. *3Biotech*. 6, 57.

Nelson P.E., Toussoun, T.A., Marasas, W.F.O. 1983. Fusarium *Species: An illustrated manual for identification*. Pennsylvania State University Press, University Park, USA.

Niessen, L., Vogel, R.F. 1998. Group-specific PCR-detection of potential trichothecene-producing *Fusarium species* in pure cultures and cereal samples. *Systemic Applied Microbiology*. 21: 618–631.

Nirenberg, H.I. 1981. A simplified method for identifying *Fusarium* spp. occurring on wheat. *Canadian Journal of Botany*. 59: 1599–1609.

O'Donnell, K., Cigelnik, E. 1997. Two divergent intragenomic rDNA ITS2 types within a monophyletic lineage of the fungus *Fusarium* are non-orthologous. *Molecular Phylogenetic Evolution*. 7: 103–116.

Pasquali, M., Marena, L., Gullino, L., Garibaldi, A. 2004. Vegetative compatibility grouping of the *Fusarium* wilt pathogen of Paris daisy (*Argyranthemum frutescens* L.). *Journal of Phytopathology*. 152: 257–259.

Patino, B., Mirete, S., Gonzalez-Jaen, T., Mule, G., Rodriguez, T.M., Vazquez, C. 2004. PCR Detection assay of Fumonisin- Producing *Fusarium verticillioides* strains. *Journal of Food Protection*. 6: 1278–1283.

Patino, B., Mirete, S., Vazquez, C., Jimenez, M., Rodriguez, T.M., Gonzalez-Jaen, T. 2006. Characterization of *Fusarium verticillioides* strains by PCR-RFLP analysis of the intergenic spacer region of the rDNA. *Journal of Science Food Agriculture*. 86: 429–435.

Pedersen, L.H., Skouboe, P., Boysen, M., Soule, J., Rossen, L. 1997. Detection of *Penicillium* species in complex food samples using the polymerase chain reaction. *International Journal of Food Microbiology*. 35: 169–177.

Schaad, N.W. and Frederick, R.D. 2002. Real-time PCR and its application for rapid plant disease diagnostics. *Canadian Journal of Plant Pathology*. 24: 250–258.

Schena, L., Nigro, F., Ippolito, A., Gallitelli, D. 2004. Real-time quantitative PCR: A new technology to detect and study phytopathogenic and antagonistic fungi. *European Journal of Plant Pathology*. 110: 893–908.

Schlacht, U.F., Moller, E.M., Geiger, H.G. 1997. Genetic diversity of *Gibberella fujikuroi* isolates from different geographic origins. Proceedings of the 5th European *Fusarium* seminar. *Cereal Research Communication*. 25: 557–600.

Shephard, G.S. 1998. Chromatographic determination of fumonisin mycotoxins. *Journal of Chromatography A*. 815: 31–39.

Smith, G.W., Constable, P.D., Bacon, C.W., Meredith, F.I., Haschek, W.M. 1996. Cardiovascular effects of fumonisins in swine. *Fundamental Applied Toxicology*. 31: 169–172.

Sreenivasa, M.Y., Dass, R.S., Charith Raj, A.P., Janardhana, G.R. 2008. PCR-based detection of genus *Fusarium* and Fumonisin-producing isolates from freshly harvested Sorghum grains grown in Karnataka, India. *Journal of Food Safety*. 28, 236–237.

Sreenivasa, M.Y., Dass, R.S., Charith Raj, A.P., Janardhana, G.R. 2011a. Mycological evaluation of Maize grains produced in Karnataka (India) for the post-harvest fungal contamination. *World Applied Sciences Journal*. 13(4), 688–692.

Sreenivasa, M.Y., Dass, R.S., Charith Raj, A.P., Prasad, M.N.N., Premila N. Achar, Janardhana, G.R. 2011b. Assessment of the growth inhibiting effect of some plant essential oils on different *Fusarium* species isolated from sorghum and maize grains. *Journal of Plant Diseases and Protection*. 118 (6), 208–213

Yao, C., Frederiksen, R.A. and Magill, C.W. 1992. Length heterogeneity in ITS2 and the methylation status of CCGG and GCGC sites in the rRNA genes of the genus *Peronosclerospora*. *Current Genetics*. 22:415–420.

Yoder, W.T. and Christianson, L.M. 1997. Species-specific primers resolve members of *Fusarium* section *Fusarium*. *Fungal Genetics Biology*. 23: 68–80.

1 Fungal Phytopathogens and the Importance of Early Detection through Conventional Methods

K. Ajithkumar, A. S. Savitha[*], *and M. Renuka*
University of Agricultural Sciences

S. Mahadevakumar
Andaman and Nicobar Regional Centre

Fungal phytopathogens are the most important microorganisms causing havoc and diseases in fields, horticulture, floriculture, ornamental and forest plant species across the globe, resulting in a considerable decline in the quantity and quality of yield. The keystone of plant pathology is the identification of the fungi that cause specific crop diseases. Only then can the disease be adequately understood and managed. The "gold standard" for identifying fungal pathogens is conventional or traditional detection methods, which include visual observation with the evidence of symptoms or signs of pathogens; laboratory techniques such as isolation, cultural and morphological identification; and the possibility of biochemical tests to confirm the pathogen. Despite this, there is still a glaring knowledge vacuum in this field of study, since no method can satisfy the expanding need for quicker, more efficient, reproducible and sensitive results. This gap can be effectively and successfully addressed by advanced molecular techniques based on nucleic acids, enzymes and proteins. However, the primary focus of this book chapter is on the different traditional techniques utilized in the early detection of fungal phytopathogens, as well as their benefits and limitations.

[*] Conceptualization and editing: A. K. and S. A. S.; Writing and review: R. M. and A. K; Supervision: A. K. and S. A. S.

DOI: 10.1201/9781003399704-1

GRAPHICAL ABSTRACT

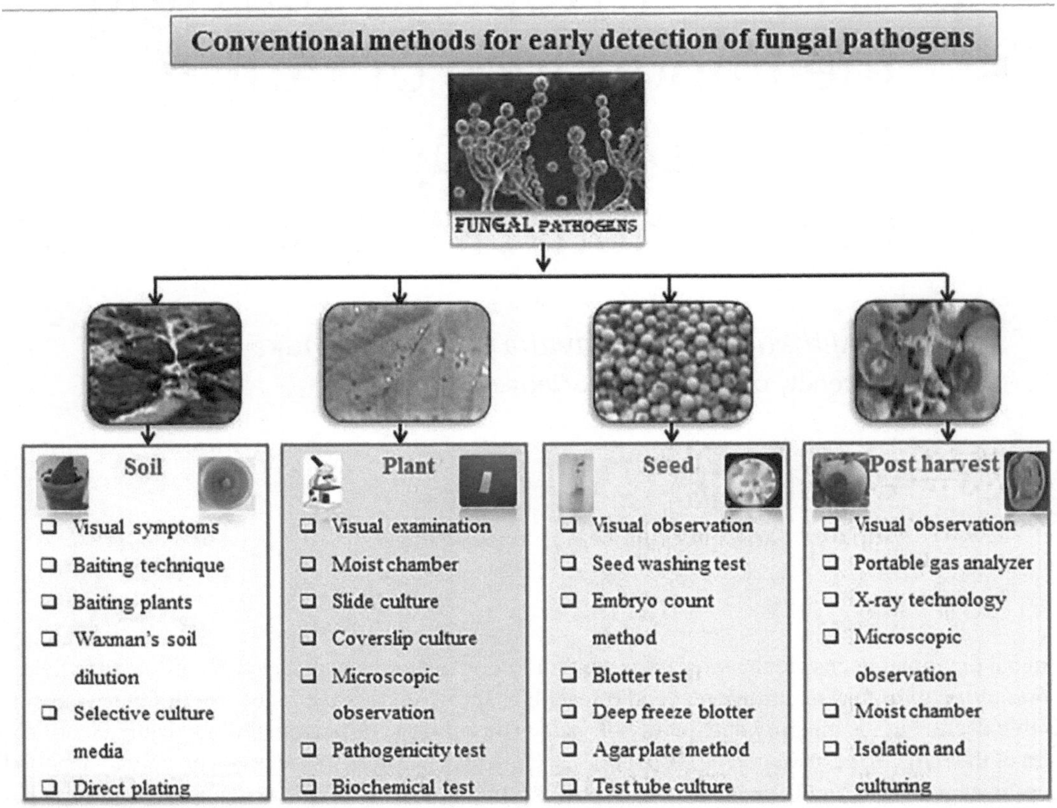

1.1 INTRODUCTION

Fungi can be defined as

> eukaryotic, unicellular or multicellular, heterotrophic, ubiquitous, non-chlorophyllus, spore-forming filamentous organisms. Fungi consists of a body (thallus) composed of microscopic tubular cells called hyphae. Its cell wall is made up of chitin, which helps in absorption of nutrients directly. They reproduce either sexually or asexually by means of spore production.

A few fungi that derive their nutrients from a living host are called biotrophs; others depend on dead and decayed organic matter for their nutrients and are referred to as saprophytes. Some fungi attack the living host, eventually killing host cells to obtain their nutrients, and are hemi-biotrophs.

Fungal phytopathogens are an important component of the biotic factor responsible for destructive devastating diseases in all crop plants across the globe, with a significant reduction in the quality as well as yield. Crop plants are attacked by various diseases causing 10%–15% losses, of which 70%–80%of diseases are caused by fungal pathogens (Kang, 2010). More than 20,000 fungal species are in parasitic habitats, which result in causing diseases in crop plants. The crop infection by the fungal pathogen starts from the seedling stage to post-harvest stages either in field, transit, or storage of particular agricultural commodities. Fungi are fully packed with the capacity to produce secondary metabolites known as mycotoxins, which readily contaminate agricultural commodities, which in turn have harmful effects on human and animal health upon consumption (Goyal et al.,

2017; Dass et al., 2007). These pathogens can cause an array of plant diseases ranging from seed rot, damping-off, dieback, blights, canker, leaf spots, mildews, rust, smuts, galls, scab, wilt, root rot, sheath rot, *etc.*, which can cause significant reduction in the quality and quantity of yield (Wenneker and Thomma, 2020; Sreenivasa et al., 2008).

The plant pathogen detection methods should ideally be sensitive,.rapid, accurate, dependable, easy to use and affordable. It should also be able to identify pathogens in complicated matrices like soil samples or plant extracts. Therefore, traditional/conventional tools came to be used in all the basic laboratories for the detection of fungal phytopathogens through the utilization of visual observation based on the presence of symptoms (oldest traditional method; Deepa & Sreenivasa, 2019a), direct microscopic examination of pathogenic propagules, culture and pathogenicity-based studies, as they are inexpensive, simple and yield results quickly (Venbrux et al., 2023).

1.2 CRITERIA AND NEED FOR EARLY DETECTION OF FUNGAL PHYTOPATHOGENS

Conventional or traditional detection methods are the "gold standard" for the identification of fungal pathogens, which consist of visual observation based on the presence of symptoms or pathogenic propagules of fungi, such as mycelia, spore and sclerotia adhered to the external surface of a plant (Nezhad, 2014). Further laboratory techniques including isolation, cultural and morphological identification, as well as biochemical tests could be done to confirm the pathogen, but the major drawback is they are time-consuming, require expertise with good handling capacity and the obtained results may not be specific and sensitive as complex diseases caused by synergy of multiple fungal pathogens therefore accuracy is less in identification (Sharma and Sharma, 2016). For those fungi which show latent infection, the detection becomes difficult, in all such kinds of situations the helping hand is the molecular tools, which overcome all these lacunae and detect fungal phytopathogens early, ensuring minimal damage to the crop, maximizing productivity and allowing agricultural sustainability (Tsui et al., 2011). The early and accurate detection of fungal phytopathogens responsible for specific crop diseases is a milestone of plant pathology as it helps (i) start preventive or curative treatments by determining the pathogen's presence and amount in the crop, (ii) ensure that seeds and planting supplies are certified for certification and quarantine programmes, (iii) calculate the population of the pathogen for a specific area in question and relate it to the resulting production loss, (iv) analyse the differences in infection by the pathogen in the germplasm sets to choose origins of the disease resistance, (v) recognize rapidly new races or strains of existing pathogens to check their entry to new area, (vi) understand the taxonomic and evolutionary association of plant pathogens, (vii) resolve the complex diseases resulted by more than one pathogen and (viii) to know the process of pathogenesis and gene function in pathogen (Narayanasamy, 2010).

A competent and effective detection frame for fungal diseases based on modern DNA-based molecular tools such as standard polymerase chain reaction (PCR), nested PCR, real-time PCR, immunocapture-PCR (IC-PCR), droplet digital PCR (dd-PCR), loop-mediated isothermal amplification (LAMP), rolling circle amplification (RCA) and nucleic acid sequence-based amplification (NASBA) is used for species identification (Cheng et al., 2020). Other methods such as magnetic-capture hybridization PCR (MCH-PCR), fluorescence *in situ* hybridization (FISH), *in situ* PCR, co-operational PCR, multiplex PCR, DNA macro and micro arrays, PCR restriction fragment length polymorphism (PCR-RFLP) and PCR denaturing-gradient gel electrophoresis (PCR-DGGE) are appropriate for genotyping (Monje and Mejia, 2020; Deepa et al., 2016a, 2016b). These provide accuracy, specificity, reliability, sensitivity, confidence and speed in detection, as well as quantification of fungal phytopathogens with minimal availability of fungal DNA in many important crops (Deepa & Sreenivasa, 2019b). They can be used to detect the fungal phytopathogens in crop growth, harvest and post-harvest stages. Their detailed studies are included in upcoming chapters. But the present book chapter mainly aims at various conventional methods utilized in early detection of fungal phytopathogens with their rewards and downside.

1.3 CONVENTIONAL OR TRADITIONAL TECHNIQUES FOR EARLY DETECTION OF FUNGAL PHYTOPATHOGENS

1.3.1 Early Detection of Fungal Phytopathogens

The aerial parts are preferentially attacked by the numerous air-borne fungal pathogens, which lead to a reduction of photosynthetic activity and finally lead to premature defoliation, leading to a decline in the growth of the plant, and even some cases it causes damage to branches and the stems. In plenty of cases, the floral parts also get affected by the fungal pathogens. Therefore, with the accurate diagnosis of these diseases at the correct time based on certain characteristic symptoms induced by the pathogen, we can detect the causative agents to some extent. Further, we can also make use of cultural and morphological characteristics for their taxonomy and classification (Janardhana et al., 2009) (Table 1.1).

TABLE 1.1
Conventional or Traditional Methods for Early Detection of Fungal Pathogens

Sl. No.	Traditional early detection techniques		Methodology	Reference
1.	**Plant**	Visual examination	Based on symptoms and signs	Raya et al. (2017)
		Moist chamber	Morphological characters	Jelusic et al. (2021)
		Slide culture technique	Morphological characters	Bisen (2014)
		Cover slip culture technique	Morphological characters	Mitchell and Britt (1981)
		Cellotape method	Morphological characters	Harris (2000)
		Pathogenicity test	Host-pathogen interaction	Bhunjun et al. (2021)
		Biochemical tests	Biochemical characters	Derbyshire et al. (2015)
2.	**Soil**	Visual examination	Based on symptoms and signs	Marquez et al. (2021)
		Waxman's soil dilution method	Cultural and morphological characters	Sodikov et al. (2022)
		Direct plating method	Cultural and morphological characters	Narciso and Parish (2007)
		Selective culture media	Based on their nutrient utilization specificity	Brito et al. (2019)
		Trapping or baiting technique	Attraction of the pathogen towards particular bait	Narayanasamy (2019)
		Indicator or bait plants	Attraction of the pathogen towards particular bait plants	Narayanasamy (2019)
3.	**Seed**	Visual examination	Based on symptoms and signs	Tsedaley (2015)
		Seed washing test	Morphological characters	Vishunavat et al. (2023)
		Sodium hydroxide seed soak method	Spore morphology	Vishunavat et al. (2023)
		Embryo count method	Mycelial segments adhered to the embryo	Cappelli and Covarelli (2005)
		Blotter test	Morphological characters	Kumar and Kumar (2022)
		Paper towel method	Morphological characters	Mahmoud et al. (2013)
		2,4-D blotter test	Morphological characters	Vishunavat et al. (2023)
		Agar plate method	Cultural and morphological characters	Kumar and Kumar (2022)
		Deep freeze blotter method	Morphological characters	Kumar and Kumar (2022)
		Test Tube Agar Method	Cultural and morphological characters	Kumar and Kumar (2022)
4.	**Post -harvest**	Visual examination	Based on symptoms and signs	Vico et al. (2014)

(Continued)

TABLE 1.1
(Continued)

Sl. No.	Traditional early detection techniques		Methodology	Reference
		Portable gas analyzer	Based on their ethylene production	Sorocki and Rydosz (2019)
		X-ray technology	Difference between the x-ray images	Dhondiram and Ashok (2017)
5.	Other methods	Optical sensor	Light's absorption or reflection capacity	Pieczywek et al. (2017)
		Image analysis	Based on the difference in the foliar features	Bock and Nutter (2011)
		Thermography	Differences in surface temperatures of plant leaves and canopies	Hillnhutter et al. (2011)
		Fluorescence imaging	Based on chlorophyll fluorescence	Burling et al. (2011)
		Hyperspectral imaging	variations in reflectance brought on by the biophysical and biochemical characteristic alterations	Delalieux et al. (2007)
		Remote sensing	measurement of reflectance of electromagnetic radiation from the vegetation	Qin and Zhang (2005)

1.3.2 Visual Examination Method

It is an age-old system for the detection of fungal phytopathogens based on the presence of disease symptoms (*e.g.*, rust, mildew, spots blight, gall, blast, blight, canker, wilt, rot or damping-off); next is the sectioning of infected plant tissues and observing such sections under the microscope for the pathogenic structures. Visual examination has its own merits as such it is consistent and perfect and has been the theme of exhaustive study and investigation because of the accessibility of comprehensive guidelines and standards necessary for evaluation (Ray et al., 2017). Identification based on morphological structure is a speedy approach to identifying fungi up to genus level but has its own limitations due to phenotypic plasticity as well as overlapping morphology as seen in the case of *Bipolaris, Colletotrichum, Diaporthe* and *Pythium* (Crous et al., 2015). Even the visual assessment is always subject to an individual's perception and experience. Hence, the outcome is not all-time decisive, and it demands an expert mycologist, which makes this detection technique unrealistic for on-site diagnosis of fungal phytopathogens as the diagnosis is done under the microscope inside the laboratory (Ray et al., 2017).

1.3.3 Moist Chamber Technique

The fungal genera or even species can be eminent only based on the characteristics of their fruiting bodies or spores as they augment the simplicity and accurateness of detection. Some fungal pathogens are reluctant spore producers as they need specific conditions for their spore production. The fruiting bodies produced by the fungi are minute and/or unusual, and they can be complicated to trace out. Therefore, mycologists have developed a laboratory method called "moist chamber technique" to enhance the possibility of recording the spores as well as rare, minute or inconspicuous fruiting bodies (Jelusic et al., 2021).

Procedure:

- A moist chamber with Petri plates with moistened blotter paper (*e.g.*, perlite) is covered with sterile blotter paper, and the lid of the plate is also lined with blotter paper. Then, the chamber is moistened with distilled water with the help of an atomizer. Additionally, the moist environment is kept in a chamber for extended periods of time.

- Section leaves or stems into small slices and surface sterilize them with 5% sodium hypochlorite for 30–60 sec; then, tape them onto a sterile slide.
- Few slices should be placed untreated because some pathogens are very sensitive to sterilization.
- Place the slide on the blotter paper and seal the lid with parafilm to facilitate holding moisture.
- Incubate the plate at 18–22 °C under alternating cycles of light and darkness (10 hrs light/14 hrs darkness).
- Open the plate 24, 48 and 72 hrs post incubation and observe the slide for the structures on the lesion.
- Even make small sections of the infected tissue and place them in the slide with a drop of water; use a cover slip to cover the slide.
- Observe the slide under the compound microscope for fungal structures such as mycelium, conidia and conidiophores.

Application: The method has been employed for the identification of fungal pathogens belonging to varied genera such as *Alternaria, Cercospora, Curvularia, Colletotrichum, Fusarium, Helminthosporium, Magnoporthe, Phoma, and Verticillium.*

Advantage:

- This technique assists in detecting mycelia, dormant spores and fruiting bodies immersed in substrata, which ease a better indulgence of a mycobiota.
- Additionally, the technique helps in stimulating the maturation of immature fruiting bodies by rehydrating the desiccated fruiting bodies as well as breaking the dormant resting spores and sclerotia in such chambers.
- Moist chambers are a rapid and direct way of inducing sporulation and help in identifying the causal agents of a few diseases. In addition, this technique has practical significance, because the pathogen shows rapid growth on the host and competes with saprophytes in moist chamber.

Disadvantage:

- There may be a problem of overgrowth of saprophytic aggressive fungi instead of target fungal growth.

1.3.4 Culturing and Plating Methods

This technique consists of the isolation and culturing of fungi on suitable artificial media under diverse conditions followed by microscopy techniques. Microscopy involves the identification of pathogens based on morphological characters such as spore shape, size, colour, sporulation patterns, production and characteristics of sporulating structures of asexual and sexual spore forms, which are considered key for taxonomic classification of fungi. Fungal pathogens can be isolated by following a standard isolation procedure where infected plant parts are cut along with adjacent healthy tissue (2–5 mm squares) by using flame-sterilized forceps; they are surface sterilized with 0.1% mercuric chloride for 30–60 sec. The sterilized pieces are aseptically transferred to three sequences of sterilized distilled water. Next, they are placed on tissue paper to absorb excess moisture, after that they are kept on Petri dishes containing potato dextrose agar (PDA) supplemented with streptomycin sulphate, at the rate of three to five pieces of tissues per Petri plate and incubated at room temperatures (25–27 °C) that may favour the pathogen development.

Application: The method is generally utilized for culturing saprophytic fungi *viz., Alternaria, Cercospora, Curvularia, Colletotrichum, Fusarium, Helminthosporium, Magnoporthe, Phomaand Verticillium*, which can depend on the organic and inorganic compounds served in a plate as food.

Advantage:

- The above technique can be used for the detection of a large group of fungal pathogens.
- It is a simple and cheapest method.

Disadvantage:

- The accurate identification of pathogens depends on the skill and experience of the person involved in the identification.
- It is a time-consuming method as well as takes a long time for identification and confirmation of pathogens and is also not suitable for rapid detection.

1.3.5 Slide Culture Technique

It is a rapid method where fungal colonies are prepared for easy examination and identification of pathogens with slight interruption. The technique was first developed by Riddell in 1950, wherein the fungus is grown on a slide poured with a thin film of agar. Hence, there is no necessity to take a segment of fungus from the culture plate for observation on the slide. Accordingly, this method is most excellent for preserving and observing the fungal structures. Here, we will identify macroscopic descriptions (colony features, growth rate, colour, texture, exudates, diffusible pigments, raised and submerged hyphae) and microscopic features (arrangement of spores and fruiting bodies). Even the conidial ontogeny includes the arrangements of conidiophores, and its dispersal can be observed by slide culture, which helps in the accurate detection of filamentous fungi (Bisen 2014).

Procedure:

- Cut a block (range: 5 × 5 mm to 1 × 1 cm) of an appropriate agar medium that has been prepared previously in a culture dish to a depth of approximately 2 mm by employing a sterile scalpel blade.
- Transfer the agar block to the surface of the sterilized microscope slide, inoculate the fungal culture, and place a sterile cover slip on the surface of it.
- Further, incubate it for 4–7 days at 30 °C and then remove the coverslip from the agar block. Transfer it to another slide and dye it with lactophenol cotton blue for proper observation of microscopic structures.

Application: The method has been found applicable for studying the arrangement and sporulation of filamentous fungi of ascomycetes (*Alternaria alternata*, *Exerohilum rostratum* and *Aspergillus fumigatus*), basidiomycetes, zygomycetes and aerial mycelia of aerobic actinomycetes like *Nocardia, Actinomyces* and *Streptomyces*.

Advantages:

- It is a cost-effective and quick method for morphometry of fungal isolates.
- As the fungus grows directly on the agar slide, there is no necessity to take a segment of fungus from the culture plate for observation on the slide. This reduces the possibility of damage to spore-bearing as well as reproductive structures.

Disadvantage:

- As the agar slides contain an inadequate nutrient source, the fungal pathogens can survive only for a short time period. Therefore, the technique is not suitable for fungi, which take more days to reach maturity, for example, members of *Xylariaceae* and many endophytic fungi.

1.3.6 Cover Slip Culture Technique

This method is easy to learn, takes less time and produces high-quality permanent mounts. It can be used to identify clinical isolates and to examine the fungi at various stages of development without disturbing the hyphal structure and spore arrangement. It is based on the aerial mycelia's propensity for adhering to glass surfaces (Mitchell and Britt, 1981).

Procedure:

- Appropriate agar medium is poured approximately 3 mm thickness in 90 mm sterilized Petri plates and allowed for solidification.
- The fungal pure culture is inoculated at a point in the media.
- Insert 4–5 sterilized 18 mm cover slips at an angle of 45°.
- Then, close the lid of Petridish and seal it with parafilm. The cultured plate is incubated at room temperature for 7 days under undisturbed conditions until examination is done.
- Finally, remove the cover slip smoothly and mount it with lactophenol or cotton blue for easy observation under a microscope.

Application: This method is employed for studying the arrangement pattern of spores of fungal genera such as *Aspergillus*, *Curvularia*, *Fusarium*, *Penicillium and Trichoderma.*

Advantages:

- It has an advantage over slide culture approach, in that if the initial preparations fail to show sufficient sporulation, there are still left to be studied at weekly intervals.
- It is a time-saving and easy-to-perform technique.

Disadvantage:

- Nutrients rapidly get exhausted.
- Despite the fact that it seems like a very easy technique to follow, it is actually challenging.
- Additionally, since splitting frequently happens when inserting cover slips underneath the flaps, it is frequently the case that the media needs to be fortified by adding extra agar.
- Sadly, it has significant limits when it comes to the study of numerous fungi that have the ability to overgrow the glass surface without sporulating or else not develop at all.

1.3.7 Staining Technique

Fungal staining is an important technique used to improve contrast in fungal samples at a microscopic level. Here, fungal chitin present in the cell wall absorbs the stain and visualizes the fungal components better under a microscope, which is required for quick and easy detection of fungal pathogens. There are different methods involved in fungal staining.

1.3.7.1 Lactophenol Cotton Blue Stain

It is a very popular method of fungal staining, in which cotton blue solution acts as a mounting medium as well as a staining agent, which stains fungal elements as blue. The components of lactophenol cotton blue have unique functions, that is, lactic acid preserves fungal structures and makes tissue clear, phenol acts as a disinfectant which kills the live organisms and prevents contamination and cotton blue mainly stains the chitin in the fungal cell wall (Suresh et al., 2016). It is also known as the "wet mount/ tease mount method" for fungal hyphae identification.

Procedure:

- Take a dirt-free and dry glass slide on which add one or two drops of lactophenol cotton blue.
- Carefully spread the fungal mycelia into the stain to make a thin preparation.

- Next, place the cover slip carefully on the fungal preparation by avoiding air bubbles, and remove the excess stain by blot drying.
- Finally, observe the slide under the microscope.

1.3.7.2 PeriodicAcid Schiff Stain

The polysaccharides in the fungal cell wall will be stained in this method. Fungal elements will stain into pink-red with blue nuclei.

Procedure:

- Take clean and dry glass slides, place the plant tissue infected with fungi on the slides and hydrate it with water.
- Oxidize the sample with 0.5% acid solution for 5 min and then wash in distilled water.
- Now apply Schiff reagent for 15 min; at this point, fungal sections become light pink in colour.
- Then, wash in lukewarm tap water for 5 min (immediately sections turn dark pink colour).
- At last counter stain with Mayer's hematoxylin for 1 min and wash under tap water for 5 min.
- Dehydrate it, and cover slip with a synthetic mounting agent, which is ready to observe under the microscope.

1.3.7.2.1 Gomari Methenamine Silver Stain

Fungi will be outlined black when stained with silver nitrate due to the silver particles precipitating the cell wall of the fungi.

Procedure:

- Take a clean and dry glass slide, place the plant tissue infected with fungi on it and fix it with methanol for 5 min.
- Wash in distilled water and dip slide in Coplin jar containing chromic acid (4%) for 45 min and again rinse in distilled water.
- Then add sodium/potassium metabisulphite (1%) for 1–2 min and rinse in distilled water.
- Dip the slide in hexamine solution for 1 hr (smear turns to dark brown), followed by washing in distilled water, if it turns to black then wash with 0.1% $FeCl_3$.
- Repeat washing with sodium thiosulphate (5%) for 2 min and rinse in distilled water.
- Again clean with the light green solution (1%) for 1 min and dry the slides; finally, observe with oil immersion microscope.

1.3.7.2.2 Trypan Blue Staining

This method was developed by Sinha and Singh (1982) for rapid detection of whip smut fungi, *Ustilago scitaminea* hyphae in growing points of nodal buds of sugarcane by using trypan blue. The final detection of fungal pathogens was completed within four hours, and the infection could be detected in buds earlier than the symptoms expression in the plant. Therefore, this could be used in quarantine and certification departments to screen the sugarcane seed material. This method also detected smut pathogen colonization in sugarcane clones which had escaped during infection in the field as they did not exhibit the symptom but carried the pathogen in latent form. Hence, it can be used for testing of a huge number of sugarcane clones within a short period (Nallathambi et al., 2001).

1.3.7.2.3 Cello Tape Method of Fungi Observation

Preparation of adhesive tape to observe fungal colonies under microscopic is a standard method (Harris 2000).

Procedure:

- A cello tape (10 cm long) is sandwiched with the thumb and forefinger.
- Firmly press the sticky side of the tape on the surface of a sporulating fungal colony cultivated in the Petri dish.
- Remove the tape, and place carefully the sticky surface of the tape carrying fungal structures over the drops of lactophenol cotton blue or aniline blue stain in the middle of a glass slide.
- Gently press the tape on the slide, if necessary extend the ends of the tape to the ends of the slide and observe under the light microscope.

Advantages:

- It is a simple, inexpensive and easy technique.
- It allows direct visualization of fungal morphology.
- We can observe intact morphology of the fungus.
- It is useful for the preparation of permanent specimens which are helpful for teaching purposes.

Disadvantage:

- The adhesive tape preparation dries easily and the wrinkles later, which makes the preparation untidy for further usage.

1.3.8 Pathogenicity Test

It is the preliminary test done to identify and detect the fungal pathogens as these organisms may be there as endophytes/epiphytes on the surface of the plant and may not result in the causation of disease (Melotto et al., 2008). Under certain inherent or environmental conditions, endophytes may become latent phytopathogens or conditional phytopathogens, which results in diseases with the age of the plant or under stress situations (Kusari et al., 2014). Fungal phytopathogens have more genetic deviation among the population because of their species' historical dynamics, topography, demography, diminutive lifetime, huge population size, and quick adaptation to environmental and human-mediated factors. Species boundary identification demarks the genetic variation that exists in nature among fungal pathogens, hence the usage of multi-gene phylogeny of nuclear, mitochondrial, protein and ribosomal coding genes to perfectly establish the phylogenetic association of fungal taxa (Bhunjun et al., 2021).

Application: Pathogenicity test is used to confirm the infection by the fungal genera such as *Alternaria*, *Cercospora*, *Curvularia*, *Colletotrichum*, *Fusarium*, *Helminthosporium*, *Magnoporthe*, *Phoma*, *Rhizoctonia*, *Sclerotium* and *Verticillium*, which further helps in studying the host-pathogen interaction.

Advantage:

- This method encourages the formation of fruiting structures that are not typically formed on standard media. Hence, it is also beneficial for the identification of fungi.
- It is the preliminary test done to detect the endophytic fungal pathogens as these later on may cause latent infection.

Disadvantage:

- Pathogenicity tests implemented on detached plant tissues might not be accurate since the host defensive processes are suppressed.
- Environmental and physiological factors such as temperature, humidity as well as plant maturity have an impact on artificial inoculation.

1.3.9 Biochemical Tests

This method is employed to detect fungal pathogens based on biochemical activities. The variation in the metabolism of fat, protein and carbohydrate, as well as their ability to production and utilization of enzymes can be taken into consideration during biochemical analysis.

1.3.9.1 Acid Phosphatase Test

The hydrolysis of phosphate esters is catalysed by this enzyme in an acidic environment. This reaction is necessary for metabolic processes in fungi. A drop of α-naphthol phosphate (1%) and acetone base solution (1%) was added to 10 ml of water agar having the fungal growth, and it turned pink to purple within 24 hrs indicating the positive response in the case of *Sclerotium rolfsii.*

1.3.9.1.1 Cytochrome Oxidase and Reductase Test

A mixture composed of 50 mg tetramethyl phenylenediamine dihydrogen chloride and 10 ml aqueous solution of ascorbic acid (15 ppm) is made. When two drops of the same reagent are added to the 4 days old culture grown on basidiomycete agar, blue pigment zone appears within 30 min, which evidences the existence of cytochrome oxidase, while incase of reductase enzyme which prevents the oxidation, there is no blue colour formation even after 2 hrs. This kind of enzymatic test generally followed in *S. rolfsii* (Derbyshire et al., 2015).

Advantage:

- Biochemical tests can be performed with basic equipment.

Disadvantage:

- These are time-consuming and require successive repitation in order to confirm the fungal pathogen.

1.3.9.2 Detection of Fungal Phytopathogens from Soil

Soil system acts as a warehouse for the habitation of various kinds of phytopathogenic fungal pathogens. These may be detrimental by infecting the plant's root, stem or aerial portions during various phases of plant growth, leading to a variety of negative effects (diseases), which weaken and eventually kill the plant. A significant challenge in identifying diseased plants before obvious symptoms is shown on the plants is posed by soil-borne fungal diseases. The majority of the time, the symptoms of infection on aerial plant parts can be observed after the establishment of fungal pathogens inside the contaminated plants. The only way to detect root infection by these diseases is through destructive sampling, which involves removing the suspect plants. It might be tough to preserve the afflicted plants at this point, but using the right management techniques might be able to stop the illness from spreading to nearby plants. Therefore, it is crucial to quickly and accurately determine their identity by using the proper techniques.

1.3.9.2.1 Visual Symptoms

The root-infecting fungal phytopathogen results in plant death by yellowing, dieback, stunting, root rot and wilting. Generally, root rot is caused by *Pythium*, *Phytophthora*, *Rhizoctonia*, *Ceratocystis*, *Fusarium* and *Cylindrocladium*, whereas *Sclerotinia* and *Sclerotium* cause crown, stem and collar rot at ground level of the plant. For example, the plants infected by *Fusarium* exhibit distinctive symptoms, such as yellowing of the foliage from bottom leaves to young leaves; further, they show drooping. Finally, the wilting of the plant may be partial or complete. If such infected plants are cut open, the infected vascular tissue (xylem) shows brown discolouration. Spore-bearing stalk masses are sometimes visible on the dead tissue, which may look like a small pink cushion. The root system remains intact without any distortion; therefore, some strength is needed to uproot the plant. While in the case of *Rhizoctonia* infection, we can notice yellowing and softening of the leaves which always hang to the stem, sloughing of cortical tissue from the stem and taproot with grey discolouration of these infected tissues due to the block of microsclerotia, which results in premature plant death, the pathogen majorly attacks on the lateral roots and root hairs and destroys them, due to that we can easily pull out the plant from soil level (Marquez et al., 2021).

Advantage:

- It is an age-old method of detection of fungal pathogens.
- Visual examination is a reliable and cost-effective method.

Disadvantage:

- It is always subject to an individual's experience, hence needs an expert mycologist, and therefore the results are not always conclusive.

1.3.9.2.2 Waxman's Soil Dilution Technique

It is used to separate the pathogenic fungi and research fungal flora of various soils and the impact of agrochemicals on the species of fungi (Sodikov et al., 2022). It is employed to detect the soil-inhabiting fungal genera *viz.*, *Ceratocystis*, *Cylindrocladium*, *Fusarium*, *Pythium*, *Phytophthora*, *Rhizoctonia*, *Sclerotinia* and *Sclerotium*. The soil dilution method was used to analyse the soil microflora in the manner described below:

- Place the soil samples collected from the field in paper bags or sterile test tubes.
- Pass the collected soil samples through a fine sieve (4 mm dia) then place 10 g of soil on the filter paper.
- Add the samples to 90 ml distilled water in a sterile flask and shake for 5 min.
- Then, transfer 1 ml of soil mixture to 9 ml of water. Mix well and then transfer again. Repeat the procedure four times, each time transferring 1 ml of the suspension from one to the next tube.
- Spread 1 ml of the suspension from the fourth test tube onto Petri plate. Pour 10 ml of the nutrient medium into the Petri plate and incubate for 8–12 days at 23–25 °C.
- Observe and count the plates for the growth of colonies of the fungi. Colonies are shifted to the test tubes having the nutrient medium. The count will be taken after 15 days.

Advantage:

- It is used to obtain the pure fungal strains from miscellaneous densities of fungal propagules present in the soil sample.
- Both quantitative and qualitative assessments of the fungal species are frequently made using this method.

Disadvantage:

- Even before the suspension is fully transferred, some of the heavier components of the soil and fungal materials immediately sink to the bottom of the vessel. These contain a few of the species that aren't shown on the plate.

1.3.10 Direct Plating Method

This is the preferred technique for identification, enumerating and separating fungi from soil, which are inoculated directly on the agar media to facilitate the dispersion of soil particles all through the agar (Narciso and Parish, 2007). An adequate dispersal of sandy soils may be attained by shaking and rotating the mixture prior to the solidification of the agar. If the soil is dry or has a high amount of clay, it is advisable to add a drop of sterile water to the soil particles in the plate, before pouring the medium. To assess the impact of soil epidemiological factors on disease incidence and transmission, it is necessary to study the soil-borne fungal pathogens both qualitatively and quantitatively that are present in the field soil. This method has been employed for isolation of *Pythium*, *Mortierella*, *Marasmius*, *Tricholoma* and *Psalliota arvensis* from the soil.

Advantage:

- It is the best technique for identifying, counting and isolating fungi from soil.

Disadvantage:

- Accuracy and sensitivity of detection of fungal pathogens present in soil are compromised.

1.3.10.1 Selective Culture Media

This method is employed to promote the growth and development of specific "selected" fungal pathogens. Only these microorganisms are capable of growing on the medium without contamination of non-targeted microorganisms. The selectivity of the pathogen can be acquired by numerous techniques such as by making a definite sugar for the lone source of carbon in the medium, as well as the addition of antibiotics, colours, salts or certain growth retarders, which impact the metabolism or enzymatic systems for the selectively inhibiting non-targeted microbes.

Kang et al. (2014) developed selective media for isolation of soil inhabitant ascomycete *Cylindrocarpon destructans* incitant of ginseng root rot, where they used minimal media supplemented with 50 mg/l of brefeldin A/radicicol. These compounds have suppressed the mycelial growth of non-targeted such as *Rhizoctonia solani*, *Botrytis cinerea* and *Alternaria panax*, without affecting the growth of *C. destructans*. Even Czapek dox agar added with fungicides such as metalaxyl and pentachloronitrobenzene acts as selective media for isolating *C. destructans*, which avoids contamination of oomycetes and zygomycetes.

The selective medium for *Fusarium graminearum* incitant of head blight of cereals contains minimal media supplemented with toxoflavin induced the growth of *F. graminearum* from rice grains by retarding other saprophytic fungi (Jung et al., 2013). A semi-selective (benomyl fortified water agar and chloramphenicol) medium is effective for isolation of *Rhizoctonia solani* incitant of root rot and bare patches of wheat and barley. A semi-selective medium containing modified V8 juice supplemented with ox-bile, antibiotics (streptomycin sulphate, tetracycline hydrochloride) and fluazinam fungicide is a competent and quick technique for identification of *Ceratocystis fimbriata* from soil (Brito et al., 2019).

Masago's *Phytophthora* selective medium was used to isolate *Phytophthora* spp. from soil, while inhibiting the growth of other oomycete fungi *Pythium* spp. with the help of 25–50µg/ml of 3-hydroxy-5-methylisoxazole incorporated to PDA medium with other antimicrobial compounds (benomyl, pentachloronitrobenzene, nystatin, rifampicin and ampicillin), which completely inhibited the growth of the bacterial colony, non-target fungi and 12 other species of *Pythium* but ineffective against *Phytophthora* spp. The NARM substrates (Nyastatin ampicillin miconazole) containing nystatin + ampicillin + miconazole + agar were utilized in inhibiting non-pythiaceous microorganisms, and encouraging mycelial growth of *Pythium* spp. obtained from field soils (Morita and Tojo, 2007).

Advantage:

- It helps in the isolation and detection of target fungal phytopathogen.

Disadvantage:

- Many extremophilic fungi are not able to be isolated and detected as the selective media may become unstable over extreme conditions.

1.3.10.2 Trapping or Baiting Technique

Incubate the plant tissue into the soil for a period of time in order to permit the growth of interested pathogens in the bait. It is typical to recover and place the plant tissue used as bait on the culture media. This method imparts the weakness of the pathogen's parasitic nature, which enables it to be distinguished from the enormous unknown microorganisms in soil by growing preferentially in the plant tissue. *Thielaviopsis basicola* can be effectively baited from the soil on the carrot discs (5 mm thickness). Collect the baited disc and transfer it to a moist filter paper-lined Petri plate with soil-side up and incubate it. Periodical checking can be done to ensure sufficient moisture in the incubated plates. Observe plates at 7 and 14 days for growth of *T. basicola*. Even the carrot baiting technique is considered a traditional method for the isolation of *C. fimbriata* but is routinely performed to isolate the fungus by placing diseased wood fragments between slices of carrot. The pepper leaf disc assay is used to bait *Phytophthora capsici*, which affects pepper and other vegetable crops. Place the pepper leaf discs in soil samples saturated with water for 24 hrs. Sterilize the leaf discs for 1 min with sodium hypochlorite solution (0.5%) followed by washing in sterile distilled water and then plate it on Masago's *Phytophthora* selective agar and then incubate it for 72 hrs (Narayanasamy, 2019). *R. soalni* AG-8 easily isolated from root systems of wheat and barley and quantified in soil samples by employing toothpick baits into soil samples. After a 2-day incubation period, transfer the toothpicks to a selective medium. Using a dissecting microscope, the developing pathogen colonies are counted 2 days after inoculation.

Advantage:

- This method makes use of the pathogen's parasitic nature, which in turn enables to distinguish it from the enormous variety of other organisms in the soil by growing preferentially into the plant tissue.
- Numerous food sources are offered by different baits, making it feasible to isolate a large variety of fungal taxa.

Disadvantage:

- It is not applicable for detecting all the phytopathogenic fungi.

1.3.10.3 Indicator or Bait Plants

Sensitive indicator (bait) plants are grown in the test soil with a thought to have infection by the target pathogen. The pathogen present in the field soil is evidenced by the development of recognizable symptoms in the indicator plants. For example, the infected root samples are added to soil for the purpose of *Phytophthora fragariae* f. sp. *fragariae* detection and then extremely susceptible plants of *Fragaria semperflorans* var. *alpine* (used as bait plants) are planted. The infected plants perish after 3–6 weeks and the stele tissue of the roots display red colouring and the pathogen's oospores (Narayanasamy, 2019).

Advantage:

- This method is helpful for detecting target fungal phytopathogen.

Disadvantage:

- This method also requires an experienced mycologist to identify symptoms.

1.3.11 Early Detection of Fungal Pathogens in Seeds

The seeds are considered as both victims and vehicles of fungal disease. Because they are easily attacked by the fungal pathogens and even harbour the fungal pathogens and carry them from one cropping season to another and help in their widespread dispersal (Sheppard, 1979). Sometimes, these infected seeds act as symptomless carriers and, in addition to this low percentage of seed infection, could cause severe crop losses, particularly by quarantine pathogens (Pellegrino et al., 2010). Many countries have regulations to limit the introduction and spread of exotic pathogens to new areas. Thus, seed health testing is carried out routinely in many countries for quality assessment, plant quarantine and domestic seed certification.

1.3.11.1 Direct Examination/Visual Observation

The dry seeds are examined for the presence of either the fungal propagules or their effects on the physical appearance to detect the seed-borne fungal pathogens. The physical changes include discolouration of seeds, and variation in size and shape, which can be visibly detected. For example, maize seeds infected with *Fusarium verticillioides* will have white starburst symptoms (Renuka, 2021). While *Acremonium*-infected sorghum seeds are completely deformed, soybean seeds infected with *Cercospora kikuchii* show signs of purple seed stain. The fungal structures may be present on or inside the seeds in the form of fungal spores, sclerotial bodies, smuted balls, resting hyphae and fruiting bodies such as perithecia, pseudothecia and pycnidia. For example, in the case of rye and bajra, the ergot fungi *Claviceps* is often mixed with seeds as sclerotia (Tsedaley, 2015).

Advantage:

- It is a basic method of detecting the fungal contamination in seeds.
- It is easy to carry out and does not require much equipment.

Disadvantage:

- It will not detect fungal pathogens producing latent infection.
- Sometimes this method is not more reliable because severely infected seeds may look healthy.

1.3.11.2 Seed Washing Test

It is a type of seed health test solely used to detect externally seed-borne fungal pathogens. Here the pathogen inoculum is loosely adhered on the seed surface, which is generally used for detecting fungal spores of obligate pathogens, which cause diseases, *viz*., downy mildew, powdery mildew, rust, smut and bunt (Vishunavat et al., 2023).

Procedure:

- Add 2 gm of seeds to the test tube containing water (10 ml), and shake the mixture with a shaker for 10 min.
- The filtered suspension can be observed as such or centrifuged for 18 min at 3000 rpm to concentrate the suspended spores.
- Reject the suspension and dissolve the spores pellet in lactophenol (2 ml).
- Observe the obtained suspension for the fungal structures under the microscope and calculate the load of spore per seed using a haemocytometer.

Advantage:

- It is used to detect the externally seed-borne fungal pathogen.

Disadvantage:

- Internally seed-borne fungal pathogens are not detected.

1.3.11.3 Sodium Hydroxide Seed Soak Technique

The kernel bunt in wheat and bunt in paddy caused by *Tilletia indica T. barclayana*, respectively, are detected by this technique. The contaminated wheat seeds exhibit black spores of the fungus at the infected portion compared to the healthy area which are yellow in colour. Whereas, in paddy NaOH increases the colour disparity between healthy and infected seeds (Vishunavat et al., 2023).

Procedure:

- Soak the seeds in NaOH (0.2%) in a beaker for 24 hrs.
- Decant the NaOH solution and thoroughly wash the seeds with tap water.
- Shade dry the seeds over a blotter paper to remove excess water on the surface of the seed.
- Observe for the formation of jet-black shiny appearance by the infected seeds.
- Crush the infected seeds in a drop of water and observe for the ooze of fungal spores.

Advantage:

- Low-cost seed health testing technology.

Disadvantage:

- Seed germination is affected because NaOH is a strong base and even further limits the genetic or phenotypic inspection of individual seeds.

1.3.11.4 Embryo Count Method (Examination of embryo)

This method is mainly used for the diagnosis of loose smut fungi (*Ustilago nuda tritici*) in wheat as well as *U. segetum* f. sp. *tritici* incitant of loose smut of barley. Here, fungal mycelium is present inside the embryo; therefore, the embryo should be separated from the rest of the seed to be observed under a microscope. Around 2000 seeds should be soaked in a sodium hydroxide (5%) solution and trypan blue (0.03%) for 12 hrs, during which the released embryo will be processed further and dehydrated by alcohol for 5 min, and then cleared with 50 ml glycerine and lactic acid (1:1).Finally, these embryos are examined under a stereo binocular microscope (Cappelli and Covarelli, 2005).

Advantage:

- It reduces the number of embryos examined per seed sample as well as avoids the use of phenols.

Disadvantage:

- It is a laborious process and requires skill.

1.3.11.5 Incubation Test

It is the generalized technique for the diagnosis of seed-inhabitant fungi, where seeds are incubated on a substrate for a certain period under certain controlled conditions to allow the growth and sporulation of fungi on seed. Then the seeds are examined under the microscope and the fungi are categorized based on the morphological features of mycelia, spores, conidiophores and shape of fruiting body. Different methods such as standard blotter plate, agar plate, 2, 4-D blotter and deep-freezing blotter are generally utilized for the identification of seed-borne fungi (Tsedaley, 2015).

1.3.11.6 Blotter Test

It is a standard and most widely used method of testing seed-borne fungal pathogens in cereals, pulses, vegetables, ornamental and forest seeds, similar to germination tests. This technique involved placing the seeds on moistened blotter paper and incubating them at congenial conditions for fungal growth. The presence of mycelium or other fungal structures will be germinated and manifested seeds with fungal infection resulting in symptom or sign development on young seedlings. This method is still more efficient in most of the seed testing laboratories for screening the health status of the seeds. This is employed for detection of *Fusarium*, *Alternaria*, *Botrytis*, *Colletotrichum*, *Ascochyta*, *Botryodiplodia*, *Cercospora*, *Bipolaris*, *Curvularia*, *Cladosporium*, *Drechslera*, *Myrothecium*, *Phoma*, *Phomopsis*, *Rhizoctonia* and *Sclerotinia* (Kumar and Kumar, 2022).

- The sterilized Petri plates are lined with two to three layers of blotter paper. The upper lid of Petri plates is lined with one layer of the blotter paper, and the papers will be soaked in sterilized distilled water.
- Surface sterilized the seeds with NaHCl (2%) for 5 min followed by washing thrice in the distilled water.
- Twenty-five seeds are arranged systematically inside the Petri plate over the moistened blotter paper.
- Incubate the Petri plates at 22 ± 2 °C for 7 days in alternating 12 hrs of darkness and light.
- Observe the incubated seeds for the growth of different seed-borne fungi under the microscope.

Advantages:

- Quick identification of the pathogen based on their morphological characters.
- Cost-effective and can be employed for the identification of pathogens of seed-borne nature.
- It combines both *in vivo* and *in vitro* techniques. Hence, it is more reliable.

Disadvantages:

- The fast growth of non-target pathogens will hamper the identification of fungi which are slow-growing ones.
- Time-consuming and difficult to prove the pathogenicity.

1.3.11.7 2,4-D (Dichlorophenoxy Acetic Acid) Blotter Test

This is similar to the blotter test, but in this method the seed germination is intentionally concealed to enhance the growth and development of seed-borne infecting fungi. For example, in the case of *Phoma lingam* infecting brassica seeds to allow the pycnidia formation, as the blotter paper is soaked in 2,4-D (2%) and seeds are placed on such treated blotter paper and allowed for incubation at 28 °C with alternates of 12 hrs light and darkness. Later, observe the seeds under microscope for the different fungal species (Vishunavat et al., 2023).

Advantage:

- It allows only the growth of the fungi and avoids the germination of seeds as a result the stored food present in the seed is solely utilized by the pathogen.

Disadvantage:

- Chemical may cause irritation.

1.3.11.8 Paper Towel or Rolled Towel Technique

Generally, 100 seeds are carefully placed on the bilayered moist paper towels with polythene sheets at the base and 10 seeds arranged per row. Cover the seeds with a moist towel and carefully rolled to prevent excess pressure on the seeds and incubate for 10 days at 25 °C. After incubation, seeds are checked for the presence of microorganisms and germination. For the identification of *Ascochyta* infection in peas and *Fusarium* spp. in cereals, this approach is used (Mahmoud et al., 2013).

Advantage:

- This technique was used to perform germination and vigour experiments on different seeds to determine the impact of seed-borne inoculum on seed quality parameters.
- It also examines how various seed treatments affect seed-borne inoculum.

Disadvantage:

- There may be chances for infection by secondary saprophytic fungi.

1.3.11.9 Agar Plate Method

It is regularly used for the identification of seed-inhabitant microorganisms, which allows the detection of viable fungi even at their initial growth phase. It is employed for quantitative determination of the colony forming units of the fungi or to determine the different species of fungi. It is an efficient technique to categorize the fungi into genera and species, as well as valuable practice for the detection of internal seed-borne fungi and subsequently constructive tool for examining the quality of seed lots (Kumar and Kumar, 2022).

Procedure:

- Surface sterilize the seeds with NaHCl (2%) for 5 min, followed by cleaning with distilled water.
- Transfer the sterilized seeds onto Petri dish, where nine seeds are arranged in the periphery with a seed at the centre.
- Incubate the Petri dish for 6–8 days at 24 °C under 12 hrs of alternate light and darkness.
- Observe the incubated seeds for the growth of different seed-borne fungi on the agar.

Advantage:

- This technique was employed for quantitative and qualitative determination of the fungal species.

Dis**advantage:**

- It may also allow the growth of secondary saprophytic fungi.

1.3.11.10 Deep Freeze Blotter Method

It is similar to the blotter method but after placing the seeds inside the Petri plates containing the moist blotter papers, they are incubated at −20 °C for a day. Further, incubate the plates at 25 °C for the remaining 6–8 days. This technique will permit the growth of few fungi, and there is no deleterious effect on the fungi which are associated with seeds by the deep freezing (Kumar and Kumar, 2022).

Advantages:

- Pathogens can be detected quickly as the food material is entirely utilized by the fungi alone.
- It is a more reliable method as it combines with both *in vivo* and *in vitro* techniques.

Dis**advantage:**

- It is a time-consuming process.

1.3.11.11 Test Tube Agar Technique

It is employed for the identification of *Septoria nodorum* infecting wheat grains. It also assists in evaluating the fungicides as well as in quarantine stations. The seeds are inoculated into the test tube slants filled with 10 ml water agar and incubated at 20 °C for 7 days. These tubes are covered with polythene or aluminium sheets to retain moisture. After 14 days of incubation, the seedlings are examined for the typical symptom development in the young developing root and shoot system.

Advantage:

- It is utilized in the quarantine unit for the detection of fungal infections.
- It also avoids contamination by other secondary fungal infections.

Disadvantage:

- It is a time-consuming process.

1.3.12 Early Detection of Fungal Pathogens from Post-Harvest Products

The freshly harvested fruits and vegetables are highly prone to post-harvest losses, which account for around 40%–50%, of which fungal pathogens alone contribute 22.5% loss. The moisture content in the horticultural products is higher at the end of their harvest period and have thin skin, which makes them an easy target to the fungal pathogens, which attack them on or before harvest and throughout their storage period. Bruising, transport, harvest damage and unfavourable conditions for storage will induce the growth and dissemination of phytopathogenic fungi in store rooms. Therefore, early detection ensures protection to the product's health and avoids further dissemination of the pathogen. Several identification methods are used for the early identification of fungi in post-harvest commodities.

1.3.12.1 Visual Observation

Penicillium expansum incitant of mould decay of apple fruits is characterized by peculiar symptoms such as light discolouration with smooth lesions on the fruit surface, which later expands and becomes soft to rotting of the tissues leading to the emission of foul smell. The infected portion separates the healthy tissue by leaving a gap in the form of a "bowl" shape. In addition, the infected area can be covered by white fungal mycelium and spores, which imparta pale blue colour and cause "nesting" symptoms on the fruits (Vico et al., 2014). In the case of grapes anthracnose, the symptom initiates as dark red spots on berries. Later spots turn into sunken circular spots with ashy-grey centre; finally, a dark margin surrounds the spots to give an appearance of "bird's-eye rot".

Advantage:

- It is a basic method of detecting fungal contamination in fruits and vegetables.
- It is easy to carry out and does not require much equipment.

Disadvantage:

It will not detect fungal pathogens producing latent infection.

1.3.13 Identification of *Penicillium Expansum* Causing Post-harvest Blue Mould Decay of Apple Fruit

1.3.13.1 Portable Gas Analyser

It is a device used to measure the fungal ethylene production (ranging from 0.1 to 100 ppm) at real time, as the fungal infection enhances the ethylene production. It is used in the detection of fungal pathogens in a wide variety of fruits and vegetables, because it is non-destructive, small, portable and accurate to measure low levels of ethylene under different levels of humidity as well as temperature. It is a quick method as it records each reading within 30 sec and is a battery-charged device, which lasts for 8 hrs and even has 16 GB storage capacity. Hence, whatever data is collected is easily stored in the device, and further this data is transferred to the scientist's computer. Here the fungi-infected fruit or vegetable sample or the fungal culture plate can be kept inside the airtight chamber connected to the needle-fitted airtight portable gas analyser through inlet and outlet tubes. A micro-pump is adjusted to circulate airflow and levels of ethylene and they were measured continuously at every second. It is generally used in the case of grapes to detect *Botrytis cinerea* infection. Here, we can notice fungal ethylene production along with mycelial growth (under the microscope) even though the external appearance of the fruit is healthy. Hence, it makes it easy for the suppliers and distributors to separate the contaminated fruits and vegetable batches and prevent the spread of the diseases to combat post-harvest losses. It can be efficiently used to sort the produce efficiently and helps in long-distance transit or extended storage (Sorocki and Rydosz, 2019).

Advantage:

- It is a non-destructive, accurate, small and portable device used to detect fungal phytopathogens.
- It is a quick method as it records each reading within 30 sec and long-lasting device.

Disadvantage:

- As it only detects the variation in ethylene production, it may even be produced by changes in the physiological condition; therefore, it may lead to wrong depictions.

1.3.13.2 X-ray Technology

This technique has been applied regularly in industry and quarantine units for detecting early infestation of mango spongy tissue. X-ray scanning is a non-destructive technique used to identify *Aspergillus niger*, *Colletotrichum gloeosporioides* and *Rhizopus stolonifer* infection in mango fruit. Where the fungus-infected fruit showed dark areas, while the uninfected fruit showed a uniform grey area (Dhondiram and Ashok, 2017).

Advantage:

- It may be useful for traders exporting bulk quantities of fruits where the infestation is detected.

Disadvantage:

- X-rays may be somewhat harmful.

1.4 OTHER METHODS FOR DETECTION OF FUNGI

1.4.1 Optical Sensor

It is an instrument used to measure the amount of light rays present in the product and convert it into an electrical signal. Finally, the signal can be interpreted with an electronic instrument or by a person. These are more sensitive, small, inexpensive and portable to use. These sensors will mainly work on the principle of light absorption or reflection. By measuring the intensity of the reflected light, the optical sensor can detect whether the plant is healthy or diseased. Pieczywek et al. (2017) reported the use of optical sensors for the diagnosis of fungal storage diseases in apples by comparing the symptoms of infected fruits with those of healthy fruits.

Advantage:

- These are more sensitive, small, inexpensive and portable to use devices for detecting the fungal pathogen.

Disadvantage:

- There may be confusing results as both physiological disorders and pathogens yield similar kinds of symptoms.

1.4.2 Lateral Flow Devices (LFD)

It is a simple, rapid, low cost and in-field paper-based dipstick assay to detect the fungal pathogens in liquid samples within 5–30 min. LFD is a suitable pre-screening method because of its high diagnostic sensitivity and specificity.

Application: In order to detect *Phytophthora ramorum* in the field, the infected sample bits are placed in a bottle containing 5 ml LFD buffer C along with five bar bearings and shaken well for 2–3 min. Then 60 µl buffer sample will run on the LFD for 5 min. The formation of two lines indicates the presence of *P. ramorum*, whereas single line depicts the absence of the fungi.

Advantage:

- It is a low-cost, rapid, sensitive and simple operated technique, which makes it a particularly suitable method for on-site fungal phytopathogen diagnosis.

Disadvantage:

- They are less precise than other nanotechnology-based methods because of the variety of potential inorganic-biological problems that could result in non-specific adsorption and target analytes being destroyed.

1.4.3 Image Analysis

It is the non-destructive and non-visible detection measure used to monitor fungal phytopathogens, where the disease plant image is captured using digital cameras or flatbed scanners and further edited by image processing software.

Application: Image analysis develops standard area diagrams, which help in the visual assessment of disease severity such as *Alternaria* leaf spot of sunflower and stem rust in ryegrass (Bock and Nutter, 2011).

Advantage:

- It is a more accurate, efficient and precise method for detection of the fungal phytopathogens.

Disadvantage:

- It is complex and challenging in terms of high variability in field conditions.
- Various fungal phytopathogens produce similar kinds of external symptoms on plant surfaces, hence difficult to detect properly.

1.4.4 Thermography

It is a potential technology to supervise the heterogeneity in the fungal infection because it enables imaging of the variation in the temperature of plant leaves and canopies. It is effectively employed for the recognition of stress. In thermography, the camera confines released infrared radiation and variation in the colour will be estimated. Here the impact of microorganisms on host plant transpiration is studied with the help of thermography. Initially, pathogenesis results in the increase or reduction in leaf temperature, impacting either inoculation sites or the entire leaf, depending on pathogen interaction with host plants. Temporary temperature decrease may occur as symptoms progress because the afflicted cells are dying and losing water. Lack of natural cooling of the necrotic tissues and later infection phases typically result in an increase in temperature (Hillnhutter et al., 2011).

Application: *Blumeria graminis* and *Puccinia striiformis* infected wheat plants exhibited low temperature because of susceptibility to evaporation in the apical mycelia. A soil-borne hemibiotrophic fungus, *R. solani*, causes necrosis of lettuce, which is observed with increased leaf temperature, the necrotrophic fungus *Cercospora beticola* displayed low temperature.

Advantage:

- It is suitable for pre-symptomatic detection of pathogen activities within plant tissue.

Disadvantage:

- It lacks diagnostic potential as various pathogens and abiotic factors affect the tissue temperature in a very similar manner.

1.4.5 Fluorescence Imaging

It measures the chlorophyll content of the fluorescence on the leaves. Pathogen contagions can be examined on the basis of photosynthetic apparatus changes as well as the electron transport system (Burling et al., 2011). The diagnosis of leaf rust and powdery mildew in wheat leaves at 470 nm, this method is used to analyse the difference in the chlorophyll fluorescence.

Advantage:

- It helps in the identification of fungal diseases in the field with more accuracy.

Disadvantage:

- The cost is high compared to other methods.

1.4.6 Hyperspectral Imaging

It is another new technology that can be used to identify plant pathogens indirectly. It can be used to gather important data about the health of the plant at a wider range of 350 and 2500 nm. In large-scale agriculture, it is increasingly employed for phenotyping plants and identifying crop diseases. This method offers quick analysis of imaging data and is very reliable. By monitoring the variations in reflectance brought on by the metabolic alterations due to infection, hyperspectral techniques are employed for plant pathogen detection.

Application: It is generally implemented in the detection of *Pyricularia oryzae* in the paddy, *Phytophthora infestans* in tomatoes and *Venturia inaequalis* in apples (Delalieux et al., 2007).

Advantage:

- It offers quick analysis of imaging data and is very reliable.
- Among the variety of plants under consideration, it aids in the more reasonable diagnosis of various diseases.

Disadvantage:

- The data may be a bit confusing as the abiotic stress may also induce a similar kind of variation in biophysical and biochemical characteristics.

1.4.18 Remote Sensing

An unusual technique for non-disparage assessing plant infection quickly, constantly and over a larger region without physical contact with the sample. This technique involves the measurement of electromagnetic radiation reflected from the vegetation (plant population) usually in the invisible to middle infrared ranges (390–2,500 nm). The potential for detecting rice sheath blight disease using multispectral remote sensing was demonstrated by Qin and Zhang (2005).

Advantage:

- It assesses the plant disease rapidly and repeatedly.
- It is an efficient way to identify diseased plants in the field.

Disadvantage:

- To analyse the image data, specialized training is required.
- The overall cost is high as it requires a series of aerial photos to analyse the data.

1.5 ADVANTAGES OF CONVENTIONAL DETECTION METHODS

- These techniques are still regarded as the best resources for finding comprehensive information on the fungi that coexist in various ecosystems across the globe.
- It is the only criterion for determining basic information about individual species, such as geographic range, host connections and ecological distribution, because DNA-based sequencing data of all fungi are not available.
- These techniques are less expensive and require less specialized equipment to conduct than an advanced one.
- These techniques make it easier to study how environmental circumstances affect fungal growth.
- Overall, these strategies are key factors for many researchers in developing countries where obtaining appropriate research funding is still a major difficulty.

1.6 LIMITATIONS OF CONVENTIONAL DETECTION METHODS

- In comparison to more sophisticated approaches, the meticulous sampling, culturing, isolation and identification processes take a lot longer.
- These approaches are not appropriate for the fungus that cannot grow or develop reproductive structures on the culture media and thus constitute a significant limitation in recognizing, classifying and delineating the fungi of a certain area.
- Identification of plant disease by the specialist with the knowledge of symptoms or signs found on the affected tissues. Many diseases have symptoms that cannot be identified visually from one another, complicating the pathogen's diagnosis and frequently making even eye exams ineffectual. Accurate identification calls for the involvement of trained experts.
- Many plant infections have a great deal of morphological similarity with one another, making them challenging to distinguish. Discriminating between populations of the same pathogen that have distinct characteristics, such as fungicide resistance, toxin production or variations in virulence, can occasionally be challenging and time-consuming.
- Additionally, the need for high accuracy as well as quick detection methods is particularly important for quarantining pathogens due to the danger of disease and the spread of the inoculum.

1.7 NEED FOR MOLECULAR DETECTION OVER CONVENTIONAL

These methods have a number of potential advantages over traditional diagnostic procedures, including the fact that they are more precise, quicker and usable by any employees without specialized taxonomical knowledge. More importantly, these methods enable the detection of microscopic organisms that are not able to culture under laboratory conditions. Additionally, molecular identification methods are helpful in identifying novel diseases with unknown aetiologies. These tools can be used to accurately quantify the pathogen's biomass in addition to detecting its existence. Genomic sequence comparison enables the reconstruction of the phylogenetic relationship of pathogens at various revolutionary stages (Ghosh et al., 2019).

1.8 SUMMARY

Fungi that are plant pathogenic inflict significant harm to a wide variety of crop plants globally. Every year, there are enormous economic losses as a result of the fungal damage, which is frequently severe to crop production. Diagnostic methods for plant fungal pathogens have greatly improved our capacity of the laboratory for detection and analysis more effectively, and unswervingly in the field. For disease surveillance and the creation of cutting-edge disease management practices, improved perceptiveness of pathogenicity and quick, precise identification of fungal infections to the level of species are prerequisites. Furthermore, the detection of fungal disease in a field will assist growers in making informed choices regarding resistance management strategies for disease prevention. Despite this, there is still a glaring knowledge vacuum in this field of study, since no method can satisfy the expanding need for quicker, more efficient, reproducible and sensitive results. This gap can be effectively and successfully addressed by advanced molecular techniques based on nucleic acids, enzymes and proteins.

1.9 FUTURE PERSPECTIVE

Conventional detection techniques are still considered the "gold standard" in the identification of fungal pathogens. We are mainly concentrated on providing an up-to-date and detailed account of the various conventional methods engaged for early identification of fungal phytopathogens along with benefits and lacunas. It acts as a key reference chapter for anyone interested in diagnosing fungal pathogens in an easy and inexpensive way. These traditional methods are considered the fundamental basements on which the molecular detection methods have been built. Even though they are basic guiding tools, there is a thrust for the development of accurate and quick detection techniques which work at a genomic level and other high-throughput methods for early detection of fungal pathogens at the field level, because if the pathogen is correctly detected at an early stage, half of the problem is solved at this stage only as it not only detects the pathogen, but it also prevents its further dissemination.

STUDY QUESTIONS

1. Write the advantages and disadvantages of conventional methods of detection of fungal pathogens.
2. Differentiate between slide culture and cover slip culture technique.
3. What are bait plants?
4. What is a portable gas analyser and how it is used in detecting fungal phytopathogens?

ACKNOWLEDGEMENT

The senior author is grateful to all the authors for the support, given on the successful completion of the chapter.

CONFLICT OF INTEREST

The authors confirm that they have no conflict of interest.

FURTHER READING

Bhunjun, C. S., Phukhamsakda, C., Jayawardena, R. S., Jeewon, R., 2021. Investigating species boundaries in *Colletotrichum*. *Fungal Diversity* 107(14), 107–127.

Kumar, J., Kumar, B., 2022. Detection and identification of seed-borne fungal pathogens associated with seed discoloration of rice. *The Pharma Innovation Journal* 11(9), 3091–3096.

Renuka, M., 2021. Variability and characterization of fumonisin produced by *Fusarium verticillioides* incitant of ear rot of maize. M. Sc. (Agri) Thesis, University of Agricultural Sciences, Raichur, p. 87.

REFERENCES

Bhunjun, C. S., Phukhamsakda, C., Jayawardena, R. S., Jeewon, R., 2021. Investigating species boundaries in *Colletotrichum*. *Fungal Diversity* 107(14), 107–127.

Bisen, P. S., 2014. *Laboratory protocols in applied life sciences*. CRC Press, Boca Ratan, FL., USA, pp. 771–888.

Bock, C. H., Nutter, F. W., 2011. Detection and measurement of plant disease symptoms using visible-wavelength photography and image analysis. *Critical Reviews in Plant Sciences* 6, 73–88.

Brito, R. A. S., Cavalcant, G. P., Borel, F. C., Maffia, L. A., 2019. Detection and isolation of *Ceratocystis fimbriata* in mango trees on semi-selective medium. *European Journal of Plant Pathology* 155: 667–669.

Burling, K., Hunsche, M., Noga, G., 2011. Use of blue green and chlorophyll fluorescence measurements for differentiation between nitrogen deficiency and pathogen infection in winter wheat. *Journal of Plant Physiology* 168, 1641–1648.

Cappelli, C., Covarelli, L., 2005. Methods used in seed pathology and their current improvements. *Phytopathologia Polonica* 35, 11–18.

Cheng, Y., Tang, X., Gao, C., Li, Z., Chen, J., Guo, L., Wang, T., Xu, J., 2020. Molecular diagnostics and pathogenesis of fungal pathogens on bast fiber crops. *Pathogens* 9 (3), 223–242.

Crous, P. W., Hawksworth, D. L., Wingfield, M. J., 2015. Identifying and naming plant-pathogenic fungi: past, present and future. *Annual Review of Phytopathology* 53, 247–267.

Dass, R. S., Sreenivasa, M. Y., Janardhana, G. R. 2007. High incidence of *Fusarium verticillioides* in animal and poultry feed mixtures produced in Karnataka, India. *Plant Pathology Journal*. 6(2), 174–178.

Deepa, N., Charith Raj, A. P., Sreenivasa, M. Y. 2016a. Nested PCR method for the early detection of fumonisin producing *Fusarium verticillioides* in pure cultures, cereal samples and plant parts. *Food Biotechnology*. 30(1), 18–29.

Deepa, N., Charith Raj, A. P., Sreenivasa, M. Y. 2016b. Multiplex PCR method for the early detection of fumonisin producing *Fusarium verticillioides*. *Food Bioscience*. 13, 84–88.

Deepa, N and Sreenivasa, M. Y.. 2019a. Molecular methods and key genes targeted for the detection of fumonisin producing *Fusarium verticillioides* - An updated review. *Food Bioscience*. 32, 100473.

Deepa, N., Sreenivasa, M. Y. 2019b. Biocontrol Strategies for Effective Management of Phytopathogenic Fungi Associated with Cereals, In: Singh, J. S., Singh, D. P.. (eds) *New and Future Developments in Microbial Biotechnology and Bioengineering*, Elsevier, pp, 177–189, ISBN 9780444641915.

Delalieux, S., Van Aardt, J., Keulemans, W., Schrevens, E., Coppin, P., 2007. Detection of biotic stress (*Venturia inaequalis*) in apple trees using hyperspectral data: Non-parametric statistical approaches and physiological implications. *European Journal of Agronomy* 27, 130–143.

Derbyshire, M. C., Michaelson, L., Parker, J., Kelly, S., Thacker, U., Powers, S. J., Bailey, A., Hammond-Kosack. K., Courbot, M., Rudd, J., 2015. Analysis of cytochrome b(5) reductase-mediated metabolism in the phytopathogenic fungus *Zymoseptoria tritici* reveals novel functionalities implicated in virulence. *Fungal Genetics and Biology* 82, 69–84.

Dhondiram, G., Ashok, C., 2017. Detection of post-harvest fungal diseases of mango by X-ray scanning non-destructive technology. *Plant Pathology and Quarantine* 7(1), 65–69.

Ghosh, R., Tarafdar, A., Chobe, D. R., Chandran, U. S. S., Sudharani, S. M., 2019. Diagnostic techniques of soil borne plant diseases: recent advances and next generation evolutionary trends. *Biological Forum-An International Journal* 11(2), 1–13.

Goyal, S., Ramawat, K. G., Merillon, J. M., 2017. Different shades of fungal metabolites: An overview. In *Fungal Metabolites; Reference Series in Phytochemistry*, Merillon, J. M., Ramawat, K., Eds., Springer, Berlin/Heidelberg, Germany, pp. 1–29.

Harris, J. L., 2000. Safe, low-distortion tape touch method for fungal slide mounts. *Journal of Clinical Microbiology* 38, 4683–4684.

Hillnhutter, C., Mahlein, A. K., Sikora, R., Oerke, E. C., 2011. Remote sensing to detect plant stress induced by *Heterodera schachtii* and *Rhizoctonia solani* in sugar beet fields. *Field Crops Research* 122, 70–77.

Janardhana, G. R., Dass, R. S., Sreenivasa, M. Y. 2009. Fumonisins – A new class of Fusarial toxins. *In Frontiers in Fungal Ecology, Diversity and Metabolites*. Ed: K. A. Sridhar et al. I. K. International Publishing House Pvt. Ltd. New Delhi.

Jelusic, A., Popovic, T., Dimkic, I., Mitrovic, P., Peeters, K., Miklavcic Visnjevec, A., Tavzes, C., Stankovic, S., Berić, T., 2021. Changes in the winter oilseed rape microbiome affected by *Xanthomonas campestris* pv. *campestris* and biocontrol potential of the indigenous *Bacillus* and *Pseudomonas* isolates. *Biological Control* 160, 1–17.

Jung, B., Lee, S., Ha, J., Park, J. C., Han, S. S., Hwang, I., Lee, Y. W., Lee, J., 2013, Development of a selective medium for the fungal pathogen *Fusarium graminearum* using toxoflavin produced by the bacterial pathogen *Burkholderia glumae*. *Plant Pathology Journal* 29(4), 446–450.

Kang, Y., Lee, S. H., Lee, J., 2014. Development of a selective medium for the fungal pathogen *Cylindrocarpon destructans* using radicicol. *Plant Pathology Journal* 30(4), 432–436.

Kang, Z. S., 2010. Current status and development strategy for research on plant fungal diseases in China. *Plant Protection* 36(3), 9–12.

Kumar, J., Kumar, B., 2022. Detection and identification of seed-borne fungal pathogens associated with seed discoloration of rice. *The Pharma Innovation Journal* 11(9), 3091–3096.

Kusari, P., Spiteller, M., Kayser, O., Kusari, S., 2014. Recent advances in research on *Cannabis sativa* L. endophytes and their prospect for the pharmaceutical industry. *In microbial diversity and biotechnology in food security*; Springer: Berlin/Heidelberg, Germany, pp. 3–15.

Mahmoud, S. Y. M., Hosseny, M. H., El-Shaikh, K. A. A., Obiadalla, A. H. A., Mohamed, Y. A., 2013. Seed borne fungal pathogens associated with common bean (*Phaseolus vulgaris* L.) seeds and their impact on germination. *Journal of Environmental Studies* 11, 19–26.

Marquez, N., Giachero, M. L., Declerck, S., Ducasse, D. A., 2021. *Macrophomina phaseolina*: General characteristics of pathogenicity and methods of control. *Frontiers in Plant Science* 12, 1–16.

Melotto, M., Underwood, W., Sheng, Y. H., 2008. Role of stomata in plant innate immunity and foliar bacterial diseases. *Annual Review of Phytopathology* 46, 101–122.

Mitchell, J. L., Britt, E. M., 1981. A cover slip culture technique for preparing permanent fungus mounts. *Mycopathologia* 76(1), 23–25.

Monje, J. D., Mejia, J. L., 2020. Botanical microbiomes on the cheap: Inexpensive molecular fingerprinting methods to study plant-associated communities of bacteria and fungi. *Applications in Plant Sciences* 8(4), 1–12.

Morita, Y., Tojo, M., 2007. Modifications of PARP medium using fluazinam, miconazole and nystatin for detection of *Pythium* spp. in soil. *Plant Disease* 91, 1591–1599.

Nallathambi, P., Padmanaban, P., Mohanraj, D., 2001. Standardization of an indirect ELISA technique for the detection of *Ustilago scitaminea* Syd., causal agent of sugarcane smut disease. *Journal of Mycology and Plant Pathology* 31(1), 76–78.

Narayanasamy, P., 2010. Detection of fungal pathogens in plants. In: *Microbial plant pathogens-detection and disease diagnosis*. Springer, pp. 5–199.

Narayanasamy, P., 2019. Detection and identification of soilborne fungal plant pathogens: Soilborne microbial plant pathogens and disease management, *Nature and Biology* 1, 1–82.

Narciso, J. A., Parish, M. E., 2007. Direct plating: a method for detecting fungal contamination in paper board cartons. *Journal of Food Safety* 21(3), 141–151.

Nezhad, A. S., 2014. Future of portable devices for plant pathogen diagnosis. *Lab on a Chip* 14(16), 2887–2904.

Pellegrino, C., Gilardi, G., Gullino, M. L., Garibaldi, A., 2010. Detection of *Phoma valerianellae* in lamb's lettuce seeds. *Phytoparasitica* 38, 159–165.

Pieczywek, P. M., Cybulska, J., Chargot, S. M., Siedliska, A., Zdunek, A., Nosalewicz, A., Baranowski, P., Kurenda, A., 2017. Early detection of fungal infection of stored apple fruit with optical sensors-comparison of biospeckle, hyperspectral imaging and chlorophyll fluorescence. *Food Control* 85, 327–338.

Qin, Z., Zhang, M., 2005. Detection of rice sheath blight for in-season disease management using multispectral remote sensing. *International Journal of Applied Earth Observation and Geoinformation* 7(2), 115–128.

Ray, M., Ray, A., Dasha, M. A., Achary, K. G., Nayak, S., Singh, S., 2017. Fungal disease detection in plants: traditional assays, novel diagnostic techniques and biosensors. *Biosensors and Bioelectronics* 87, 708–723.

Renuka, M., 2021. Variability and characterization of fumonisin produced by *Fusarium verticillioides* incitant of ear rot of maize. M. Sc. (Agri) Thesis, University of Agricultural Sciences, Raichur, p. 87.

Riddell, R. W., 1950. Permanent stained mycological preparations obtained by slide culture. *Mycologia* 42, 265–270.

Sharma, P., Sharma, S., 2016. Paradigm shift in plant disease diagnostics: A journey from conventional diagnostics to nano-diagnostics. In *Current Trends in Plant Disease Diagnostics and Management Practices*. Eds. Kumar, P., Gupta, V., Tiwari, A., Kamle, M., (Cham: Springer), pp. 237–264.

Sheppard, J. W., 1979. Methods for routine detection of seedborne fungal pathogens. *Seed Technology Journal* 4(2), 74–77.

Sinha, O. K., Singh, K., 1982. Stain technique for detection of smut hyphae in buds of sugarcane. *Plant Disease* 66(10), 932–933.

Sodikov, B., Khakimov, A., Rakhmonov, U., Omonlikov, A., Gulmatov, R., Utaganov, S., 2022. Soil-borne plant pathogenic fungi biodiversity of sunflower, sustainable management of earth resources and biodiversity, *IOP Conference Series Earth and Environmental Science* 1068(1), 1–9.

Sorocki, J., Rydosz, A., 2019. A prototype of a portable gas analyzer for exhaled acetone detection. *Applied Sciences* 9(13), 1–13.

Sreenivasa, M. Y., Dass, R. S., Charith Raj, A. P., Janardhana, G. R. 2008. PCR-based detection of genus *Fusarium* and Fumonisin-producing isolates from freshly harvested Sorghum grains grown in Karnataka, India. *Journal of Food Safety* 28, 236–237.

Suresh, S., Arumugam, D., Zacharias, G., Palaninathan, S., Vishwanathan, R., Venkatraman, V., 2016. Prevalence and clinical profile of fungal rhinosinusitis. *Allergy Rhinology Journal* 7(2), 1–7.

Tsedaley, B., 2015. Review on seed health tests and detection methods of seedborne diseases. *Journal of Biology, Agriculture and Healthcare* 5(5), 1–10.

Tsui, C. K., Woodhall, J., Chen, W., Lévesque, C. A., Lau, A., Schoen, C. D., Baschien, C., Najafzadeh, M. J., de Hoog, G. S., 2011. Molecular techniques for pathogen identification and fungus detection in the environment. *IMA Fungus* 2(2), 177–189.

Venbrux, M., Crauwels, S., Rediers, H., 2023. Current and emerging trends in techniques for plant pathogen detection. *Frontiers in Plant Science* 14, 1–25.

Vico, I., Duduk, N., Vasic, M., Nikolic, M., 2014. Identification of *Penicillium expansum* causing post-harvest blue mold decay of apple fruits. *Journal of Pesticides and Phytomedicine* 29(4), 257–266.

Vishunavat, K., Prabakar, K., Anand, T., 2023. Seed health: testing and management. In: Dadlani, M., Yadava, D. K., (eds) *Seed science and technology*. Springer, Singapore. pp. 335–364.

Wenneker, M., Thomma, B. P. H. J., 2020. Latent post-harvest pathogens of pome fruit and their management: from single measures to a systems intervention approach. *European Journal of Plant Pathology* 156(3), 663–681.

2 Fungal Phytopathogen Diagnostic Approaches
A Historical Perspective

H. V. Girish
University of Mysore

M. P. Raghavendra
Maharani's Science College of Women

Fungi are a fascinating world of microorganisms placed in a separate kingdom due to their characters being distinct from all other living organisms. The first report of fungal spore observation dates back to the fifteenth century, mostly the first recorded document of fungal spore description by Giambattista delia Porta. Later, Pier Antonio Micheli, Christiaan Hendrik Persoon, Elias Magnus Fries and others contributed to the understanding of mycology in general. Fungal taxonomy is an important branch wherein the classification of fungi, initially, was done on the basis of spore colour and microscopic observations. Advances in genetics, molecular biology, biotechnology, DNA sequencing, biochemistry and phylogenetic analysis in due course have revolutionized fungal identification from the gold standard classical methods, viz., microscopic, morphological, histopathological, symptom-based, incubation, standard blotter, agar plate and cultural characteristics to more advanced molecular diagnostic approaches. Closely related organisms at different taxonomic levels, biotypes, vegetative incompatible/compatible groups, different sexual morphs with one or more asexual states, complexity of lineage-dependent evolutionary processes and non-availability of thorough and experienced fungal taxonomists to evaluate data are the challenges faced at different levels of fungal phytopathogen diagnosis. This chapter describes briefly the different techniques which laid the foundation for diagnosing phytopathogens and the need for holistic diagnosis to increase sensitive and specific rapid detection.

DOI: 10.1201/9781003399704-2

GRAPHICAL ABSTRACT

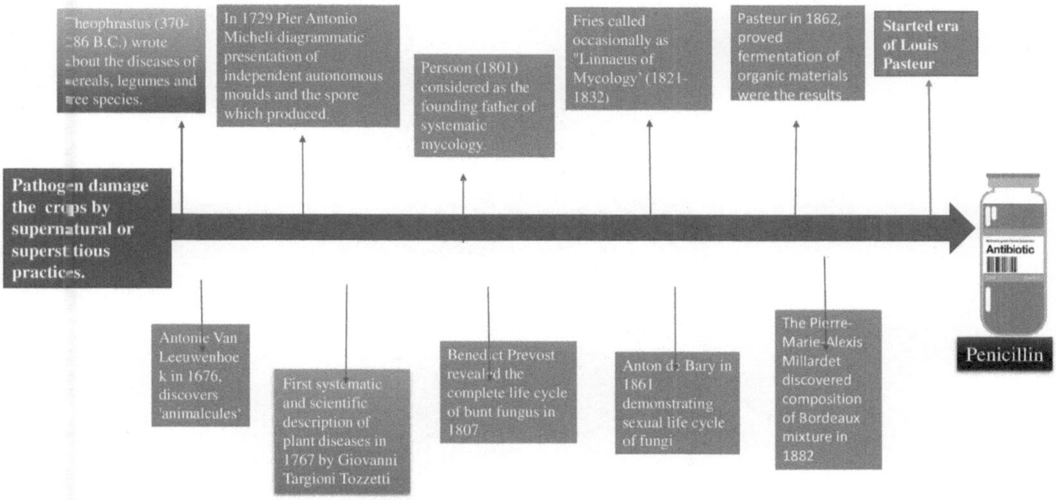

Graphical representation of fungal phytopathogens diagnostic: A historical perspective

2.1 INTRODUCTION

There are proportionate evidences to rely on that microbes were the earliest of the group of biological forms to appear on this earth. The other biological forms, together with all plants and animals, are believed to have appeared on this planet a few million centuries later (Agrios, 2005; Rangaswamy and Mahadevan, 2006). About ten thousand years ago, by the time humans were knowledgeable enough to grow plants for their well-being, microbes had already increased exponentially in total numbers and microbial biodiversity had amplified and rolled out comprehensively throughout this planet. Presently, microorganisms are affiliated with both living organisms and non-living substances (Sreenivasa et al., 2008; 2010). Their continuous association with numerous matters, predominantly with other biological organizations, has given rise to particular adaptations and distinct features between them. Such particular adaptation features have affected evolutionary variations to result in dissimilar types and forms of organisms, which we now distinguish into species and genera. Plant and microbe association are consequently as long as the plants themselves (Rangaswamy and Mahadevan, 2006). With time, the association between microorganisms and plants became increasingly multifaceted. In this evolutionary process, microorganisms which were not virulent enough to compete with others were wiped out, whereas virulent organisms persisted (Sreenivasa et al., 2011; 2011). Such variations continued to take place in an everlasting succession. Though humans were not conscious of the role of microbes in producing destructive and favourable effects on crop varieties, they could, from the very initial days, distinguish between a diseased and a normal plant (Nagaraja et al., 2016).

2.2 PREHISTORIC RATIONAL

Knowledge of plant diseases is a few thousand centuries old, and it is not acknowledged precisely when humans came to realize their predominance. However, there are definitely no written documents of the earliest observations made, paleontological data have publicized the origins of some

FIGURE 2.1 Bust of Giovanni Targioni Tozzetti.

plant specimens, predicted to be around 230 million years old, comprising some fossil fungi in the cortex. At Florence Botanical Garden in 1767 the discipline was established by Giovanni Targioni Tozzetti (Figure 2.1) investigated the reason for a famine and discovered *Puccinia gramini*, agent of "rust of wheat", in central Italy (Martinelli et al., 2015), there is further direct proof of the knowledge of disease in plant, which is documented in the Ancient Testament. Mildews and Blight crop varieties were more common during earlier times (Egerton, 2012). The Romans acknowledged rust disease in wheat, as it affected their lives to a substantial magnitude. The destruction caused by some diseases was known to such a level that they celebrated distinct festivals to ward them off. For instance, the Rust God Robigo was honoured by the Romans at a distinct festival called 'Robigalia', when their wheat crop reached maturity (Rangaswamy and Mahadevan, 2006). There are details of blight mildew and rust diseases of different crop varieties of that period in the writings dating back to the first century B.C. Theophrastus, Varro, Secundus, Pliny, Columella and others transcribed methods to check these diseases. Though no novel knowledge on diseases in plants was established in the course of the Dark Era, numerous new crop diseases of plants became known in the course of the Shakespearean Era. Nevertheless, the old impression, that diseases were God-sent, continued unchallenged in the seventeenth century also.

The most primitive well-known writings make allusions to the ravages of diseases in plants. There are many more references in the religious holy literature about the incidence of diseases in plants. Artha Shastra of Kautilya (321–186 BC), Atharveda (1500–500 BC), Vishnu Puran (500 AD) and the Agnipurana (500–700 AD), Rigveda, etc., are the few ancient religious holy books in which they have pointed out on the diseases in plants and causative agents along with their management strategies. The learned humans in the Vedic era were acquainted with the fact that these plant diseases were affected by microbes. Plant disease symptoms were also revealed in Shakespeare's poems, dramas and Holy Bible (Agrios, 2005). The book "Vrikshayurveda" written by Surapal, roughly around 1000 years ago about prehistoric Indian science of plant life, contained evidence on diseases in plants (Rangaswamy and Mahadevan, 2006). This is the first Indian book, which gave foremost information on diseases in plants. Surapal divided diseases in plants into two major groups: viz., external and internal. Diseases in plants like downy mildew, rust, smut, blight and powdery mildew were revealed in the Holy Bible. Theophrastus, a famous Greek philosopher (370–286 B.C.), was the first person to analyse and write about the diseases of cereals, legumes and tree species. Theophrastus in his book *Enquiry into Plants* has documented his observations, thoughts and experiences, but these are not built on any experimental facts. He stated that plant disease of dissimilar groups has different diseases, which are spontaneous or it may be autonomous.

The Dutch scientist Antonie van Leeuwenhoek initiated a new era in biological science in 1676 with the discovery of microorganisms. Hesse, in 1690, wrote on diseases in plants subscribing to the predominant misconceptions as their causes. In 1705, J.P. de Tournefort of France revealed the internal and external causes of diseases in plants, a substantial stage forward for that period. Numerous new concepts were set forth, and suggestions were made by the scientists of the date on the possible involvement and role of microbes in natural progressions. The strong concept that life arose spontaneously through the action of God was questioned by few philosophers or researchers, and intense arguments and discussions on the concept of spontaneous generation were soon established among the important researchers of that period. The inventive experiments conducted by a few researchers, namely Redi (1668), Spallanzani (1775), Schwaan (1836), Cogniard-Latour (1836) and Schroeder and Dusch (1856), provided proof and evidence in contrast to the theory of spontaneous generation. Nevertheless, the protagonists of the concept were highly powerful and had abundant impact on the researchers of the period, so the controversy persisted notwithstanding the proofs and evidences against it, accumulated through numerous meticulously planned experiments by other scientists of the period. To conclude, the experiments and contributions of the great eminent French scholar Louis Pasteur (1822–1895) presented the theory as a death blow. Pasteur, in 1862, proved beyond doubt that putrefaction and fermentation of organic materials were the results of microorganism's action. His successive experiments proceeded in the finding of novel Germ theory and the transmissible nature of animal and human diseases and in the discovery of vaccination procedures for managing them. These significant discoveries in biological science greatly prejudiced the growth of plant pathology as a separate branch.

2.3 PLANTS AND DISEASES

Plants are a composition of the mainstream of the globe's living environment as herbs, shrubs, trees and so on. Incidentally, plants also make up all the nutrition on which animals and humans depend. Even the milk, eggs and meat that carnivore animals and humans consume are derived from animals, which are dependent on plants for their nutrition. Plants have a natural ability to translate the energy from sunlight into usable, stored chemical energy in the form of proteins, fats and carbohydrates (Schumann, 1991). All the existing creatures comprising humans depend on these plants and their products for existence, and so are the consumers in the ecosystem. Plants, either wild or cultivated, develop and produce good yields on condition that the soil supplies them with adequate moisture and nutrients, necessary light reaches their leaf surface, and, most importantly, the temperature remains in the normal range. No matter how efficient they are plants will also develop sickness. Diseased plants produce poorly; they show several types of signs or symptoms and frequently plant parts or entire plants will die. It is not acknowledged whether diseased parts of the plants sense discomfort or pain (Manoharachary, 2017).

The causative agents that cause disease in plants are very similar to those causing sickness in animals and humans. They comprise pathogenic microbes such as bacteria, viruses, protozoa, nematodes and fungi along with abiotic factors in the form of adverse environmental conditions such as excess or lack of mineral nutrients, sunlight and its wavelength, moisture and the existence of toxic chemicals in soil or air. Plant pathology as a branch of science makes an effort to enhance our knowledge and understanding of diseases in plants. Simultaneously, knowledge of plant pathology endeavours to develop methodologies, tools and equipment, and materials through which diseases in the plant can be managed or controlled (Mancini et al., 2016).

Plant diseases, if uncontrolled, lead to several consequences such as shortage of food and increase the food prices or end up with poor quality of food. Even diseased plant products may occasionally be toxic and unhealthy for consumption. Some of the deadly plant diseases might eradicate entire plant communities or particular species, and many disturb the ecological food chain in the environment (Deepa et al., 2016b; 2016c). Plant disease management strategies will increase the superior quality of food and more aesthetically pleasing habitat, but on the other hand the end user may have to pay the penalty for the usage of equipment, materials and exertion used to manage diseases in plants. Sometimes pollution in the environment also leads to diseases in crop plants (Deepa & Sreenivasa et al., 2019a; 2019b).

2.4 ROLE OF FUNGI

Fungi are the principal and major constituent of biodiversity, crucial for the survival of other creatures and also in universal ecological and biochemical processes. Fungal species are ubiquitous microorganisms available in all forms of habitats and are the utmost versatile organisms (Bajinka et al., 2017). Soil is the most significant habitat for microbes like yeasts, bacteria, nematodes, and fungi. Microorganisms such as fungi are the chief contributors to soil biomass (Janardhana et al., 2009; Alexander, 1977). The development, growth, diversity and distribution of fungal species are affected by diverse environmental conditions such as moisture, temperature, degree of aeration, pH, quantity and kind of available nutrients (Gaddeyya et al., 2012). Fungal organisms which are available in the soil, play a significant and dynamic part in sustaining fertility and productivity of the soil and are governed by numerous factors comprising soil parameters and actions of human beings (Dass et al., 2007; Bao et al., 2012). Fungal species are the most predominant microorganisms that inhabit the soil. They play a key role in ecological recycling and progressions that induce enhancement of the healthiness and overall improvement of the plants (Mulani and Turkmane, 2014). However, the aerial surface of the plant parts is colonized by numerous microbes, comprising fungi and yeast that cause the spoilage and deterioration of fruits and vegetables. These make fungi a very crucial concern, since plants are primary food sources for all the ecological food chain and end users, and, moreover, of high economic importance to farmers (Prabakaran et al., 2011; Alsohaili et al., 2018).

2.5 TRADITIONAL TAXONOMIC METHODS

In the initial stage, philosophers categorized topics into three different types: viz., vegetable, mineral and animal. Fungal species remained positioned in the group of vegetable kingdom since particular resemblances to plants, for instance, nutrition absorptive capability, absence of mobility and reproduction through spores (Rangaswamy and Mahadevan, 2006; Mancini et al., 2016; Raya et al., 2017). In fact, it was believed at one stage that fungal organisms had evolved from algal species by losing photosynthetic apparatus. These imply the practice of analogous taxonomic categories like phycomycetes, which accurately means "algal fungi". Algal fungi, also known as lower fungi, are nowhere used since they comprise taxa not now thought to be correlated to one another (mainly Chytridiomycota, Zygomycota and Oomycota,). The initial traditional taxonomic classification system was established on external morphological resemblance. However, the light microscope discovery revealed that the arrangement of fungal species such as fruiting bodies, which look identical, could be structurally dissimilar and replicate in inherently diverse ways, leading to categorized distance apart. Till the 1980s, the initial fungal classification was completely established primarily on light microscopic investigation of distinctive morphological characters, leading to classification systems. Numerous instances of unusual categories were observed by comparing the current system of classification with the earlier (Webster and Weber, 2007). Few instances of previous traditional classifications comprising the absence or presence of septa in hypha, adequate particulars of the type, development and spore liberation mechanisms (Kendrick, 1971; Guarro et al., 1999), or detailed aspects of the ecology and biology of fungi. Transmission electron microscope provided useful ultrastructural details such as septal pore properties, mitochondrial appearance, details of the cell wall during spore germination or development and secretory vesicle organization at the tip of developing hypha. Appreciable contributions were also made through biochemical approaches, particularly in illustrating higher or advanced taxonomic levels. Examples include the cell wall chemical composition, alternate pathways of biosynthesis of lysine, the pigments distribution and occurrence (Gill & Steglich, 2016) and the types and magnitudes of sugars or polyols used in metabolism (Pfyffer et al., 1986; Rast & Pfyffer, 1989). Microscopic observed characters are quite significant at present for distinguishing fungal species, and making a preliminary identification subsequently, if essential, is supported by advanced molecular methodologies. Definitely, the DNA sequence evaluation attained from fungal organisms is relevant only if these fungal species have been characterized earlier and entitled by normal conventional methodologies. Hence, it is very much essential today to elaborate art of the conventional way of examining and identifying fungi for all mycology researchers and students (Berbee and Taylor, 1992).

2.6 HOW OLD ARE FUNGI?

Numerous literature support that fungal species are the earliest class of microorganisms. Berbee and Taylor (2001) tried to make an effort to construct a timeframe of phylogenetic trees by applying "molecular clock" concept, that is, the speculation that the degree of mutation leads to an increase in phylogenetic diversity is persistent over a span of time and in numerous classes of microorganisms. Through the calibration of their molecular clock in contrast to the fossil record, Berbee and Taylor (2001) predicted that fungal species might have detached from animal group around 900 million years ago, that is, prior to the evolution of land-dwelling organisms. This estimation is reliable with the discovery of fossilized connecting links of hyphae-like arrangements in sediments around one billion years old (Butterfield, 2005). Fungal organisms distinguishable as Ascomycota, Chytridiomycota and Zygomycota have been revealed among the fossil records of early land-dwelling plants from the Rhynie chert is a lower Devonian, established around 400 million years ago (Taylor et al., 1992, 1999, 2005). This is more evident that these early old terrestrial plant species previously engaged in a symbiotic relationship of mycorrhizal with glomalin groups of Zygomycota (Varma and Kharkwal, 2009).

2.7 EARLY DEVELOPMENTS

Amongst the several causative agents of diseases in plants, fungal organisms stood the earliest to be known. This originated mainly through the result of three eminent researchers in Europe during the second half of the eighteenth century: extensive study of Tillet (1755) on the bunt disease of wheat and its life cycle and Fontana (1767) and Torgioni (1768) demonstrated that wheat rust is produced by a tiny microscopic organism (Rangaswamy and Mahadevan, 2006).

During the seventeenth and eighteenth centuries, quite a few biologists studied and designated fungal organisms; very significant is the group of Pier Antonio Micheli (1679–1737) in Italy; Christiaan Hendrik Persoon (1761–1836) in Europe; Joseph-Henry Leveille (1796–1870) in France; Edmond Tulasne and Charles Tulasne (1815–1885) in France; Lewis David von Schweinitz (1780–1834) in USA; Elias Magnus Fries (1794–1878) in Sweden; Christian Gottfried Daniel Nees Von Esenbeck (1776–1858) in Germany; Miles Joseph Berkeley (1803–1889) in England. Further few more biologists who contributed to the growth of plant pathology as a branch of science were Johann Christian Fabricius (1774) in Denmark; Johann Baptista Zallinger (1779) in France; Franz Unger (1800–1870) in Germany and Meyen (1841) in Germany. In 1807 Prevost showed the transmissible nature of wheat smut. Based on this research analysis, several other research reports were published on different crop plants caused by fungal organisms in the early part of the nineteenth century (Agrios, 2005; Rangaswamy and Mahadevan, 2006). Nevertheless, recognition for the original systematic examination of the pathogenic nature of plant pathogens, the role of fungal organisms in triggering the diseases and for a complete interpretation of the life cycle of the pathogens involved goes to eminent German researcher Anton de Bary (1831–1888), who is famously known as the Father of Plant Pathology (Rangaswamy and Mahadevan, 2006).

The ancient progression in mycology during the latter half of the nineteenth century focused mostly on Germany. During this time, leading researchers such as Julius Kuhn (1825–1910), Julius Oscar Brefeld (1839–1925), Heinrich Julius Adolph Robert Hartig (1839–1901), Albert Bernhard Frank (1839–1900), Frederik George Emil Rostrup (1831–1907), Georg Klebs (1857–1918), Woronin Michael Stephanovich (1838–1903) in Germany, Thomas Jonathan Burrill (1839–1916) in the United States of America, Savastano (1853–1910) in Italy and Harry Marshall Ward (1861–1906) in England mark substantial contributions, not merely through their innovative methodologies on various difficulties in plant pathology but also along with their skilled training of young researchers in the group and along with their documents and writings which blow out to different regions of the world (Webster and Weber, 2007).

In the later part of nineteenth-century British and Canadian Arthur Henry Reginald Buller (1874–1944), Jacob Eriksson (1896) in Sweden and Leon Hatchig Leonian-Mycotree (1888–1945) in the USA also ensured innovative research work for the development of plant pathological aspects. Pierre-Marie-Alexis Millardet discovered that the composition of the Bordeaux mixture in 1882 in France was historic and remarkable and led eventually to the management of several diseases in plants. The loss and damage originating from plant diseases have affected humans since immemorial (Agrios, 2005). Moreover, with the reappearance of mildew and rust in crop varieties, there have been widespread outbreaks of plant diseases, which caused an enormous increase in death rate and led to famine. The severe pandemic incidence of the potato late blight in Europe during 1830–1840 and far ahead in Ireland in 1845 caused the Irish Famine, which resulted in the migration of people to other countries, particularly to North America, and even the demise of about a million individuals. The periodic epidemics of ergot of rye in Europe grabbed numerous thousands of individuals during the eighteenth century. The severe epidemic of *Helminthosporium* in rice varieties, particularly in North East India, caused substantial loss of paddy grains, which is recorded in history as the Bengal Famine (1943), resulted in the death of a million individuals and also caused administrative conflict in the nation (Rangaswamy and Mahadevan, 2006).

2.8 DIVERSITY OF FUNGAL ORGANISMS

It has been a big debate in the group of mycologists for more than 200 years about which organisms must be counted as fungi. Elias Magnus Fries, in 1825, anticipated that the fungal organisms beyond doubt are the biggest in the group of the vegetable kingdom, equivalent to the insects in the kingdom of animals (Bing et al., 2019). Fungal species are one of the chief microbial constituents of all types of habitats. Since the 1860s, a lot of research has been carried out on the fungi of different habitats, such as soils of forest, marine, grasslands, desert, polar region, and mangrove habitats and coastal sand belt from several fragments of the biosphere (Rabosky and Daniel, 2009). Microbial communities predominantly fungi and bacteria are vital constituents of biological pyramids in all types of ecosystems. By 1960, hardly a few fungal biologists were hypothesizing that conceivably several fungi as plant species, but more or less no efforts were made to estimate them from the available data. Hawksworth published a paper in 1991, maybe the starting point in which he estimated roughly 1.5 million fungal organisms existing in natural ecosystem, but hardly 5–10% has been described precisely (Figure 2.2). During the past decade, because of new advanced DNA sequencing knowledge which transformed studies on diversity and taxonomy of fungi, leading to deviations in the estimations of fungal organisms' statistics based on host relationship extending from 2.2 to 3.8 million and based on high-throughput technology around 11.7 to 13.2 million species of fungi (Bing et al., 2019).

Nevertheless, only about 150,000 fungal organisms have been designated to date, and it is believed that these fungi encompass a considerably greater diversity. Given the crucial prominence

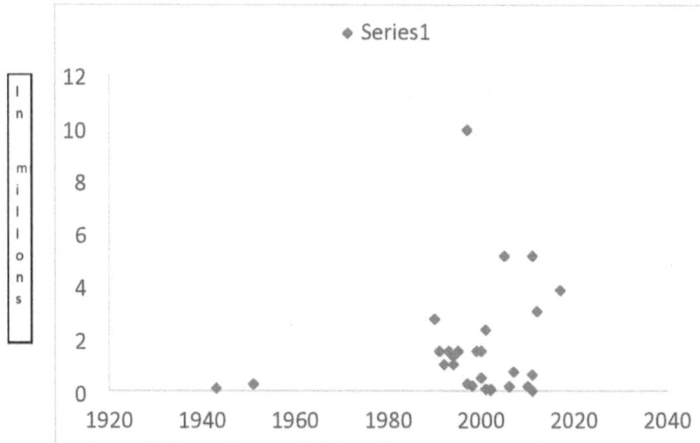

FIGURE 2.2 Diversity of the fungi.

of fungal species for ecosystem function, it is very much essential to examine diversity of fungal organisms, utilizing the most varied approaches and practical demonstrations of fungal diversity and to address their roles in the diverse ecosystem. Schmit and Mueller (2007) assessed that there are at least 712,000 fungi distributed universally. The actual number of fungal species is still mysterious, though, only 5–13% of the total estimation of global fungi was estimated to be described.

2.9 HISTORY OF FUNGAL DISEASE IDENTIFICATION IN PLANTS

Humans have frequently considered fungal species with impression of suspicion. These peculiar microorganisms become visible unexpectedly and for no evident reason. These fungal spécies were associated with deterioration, and few of them remained exceptionally poisonous (Kendrick, 2011; Deepa & Sreenivasa 2017a; 2017b). As typical, this undesirable assertiveness appeared basically from unawareness (earliest and utmost well-known mycological instances involving in assassination of the emperor "Claudius" by Agrippina, his wife, with noxious mushroom juice of *Amanita phalloides*). More conventional but very significant and mainly positive events, like the rising of bread doughing and the fermented beer and wine, remained well known for more than a thousand years earlier, before anybody associated with fungal species (Kendrick, 2011; Webster and Weber, 2007).

The research on fungal species properly commenced efficiently at around the same period, with the discovery of the microscope. In 1667, Robert Hooke was the first person to observe a fungal organism using the compound microscope. He anticipated that they appeared on decomposed parts of the plants by spontaneous generation, but he was suspiciously inclined to lay down principles as undeniably true about this. In 1689, Leeuwenhoek observed the budding cells of yeast (Rangaswamy and Mahadevan, 2006).

An Italian botanist in 1729, Pier Antonio Micheli (Figure 2.3), described 900 and plus moulds of *Botrytis* and *Aspergillus* with beautifully well-illustrated diagrammatic presentation. He knew that, before anyone else, they remained independent autonomous organisms, and he also knew in advance the role of the spore produced by the fungi. So, it is not exaggerated to call him the Father of mycology (Agrios, 2005).

Pier Antonio Michelli: an Italian botanist. He was the founder and father of mycology. He remained the first individual who witnessed fungal spores' identification for the first time and conducted numerous fungal spore germination experiments. He also confirmed that if spores of these fungi are positioned on the portions of the fruits, they develop into fresh thalli of the fungus (grown fungal organisms on cut portions of pears and melons). He was the first well-known individual to detect *Cystidia* on the lamellar edge or hymenial layer of Agaricales. He brought out the novel book

FIGURE 2.3 Pier Antonio Michelli.

Nova Plantarum Genera in 1729, in which he provided explanations about 1900 species in Latin out of which 900 plus were fungal species. The key genera are *Botrytis* sps., *Aspergillus niger*, *Polyporus* sps., etc.

After the demonstration of Antonio Michelli's experiments, subsequent prominent studies on fungal species were associated with plant diseases. In 1750, "The Bordeaux Academy of Arts and Science" announced an award for the best experiment and publication for the treatment of blackening of wheat, an infection called stinking bunt. Tillet grabbed the opportunity and taken as a challenge. Though he had no proper knowledge and preparation in botany and agriculture, he constructed an experimental model that efficiently tested all the concepts in his days (Rast and Pfyffer, 1989). Tillet separated a portion of land into four different small plots successively one way, and five small plots in succession crosswise at right angles – total of 20 strips – each in receipt of a diverse set of management (dissimilar type of treatment for seeds, manure and control strategies). He partitioned each small plot into six strips to examine the probable efficacy of altered planting times. He had a total of 120 dissimilar altered combinations, thus demonstrating a natural ability for investigational strategies (Kendrick, 2011).

Seeds contaminated by the black bunt dust continuously contribute an increased percentage of bunted ears. Lime-treated seeds sown in the plots had always less bunted ears (Deepa et al., 2016). Tillet (Figure 2.4) showed that the kind of manure and time of sowing had no significant effect by this he answered many rumours to the contrary. He is not aware of the fact that bunt disease was seed-borne; however, he had no complete knowledge of how it functioned, because plants developed from spore-contaminated seeds looked robust till they developed their own seeds (fungi grow inside the particular host for most of the period devoid of showing any kind of symptom) (Webster and Weber, 2007). Through well-organized experimentations, he demonstrated that wheat seeds with black powder spores on their superficial surface develop more diseased plants than uncontaminated seeds. However, this was prior to the period when the "germ theory of disease" was established, he ensured that the fungal spores were related to the disease in wheat varieties. However, he was convinced that it was a "poisonous entity" related to the black fungal spore masses that really caused the disease in plants, not the fungus itself. He experimentally demonstrated that the chemical treatment of seeds with lime and common salt suppressed the transmissible activity. The disease-causing fungus was future called *Tilletia* to honour his name. His way of demonstration and experimental strategy functioned beautifully, and he deserved to win the award (Kendrick, 2011).

Before the nineteenth century, the study of fungal organisms was not very well illustrated and did not gain respectability and urgency. A Swiss Professor of Philosophy, Benedict Prevost, at the Academy of France, revealed the complete life cycle of bunt fungus in 1807. Prevost did extensive

FIGURE 2.4 Tillet.

research on the germination of bunt fungal spores on wheat and came up with the idea that these fungal organisms entered the immature wheat plants, and this was the authentic reason for the wheat plants to get disease. Prevost's experimentations provided the primary evidence and understanding of the role of microbes in disease development. For more than 10 years, he extensively studied wheat bunt disease. He published his discoveries in the article "Memoir on the Immediate Cause of Bunt or Smut of Wheat and of Several Other Diseases of Plants and on Preventives of Bunt". He showed that the fungus *Tilletia caries* is the causative agent for the bunt of wheat. He also demonstrated management strategies to control smut by immersing seeds in the copper sulphate solution.

Christiaan Hendrik Persoon is commonly considered as the founder father of systematic mycology. He is recognized for his publications, particularly in mycology, which concluded in the Synopsis Fungorum (1801), which remained as the initial point for the naming of fungi in the orders Gasteromycetes, Ustilaginales and Uredinales and laid the foundation for modern mycology. Later, his classification was magnified by P. A. Saccardo and E. M. Fries. A minor adaptation of his classification system of the "macromycetes," presently identified as the Friesian tradition, still plays a significant role in mycology, though it was steadily being substituted by a thoroughly diverse system constructed mainly on microscopic characters. With the inspiration of the synopsis available during that period and subsequent to his extensive research publications, it was made possible to achieve unparalleled progress on the number of designated genera and species of fungi. Persoon initiated a prominently revised version of the synopsis under the entitled "Mycologia europaea". There were three volumes published (1822–1826), and he gave the label to rust pathogen of wheat as *Puccinia graminis* (Alfen, 2001).

Elias Magnus Fries (Figure 2.5) was a Swedish botanist and mycologist. He is called occasionally as "Linnaeus of Mycology". In his research work, he designated and gave scientific botanical names to hundreds of fungal species and lichens, several of which continue to be authoritative today. Originator of the first system used to classify fungal species, Fries had initiated to accumulate and designate well-known organisms for his Systema Mycologicum in three volumes (1821–1832), in which Fries established a novel system for the classification of fungal species.

FIGURE 2.5 Elias Magnus Fries.

With the exemption of limited variations with respect to microscopic identification, the classification system is still effective and valid for various classes of fungi even today. Fries also established a different system for the classification of lichens constructed on the morphological characters of their fruiting bodies (the organ that develops reproductive spores). This novel system of classification documented in his Lichenographia Europaea Reformata (1831) was broadly acknowledged until the usage of microscopic knowledge which revolutionized the field of mycology. Fries remained the first individual to differentiate between lichens with external morphological cover on the fruiting body and those devoid of. Regardless of Fries's frequently established brief explanations, thousands of drawings and paintings have been documented accurately to represent Fries's concepts of fungal taxonomy.

Diseases caused by fungal species in plants have been there since the beginning of agriculture, but not until the 1840s, after the devastating late blight of potato widespread in Ireland, the ultimate, modern era eruption on mycology along with the history of late blight of potato in the middle of the past century led to an emergence of a separate branch of plant science called as plant pathology. Plant pathologists and mycologists came to mean the same thing, and continued to do so for where they associated to their causative organisms. Rev. Berkeley MJ, England's leading mycologist, was persuaded that an oomycetous fungal species, now recognized as *Phytophthora infestans*, was the causative agent for the disease. The remnants of the British scientific community believed Berkeley was too preoccupied with fungal species, and they desired the faultiness of the disease on an additional supply of water content in the potato field, or steam locomotives generated by electricity or even on miasmas arising from subterranean volcanoes (Prasad et al., 2010). Ultimately, Berkeley was demonstrated experimentally right, and the special branch of plant pathology was started.

At the age of 22, Anton de Bary (Figure 2.6), in 1853, published his research work on smut and rust diseases in cereals. This publication was the first of his various outstanding contributions to the field of mycology. In 1861 he revealed that the motile flagellated zoospores from the fungus, which are chiefly responsible for the transmission of the potato blight disease. He furthermore predicted that the fungal species produces oospores sexually, and it is very difficult to find them. Later, the sexual phase was finally identified in Europe in 1910. At the time of 1863, Anton de Bary had established the complete life cycle of powdery mildew fungi. He ensured, by 1865, to solve the multifaceted life cycle of wheat rust fungus, and in 1870 de Bary evidenced that a plant-parasitic powdery mildew was continued to be parasitized by another fungal species. Understanding of diseases in plants was then far more progressive compared to human diseases knowledge. By 1866, he was

FIGURE 2.6 Heinrich Anton de Bary.

FIGURE 2.7 Louis Pasteur.

capable of publishing a fungi classification slightly similar to that acknowledged until very recent times (after molecular diagnosis started to make an impact). Later started the era of Louis Pasteur (Figure 2.7) (Table 2.1), an eminent microbiologist and chemist, who was responsible for the utmost significant scientific revolutions of the nineteenth century in the different areas of medicine, agriculture and hygiene. He soon embarked on a journey filled with lots of innovative discoveries, which led him to develop the vaccination.

2.10 SUMMARY

Fungal organism belongs to the kingdom that since antiquity has stood performing a very significant role in the ecosystem through the breakdown of dead organic substances to keep and balance the life cycle. In addition, fungal species are of abundant economic significance. They can be a source of causing considerable diseases in both plants and animals. They are occasionally however dependent

TABLE 2.1
Enlists the Glimpses of Some Historical Discoveries and Contributions of Mycology

Year	Name	Contribution
1676	Antonie van Leeuwenhoek	Developed the first microscope
1729	Pier Antonio Micheli	Well-illustrated diagrammatic presentation of independent autonomous moulds and the spore
1755	Mathieu Tillet	Bunt disease of wheat and its life cycle
1767	Giovanni Targioni Tozzetti	First systematic and scientific description of plant diseases
1801	Christiaan Hendrik Persoon	Considered as the founding father of systematic mycology
1807	Prevost	Primary evidence and understanding of the role of microbe in the relationship to cause the disease
1821	Elias Magnus Fries	Novel system for classification of fungal species – Linnaeus of mycology
1846	Miles Joseph Berkeley	They believed that late blight of potato was caused by the fungus
1866	Anton de Bary	First to indicate the nature of obligate and facultative forms, regarded as the "founder of modern mycology"
1839–1925	Oscar Brefeld	Discovered the methods of artificial culture of microorganisms
1861–1906	Harry Marshall Ward	Emphasized the role of environment in the epidemiology of coffee rust
1867	Louis Pasteur	Pasteurization – Later started era of Louis Pasteur
1885	Pierre Marie Alexis Millardet	Discovered the Bordeaux mixture for the control of downy mildew of grapes

on each other; an instance can be the synergetic relationship revealed by mycorrhizae (symbiotic relationship between roots of vascular plants and fungus).

Regardless of the economic loss caused by fungal species by causing diseases in plants, till today fungal species have been provided the very important medicinal drugs without number, and many more secrets may be hidden in fungal organisms; a few common ones comprise antibiotics and penicillin. The incidence of evolving plant pathogenic organisms continues to become a chief threat to food security, ecosystems, and the worldwide economy. Besides, key factors such as variations in climate, globalization, mutations and pathogen evolution have stimulated the spread of non-native plant pathogens. Perfect identification of fungal phytopathogens to the species or strain level is very crucial for fundamentally all phases of plant pathology, from fundamental research on the biology of pathogens to prerequisite for disease surveillance and improvement of innovative disease-controlling strategies. Definite and well-organized diagnostic tools have been established globally to diminish and avoid the phytopathogenic threat. The capability to identify, classify and quantify phytopathogens precisely is the landmark for the branch of phytopathology. The identification of the microorganisms through reliable means is very essential for the application of plant disease management strategies. Hence, in the phytodiagnosis of microbial infections, particularly of fungal species, it is very much a prerequisite to accomplish an accurate disease detection method. Since numerous fungal pathogenic organisms produce similar morphological symptoms and therefore it is very essential to be able to differentiate among different species. Thereby, precise and primary diagnoses of the plant pathogen are crucial to providing management strategies. The current chapter has given an in-depth exploration of historical account of mycology as contributed by eminent world and Indian mycologists.

2.11 FUTURE PERSPECTIVES

It is well known that fungi serve as mini-factories to synthesise several biomolecules of interest and fungal-derived biotechnological products, which are very important in major sectors, viz., healthcare, agriculture, pharmaceutical, food security, waste management and environmental stabilization. Proper identification and strain improvement strategies will pave the way for the development of efficient fungal strains of industrial importance.

STUDY QUESTIONS

1. Important discoveries in mycology
2. Role of fungi

ACKNOWLEDGEMENT

The authors are thankful to DOS in Botany, University of Mysore, Manasagangothri, Mysuru and PG Department of Microbiology, Maharani's Science College of Women (Autonomous), JLB Road, Mysuru, Karnataka, India. We thank Ms. Pushpa and Prasad Murthy for the timely sketch of all the scientists' photos we incorporated in this chapter.

FURTHER READING

Manoharachary, C. 2017. History of Mycology from India—Some Glimpses Springer Nature Singapore Pte Ltd. T. Satyanarayana et al. (eds.), *Developments in Fungal Biology and Applied Mycology*. DOI: 10.1007/978-981-10-4768-8_3.

REFERENCES

Agrios, G. N. 2005. *Plant pathology*. 5 Edition. Elsevier, Academic Press, Amsterdam, The Netherlands.

Alexander, M. 1977. *Introduction to Soil Microbiology*. John Wiley, New York.

Alfen, N.K.V. 2001. Fungal pathogens of plants. *Encyclopedia of Life Sciences*. DOI: 10.1038/npg.els.0000362

Alsohaili, S.A., Bayan, M., Bani-Hasan, 2018. Morphological and Molecular Identification of Fungi Isolated from Different Environmental Sources in the Northern Eastern Desert of Jordan. *Jordan Journal of Biological Sciences*, 11(3), 329–337.

Bajinka, O., Terzi, Y., Ucar, F. 2017. The Development of Diagnostics Tools and Techniques in the Isolation and Detection of Fungal Pathogens. *Journal of Infectious Diseases and Medicine* 2, 119. DOI: 10.4172/2576-1420.1000119.

Bao, Z., Ikunaga, Y., Matsushita, Y., Morimoto, S., Takada-Hoshino, Y. 2012. Combined analyses of bacterial, fungal and nematode communities in Andosolic agricultural soils in Japan. *Microbes and Environments*, 27,72–79.

Berbee M.L., Taylor, J.W. 1992. Two ascomycete classes based on fruiting-body characters and ribosomal DNA sequence. *Molecular Biology and Evolution* 9, 278–284.

Berbee, M.L., Taylor, J.W. 2001. Fungal Molecular Evolution: Gene Trees and Geologic Time, *The Mycota VII Part B Systematics and Evolution* McLaughlin/McLaughlin/Lemke (Eds.) Springer-Verlag Berlin Heidelberg.

Bing, W.U., Hussain M., Zhang W., Stadler M., Liu X., Xiang M.. 2019. Current insights into fungal species diversity and perspective on naming the environmental DNA sequences of fungi, *Mycology* DOI: 10.1080/21501203.2019.1614106.

Butterfield, N. J. 2005. Probable Proterozoic fungi. *Paleobiology*, 31, 165–182.

Dass, R.S., Sreenivasa, M.Y., Janardhana, G.R. 2007. High incidence of *Fusarium verticillioides* in animal and poultry feed mixtures produced in Karnataka, India. *Plant Pathology Journal* 6(2), 174–178.

Deepa, N., Charith Raj, A. P., Sreenivasa, M.Y. 2016b. Nested PCR method for the early detection of fumonisin producing *Fusarium verticillioides* in pure cultures, cereal samples and plant parts. *Food Biotechnology* 30(1), 18–29.

Deepa, N., Charith Raj, A.P., Sreenivasa, M.Y. 2016c. Multiplex PCR method for the early detection of fumonisin producing *Fusarium verticillioides*. *Food Bioscience* 13, 84–88.

Deepa, N., Nagaraja, H., Sreenivasa, M.Y. 2016. Prevalence of fumonisin producing *Fusarium verticillioides* associated with cereals grown in Karnataka (India). *Food Science and Human Wellness* 5, 156–162. DOI: 10.1016/j.fshw.2016.07.001

Deepa, N., and Sreenivasa, M.Y. 2017a. Fumonisins: A Review on its Global Occurrence, Epidemiology, Toxicity and Detection. *Journal of Veterinary Medicine and Research*. 4(6), 1093.

Deepa, N., and Sreenivasa, M.Y. 2017b. *Fusarium verticillioides*, a Globally Important Pathogen of Agriculture and Livestock: A Review. *Journal of Veterinary Medicine and Research*. 4(4), 1084.

Deepa, N and M Y Sreenivasa. 2019a. Molecular methods and key genes targeted for the detection of fumonisin producing *Fusarium verticillioides* - An updated review. *Food Bioscience* 32, 100473.

Deepa N and Sreenivasa, M.Y. 2019b. Biocontrol Strategies for Effective Management of Phytopathogenic Fungi Associated with Cereals, In: Singh JS, Singh DP. (eds) *New and Future Developments in Microbial Biotechnology and Bioengineering*, Elsevier, pp, 177–189, ISBN 9780444641915.

Egerton F.N. (2012) History of Ecological Sciences. Part 44: Phytopathology during the 1800s. *Bulletin of the Ecological Society of America* 93, 303–339.

Gaddeyya, G., Niharika, P. S., Bharathi, P., Kumar, P. K. R. 2012. Isolation and identification of soil mycoflora in different crop fields at Salur Mandal. *Advances in Applied Science Research*, 3,2020–2026.

Gill, M. and Steglich, W. 2016. Pigments of fungi (Macromycetes) in: Herz W., editor. *Progress in the Chemistry of Organic Natural Products*. New York, NY: Springer.

Guarro, J., Gene, J., Stchigel, A.M. 1999. Developments in fungal taxonomy. *Clinical Microbiology Reviews* 12, (3) 454–500. DOI: 10.1128/CMR.12.3.454.

Hawksworth, D.L. 1991. The fungal dimension of biodiversity: magnitude, significance, and conservation. *Mycological Research*, 95(6), 641–655.

Janardhana, G. R., Dass, R.S., Sreenivasa, M.Y. 2009. Fumonisins – A new class of Fusarial toxins. In *Frontiers in Fungal Ecology, Diversity and Metabolites*. Ed: K. A. Sridhar et al. I. K. International Publishing House Pvt. Ltd. New Delhi.

Kendrick. 1971. International Specialists' Workshop-Conference on Criteria and Terminology in the Classification of Fungi Imperfecti (1st: 1969: University of Calgary), University of Toronto Press.

Kendrick, B. 2011. *Fungi and the History of Mycology. eLS*. John Wiley & Sons, Ltd: Chichester. DOI: 10.1002/9780470015902.a0002320.pub2.

Mancini, V., Murolo, S., Romanazzi, G. 2016. Diagnostic methods for detecting fungal pathogens on vegetable seeds. *Plant Pathology* 65, 691–703.

Manoharachary, C. 2017. History of Mycology from India—Some Glimpses. T. Satyanarayana et al. (eds.), *Developments in Fungal Biology and Applied Mycology*, Springer Nature Singapore Pte Ltd. DOI: 10.1007/978-981-10-4768-8_3.

Martinelli, F., Scalenghe, R., Davino, S., Stefano Panno, S., Scuderi, G., Ruisi, P., Villa, P., Stroppiana, D., Mirco Boschetti, M., Goulart, L.R., Davis, C.E., Dandekar, A.M. 2015. Advanced methods of plant disease detection. A review, *Agronomy for Sustainable Development*. 35, 1–25.

Mulani, R.M., Turkmane, K.L. 2014. Diversity of hizospheric fungi of Ceropegia bulbosa Var. bulbosa Roxb. *Journal of Global Bioscience*. 3(4), 1089–1093.

Nagaraja, H., Chennappa, G., Poorna Chandra Rao, K., Mahadevprasad, G., Sreenivasa, M.Y. 2016. Diversity of toxic and phytopathogenic *Fusarium* species occurring on cereals grown in Karnataka state, India. *3 Biotech*. 6, 57.

Pfyffer, G.E., Pfyffer, B. U. & Rast, D. M. 1986. The polyol pattern, chemotaxonomy, and phylogeny of the fungi. *Sydowia* 39, 160–201.

Prabakaran, M., Merinal, S., Panneer, S.A. 2011. Investigation of *phylloplane mycoflora* from some medicinal plants. *European Journal of Experimental Biology*, 1, 219–225.

Prasad, M.N.N., Shankara Bhat, S., Sreenivasa, M.Y. 2010. Antifungal activity of essential oils against *Phomopsis azadirachtae*- the causative agent of die-back disease of neem. *Journal of Agricultural Technology*. 6(1), 127–133.

Rabosky, Daniel, L. 2009. Ecological limits and diversification rate: alternative paradigms to explain the variation in species richness among clades and regions. *Ecology Letters* 12(8), 735–743.

Rangaswamy, G., Mahadevan, A. 2006. Disease of crop plants in India, 4h Eds, Prentice Hall of India Private Limited, New Delhi, ISBN: 9788120312470.

Rast, D.M., Pfyffer, G.E. 1989. Acyclic polyols and higher taxa of fungi. *Botanical Journal of the Linnean Society* 99, 39–57. DOI: 10.1111/j.1095-8339.1989.tb00391.x

Raya, M., Raya, A., Dasha, S., Mishraa, A., Acharyb G.K., Nayaka, S., and Singh, S. 2017. Fungal disease detection in plants: Traditional assays, novel diagnostic techniques and biosensors. *Biosensors and Bioelectronics* 87, 708–723.

Schmit, J.P., Mueller, G.M. 2007. An estimate of the lower limit of global fungal diversity. *Biodiversity and Conservation* 16:99–111.

Schumann, G.L. 1991. *Plant Diseases: Their Biology and Social Impact*. St Paul, MN: APS Press, The American Phytopathological Society.

Sreenivasa, M.Y., Dass, R.S., Charith Raj, A.P., Janardhana, G.R. 2008. PCR-based detection of genus *Fusarium* and Fumonisin-producing isolates from freshly harvested Sorghum grains grown in Karnataka, India. *Journal of Food Safety* 28, 236–237.

Sreenivasa, M.Y., Dass, R.S., Charith Raj, A.P., Janardhana, G.R. 2011. Mycological evaluation of Maize grains produced in Karnataka (India) for the post-harvest fungal contamination. *World Applied Sciences Journal* 13(4), 688–692.

Sreenivasa, M.Y., Dass, R.S., Charith Raj, A.P., Prasad, M.N.N., Achar, P.N., Janardhana, G.R. 2011. Assessment of the growth inhibiting effect of some plant essential oils on different *Fusarium* species isolated from sorghum and maize grains. *Journal of Plant diseases and Protection*. 118(6), 208–213.

Sreenivasa, M.Y., Dass, R.S., Janardhana, G.R. 2010. Post harvest fungi associated with sorghum grains produced in Karnataka (India). *Journal of Plant Protection Research*. 50(3), 335–339.

Taylor T. N., Hass, H., Kerp, H. 1999. The oldest fossil ascomycetes. *Nature*, 399, 648.

Taylor T. N., Hass, H., Kerp, H., Krings, M., Hanlin, R. T. 2005. Perithecial ascomycetes from the 400 million year old Rhynie chert: an example of ancestral polymorphism. *Mycologia*, 97, 269–285.

Taylor T.N., Remy, W., Hass, H. 1992. Fungi from the Lower Devonian Rhynie chert: Chytridiomycetes. *American Journal of Botany*, 79, 1233–1241.

Varma, A., Kharkwal, A.C. 2009. *Symbiotic Fungi, Soil Biology*. Berlin: Springer. 18. DOI: 10.1007/978-3-540-95894-9_1

Webster, J., Weber, R. 2007. *Introduction to Fungi*, Cambridge University Press, Cambridge, New York, Melbourne, Madrid, Cape Town, Singapore, São Paulo.

3 Methodology for the Isolation and Identification of Fungal Pathogens

S. Mahadevakumar
Andaman and Nicobar Regional Centre

J. Joy and M. Mahesh
University of Mysore

K. R. Sridhar
Mangalore University

New fungal diseases and emerging fungal pathogens have been described on a regular basis in recent years, and the majority of known phytopathogens include species complexes, which have always been difficult to identify based on appearance alone. Many of the traditional diseases reported are mostly diagnosed based on symptoms and morphology. Without proper isolation protocols, the identification and diagnosis of plant fungal pathogens is always challenging. So far, the successful diagnosis of fungal disease relies on the establishment of pure cultures. A successful and precise diagnosis needs accurate details of hosts or substrates, symptoms of hosts, and a pure culture with fruiting structures. This process involves host specificity, the nature of the fungal pathogens being targeted, media requirements and some cases where biotrophic fungal pathogens cannot be cultivable on any artificial media. In such cases, the method of isolation plays an important role. Isolation techniques are plentiful, time-consuming and often intricate. Traditionally, identification of fungal pathogens was mainly based on visible and microscopic phenotypic characteristics. Since there are distinct micro- and macro-forms of fungal pathogens, identification based on phenotype is highly grim, and phenotypic observations recorded after the establishment of pure culture would be helpful to the identification process in investigating and establishing the correct and authentic identity as well as describing a new fungal species by comparing it with previously described species. This is a generalized procedure developed for the isolation of fungi, especially fungal pathogens, and this procedure differs for other groups that are in need of specialized media, nutrient enrichment of the medium, or are also influenced by the source we are using for isolation purposes. The collection of the host or substrate is crucial, and after that, isolation is very important, which forms the basis for further morphological as well as molecular analysis. This chapter addresses the detailed procedures available for the isolation and identification of fungal phytopathogens to understand the significance of the host-pathogen relationship, the importance of nutrients for growth and development and how the identification process will be assisted by good isolation methods.

DOI: 10.1201/9781003399704-3

GRAPHICAL ABSTRACT

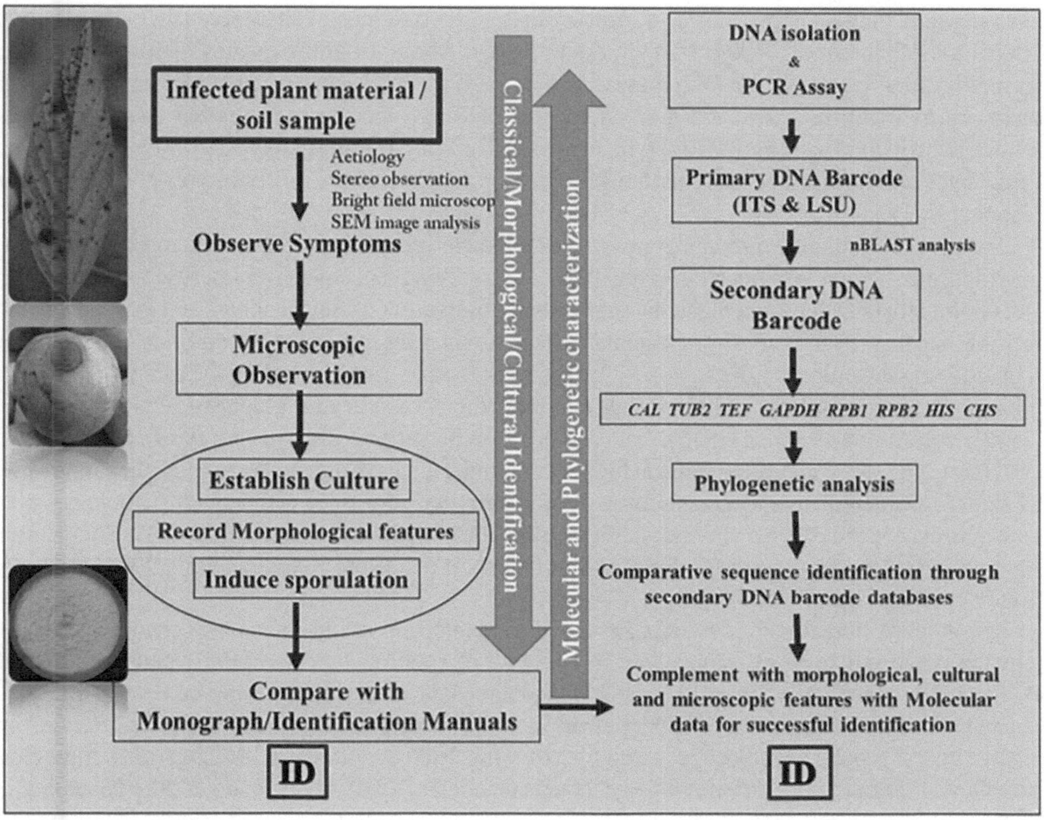

3.1 INTRODUCTION

The majority of life on earth is made up of plants, which give animals and people the nourishment they need. Plants use photons to create stored chemical energy, such as lipids, proteins and carbohydrates. Plant materials are essential to the lives of animals (Agrios, 2005). For plants to flourish, the soil must supply enough moisture, nutrients and light to reach the leaves while also keeping the temperature within a typical range. Plant illness results in both low growth and poor yield. Plants can become infected due to a variety of agents, which include bacteria, fungi, nematodes, viruses and viroids (Mahadevakumar and Sridhar, 2020, 2021). It is also recognized that plants might suffer from insect damage and competition from weeds. Fungal plant pathogens are responsible for a significant loss of output and a varied range of diseases in important crops around the world (Anderson et al., 2004; Strange and Scott, 2005; Dass et al., 2007; Sreenivasa et al., 2008).

In the past, numerous phytopathogenic fungi have posed grave risks to agriculture. Since plant diseases have a great impact on the food security of a nation, research on crop protection, diagnosis of crop plant diseases and management of disease has gained prominence. A successful diagnosis involves the isolation and proper identification of the associated pathogen (Deepa & Sreenivas 2019a; 2019b). The source could vary; it may be associated with any part of the plant. Mycology and plant pathology research in India have evolved in parallel lines, and E.J. Butler laid a strong foundation in mycology and phytopathology (Mahadevakumar et al., 2021). Berkeley (1839) is credited with starting the scientific study of fungi (Ainsworth, 1976).

Previously, taxonomists believed that because of their comparable morphology and growing environments, fungi and plants were closely linked. Fungi were later found to belong to a different kingdom that separated about a billion years ago (Baldauf and Palmer, 1993; Bruns et al., 2006; Janardhana et al., 2009; Parfrey et al., 2011). About 144,000 species of fungi have been formally documented so far (Wijayawardene et al., 2020). However, it has been predicted that there are as many as 2.2 to 3.8 million species of fungi and the real number is still not clear (Hawksworth and Lucking, 2017; Hyde et al., 2020). Fungal species have historically been identified by various theories and methods involving biochemistry, morphology, physiology and responses to analytical tests.

The predominant conventional approach for characterizing fungi is classification based on phenotypic traits, which was boosted in the 19th century by advancements in microscope technology. But Giambattista Della Porta made the first known observation of fungal spores and their germ tubes in 1588. Further, Persoon (1794) created a basic taxonomy for mushrooms, and Fries (1873) refined it by utilizing the colour of the spores as well as microscopic traits, including the arrangement of the gills, pores, hymenophore and teeth. Various fundamental studies by Hughes (1953), Carmichael (1962), von Arx and Müller (1975), Barr (1978), Kohlmeyer and Kohlmeyer (1979), Sutton (1980) and Barr (1987) have classified fungi at higher taxonomic levels using morphology. Other approaches to study considered fungal biochemistry, bioremediation, physiology and plant pathology, along with morphology-based taxonomic studies (Ali, 1962; Hyde and Alcorn, 1993; Ali-Shtayeh and Jamous, 2000; De Souza and Declerck, 2003; Duong et al., 2008; Evidente et al., 2008; Hyde and Soytong, 2008).

The modern fungal taxonomy has advanced by exploring various areas like chemical profiling (chemotaxonomy), ecology, molecular biology with morphology and secondary metabolites (Hyde et al., 2020; Manawasinghe et al., 2019; Phukhamsakda et al., 2020; Samarakoon et al., 2020; Senanayake et al., 2017, 2018; Wibberg et al., 2020). The richness and intra- and inter-relationships among fungal groups have been revealed by assessing 2680 evolutionary investigations, biological, genetic and sequence data for phylogeny (Hongsanan et al., 2017; Hyde et al., 2020). These studies have challenged but also complimented, traditional morphology-based fungal classification (Ekanayaka et al., 2017). The procedures for morphological studies of fungi are dispersed over numerous publications, and different researchers employ varied techniques. Because of the disparities in the approaches used, it is challenging to compare the outcomes.

3.2 BASIC OUTLINE OF THE DIAGNOSIS OF PLANT PATHOGENS

The diagnosis of plant diseases involves identifying the cause of abnormal plant growth or symptoms. It's a crucial step in plant pathology, as an accurate diagnosis allows for effective disease management strategies. Important steps involved in disease diagnosis are field observation, followed by sampling, and isolation and characterization in the laboratory. Here are some of the common steps and methods involved in the diagnosis of plant diseases:

- Field Inspection:
  - Symptom Observation: Careful examination of the plant for any abnormal symptoms such as leaf spots, wilting, discolouration, stunting or deformities.
  - Environmental Conditions: Consideration of environmental factors like weather, soil conditions, and cultural practices that may contribute to disease development.
- Sample Collection:
  - Collecting Samples: Choose plant samples that exhibit typical disease symptoms. Include both healthy and affected plant parts (leaves, stems and roots) for a comprehensive analysis.
  - Proper Handling: Handle samples with care to avoid contamination, and submit them to a diagnostic laboratory as soon as possible.

- Laboratory Analysis:
  - Microscopy: Microscopic examination of plant tissues for the presence of pathogens or characteristic structures like spores, mycelium or hyphae.
  - Cultural Techniques: Growing pathogens on suitable media to isolate and identify them. This may involve observing colony characteristics, spore production and growth patterns.
  - Molecular Techniques: PCR (Polymerase Chain Reaction) and other molecular methods can be employed for the precise identification of pathogens based on their DNA or RNA.
  - Serological Tests: ELISA (Enzyme-Linked Immunosorbent Assay): A technique used to identify or characterize specific plant pathogens by utilizing antibodies. It's particularly useful for viral and bacterial diseases.
  - Nematode Analysis: Nematode Extraction: If nematodes are suspected to be the cause, soil or root samples can be processed to extract and identify nematodes under a microscope.
- Biological Testing:
  - Inoculation Studies: Inoculating healthy plants with suspected pathogens to observe if they develop similar symptoms confirms the presence of the pathogen.
- Symptomatology and Pathogen Identification:
  - Comparative Symptomatology: Comparing observed symptoms with known symptoms caused by different pathogens.
  - Taxonomic Identification: Identification of the pathogen based on its taxonomy and characteristics.
- Data Integration:
  - Field History: Integration of field history, including crop rotation practices, previous disease occurrences and pest management practices, to understand the context of disease development.
- Consultation:
  - Expert Opinion: Seeking advice from plant pathologists or agricultural extension services for complex cases.
- Disease Reporting:
  - Documentation: Recording and reporting the diagnosis, which may include recommendations for disease management and prevention.

Plant disease diagnosis is a multi-step process that often involves a combination of field observations, laboratory analyses and expert interpretation. It requires a comprehensive approach to ensure accurate identification and effective management strategies.

3.3 ISOLATION OF FUNGI

Isolating fungi from various sources is a fundamental step in studying their biology, ecology and potential applications. Several isolation methods are commonly used in mycology to obtain pure cultures of fungi. Here are some general steps and techniques for isolating fungi:

- Sample Collection: Collect samples from the substrate where fungi are suspected to be present. This could include soil, plant material, air, water or other environmental sources.
- Sterilization of Equipment: Sterilize all equipment (including containers, tools and media) to prevent contamination by other microorganisms.
  - **Dilution and Spread Plate Technique**: Dilute the sample by serial dilution, and then spread a small portion of each dilution onto the surface of a solid agar medium. Use a sterile spreader to evenly distribute the dilution on the agar surface.
  - **Pour Plate Technique**: Mix the sample with a sterile, molten agar medium before it

solidifies. Pour the mixture into a sterile Petri dish and allow it to solidify. Colonies will develop both on the surface and within the agar.

- **Direct Plating**: Place a small portion of the sample directly onto the surface of a solid agar medium. Streak the material using a sterile loop to separate and isolate individual fungal colonies.
- **Moist Chamber Technique**: Place a sample in a moist chamber (a closed container with moist filter paper) to promote the germination of spores. After germination, transfer the developing hyphae to a culture medium.
- **Baiting Techniques**: Use specific substrates or baits that attract certain fungi. For example, pieces of plant material can attract plant-pathogenic fungi. Incubate the baited material and then transfer any emerging fungi to a culture medium.
- **Selective Media**: Use media that specifically promote the growth of certain fungi or inhibit the growth of others. This can help in isolating fungi with specific characteristics.

- Examples include media with antibiotics to inhibit bacterial growth or media with specific nutrients for particular fungal groups.
  - **Incubation Conditions**: Incubate the cultures under controlled environmental conditions (temperature, humidity and light/dark cycles) to encourage the growth of the isolated fungi.
  - **Sub-culturing**: Transfer individual fungal colonies from the primary isolation plates to fresh media to obtain pure cultures. Subculture is done until a pure culture with consistent characteristics is achieved.
  - **Identification**: Once pure cultures are obtained, the fungi can be identified using various methods, such as microscopy, molecular techniques and morphological characteristics.

It is pertinent to note that the specific isolation method chosen depends on the type of sample, the objectives of the study, and the characteristics of the fungi being targeted. Aseptic techniques are critical throughout the process to avoid contamination and ensure the purity of isolated cultures.

3.4 MEDIA FOR THE ISOLATION OF FUNGAL PATHOGENS

Isolation of fungal pathogens requires a suitable medium for their successful isolation, which can be maintained for the long term for storage as well as further utilization for research and development activities. Artificial media developed for the isolation of fungi provide raw materials for the growth of fungi, which will be used by the fungi to establish colonies and produce sporulating structures. However, all fungi cannot be cultured on one common medium, mainly because their life cycle and requirements for growth and development differ. Therefore, there are generalized media on which most of the fungi were isolated and specialized media with different chemical compositions as per the requirements of the specialist group. Some of the common media for the isolation of fungi are provided below:

- **Potato Dextrose Agar (PDA)**: This is a general-purpose medium widely used for the isolation and cultivation of various fungi. It contains potato infusion, dextrose (glucose) and agar. PDA provides a balanced nutrient source and is suitable for observing colony morphology.
- **Czapek-Dox Agar**: This medium is designed to support the growth of many fungi. It contains sodium nitrate, potassium chloride, magnesium sulphate, ferrous sulphate, dipotassium phosphate and sucrose, among other components. Czapek-Dox Agar is often used for the cultivation of fungi for taxonomic and physiological studies.
- **V8 Juice Agar**: V8 juice agar is prepared using V8 vegetable juice as a base. It provides a

rich source of nutrients, including vitamins and minerals. This medium is particularly useful for the isolation and cultivation of *Fusarium oxysporum*.

- **SNA (Spezieller Nährstoffarmer) Medium**: SNA medium is specifically designed for the cultivation of Fusarium species, especially those associated with plant roots. It contains specific nutrients that mimic the root exudates of plants, providing an environment suitable for the growth of root-associated fungi.
- **Corn Meal Agar**: This medium contains cornmeal, dextrose and agar. It is used for the cultivation of fungi and is particularly suitable for observing the sporulation and morphology of Fusarium species.
- **Water Agar**: Simple water agar, which contains agar and distilled water, can be used for the isolation of Fusarium from environmental samples. It allows the fungi to grow and produce mycelium without the interference of complex nutrients.

It is important to note that the selection of medium depends on the precise objectives of the study, the fungal species being investigated, and the desired characteristics of the cultured isolates. Additionally, some researchers may modify or customize these media based on their experimental requirements. Always follow proper laboratory protocols and guidelines when working with fungal cultures.

Isolation means obtaining a pure colony of a microbe. The source may be any plant part, soil, water, air or any surface from which we wish to isolate. There are several reasons to consider that isolation is very important in disease diagnosis. In addition to providing the starting material for DNA extraction used for molecular research, cultures can yield significant morphological features relevant for identification. The taxon is represented by ex-type cultures (Gams, 2015). One of the main resources for researching and comprehending a species' biology is its culture. According to Zeng et al. (2019), sequencing data from a few genes to entire genomes and transcriptomes under various circumstances can be obtained using extracted DNA from the culture. Furthermore, several taxa produce secondary metabolites that are useful as biocontrol agents, biopesticides, nematicides and enzymes (Chandler et al., 2011). These metabolites can be isolated from cultures (Strobel et al., 1996, Choi et al., 1999). On an industrial scale, the final output may be impacted by other fungi contaminating the cultures (Noman et al., 2018). In mating research, isolates' + and – hyphae are utilized to examine inbreeding. These breeding cultures are employed in biomedicine, agriculture, and the production of fungal inoculum for mushrooms (Chandler et al., 2011).

It has never been easy to isolate taxa from hosts or substrates and produce pure cultures (Goh and Hanlin, 1997). According to the majority of mycologists, a single spore that has germinated should yield a pure culture (Kirsop and Doyle, 1991; Goh and Hanlin, 1997). As a result, pure cultures are frequently obtained by single-spore isolation (Goh and Hanlin, 1997). It is preferable to get isolates from single spores in order to establish links between sexual and asexual morphs and to confirm that they are the same fungus (Choi et al., 1999; Goh and Hanlin, 1997; Chomnunti et al., 2011; Noman et al., 2018).

Easy, affordable and efficient isolation techniques are what we need (Choi et al., 1999). Certain species yield a large number of fruit bodies, while others only produce a few (Goh and Hanlin, 1997). Occasionally, fruit bodies are not developed to allow for identification. It is necessary to collect fully developed fruiting bodies or a sufficient quantity of spores (Goh and Hanlin, 1997). Along with spores or conidia, contaminants' conidia on the specimens may also be transmitted (Bernadovičová and Ivanová, 2011). Bacterial or fungal contamination may outgrow slow-growing species, leading to the identification of incorrect species (Goh and Hanlin, 1997; Smith, 1969). Even when the culture is produced in media that contains antibiotics, some bacteria can still limit development (Goh and Hanlin, 1997; Smith, 1969).

Other ways have to be developed because not all spores germinate in artificial media (Choi et al., 1999). To obtain cultures, numerous spores or spores clinging to the walls of the fruit are isolated in addition to single spores. To sustain their metabolic processes, certain species require certain culture media (Leyronas et al., 2012; Smith and Onions, 1994). According to Noman et al. (2018), pure cultured fungus mycelia are also necessary for the chemical and physiological profiling of the strain of interest.

Purification is necessary to produce a pure culture if the isolated taxon is mixed with other fungi or bacteria. Methods such as single colony subculture onto the new media or hyphal tip culture (Choi et al., 1999; Noman et al., 2018). When the taxon's growth rate exceeds that of the contaminant, hyphae tipping may be employed (Choi et al., 1999; Goh and Hanlin, 1997; Rangaswami & Mahadevan, 2006). As opined by Zeng et al. (2018), it is advised to directly harvest DNA from fruit bodies for certain taxa that are not able to thrive in artificial media. Goh and Hanlin (1997) proposed the glass needle method as a means of spore purification prior to isolation. Here, the spores are rolled by dragging them across the agar surface, and then by rolling again, impurities on the spore's surface are eliminated.

3.5 ISOLATION OF SOME SPECIAL GROUPS OF FUNGI

3.5.1 Isolation of Hyphomycetous Fungal Pathogens

Because hyphomycetes are exposed to the outside world, isolating them might be challenging (Subramanian, 1983). Bacteria and other fungi can readily contaminate these fungi. The most crucial tactic for obtaining a pure culture of targeted hyphomycetes is to stop non-targeted fungi from growing (Ellis, 1971). Nonetheless, hyphomycetes are isolated using a variety of techniques, some of which are listed here. Due to their exposure to the outside world, hyphomycetes can be challenging to isolate (Subramanian, 1983). Bacteria and other fungi can readily contaminate these fungi. The most crucial tactic for obtaining a pure culture of targeted hyphomycetes is to stop non-targeted fungi from growing (Ellis, 1971). Nonetheless, hyphomycetes are isolated using a variety of techniques.

- **Direct Streaking Method**: The direct streaking method is a technique for isolating pure cultures from hyphomycetes or spore-rich fungi. Here, spores are not suspended in spore solution; instead, they are immediately streaked over culture media in an aseptic manner. This approach saves time and just requires a few tools (Andrews and Pitt, 1986).
- **Soil Plate Method for the Isolation of Soil Hyphomycetes**: The manufacturer's instructions must be followed in order to prepare the media for the best possible conditions for fungal growth and sporulation when using this approach to isolate soil hyphomycetes (Barron, 1972).

3.5.2 Isolating Techniques for Coelomycetes

Fungal species recognized as coelomycetes bear conidia, which develop inside a cavity surrounded by either fungal, host or a combination of both tissues (Li et al., 2020; Sutton, 1980; Wijayawardene et al., 2016). Coelomycetes' conidia are frequently extruded from their conidiomata, and it is best to isolate each one separately using a different method.

- **Single Conidial Isolation**: According to Davis (1930), single conidial isolation is the act of transferring a single, germinating conidium to a pure medium plate. By doing this, it is ensured that the colony was formed from a single, pure conidium and that no genetic material from several conidia was combined. A simple and reliable technique for obtaining a pure culture is single-conidium isolation. Pour plating, spread plating or streaking on the plate are the methods used for the single conidial approach (Zhang et al., 2003). A pure conidia suspension that has been diluted enough is employed because it effectively separates the conidia and makes choosing them easier. One specimen can harbour more than one fungal taxon, and fast-growing species frequently inhibit other species during isolation;

thus, it is critical to validate that only target conidia have been separated using this procedure. Along with the targeted fungi, endophytes are often also isolated (Ju et al., 2009). Thus, 70% ethanol should be used to first sterilize the specimen's surface. Under a dissecting microscope, a single conidium is acquired using a capillary pipette from a conidial suspension or directly picked with forceps (Booth, 1971).

- **Spore Suspension Method by Vortex**: This technique uses a vortex mixer to create a spore suspension (Fei et al., 2019; Leyronas et al., 2012). The vortex mixer makes it easier for the ascus wall to break down and release spores into the suspension.

3.5.3 Isolating Techniques for the Sexual Morph of Ascomycetes

According to Pöggeler et al. (2006), ascomycetes generate apothecia or cleistothecia, perithecia and pseudothecia as sexual fruit bodies. The surface of the host may briefly come into contact with these reproductive organs. Ascospores are removed from fruiting bodies by opening them during isolation. Several pertinent strategies are covered in this article.

- **Spotted Pour Technique**: To isolate a single spore that is growing, the spotted pour technique is frequently employed (Choi et al., 1999). As previously mentioned, a spore suspension is made. A marker is used to mark 16 squares on the Petri dish's bottom. Using a sterile pipette, one drop of uniform spore suspension is placed onto each square of media (Choi et al., 1999). To ensure that the right fungus has been chosen, individual drops are examined under a stereo microscope (Hildebrand, 1938). To avoid water buildup on the agar surface, parafilm is used to encapsulate the plates and they are kept upside down. For 12–24 hours, plates are incubated at 18 °C to 25 °C. Within 12 to 24 hours, spore germination is assessed. When a spore begins to germinate, a very tiny piece of agar surrounding it can be cut with a sterilized, ultra-fine needle and moved to another medium (Booth, 1971). Transplanting too much agar is avoided by choosing germinated spores with fairly small germ tubes. When seen using a compound microscope with lenses 10 to 20 times magnification or a stereo microscope with the greatest magnification, germinating ascospores are plainly identifiable. Transferring spores to multiple plates is a better way to maintain duplicates. After germination, spores must be moved to fresh media within 1 hour. If not, spores that are germinating form colonies and may absorb spores from other species. Additionally, when the germ tube transforms into mycelia, the spore wall vanishes.
- **Multiple Spore or Asci Isolation**: Multiple spore isolation, also known as asci isolation, is possible for taxa that generate a lot of spores. For fungi that produce small spores and whose germinating spores are not easily visible, this is an ideal isolation technique (Stadler et al. 2014). Furthermore, it can be challenging to see hyaline spores on culture media plates. Thus, in this case, repeated spore isolation might be appropriate (Bills et al., 2012). This method involves slicing the fruiting body on a sterile host horizontally with a sterile blade, and then streaking the perithecial content onto a media plate (Wongkanoun et al., 2019). On media plates, asci or many spores may germinate.
- **Direct Isolation from Fungal Tissues**: Spores may occasionally be overly developed or sparsely distributed in specimens that were not in excellent condition when they were obtained. In these situations, a culture can be obtained by using the living portions of the peridium, hamathecium and vegetative hyphae (Rozas et al., 2011). Mature spores that are adhered to peridium or hamathecium content have the potential to germinate and form colonies. Tissues are rehydrated for this procedure prior to being added to the medium (Liu et al., 2015).
- **Spore Shooting or Drop Method**: Ascomycetes can be isolated using the spore drop method or spore shot technique (Choi et al., 1999). Huge spore masses are produced by certain fungi, and the spores are discharged by lengthy necks. Spore-firing or dropping

methods can be used to deal with species that expel their spores. According to Choi et al. (1999), ascospores for ascomycetes should be removed from the fruiting body.

3.6 ISOLATING TECHNIQUES FOR PLANT PATHOGENS

The interphase or transitional region between healthy and infected tissue typically makes up the specimen or tissue sections utilized in plant pathogen isolation (Melotto et al., 2008). The mycologist's experience may influence the material selection for isolation. Various methods may be needed for isolation, depending on the pathogens (Abdulkhair and Alghuthaymi, 2016; Deepa et al., 2016a, 2016b). Generally speaking, commonly occurring fungi are isolated using general techniques, including tissue and single-spore isolation. Selective media or baits are two methods that can be used to isolate primary pathogens.

The Petri dishes used in all isolation techniques are kept either at ambient temperature or in an incubator (25–30°C). These cultures are checked daily for microbial growth and are usually kept alive for 3–10 days. An essential consideration in the isolation of plant pathogens is the choice of culture medium. The standard culture medium that many plant pathogens can use is PDA. While fungi can thrive, acidified PDA inhibits the growth of bacteria (Trigiano and Ownley, 2017). Nonetheless, certain fungal infections could necessitate specific growth media with a high nitrogen and carbohydrate content, a pH range of 5–7 and a temperature range of 15–37 °C (Basu et al., 2015).

3.6.1 Isolation of Soil-Borne Plant-Pathogenic Fungi (Baiting method)

According to Kariman et al. (2018), soil-borne plant-pathogenic fungi can be classified into two groups: specialists that grow in root-free soil and those that are root-inhabiting and cannot grow in root-free soil. In every instance, roots are taken, carefully cleaned, surface sterilized, dried, dissected and then plated on antibiotic-containing media. Plant pathogens are captured and isolated using baits. A favourable plant host that can promote the growth of a disease is typically used as a bait. For instance, some Phytophthora species generate mobile zoospores, which serve as a means of dissemination and introduce fresh infections. Leaf disc fragments can float on the water's surface and both the infected plant root and the surrounding soil are submerged in water to isolate Phytophthora species. The leaf discs will attract Phytophthora zoospores, which are then carefully taken from the water, blotted dry, and placed on a semi-selective medium after a few days. Phytophthora species will be seen producing mycelium from the leaf disc if they are present in the infected root or soil.

3.6.2 Direct Isolation of Plant-Pathogenic Fungi

Plant material is incubated for 24–48 hours in a wet chamber, and single-spore isolation is utilized to obtain a pure culture if fruiting bodies are clearly evident on symptomatic plant tissues (Ampt et al., 2019). With certain adjustments, plant pathologists employ this technique. For instance, plant tissue portions that contain multiple mature fruit bodies can be plated out by immersing them in a tube of sterile water until a sufficient number of spores are observed in the suspension (Schuck et al., 2014).

3.6.3 Indirect Isolation of Plant-Pathogenic Fungi

Plant-pathogenic fungi must be indirectly isolated from symptomatic plant tissues if they are poorly developed or do not sporulate. In order to reduce the possibility of contaminants or other saprobes being present, these plant tissues are often surface sterilized. The virus responsible for the sickness would be found in the injured and infected interior tissue, which explains why (Bolton and Thomma, 2012).

- **Indirect Isolation from Fruiting Bodies**: Direct agar transfer of mycelium pieces or poorly sporulating fruiting bodies is another option (Zhang et al., 2013). Fruiting bodies are crushed in this way, and the shards are then streaked on a brand-new culture plate.

3.6.4 Isolation Techniques for Soil Fungi

To quantify fungal diversity, fungi have been isolated from soil using a variety of techniques (Warcup, 1960). Active mycelial forms in soil have been evaluated using the hyphal separation approach (Warcup, 1955; Parkinson and Williams, 1960). A total of 2721 microfungi on decomposing wood, leaf litter, root surfaces and organic materials have been isolated using the soil-washing approach (Harley and Waid, 1955; Robertson, 1954; Simmonds, 1930; Thorn et al., 1996). Nonetheless, this technique eliminated zygomycetous and ascomycetous fungal spores. As a result, ascomycetous and zygomycetous species have been isolated from soil using the immersion-tube technique (Chesters, 1940). According to Chesters (1940), soil fungus can be extracted straight from dispersed soil particles, mycelial pieces suspended in soil and spores. Given that saprophytic fungi are the main recyclers of soil and litter, soil has a high diversity of fungi (Gams, 1992; Thorn et al., 1996). The most effective isolation technique for saprophytic soil Zygomycetes is Warcup's soil plate method (Steiman et al., 1995; Thorn et al., 1996).

- **Soil Dilution Plating Method**: Waksman (1927) used this dilution technique, which Aziz and Zainol (2018) improved. To lower the number of fungi in each dilution tube, soil samples are diluted.
- **Soil Plate Method**: In order to separate the soil fungi, Warcup (1950) devised the soil plate method (Kamal and Bhargava, 1971). This is a basic plating method where a culture medium is covered with a small amount of dirt (Warcup, 1950).
- **Immersion-Tube Method**: Chesters (1940, 1948) described the use of the immersion-tube method as a means of isolating a diverse range of microflora from soil.

3.7 ISOLATION OF OBLIGATE PATHOGENIC FUNGI

There are several obligate pathogenic fungi known to cause devastating diseases on various economically important crop plants. Important fungi represented in this group include rusts, smuts, powdery mildews and chytridiomycetes members with representations of *Pythium*, *Phytophthora* and *Phytopythium*. Though recent technological advancements have resulted in the devising of specialized media for some obligate parasites like Phytophthora and related genera, there is no media developed for successful cultivation or isolation. In such cases, direct isolation for microscopy purposes or maintaining them on a living host is the most feasible method. Whenever there is a need for inoculum, the fungus can be harvested from the field and maintained exclusively for the same purpose. For proper identification of obligate parasitic fungal pathogens, depending on key manuals and micro-morphological features will certainly help and further application of molecular barcoding or sequencing using specific primers will be a more suitable method for identification. However, efforts are made to explore and understand the complexity of their lifestyle and in the future, there may be different sorts of outcomes we may expect to culture and isolate such pathogenic organisms.

3.8 RESOURCES FOR THE IDENTIFICATION OF FUNGAL PATHOGENS

It can be quite challenging to accurately identify fungal diseases based on their morphology and the appropriate species. Careful examination of important morphological traits, however, can quickly rule out a large number of species and enable the identification of a diagnostically significant taxon that may or may not be representative of a true species. A combination of morphological and molecular techniques may be needed for the accurate identification of species and *formae speciales*. Before the invention of molecular tools and sequencing platforms, identification was solely based on morphology, cultures, biochemical responses and physiological responses. Some of the key

TABLE 3.1
Details of Identification Manuals and Information Available for Phytopathogenic Fungi

Fungus	Monograph	Reference
Ascomycetes	*A Reevaluation of the Bitunicate Ascomycetes with Keys to Families and Genera*	von Arx and Müller, 1975
Coelomycetes	*Coelomycetes*	Sutton, 1980
Coelomycetes	*Coelomycetous Anamorphs with Appendage-Bearing Conidia*	Nag Raj, 1993
Deuteromycetes	*Illustrated Genera of Imperfect Fungi*	Barnett and Hunter, 1997
Fusarium	*Fusarium Species: An Illustrated Manual for Identification*	Nelson et al., 1983
Fusarium	*Toxigenic Fusarium Species, Identity and Mycotoxicology*	Marasas et al., 1984
Fusarium	*The Fusarium Laboratory Manual*	Leslie and Summerell, 2006
Hyphomycetes	*Dematiaceous Hyphomycetes*	Ellis, 1971
Hyphomycetes	*More Dematiaceous Hyphomycetes*	Ellis, 1976
Hyphomycetes	*Genera of Hyphomycetes*	Carmichael et al., 1980
Loculoascomycetes	*Prodromus to Class Loculoascomycetes*	Barr, 1987
Phoma	*Phoma Identification Manual*	Boerema et al., 2004
Rusts	*Illustrated Genera of Rust Fungi*	Cummins and Hiratsuka, 2003
Rusts	*Rust Fungi of Australia*	Shivas et al., 2014
Rusts	*Rust Fungi of South Africa*	https://www.daf.qld.gov.au/our-organisation/research/biological-collections/identification/rust-fungi-of-southern-africa
Smuts	*Smut Fungi of Thailand*	Shivas et al., 2013
Smuts	*Smut Fungi of Australia* *https://www.daf.qld.gov.au/our-organisation/research/biological-collections/identification/smut-fungi-of-australia*	Shivas et al., 2014
GOPHY1	*Genera of Phytopathogenic Fungi*	Marin-Felix et al., 2017
GOPHY2	*Genera of Phytopathogenic Fungi*	Marin-Felix et al., 2019
GOPHY3	*Genera of Phytopathogenic Fungi*	Marin-Felix et al., 2019
GOPHY4	*Genera of Phytopathogenic Fungi*	Chen et al., 2022

literature that played an important role in the progress of the identification and systematics of fungi and plant pathology is presented in Table 3.1. Important key manuals are described here and are still a valid reference for understanding the identification process. Further, a comprehensive review of the isolation and identification of fungi is provided by Senanayake et al. (2020). Likewise, important monographs and information related to genera and family are available online from journal websites, including studies in fungi, Persoonia, fungal diversity and other taxonomy-related journals. However, specific information related to plant pathogens can be retrieved from reputed plant pathology journals like *Plant Disease*, *Plant Pathology*, *Phytopathology*, *New Disease Reports*, *Australasian Plant Pathology*, *Mycological Progress*, *Molecular Plant Pathology*, *Indian Phytopathology*, *Journal of Mycology and Plant Pathology* and many others.

The manual "Illustrated genera of rust fungi" by Cummins and Hiratsuka (2003) contains identification keys, illustrations and photographs for 13 anamorph and 120 holomorph genera of rusts that will aid in identification. It also presents life cycles, morphological diversity and the spore states of the Uredinales. A key to 13 families is included, along with extensive information on rust fungi, including their economic significance, symptoms, collection and identification, phylogeny and relationships with other groups of fungi (cytology, host-rust relationships, life cycles, morphology of spore states and vegetative structures).

Many genera of phytopathogenic fungi are poly- and paraphyletic, and the majority of species comprise species complexes (Crous et al. 2015). Therefore, determining the identity of these genera and species is crucial for both plant trade and health. Consequently, in order to update the approved generic names of fungal genera, the project genera of phytopathogenic fungi (www.Genera Of Fungi.org) concentrated on fungal genera whose members cause illness in plants (Crous et al., 2014; Kirk et al., 2013). Up to the present, around 18,000 genera have been described; we currently recognize about 8000 of them. But most of these were described before the era of DNA. In order to confirm the traceability of the nomenclatural act, the type species of these names must be recollected, their epitypes or neotypes specified, and they must be assigned an MBT number (Robert et al., 2013). Furthermore, it is necessary to validate the sexual–asexual relationships across fungi in order to transition to a single nomenclature (Crous et al. 2015). It is one of the major initiatives of the International Committee for the Taxonomy of Fungi. The International Subcommission for the Taxonomy of Phytopathogenic Fungi was co-chaired by Pedro Crous and Amy Rossman (www.fungaltaxonomy.org/). So far, there are four series of GOPHY publications providing a detailed, up-to-date overview of each genus, thereby considering the application of the name of the genus wherever possible, and providing an overview of the presently accepted species (as well as their DNA barcodes) and relevant literature (Chen et al., 2022; Marin-Felix et al., 2017, 2019, b).

3.9 SUMMARY

Successful diagnosis relies on the establishment of pure cultures and recording each and every minute character that can be documented through proper microscopic and visual (cultures or pigments) examination. Therefore, the techniques we follow for isolation and identification play a key role. Further, these techniques vary based on the substratum and/or type of sample we are examining and if any special assistance is required, that needs to be followed.

3.10 FUTURE PERSPECTIVE

Various techniques have been devised and applied successfully to isolate plant pathogens from various sources, viz., plant tissues (leaves, stems, inflorescence and roots), fungi associated with soil (root-associated or rhizosphere-associated), or directly from soil, living tissues (endophytes) and air. Many of the pathogenic fungi (general fungi) can grow on a common medium, but it is not good for all. Since the biology and growth of every organism are unique, specialized media play a key role. Yet there are many fungi that have yet to be cultivated on artificial media. So, research and development are constantly trying to explore the nature of the media composition on which sporulation can be induced and how sexual and asexual forms can be connected, which will greatly help in understanding the biology of pathogens, which in turn helps to combat the diseases caused by them.

STUDY QUESTIONS

1. Add a note on media for fungi isolation.
2. Different techniques for isolation of fungi.
3. Identification of fungal pathogens.

FURTHER READING

Crous P. W., Hawksworth D. L., and Wingfield M. J., 2015. Identifying and naming plant-pathogenic fungi: past, present, and future. *Annu Rev Phytopathol*, 53, 247–267. https://doi.org/0.1146/annurev-phyto-080614-120245

Senanayake, I. C., Rathnayaka, A. R., Marasinghe, D. S., Calabon, M. S., Genteckaki, E., et al., 2020. Morphological approaches in studying fungi: collection, examination, isolation, sporulation and preservation. *Mycosphere*, 11(1): 2678–2754. https://doi.org/10.5943/mycosphere/11/1/20

REFERENCES

Abdulkhair, W. M. and Alghuthaymi, M. A., 2016. Plant Pathogens, *Plant Growth, Everlon Cid Rigobelo, IntechOpen*, https://doi.org/10.5772/65325

Agrios, G. N., 2005. *Plant pathology*, 5th Edn. Academic, London

Ainsworth, G. C., 1976. *Introduction to the History of Mycology*. Cambridge University Press.

Ali, M. M., 1962. Comparison of the physiology of three isolates of Colletotrichum graminicola. *Mycopathologia et Mycologia Applicata* 17, 261–268. https://doi.org/10.1007/BF02279299

Ali-Shtayeh, M. S. and Jamous, R. M. F., 2000. Keratinophilic fungi and related dermatophytes in polluted soil and water habitats. *Revista Iberoamericana de Micologia* 17, 51–59.

Ampt, E. A., van Ruijven, J., Raaijmakers, J. M., Termorshuizen, A. J. and Mommer, L., 2019. Linking ecology and plant pathology to unravel the importance of soil-borne fungal pathogens in species-rich grasslands. *Eur J Plant Pathol* 154, 141–156. https://doi.org/10.1007/s10658-018-1573-x

Anderson, P. K., Cunningham, A. A., Patel, N. G., Morales, F. J., Epstein, P. R. and Daszak, P., 2004. Emerging infectious diseases of plants: pathogen pollution, climate change and agro-technology drivers. *Trends Ecol Evol* 19:535–544. https://doi.org/10.1016/j.tree.2004.07.021

Andrews, S. and Pitt, J. I., 1986. Selective medium for isolation of Fusarium species and dematiaceous hyphomycetes from cereals. *Appl Environ Microbiol* 51, 1235–1238. https://doi.org/10.1128/aem.51.6.1235-1238.1986

von Arx, J. A. and Müller, E., 1975. A re-evaluation of the bitunicate ascomycetes with keys to families and genera. *Stud Mycol* 9, 1–159.

Aziz, N.H. and Zainol N. 2018. Isolation and identification of soil fungi isolates from forest soil for flooded soil recovery. *IOP Conf. Series: Mater Sci Eng* 342, 012028.

Baldauf, S. L. and Palmer, J. D., 1993. Animals and fungi are each other's closest relatives: congruent evidence from multiple proteins. *Proc Natl Acad Sci* 90, 11558–11562. https://doi.org/10.1073/pnas.90.24.11558

Barnett, H. L. and Hunter, B. B., 1997. *Illustrated genera of imperfect fungi*. 4th Edn., APS Press, St. Paul, 218 p.

Barr, M. E., 1978. The Diaporthales in North America: with emphasis on Gnomonia and its segregates. *Mycol Mem* 7, 1–232.

Barr, M. E., 1987. *Prodromus to Class Loculoascomycetes*. M.E. Barr Bigelow, Amherst, Massachusetts.

Barron, G. L., 1972. *The genera of Hyphomycetes from soil*. Robert E. Krieger Publishing Co. Inc., New York.

Basu, S., Bose, C., Ojha, N., Das, N., et al., 2015. Evolution of bacterial and fungal growth media. *Bioinformation* 11, 182–184. https://doi.org/10.6026/97320630011182

Berkeley, M. J., 1839. Descriptions of exotic fungi in the collection of Sir W. J. Hooker, from memoirs and notes of J. F. Klotzsch, with additions and corrections. *Ann Nat Hist* III, 375–401.

Bernadovičová, S. and Ivanová, H., 2011. Hyphomycetes and Coelomycetes fungi isolated from affected leaves and twigs of cherry laurel trees. *Folia Oecologica*, 38, 137–145.

Bills, G. F., González-Menéndez, V., Martín, J., Platas, G., Fournier, J., Peršoh, D. and Stadler, M., 2012. Hypoxylon pulicicidum sp. nov. (Ascomycota, Xylariales), a Pantropical Insecticide-Producing Endophyte. *PLoS ONE* 7, e46687. https://doi.org/10.1371/journal.pone.0046687

Boerema, G. H., De Gruyter, J., Noordeloos, M. E. and Hamers, M. C. E., 2004. *Phoma identification manual differentiation of specific and intra-specific taxa in culture*. CABI Publishing, Cambridge MA, USA, Wallingford, OX, UK, https://doi.org/10.1079/9780851997438.0000

Bolton, M. D. and Thomma, B. P. H. J., 2012. *Plant Fungal Pathogens: Methods and Protocols*. (Methods in Molecular Biology; No. 835). Humana Press, New York, USA. https://doi.org/10.1007/978-1-61779-501-5

Booth, C., 1971. *Methods in Microbiology*. Academic Press INC LTD, London.

Bruns, J. A., Zehr, J. P., Montoya, J. P., Kustka, A. B. and Capone, D. G., 2006. Effect of EDTA additions on natural Trichodesmium spp. (Cyanophyta) populations. *J Phycol* 42, 900–904. https://doi.org/10.1111/j.1529-8817.2006.00239.x

Carmichael, J. W., 1962. Viability of mold culture stored at-20 C. *Mycologia* 54, 432–436. https://doi.org/10.2307/3756573

Carmichael, J. W., Kendrick, W. B., Conners, I. L. and Sigler L., 1980. *Genera of Hyphomycetes*. University of Alberta Press, Edmonton.

Chandler, D., Bailey, A. S., Tatchell, G. M., Davidson, G., Greaves, J. and Grant, W. P., 2011 The development, regulation and use of bio-pesticides for integrated pest management. *Philos Trans R Soc B: Biol Sci* 366, 1987–1998. https://doi.org/10.1098/rstb.2010.0390

Chen, Q., Bakhshi, M., Balci, Y., Broders, K. D., Cheewangkoon, R., Chen, S. F., et al., 2022. Genera of phytopathogenic fungi: GOPHY 4. *Stud Mycol.* 101, 417–564. https://doi.org/10.3114/sim.2022.101.06

Chesters, C. G. C., 1940. The Arundel Foray. *Trans Br Mycol Soc* 24, 1–3.

Chesters, C. G. C., 1948. A contribution to the study of fungi in the soil. *Trans Br Mycol Soc* 30, 100–117.

Choi, Y. W., Hyde, K. D. and Ho, W. H., 1999. Single spore isolation of fungi. *Fungal Divers* 3, 29–38.

Chomnunti, P., Schoch, C. L., Aguirre-Hudson, B., Ko-Ko, T. W., Hongsanan, S., Jones, E. B. G., et al., 2011. Capnodiaceae. *Fungal Divers* 51, 103–134. https://doi.org/10.1007/s13225-011-0145-6

Crous, P. W., Shivas, R. G., Quaedvlieg, W., Van der Bank, M. and Zhang, Y. 2014. FRungal planet description sheets 214-280. *Persoonia* 32, 184–306.

Cummins, G. B. and Hiratsuka, Y. 2003. *Illustrated genera of rust fungi.* American Phytopathological Society, St. Paul.

Dass, R.S., Sreenivasa, M.Y. and Janardhana, G.R. 2007. High incidence of *Fusarium verticillioides* in animal and poultry feed mixtures produced in Karnataka, India. *Plant Pathol J* 6(2), 174–178.

Davis, W. H., 1930. Single Spore Isolation. *Proc Iowa Acad Sci* 37, 151–159.

De Souza, F. A. and Declerck, S., 2003. Mycelium development and architecture, and spore production of *Scutellospora reticulata* in monoxenic culture with Ri T-DNA transformed carrot roots. *Mycologia* 95, 1004–1012.

Deepa, N., Charith Raj, A. P. and Sreenivasa, M.Y. 2016a. Nested PCR method for the early detection of fumonisin producing *Fusarium verticillioides* in pure cultures, cereal samples and plant parts. *Food Biotechnology*. 30(1), 18–29.

Deepa, N., Charith Raj, A.P. and Sreenivasa, M.Y. 2016b. Multiplex PCR method for the early detection of fumonisin producing *Fusarium verticillioides*. *Food Bioscience*. 13, 84–88.

Deepa, N and M Y Sreenivasa. 2019a. Molecular methods and key genes targeted for the detection of fumonisin producing *Fusarium verticillioides* - An updated review. *Food Bioscience*. 32, 100473.

Deepa N and M.Y. Sreenivasa. 2019b. Biocontrol Strategies for Effective Management of Phytopathogenic Fungi Associated with Cereals, In: Singh JS, Singh DP (eds) *New and Future Developments in Microbial Biotechnology and Bioengineering*, Elsevier, pp, 177–189, ISBN 9780444641915.

Duong, L. M., McKenzie, E. H. C., Lumyong, S. and Hyde, K. D., 2008. Fungal succession on senescent leaves of Castanopsis diversifolia in Doi Suthep-Pui National Park, Thailand. *Fungal Divers* 30, 23–36.

Ekanayaka, A. H., Ariyawansa, H. A., Hyde, K. D., Jones, E. B., Daranagama, D. A., Phillips, A. J. L., et al., 2017. Discomycetes: the apothecial representatives of the phylum Ascomycota. *Fungal Divers* 87, 237–298. https://doi.org/10.1007/s13225-017-0389-x

Ellis, M. B., 1971. *Dematiaceus Hyphomycetes.* Commonwealth Mycological Institute, Kew, Surrey, UK.

Ellis, M. B., 1976. *More dematiaceous Hyphomycetes.* Commonwealth Mycological Institute, Kew, Surrey, England, pp. 507.

Evidente, A., Cimmino, A., Andolfi, A., Vurro, M., Zonno, M. C. and Motta, A., 2008. Phyllostoxin and phyllostin, bioactive metabolites produced by Phyllosticta cirsii, a potential mycoherbicide for Cirsium arvense biocontrol. *J Agric Food Chem* 56(3), 884–888. https://doi.org/10.1021/jf0731301

Fei, L. W., Lu, W. B., Xu, X. Z., Yan, F. C., Zhang, L., Liu, J., et al., 2019. A rapid approach for isolating a single fungal spore from rice blast diseased leaves. *J Integr Agric* 18, 1415–1418. https://doi.org/10.1016/S2095-3119(19)62581-5

Fries, E. M., 1873. Observations on Gonnermann and Rabenhorst's Mycologia Europaea. *Grevillea* 2 (14), 1–27.

Gams, W., 1992. The analyses of communities of saprophytic microfungi with special reference to soil fungi. In: Winterhoff W (Ed) *Fungi in vegetation science.* Kluwer, Dordrecht.

Gams, W., 2015. An ex-type culture cannot always tell the ultimate truth. *IMA Fungus* 6(2), 69. https://doi.org/10.1007/BF03449357

Goh, T. K. and Hanlin, R. T. 1997. Nuclear divisions in the ascus and ascospores of Melanospora zamiae. *Mycol Res.* 101, 1511–1514.

Harley, J. L. and Waid, J. S., 1955. A method of studying active mycelia on living roots and other surfaces in the soil. *Trans Br Mycol Soc* 38, 104–118. https://doi.org/10.1016/S0007-1536(55)80022-8

Hawksworth, D. L. and Lucking, R., 2017. Fungal diversity revisited: 2.2 to 3.8 million species. *Microbiol Spectr* 5, 79–95. https://doi.org/10.1128/microbiolspec.FUNK-0052-2016

Hildebrand, E. M., 1938. Techniques for the Isolation of Single Microorganisms. *Bot Rev* 4, 627–664. https://doi.org/10.1007/BF02869844

Hongsanan, S., Maharachchikumbura, S. S., Hyde, K. D., Samarakoon, M. C., Jeewon, R., Zhao, Q., et al., 2017. An updated phylogeny of Sordariomycetes based on phylogenetic and molecular clock evidence. *Fungal Divers* 84(1), 25–41. https://doi.org/10.1007/s13225-017-0384-2

Hughes, S. J., 1953. Conidiophores, conidia, and classification. *Can. J. Bot.* 31, 577–659. https://doi.org/10.1139/b53-046

Hyde, K. D. and Alcorn, J. L., 1993. Some disease associated microorganisms on plants of Cape York Peninsula and Torres Strait Islands. *Australas Plant Pathol* 22, 73–83. https://doi.org/10.1071/APP9930073

Hyde, K. D., Dong, Y., Phookamsak, R., Jeewon, R., Bhat, D. J., Jones, E. B. G., et al. 2020. Fungal diversity notes 1151–1276: taxonomic and phylogenetic contributions on genera and species of fungal taxa. *Fungal Diversity* 100, 5–277. https://doi.org/10.1007/s13225-020-00439-5

Hyde, K. D., Norphanphoun, C., Maharachchikumbura, S. S. N., Bhat, D. J., Jones, E. B. G., Bundhun, D., et al. 2020. Refined Families of Sordariomycetes. *Mycosphere* 11, 305–1059. https://doi.org/10.5943/mycosphere/11/1/7

Hyde, K. D. and Soytong, K., 2008. The Fungal endophyte dilemma. *Fungal Divers* 33, 163–173.

Janardhana, G. R., Dass, R.S. and Sreenivasa, M.Y. 2009. Fumonisins – A new class of Fusarial toxins. *In Frontiers in Fungal Ecology, Diversity and Metabolites.* Ed: K. A. Sridhar et al. I. K. International Publishing House Pvt. Ltd. New Delhi.

Ju, Z., Wang, J., Pan, S. L., 2009. Isolation and preliminary identification of the endophytic fungi which produce Hupzine A from four species in Hupziaceae and determination of Huperzine A by HPLC. *J Med Sci* 36, 445–449.

Kamal, S. and Bhargava, K. S., 1971. Studies on Soil Fungi from Teak Forests of Gorakhpur IV. Isolation of fungi by soil plate and soil dilution plate methods. *Sydowia Ann Mycol*, 24: 233–240.

Kariman, K., Barker, S. J. and Tibbett, M., 2018. Structural plasticity in root-fungal symbioses: diverse interactions lead to improved plant fitness. *Peer J* 6, e6030. https://doi.org/10.7717/peerj.6030

Kirk, P. M., Staplers, J. A., Braun, U., Crus, P. W. and Hansen, K. 2013. A without prejudice list of generic names of fungi for protection under the international code of nomenclature for algae, fungi and plants. *IMA Fungus* 4, 381–443.

Kirsop, B. E. and Doyle, A., 1991. *Maintenance of microorganisms and cultured cells, a manual of laboratory methods.* 2nd edn. Academic Press, London.

Kohlmeyer, J. and Kohlmeyer, L., 1979. *Marine mycology: the higher fungi.* Academic Press, New York.

Leslie, J. F. and Summerell, B. A., 2006. *The Fusarium Laboratory Manual.* Blackwell Publishing, Iowa, USA.

Leyronas, C., Duffaud, M. and Nicot, P. C., 2012. Compared efficiency of the isolation methods for Botrytis cinerea. *Mycology* 3, 221–225. https://doi.org/10.1080/21501203.2012.727484

Li, W. J., Mckenzie, E. H., Liu, J. K., Bhat, J. D., Dai, D. Q., Camporesi, E., et al., 2020. Taxonomy and phylogeny of hyaline-spored coelomycetes. *Fungal Divers* 100, 279–801. https://doi.org/10.1007/s13225-020-00440-y

Liu, L., Wang, C. L., Peng, W. Y., Yang, J., Lan, M. Q., Zhang, B., et al., 2015. Direct DNA extraction method of an obligate parasitic fungus from infected plant tissue. *Genet Mol Res* 14, 18546–18551. https://doi.org/10.4238/2015.December.28.1

Mahadevakumar, S., Amruthesh, K. N. and Sridhar, K. R., 2021. Historical Perspectives of Rusts in India. In: *Progress in Mycology – India perspectives.* Ed. Satyanarayana, T. et al. Springer Nature. https://doi.org/10.1007/978-981-16-2350-9_12

Mahadevakumar, S. and Sridhar, K. R., 2020. Diagnosis of Pythium by Classical and Molecular Approaches. In: *The genus Pythium – Diagnosis, Diseases and Management.* Eds. Rai, Abd-Esalam and Ingle. CRC Press. pp 195–219. https://doi.org/10.1201/9780429296406

Mahadevakumar, S. and Sridhar, K.R., 2021. Diversity of Plant Pathogenic Fungi in Agricultural Crops. In: *Plant, Soil and Microbes in Tropical Ecosystems.* Dubey, S.K., Verma, S.K. (eds.). Springer Nature, pp. 101–149. https://doi.org/10.1007/978-981-16-3364-5_6

Manawasinghe, I. S., Dissanayake, A. J., Li, X., Liu, M., Wanasinghe, D. N., Xu, J., et al., 2019. High Genetic Diversity and Species Complexity of Diaporthe Associated with Grapevine Dieback in China. *Front Microbiol* 10, 1936. https://doi.org/10.3389/fmicb.2019.01936

Marasas, W. F. O., Nelson, P. E. and Toussoun, T. A., 1984. *Toxigenic Fusarium species: identity and mycotoxicology.* University Park, Pennsylvania: Pennsylvania State University Press. 328 p.

Marin-Felix, Y., Groenewald, J.Z., Cai, L., Chen, Q., Marincowitz, S., Barnes, I., et al., 2017. Genera of phytopathogenic fungi: GOPHY 1. *Stud Mycol.* 86:99–216. https://doi.org/10.1016/j.simyco.2017.04.002

Marin-Felix, Y., Hernández-Restrepo, M., Iturrieta-González, I., García, D., Gené, J., Groenewald, J. Z., et al., 2019. Genera of phytopathogenic fungi: GOPHY 3. *Stud Mycol.* 94:1–124. https://doi.org/10.1016/j.simyco.2019.05.001

Marin-Felix, Y., Hernández-Restrepo, M., Wingfield, M. J., Akulov, A., Carnegie, A. J., Cheewangkoon, R., et al., 2019. Genera of phytopathogenic fungi: GOPHY 2. *Stud Mycol.* 92:47–133. https://doi.org/10.1016/j.simyco.2018.04.002

Meloto, M., Underwood, W. and He, S. Y., 2008. Role of Stomata in Plant Innate Immunity and Foliar Bacterial Diseases. *Annu Rev Phytopathol* 46, 101–122. https://doi.org/10.1146/annurev.phyto.121107.104959

Nag Raj, T. R., 1993. *Coelomycetous anamorphs with appendage-bearing conidia.* Mycologue Publications, Waterloo, Ontario.

Nelson, P. A., Toussoun, T. A. and Marasas, W. F. O., 1983. *Fusarium species-An illustrated manual for identification.* Pennsylvania State University Press, USA.

Noman, E., Al-Gheethi, A. A., Rahman, N. K., Talip, B., Mohamed, R., et al., 2018. Single spore isolation as a simple and efficient technique to obtain fungal pure culture (IConCEES 2017). *IOP Conf Ser*, 140, 012055. https://doi.org/10.1088/1755-1315/140/1/012055

Parfrey, L. W., Lahr, D. J. G., Knoll, A. H. and Katz, L. A., 2011. Estimating the timing of early eukaryotic diversification with multigene molecular clocks. *Proc Natl Acad Sci USA* 108(33), 13624–13629. https://doi.org/10.1073/pnas.111063310

Parkinson, D. and Williams, S. T., 1960. A method for isolating fungi from soil microhabitats. *Plant and Soil* 13, 347–355. https://doi.org/10.1007/BF01394646

Persoon, C. H., 1794. Dispositio methodica fungorum: A methodical classification of the fungi. *Neues Magazin für die Botanik* 1, 81–128.

Phukhamsakda, C., McKenzie, E. H., Phillips, A. J. L., Jones, E. B. G., Bhat, D. J., Stadler, M., et al., 2020. Study of fungi associated with Clematis L. (Ranunculaceae) and integrated approach to delimiting species boundaries. *Fungal Divers* 102, 1–203. https://doi.org/10.1007/s13225-020-00448-4

Pöggeler, S., Nowrousian, M. and Kück, U., 2006. Fruiting-Body Development in Ascomycetes. In: Fischer K (Ed.). *The Mycota I: Growth, Differentiation and Sexuality.* Springer-Verlag, Berlin Heidelberg 325–355.

Rangaswami, G. and Mahadevan, A., 2006. *Diseases of crops plants in India*, 4th Ed., New Delhi Prentice Hall of India.

Robert, V., Vu, D., and Amor, A. B. 2013. MycoBank gearing up for new horizons. *IMA Fungus* 4, 371–379.

Robertson, N. F., 1954. Studies on the mycorrhiza of *Pinus sylvestris*. I. The pattern of development of mycorrhizal roots and its significance for experimental studies. *New Phytol* 53, 253–283. https://doi.org/10.1111/j.1469-8137.1954.tb05239.x

Rozas, E. E., Albano, R. M., Lôbo-Hajdu, G., Müller, W. E. G., Schröder, H. C. and Custódio, M. R., 2011. Isolation and cultivation of fungal strains from in vitro cell cultures of two marine sponges (Porifera: *Halichondrida and Haplosclerida) Braz J Microbiol* 42(4), 1560–1568. https://doi.org/10.1590/S1517-83822011000400004 3

Samarakoon, M. C., Thongbai, B., Hyde, K. D., Brönstrup, M., Beutling, U., Lambert, C., et al., 2020. Elucidation of the life cycle of the endophytic genus Muscodor and its transfer to Induratia in Induratiaceae fam. nov., based on a polyphasic taxonomic approach. *Fungal Divers* 101, 177–201. https://doi.org/10.1007/s13225-020-00443-9

Schuck, S., Weinhold, A., Luu, V. T. and Baldwin, I. T., 2014. Isolating Fungal Pathogens from a Dynamic Disease Outbreak in a Native Plant Population to Establish Plant-Pathogen Bioassays for the Ecological Model Plant *Nicotiana attenuata*. *PLoS ONE* 9(7), e102915. https://doi.org/10.1371/journal.pone.0102915

Senanayake, I. C., Crous, P. W., Groenewald, J. Z., Maharachchikumbura, S. S. N., Jeewon, R., Phillips, A. J. L., et al. 2017. Families of Diaporthales based on morphological and phylogenetic evidence. *Stud in Mycol* 86, 217–296. https://doi.org/10.1016/j.simyco.2017.07.003

Senanayake, I. C., Jeewon, R., Chomnunti, P., Wanasingha, D. N., Norphanphoun, C., Karunarathna, A., et al., 2018. Taxonomic circumscription of Diaporthales based on multigene phylogeny and morphology. *Fungal Divers* 93(1), 241–443. https://doi.org/10.1007/s13225-018-0410-z

Shivas, R. G., Athipunyakom, P., Beasley, D. R., McTaggart, A. R. and Vánky, K., 2013. Smut Fungi of Thailand, https://collections.daff.qld.gov.au/thaismutfungi/

Shivas, R. G., Beasley, D. R. and McTaggart, A. R., 2014. Online identification guides for Australian smut fungi (Ustilaginomycotina) and rust fungi (Pucciniales). *IMA Fungus* 5(2), 195–202. https://doi.org/10.5598/imafungus.2014.05.02.03

Simmonds, P. M., 1930. A washing device for isolation work with plant material. *Phytopathology* 20, 911–913.

Smith, G., 1969. *An Introduction to Industrial Mycology*, 6th ed., Arnold Ltd, London, UK.

Smith, D. and Onions, A. H. S. 1994. *Preservation and maintenance of living fungi.* CAB International, UK.

Sreenivasa, M.Y., Dass, R.S., Charith Raj, A.P., Janardhana, G.R. 2008. PCR-based detection of genus *Fusarium* and Fumonisin-producing isolates from freshly harvested Sorghum grains grown in Karnataka, India. *J Food Saf*, 28, 236–237.

Stadler, M., Læssøe, T., Fournier, J., Decock, C., Schmieschek, B., Tichy, H. V., Peršoh, D., 2014. A polyphasic taxonomy of Daldinia (Xylariaceae). *Stud Mycol* 77, 1–143. https://doi.org/10.3114/sim0016

Steiman, R., Guiraud, P., Sage, L., Seigle-Murandi, F., Lafond, J. L., 1995. Mycoflora of soil around the Dead Sea I – Ascomycetes (including Aspergillus and Penicillium), Basidiomycetes, Zygomycetes. *System Appl Microbiol* 18, 310–317. https://doi.org/10.1016/S0723-2020(11)80402-9

Strange, R. N. and Scott, P. R., 2005. Plant disease: a threat to global food security. *Annu Rev Phytopathol.* 43, 83–116. https://doi.org/10.1146/annurev.phyto.43.113004.133839

Strobel, G., Yang, X., Sears, J., Kramer, R., Sidhu, R. S. and Hess, W. M., 1996. Taxol from Pestalotiopsis microspora, an endophytic fungus of Taxus wallachiana. *Microbiology* 142(2), 435–440. https://doi.org/10.1099/13500872-142-2-435

Subramanian, C. V., 1983. *Hyphomycetes taxonomy and biology*. London Academic Press.

Sutton, B. C., 1980. *The Coelomycetes. Fungi imperfecti with pycnidia, acervuli and stromata*. CMI, Kew, 1–696.

Thorn, R. G., Reddy, C. A., Harris, D. and Paul, E., 1996. Isolation of saprophytic Basidiomycetes from soil. *Appl Environ Microbiol* 62(11), 4288–4292. https://doi.org/10.1128/aem.62.11.4288-4292.1996

Trigiano, R. N. and Ownley, B. H., 2017. Identifying obligate, biotrophic fungi (and hosts) using the sequence of the internal transcribed spacer (ITS) region. In: Trigiano RN, *Ownley BH 2017 – Plant Pathology Concepts and Laboratory Exercises*, 3rd Ed., Taylor and Francis, Boca Raton, FL, 483–494.

Waksman, S. A., 1927. *Principles of soil microbiology*. Williams & Wilkins, Baltimore.

Warcup, J.H. 1950. The soil-plate method for isolation of fungi from soil. *Nature* 166, 117–118.

Warcup, J. H., 1955. Isolation of fungi from hyphae present in soil. *Nature* 175, 953–954. https://doi.org/10.1038/175953b0

Warcup, J. H., 1960. *Methods for isolation and estimation of activity of fungi in soil*. The Ecology of Soil Fungi, Liverpool University Press.

Wibberg, D., Stadler, M., Lambert, C., Bunk, B., Spröer, C., Rückert, C., et al., 2020. High quality genome sequences of thirteen Hypoxylaceae (Ascomycota) strengthen the phylogenetic family backbone and enable the discovery of new taxa. *Fungal Divers* https://doi.org/10.1007/s13225-020-00447-5

Wijayawardene, N. N., Hyde, K. D., Al-Ani, L. K. T., Tedersoo, L., Haelewaters, D., et al., 2020. Outline of Fungi and fungus-like taxa. *Mycosphere* 11(1), 1060–1456. https://doi.org/10.5943/mycosphere/11/1/8

Wijayawardene, N. N., Hyde, K. D., Wanasinghe, D. N., Papizadeh, M., Goonasekara, I. D., Camporesi, E., et al., 2016. Taxonomy and phylogeny of dematiaceous coelomycetes. *Fungal Divers* 77, 1–316. https://doi.org/10.1007/s13225-016-0360-2

Wongkanoun, S., Wendt, L., Stadler, M., Luangsa-Ard, J. and Srikitikulchai, P., 2019. A novel species and a new combination of Daldinia from Ban Hua Thung community forest in the northern part of Thailand. *Mycol Progress* 18, 553–564. https://doi.org/10.1007/s11557-019-01469-3

Zeng, X., Jeewon, R., Wen, T., Hongsanan, S., Boonmee, S. and Hyde, K. D., 2018. Simplified and efficient DNA extraction protocol for Meliolaceae specimens. *Mycol Progress* 17, 403–415. https://doi.org/10.1007/s11557-017-1361-6

Zeng, L., Tu, X. L., Dai, H., Han, F. M., Lu, B. S., Wang, M. S., et al., 2019. Whole genomes and transcriptomes reveal adaptation and domestication of pistachio. *Genome Biol* 20, 79. https://doi.org/10.1186/s13059-019-1686-3

Zhang, K., Su, Y. Y. and Cai, L., 2013. An optimized protocol of single spore isolation for fungi. *Cryptogam Mycol* 34(4), 349–356. https://doi.org/10.7872/crym.v34.iss4.2013.349

Zhang, W., Wolf, T. M., Bailey, K. L., Mortensen, K. and Boyetchko, S. M., 2003. Screening of adjuvants for bioherbicide formulations with Colletotrichum spp. and Phoma spp. *Biol Control* 26, 95–108. https://doi.org/10.1016/S1049-9644(02)00133-0

4 Polymerase Chain Reaction and Its Variants as Detection Tools for Fungal Phytopathogens

Deepa Nagaraju and Sreenivasa Marikunte Yanjarappa
University of Mysore

Nearly, more than 1.5 million species of fungi have been estimated, among which less than 5% of the fungi have been described so far. Worldwide, many of the fungal species cause diseases in plants, leading to major economic losses in the agricultural field. Plant ill health needs to be monitored regularly by diagnosing the fungal phytopathogens at early stages. Even though these molecular diagnostics have grown to a great extent over a decade yet there is a long way for its development and application in plant disease diagnostics. Many molecular diagnostic tools such as PCR variants like nested PCR, multiplex PCR, Q-PCR, RT-PCR, next-generation PCR, conventional PCR, end-point PCR and bio-PCR; PCR coupled methods like MCH-PCR, PCR-ELISA; in-situ PCR, PCR-DGGE, PCR-based methods like RFLP, AFLP, RAPD; some PCR post amplification methods such as microarray, macroarray, LAMP; and advance PCR methods like FISH, NASBA, SAGE, RNAi, Northern blot and DNA/RNA probe are in practice to reduce the spread of disease and minimising the yield and crop loss. Currently, in the present chapter, we will be discussing PCR and its variants for the detection of fungal phytopathogens. As they are reliable, easy, inexpensive, robust and compete with traditional techniques.

DOI: 10.1201/9781003399704-4

GRAPHICAL ABSTRACT

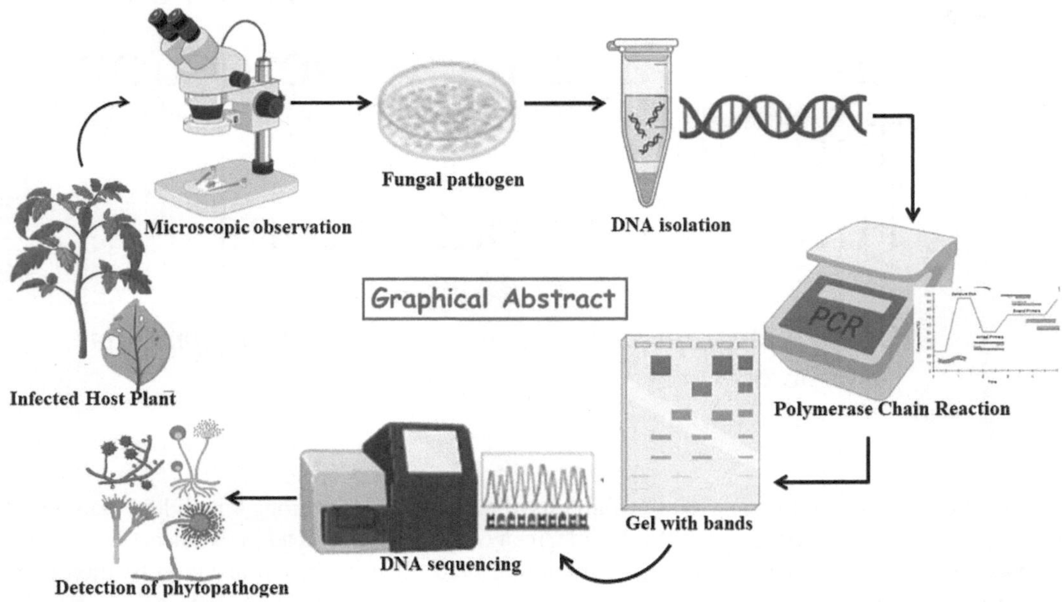

4.1 INTRODUCTION

Fungi perform dual roles as both foes and friends. Many of the fungi due to their medical, economic and commercial uses are beneficial to animals, humans and vegetables. Its other role is harmful due to its capability to cause disease and act as a pathogen (Aslam et al., 2017). Phytopathogenic fungi mainly infect plants in between the seedling to seed maturing stage individually or synergistically with any other kind of fungal pathogens under natural environmental conditions (Nagaraja et al., 2016; Narayanasamy 2011). Powdery mildew, blight, anthracnose, canker, damping off, blight, gall, dieback, root rot, leaf spot, rust, scab and wilt, and anthracnose are the most common fungal phytopathogenic diseases (Jain et al., 2019). These diseases cause massive losses for crop yields in both quantity and quality. This serves to be the major issue in the agricultural sector, horticultural, agronomical, ornamental, floricultural, forest and plant species worldwide (Hariharan and Prasannath, 2021; Wenneker and Thomma 2020). Approximately 70% to 80% of cereal crops, vegetables and fruits are infected by fungal phytopathogens, which leads to substantial reduction in yield and quality of crops (Deepa et al., 2021). Even though the control of phytopathogenic fungi is a topmost priority with advanced strategies, its early detection fetches great initiative (Deepa and Sreenivasa, 2019b; Sreenivasa et al., 2011a; Deepa and Sreenivasa, 2019b; Chennappa et al., 2014).

Identification and detection of phytopathogenic fungi has become the cornerstone (Borman et al., 2008). The main agenda for the diagnostic field is reliable, fast and accurate detection of the origin of a disease. Conventional methods like morphological, microscopic and cultural techniques mainly work on Phyto fungal infections like propagules of fungi such as mycelia, sclerotia, and conidia present on the external surface require prolonged time, taxonomy specialists and extensive labour (Nilsson et al., 2011; Tör and Woods-Tör, 2017; Janardhana et al., 2009; Dass et al., 2007). Even though these approaches form the cornerstone for phytopathogen detection, they may further lead to variable results during identification on requiring the specialist or experts with fungal identification knowledge for actual results (Chalupova et al., 2014; Deepa and Sreenivasa, 2017a; 2017b). Hence, the development and implementation of effective and novel diagnostic methods for fungal phytopathogens are urgent. As the earlier methods depend on phenotypic characters, the latter methods

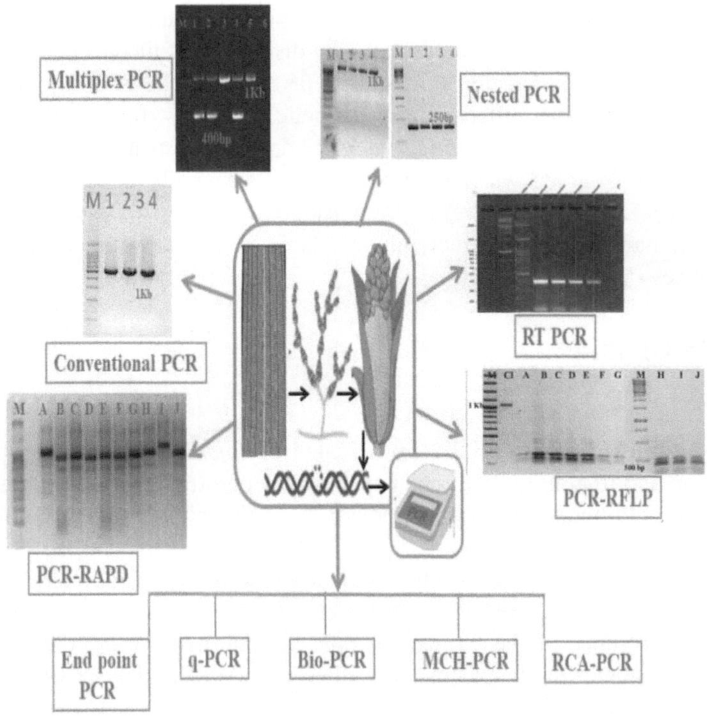

FIGURE 4.1 PCR and its variants for detection of fungal phytopathogens.

mainly rely on the genotypic characters exhibiting effective, fast, potentially accurate and highly specific and accurate results.

Currently, based on the limitations of conventional methods, plant-fungal diagnoses have further moved to molecular approaches, which have become valuable tools for pathogen recognition and quantification. Various molecular methods have been in trend in recent days, but we hereby discuss variants of polymerase chain reaction (PCR) on the detection of fungal phytopathogens (Deepa and Sreenivasa, 2023). PCR is a molecular technology that results in the exponential amplification of the desired region in the DNA molecule. The method was invented by Karry Mullis at a California biotech company in 1983 and was awarded a Nobel Prize for this discovery in 1993. It is applicable to identify and manipulate DNA, detect infectious pathogens and diagnose genetic variations including mutations. The methodology involves an enzyme Taq DNA polymerase (1–2.5 units) isolated from thermophilic bacteria *Thermus aquaticus* to catalyse the reaction, target DNA (<1 µg) having sequence to be amplified, primer pairs (1.5 µL) oligonucleotides defines sequence to be amplified, dNTPs (10 Mm – 0.4 µL), Mg++ ions (50 Mm) cofactor of the enzyme, and buffer (2µL with Ph 8.3–8.8) solution maintains pH and ionic strength of the reaction for the activity of the enzyme. Mainly follows the steps of denaturation which heats up the machine up to 94° C to denature DNA strands. Annealing cools the reaction to 54 °C to anneal the primers required to template. Extension warms the reaction to 72 °C on activating the Taq polymerase enzyme that extends primers and replicates the DNA. The entire cycle will be repeated 30–35 times (Deepa et al., 2016a; Sreenivasa et al., 2010, 2011b).

Standard PCR, conventional PCR, real-time PCR, end-point PCR, bio-PCR, polymerase spiral reaction, nested PCR, rolling circle amplification (RCA), nucleic acid sequence-based amplification (NASBA), multiplex PCR, quantitative PCR (Q-PCR), magnetic capture hybridization (MCH-PCR), co-operational PCR, PCR denaturing-gradient gel electrophoresis (PCR-DGGE), PCR restriction fragment length polymorphism (PCR-RFLP), next-generation PCR, PCR-ELISA, amplified length polymorphism (PCR-AFLP) and random amplified polymorphic DNA (PCR-RAPD) are the types of PCR variants involved in fungal phytopathogens (Deepa et al., 2018) (Figure 4.1).

Diagnosis and detection of fungal phytopathogens with molecular-based technologies need to follow certain basic stages like extraction of DNA by disruption of the chitin cell wall, removal of proteins and lyophilization of mycelia. Purification of extracted DNA assay is carried by silica membrane, pelleting and spin filter using magnetic particles for silica-coated separation. Quantification of the concentration of fungal DNA is determined using a UV spectrophotometer further diluting with PCR-grade water (Hariharan and Prasannath, 2021; Mancini et al., 2016; Abdullah et al., 2018). Even though several protocols are in practice for the isolation of DNA from pathogens currently, many kits are commercially available for the isolation of DNA in plant disease diagnostic approaches (Yang et al., 2016).

4.2 VARIANTS OF PCR

4.2.1 End-Point PCR

DNA template is amplified exponentially by repeated cycles of denaturation, annealing, extension, final extension and final hold reaction at specific temperatures with particular primers, thermostable Taq DNA polymerase, deoxyribonucleotide triphosphates (dNTPs) in buffer solution (Griffiths, 2014).

Mechanism: In this method, particular oligonucleotides were designed to target specific fungi or primers to amplify the pathogens tracked by sequencing for the detection of fungal phytopathogens and by assuring the presence of the targeted sample through agarose gel electrophoresis, which assures the phytopathogenic fungal existence. Further, each isolate identity was determined using Basic Local Alignment Search Tool (BLAST) in NCBI Genbank database (Mirmajlessi et al., 2015).

Application: End-point PCR was applied by Sikdar (2014) to detect Sphaeropsis rot and speck rot causing phytopathogens *Sphaeropsis pyriputrescens* and *Phacidiopycnis washingtonensis* in apples, respectively. Hemp disease in *Cannabis sativa* causing pathogen *Cercospora flagellaris* (Hemp powdery mildew) and *Golovinomyces cichoracearum* (Hemp leaf spot) was diagnosed through end-point PCR (Pepin et al., 2018; Doyle et al., 2019). *Cercospora tezpurensis* was diagnosed by end-point PCR causing leaf spot disease in *Capsicum assamicum* (Meghvansi et al., 2013). Fruit spot and leaf spot disease was detected in Blueberry and Macadamia caused by *Exobasidium maculosum*,

TABLE 4.1
PCR Variants to Diagnose Fungal Phytopathogens

PCR Variant	Fungal phytopathogen	Host	Disease	Reference
End-point PCR	*Cercospora flagellaris*	*Cannabis sativa*	Hemp leaf spot	Doyle et al., 2019
	C. tezpurensis	*Capsicum assamicum*	Leaf spot	Meghvansi et al., 2013
	Exobasidium maculosum	*Blueberry*	Fruit spot	Brewer et al., 2014
	Golovinomyces cicharacearum	*Cannabis sativa*	Hemp powdery mildew	Pepin et al., 2018
	Neonestalationsis clavispora *Colletotrichum siamense*	*Macadamia*	Leaf spot	Prasannath et al., 2020
Nested PCR	*F. verticillioides*	Maize, paddy, sorghum	Stalk rot, ear rot	Deepa et al., 2016c
	Puccinia striiformis	Wheat	Stripe rust	Wang et al., 2009
	Phytophthora cactarum	Strawberry	Crown rot	Bhat and Browne, 2010
	Colletotrichum gleosporioides	Dioscorea	Yam anthracnose	Raj et al., 2013
	Pilidiella granati	Pomegranate	Crown rot and twig blight	Yang et al., 2017

(*Continued*)

TABLE 4.1 (Continued)

PCR Variant	Fungal phytopathogen	Host	Disease	Reference
Multiplex PCR	*F. verticillioides; F. subglutinans; F, oxysporum*	Maize, paddy, musa, sorghum	Stalk rot, ear rot, *Fusarium* head blight, beer bananas	Deepa et al., 2016b; Faria et al., 2012
	Neofabraea alba; N. perennans; N. keinholzii	Apple	Bulls eye rot	Michalecka et al., 2016
Quantitative PCR (qPCR) / Real time PCR	*Fusarium guttiforme; F. culmorum*	Pineapple, cereals	Foot rot; root rot; *Fusarium* head blight; Fusariosis	Bilska et al., 2018; Carnielli-Queiroz et al., 2019
	Rhizoctonia solani	Tobacco	Target spot	Zhao et al., 2014
	Magnanarthe oryzae	Rice	Rice blast	Sun et al., 2015
	Verticillium longisporum	Brassica napus	Wilt stripe	Depotter et al., 2017
	Parastagonaspora nodorum	Wheat	Septoria blotch	Abdullah et al., 2018
	Didymella bryaniae	Cucurbits	Stem blight	Ling et al., 2010
	Aspergillus versicolor, Cladosporium cladosporoiodes, Stachybotrys chartarum, A. alternate	Cereals	Anthracnose	Badali and Nabili 2012
Bio PCR	*Colletotrichum lupini Alternaria alternata, A. dauci, A. radicina*	Cereals	Anthracnose	Pecchia et al., 2019
MCH PCR	*Acidovorax avenae Didymella bryoniae*	Cucurbit	Gummy Stem blight	He et al., 2012
RCA	*N. vangabunda, N. kienholzin, N. perennans, N. malicorticis*	Apple	Bulls eye rot	Lin et al., 2018
	F. oxysporum, F. graminearum, F. tricinctum, F. incarnatum	Cereals	*Fusarium* head blight	Davari et al., 2012
RT-PCR	*F. graminearum*	Maize, wheat, barley, oat, rye	*Fusarium* head blight	Capote et al., 2012
	Oidium neolycopercici, M. graminicola	Wheat, tomato	blight	Matsuda et al., 2005
Conventional PCR	*Mycosphaerella finiensis, M. musicolaFusarium* species	Cereal samples	Rot, blight	Deepa et al., 2016a
SAGE	*B. graminis*	Barley	Powdery mildew	Velculescu, 1995
RNAi	*F. graminearum, A. flavus*	Zea mays, *Hordeum vulgare*	Seedling blight	
PCR-DGGE	*Arthrobotrys oligospora, Trichoderma harzianum*	Tomato	Foot rot, seedling damp	Rytkonen et al., 2011
PCR-RFLP	*Pythium myriotylum, Fusarium* species	Cereals, maize, paddy, sorghum, wheat, barley	Stem rot, root rot, blight	Gómez-Alpizar et al., 2011; Deepa et al., 2018; Visentin et al., 2012, Patino et al., 2006
PCR-RAPD	*Fusarium* species	Maize kernels, cereal samples	*Fusarium* head blight	Moeller et al., 1999; Deepa et al., 2018

Neopestalotiopsis clavisopra and *Colletotrichum siamense*, respectively, by this method (Brewer et al., 2014; Prasannath et al., 2020) (Tables 4.1 and 4.2).

Advantages: End-point PCR is a traditionally used method. It is less sensitive than real-time PCR.

Limitations: The method is cost effective and time consuming as it requires to design primer for closely related phytopathogens.

TABLE 4.2
Key Genes Targeted on Detecting Pathogenic Fungi through PCR-based Approaches

Assay	Targeted genes	References
Conventional PCR	ITS, VERT, FUM	Deepa et al., 2016a; Henson and French 1993
Nested PCR	ITS, SSU-rDNA, VERT, FUM	Yang et al., 2017; Raj et al., 2013; Deepa et al., 2016c
Multiplex PCR	TUB2, gaoB, VERT, FUM, RPC2	Faria et al., 2012; Deepa et al., 2016b
End point PCR	TEF-1a, LSU-rDNA, ITS, ACT, CAL, HIS, TUB-2. GAPDH	Prasannath et al., 2020; Meghvansi et al., 2013; Pepin et al., 2018; Doyle et al., 2019
BIO-PCR	ITS rDNA	Kumar et al., 2020
Quantitative PCR	ITS, 18S-28S, TUB2, COX2, TEF-1a	Zhao et al., 2014; Bilska et al., 2018
RCA	TEF-1α	Lin et al., 2018
PCR-RFLP	ITS, Xho I, EcoR I, Hind III, Kpn I, Hinf I, Hae III	Visentin et al., 2012; Deepa et al., 2018
PCR-RAPD	R16, R14, R1	Moeller et al., 1999; Deepa et al., 2018

4.2.2 Multiplex PCR

The method uses single reaction mixture with several species-specific primers simultaneously to detect various phytopathogenic fungi.

Mechanism: Single PCR run is appropriate for detecting more than one gene at a time. The amplicons were further separated and analysed with agarose gel electrophoresis, further subjecting to sequencing and obtaining the accession numbers through BLAST for detection of phytopathogens (Deepa and Sreenivasa, 2017b).

Application: Phytopathogens *F. oxysporum*, *Phytophthora cactorum*, *Bipolaris cactivora and Phytophthora nicotianae* causing grafted cacti in the cactus industry were diagnosed using mutiplex PCR assays (Cho et al., 2016 – 2021 review). Fruit rot disease in cranberry detected 12 fungal phytopathogens *A. lycopodina*, *Colletotrichum fiorinae*, *Phyllosticta elongate*, *C. Fructivorum*, *Coleophoma empetri*, *Fusicoccum putrefaciens*, *Phomopsis vaccinia*, *Monilinia oxycocci*, *Phyllosticta vaccinia*, *Strasseria geniculate*, *Physalospora vaccinia* in a single test run (Conti et al., 2019). *F. verticillioides* associated with cereals was diagnosed by mPCR using one forward and two reverse primers (Deepa et al., 2016b). *F. subglutinans* causing stalk and ear rot in maize was detected in Brazil by gaoB gene coding for galactose oxidase (Faria et al., 2012). *Aspergillus* species were identified from meju using primers set I (omtB-1,2, aflR-1,2 and omtA-1,2) and set II (aflR-1,2, ver-1-F-R and omtA-1,2) (Kim et al., 2010). Dessert bananas disease in Musa caused by *F. oxysporum* was detected using multiplex PCR targeting the genes TEF-1 and RPC2 in a single run (Ndayihanzamaso et al., 2020). TUB2 gene in Apple causing Bull's eye rot infected with *Neofabraea alba, N. Keinholzii and N. perennans* was detected by Michalecka et al., (2016) (Tables 4.1 and 4.2).

Advantages: Quick, reliable and accurate.

Limitations: Primer designing is crucial as primers need to have similar annealing temperatures for amplification. It is potentially expensive, sensitive and resource-intensive.

4.2.3 Nested PCR

The method is the altered version of end-point PCR involving two rounds of PCR amplification with two sets of primers that enhance the sensitivity and specificity. The sensitivity of the method is 1000 times greater than any other PCR.

Mechanism: Initial round of PCR involves non-specific PCR primers, and pathogen-specific primers particular to the respective gene are used in the second round of PCR (Bhat and Browne

2010). Product of first round of PCR is used as template in the second round of PCR followed by agarose gel electrophoresis for confirmation on subjecting to sequencing and obtaining the accession numbers through BLAST for detection of phytopathogens (Deepa et al., 2016c) (Tables 4.1 and 4.2).

Application: Crown rot of pomegranate was detected with low concentration of DNA of 10pg detecting the pathogen *Pilidiella granati* (Yang et al., 2017). *Colletotrichum gloeosporioides* and *Cylindrocladium scoparium* were diagnosed causing great yam disease and eucalyptus dieback disease, respectively (Qiao et al., 2016). Fumonisin-producing *F. verticillioides* associated with cereals was detected using one forward (VERF-1) and two reverse primers (VERTR and VERTF-2). Species-specific primer was used in the first round and fum-specific primer was used in the second test run (Deepa et al., 2016c). Klemsdal and Elen (2006) detected *Fusarium culmorum* in cereal crops, where FculAF/FculAR external primer pair was diluted 10,000 times when compared to FculBF/FculBR internal primer pair with detection limit of 5–50 fg of target DNA. *Puccinia striiformis* was detected in wheat causing stripe rust by nested PCR (Wang et al., 2009). Similarly, crown rot disease-causing pathogen *Phytophthora cactorum* in strawberries was diagnosed by the nested PCR method (Bhat and Browne 2010).

Advantages: Enhanced sensitivity from 10 to 1000-fold on comparison to end-point PCR. Accurate and avoids frequent DNA isolation for each test run.

Limitations: Time consuming, risk of contamination is high as it involves a step as manipulation of earlier amplified samples, which may lead to false-positive results which cannot be used as reliable diagnostic methods.

4.2.4 Real-Time PCR

The method is also known as (Quantitative PCR) qPCR. It mainly involves quantification and detection of phytopathogenic fungal DNA or RNA sequences. It is an improved version of conventional PCR (Tables 4.1 and 4.2).

Mechanism: The method can also be represented by the name real-time PCR. Here, copy number of target DNA and RNA sequences is estimated by analysing the cycle threshold (Ct) value of the fungal pathogen on applying the specific primers (Balodi et al., 2017). The reaction requires sequence-specific fluorescent labelled probes like TaqMan, Eva Green, and SYBR Green I to maintain the amplification. Care to be taken so that the fluorescent signal will be proportional to the quantity of amplicon produced during each run of the cycle and also by breaking down of the dye-labelled probe or intercalating dye at the time of amplification (Badali and Nabili, 2012; Alemu 2014).

Application: Cryphonectria parasitica serves to be an emerging fungal phytopathogen causing wilting, blight, cankers and dieback diseases in chestnut trees (Jain et al., 2019). Cereals, barley and rice were infected with foot rot, root rot, head blight, leaf spot, and blast diseases by *Fusarium culmorum* for COX2, *Ramularia collo-cygni*, *Magnaporthe oryzae* for 18S-28S rDNA, respectively (Bilska et al., 2018; Sun et al., 2015). *Fusarium guttiforme* causes fusariosis in pineapple was detected by quantitative PCR for gene TEF-1 α and TUB2 (Carnielli-Queiroz et al., 2019). Tobacco plant was affected by target spot disease by *Rhizoctonia solani* and was diagnosed using the ITS primer by qPCR (Zhao et al., 2014). Gummy stem blight disease was observed in cucurbits by *Didymella bryoniae*, and stem wilt stripe disease was observed in Brassica napus by *Verticillium longisporum* on detecting TUB2 gene by qPCR and reported as a novel fungal pathogen (Depotter et al., 2017; Ling et al., 2010). Quantification of *Diaporthe helianthi* and *D. gulyae* phytopathogens causing Phomopsis stem canker was distinguished in sunflower by qPCR assay (Elverson et al., 2020). Dual labelled probes with inimitable fluorogenic reporters from *Parastagonospora nodorum*, and *Pyrenophora tritici-repentis* were custom designed as a duplex qPCR resulting in appropriate and accurate differentiation screening of multiple pathogens, which caused yellow spot and Septoria blotch in wheat (Abdullah et al., 2018). Mainly used for the quantification and identification of disease-causing fungi like *Aspergillus versicolor*, *Stachybotrys chartarum*, *Cladosporium cladosporioides*, and *Alternaria alternate*. The method detects even the lowest concentration of 300 fg DNA.

Advantages: The method is sensitive and fast, provides reliable information regarding pathogens, high specificity fluorogenic probes are used, cost effective and time efficient.

Limitations: The dye used is non-specific in nature, produces false results by forming primer dimer, instrument is costly. TaqMan probe used in the reaction is with high specificity.

4.2.5 BIO-PCR

It is a modified technology of end-point PCR to amplify specific DNA segments.

Mechanism: It follows a pre-assay incubation step for infected sample that increases the biomass of the phytopathogen. The method is mainly used to concentrate the target pathogen on growing the respective target pathogen in media, which prevents other non-target microbial growth (Schaad et al., 1995). It effectively detects seed-borne fungal pathogens (Kumar et al., 2020).

Applications: Colletotrichum lupini was enriched in yeast malt broth and was detected using species-specific pair primer based on intergenic spacer (IGS) sequence rDNA through BIO-PCR, which caused anthracnose disease. Species of *Alternaria alternata, A. dauci, A. radicina* were diagnosed with specific ITS rDNA using deep freezing blotter method (Kumar et al., 2020) (Tables 4.1 and 4.2).

Advantages: Eliminates PCR inhibitors, high sensitivity, detects living cells to avoid false positives.

Limitations: Time consuming, costly on procuring selective media for the assay.

4.2.6 Magnetic-Capture Hybridization PCR

The method was established to deal with PCR inhibitors in phytopathogens at the time of DNA isolation.

Mechanism: MCH-PCR follows the steps of DNA isolation, purification with ssDNA probe on magnetic beads and PCR amplification with target DNA. The magnetic beads are of biotinylated coated oligonucleotide specific to the target DNA of the phytopathogen. Here, dsDNA and magnetic beads hybridize to separate the inhibitors (Chalupova et al., 2014). Probe used in the method needs selection of oligonucleotide probe from highly conserved region of phytopathogens, taking care that 5' end is biotin labelled on attaching streptavidin-coated magnetic beads. Further, the sequence is subjected to BLAST by in silico specificity (Johnson et al., 2013) (Tables 4.1 and 4.2).

Applications: Acidovorax avenae and *Didymella bryoniae* causing gummy stem blight in cucurbit seeds were detected by this method.

Advantages: It decreases the detection time compared to other methods; inhibitors and non-target DNA are removed during amplification.

Limitations: Highly sensitive, requires use of magnetic beads which is cost effective.

4.2.7 Rolling Circle Amplification

It is an enzymatic isothermal method that exploits either DNA or RNA polymerase to generate ssDNA or RNA. Requires DNA polymerase, circular template, short DNA/RNA primer, buffer and deoxynucleotide triphosphate.

Mechanism: Amplification of DNA with phi29 DNA polymerase by strand displacement for extension of single or multiple primers that anneal to circular DNA template is conducted. Strand displacement in the sense that newly synthesized long ssDNA replaces formerly synthesized DNA on releasing ssDNA (Tables 4.1 and 4.2).

Application: Bull's eye rot disease in apples caused by *Neofabraea vangabunda, N. Kienholzin, N. perennans, N. malicorticis* phytopathogen is diagnosed on TEF-1α segments with four padlock probes PLP-Nv, PLP-Nk, PLP-Np, PLP-Nm by RCA method (Lin et al., 2018). *Fusarium oxysporum, F. graminearum, F. incarnatum, F. tricinctum* causing Fusarium head blight disease were detected by performing RCA technology using padlock probes based on elongation factor TEF-1α (Davari et al., 2012). The method is also applied to analyse single nucleotide polymorphism, gene expression, protein post-translational modification and mRNA splicing (Gao et al., 2019).

Advantages: Efficient, simple, temperature cycling device is not required.

Limitations: It gives contamination and false-positive results due to supersensitive RCA assay.

4.2.8 Nucleic Acid Sequence-Based Amplification

NASBA is a technique that amplifies nucleic acids with isothermal conditions.

Mechanism: The method requires RNA for amplification purposes. ssRNA template will be converted to cDNA using the enzyme reverse transcriptase. It generates nearly 10^8 copies of nucleic acid sequence at a gap of 30 min and the amplicons are further detected through electrochemiluminescence probe-capture hybridization (Heo et al., 2019).

Application: The method is very rare for fungal plant pathogen detection. Mainly utilizes virus reverse transcriptase, T7 RNA polymerase and RNase Hat temperature around 41°C (Chang et al., 2012) (Tables 4.1 and 4.2).

Advantages: Highly sensitive, does not require thermal cycler, better than real-time based PCR and method is well applicable in laboratory on chip devices.

Limitations: It cannot be used for all the phytopathogens where only DNA is targeted.

4.2.9 RT-PCR

Mechanism: In this method, RNA is reversely transcribed into cDNA using primers and Reverse transcriptase enzyme and further amplified by PCR variant method.

Applications: *F. graminearum* causing *Fusarium* head blight is quantified in cereals like maize, wheat, oat, rye and barley. Applied mainly to diagnose the RNA containing viruses such as retroviruses by developing antimicrobial vaccines as a therapy (Capote et al., 2012). The method detects phytopathogens *Oidium neolycopersici* and *M. graminicola* in tomato and wheat, respectively (Matsuda et al., 2005) (Tables 4.1 and 4.2).

Advantages: Ability to differentiate between live and dead fungi, obtains qualitative and quantitative data, simple, rapid, cost effective, highly specific.

Limitations: Expertise I required, more contamination, extremely sensitive.

4.2.10 Conventional PCR

Initially, the invention of PCR was a revolution in the field of plant pathology.

Mechanism: Synthesizes a million copies of DNA using specific primers followed by denaturation, annealing and elongation steps. PCR mainly depends on deoxynucleoside triphosphate concentration and efficacy of DNA extraction. To avoid the inhibitors in the sample, cetyl trimethyl ammonium bromide is used for plant sample detection (Tables 4.1 and 4.2).

Application: Earlier the method was used for specific detection of bacteria and viruses. Currently, it is used widely for the detection of phytopathogen. *Mycosphaerella finiensis* and *M. musicola* are detected by the method. Cereal sample collected from all over Karnataka was screened by this method diagnosing 69 samples infected with *Fusarium* species and 51 samples infected with *F. verticillicides* (Deepa et al., 2016a).

Advantages: Gives highly specific results, rapid and precise.

Limitations: Requires much labour, too costly

4.2.11 Serial Analysis of Gene Expression

SAGE is a sequence-based method for the expression of quantitative genes identifying multiple transcripts.

Mechanism: The method uses two different samples as ligated samples and labelled samples with primers and amplified later. It involves quantification and sequencing of 15bp oligonucleotides to find the corresponding expressed genes against the available genome sequences (Velculescu, 1995). After amplification, primers are removed forming concatemers providing sticky ends and later cloned into vector trailed by computational analysis.

Applications: Used to identify *B. graminis* in barley leaves.

Advantages: Precise knowledge is not necessary.

Limitations: Requires large quantity of mRNA, sometimes requires even long tag than 15bp for complex genomes.

4.2.12 PCR-DGGE

Mechanism: PCR-Denaturing Gradient Gel Electrophoresis is a molecular technique that involves fungal phytopathogen DNA amplification and further subjected to electrophoresis.

Application: Fungal pathogens are detected in plants. *Arthrobotrys oligospora* and *Trichoderma harzianum* spores were detected from the technique (Rytkonen et al., 2011) (Tables 4.1 and 4.2).

Advantages: Studies on diversity

Limitations: Time consuming, does not separate properly.

4.2.13 PCR-RFLP

PCR restriction fragment length polymorphism is a method that involves the PCR for targeted region amplification, and further restriction enzyme activity of digestion of PCR products occurs.

Mechanism: It is generated either by presence or absence of the recognition site at the same location of the chromosome for the restriction endonuclease among different individuals of the species. Particular restriction enzyme from the fragments of dissimilar length signifies the same chromosome region of diverse individuals. The particular differences are further diagnosed by gel electrophoresis in combination with hybridization of labelled probe specifically for the chromosome region (Tables 4.1 and 4.2).

Applications: Applied for differentiation of phytopathogenic fungi from non-pathogenic strains like *Pythium myriotylum* (Gómez-Alpizar et al., 2011). Xho I, EcoR I, Hind III, Kpn I, Hinf I and Hae III enzymes are used for the study of genetic variations among 10 strains of *F. verticillioides* analysing strain level differentiation in Xho I, EcoR I, Hind III and Kpn I and identical pattern with other two enzymes (Deepa et al., 2018). PCR-RFLP with ITS primers was conducted to discriminate the *Fusarium* species (Visentin et al., 2012). RFLP analysed the telomorphs and DNA mating population of *F. verticillioides* (Schlacht et al., 1997). Patino et al., (2006) diagnosed and discriminated variations of 33 mycotoxin-producing *Fusarium* species irrespective of their host and origin.

Advantages: Simple, quick, reproducible, accurate in population identification, stable.

Limitations: High labour required, radioisotope is required, costly

4.2.14 PCR-RAPD

Mechanism: PCR-Random amplified polymorphic DNA is a method with 10 bases long oligonucleotide of random sequences used as primers in the reaction. Genomic DNA sequence if complementary to oligonucleotide primer, the amplified product will be further detected in gel electrophoresis. The fragments diagnosed in the profile will be scraped or removed from the gel and later sequenced to obtain sequence characterized amplified region (SCAR) where two particular primers are designed for precise detection (Tables 4.1 and 4.2).

Application: The method is applied to differentiate fungal phytopathogens within the species. Infected maize kernels analysis was analysed by PCR-RAPD fragment sequences (Moeller et al., 1999). *F. verticillioides* causing *Fusarium* head blight in cereal samples were analysed for genetic variation using three different RAPD primers R16, R14 and R1 (Deepa et al., 2018).

Advantages: Simple, quick, probes not required, efficient.

Limitations: Lacks reproducibility, mismatch occurs.

4.2.15 PCR-AFLP

Mechanism: PCR Amplified fragment length polymorphism is a method mainly used for the detection of symptoms developed on detached leaves, seedling development and seed germination.

Application: Genetic variability among the phytopathogens is investigated (Silva et al., 2017).

Advantages: Different genome part is analysed; reproducibility is high.

Limitations: Complex and lengthy protocol.

4.3 SUMMARY

Fungal phytopathogens cause major losses to the agricultural sector by reducing the quality and production of food crops. Rapid and accurate identification and detection of these pathogens is the most important criterion. Presently, diagnostic laboratories are in use of many advanced molecular techniques to detect and diagnose the emerging, novel previously reported and reemerging disease caused by plant pathogenic fungi. In the present chapter, we have discussed the PCR and its variants' application as a significant tool for identification, diagnosis and characterization of fungal pathogens. These approaches have successfully diagnosed both symptomatic and asymptomatic diseases of fungal pathogens in agriculturally important fields. The challenges are for researchers and scientists to develop in-time detection of phytopathogenic fungi to help the farmers formulate proper managemental practices to control plant disease infection.

4.4 FUTURE PERSPECTIVES

Molecular technologies using molecular markers need to be applied to a large extent for the analysis and characterization of genetic variability in fungal phytopathogens causing diseases in the agricultural crops that pose a great threat to the agricultural sector.

STUDY QUESTIONS

1. Best PCR variant for detection of fungal phytopathogen
2. Precedence and hindrance of PCR variants
3. Discovery and mechanism of PCR

ACKNOWLEDGEMENT

We wish to thank the Indian Council of Medical Research (ICMR) for providing an RA fellowship to Dr. Deepa N to carry out research work on "Potential antifungals from PGPR bacteria against fumonisin".

CONFLICT OF INTEREST

The authors confirm that they have no conflict of interest.

FURTHER READING

Aslam, S., Tahir, A., Aslam, M.F., Alam, M.W., Shedayi, A.A., Sadia, S. 2017. Recent advances in molecular techniques for the identification of phytopathogenic fungi – a mini review. *Journal of plant interactions*. 12(1), 493–504.

Hariharan, G., and Prasannath, K. 2021. Recent advances in molecular diagnostics of fungal plant pathogens: A mini review. *Frontiers Cell Infectious Microbiology*. 10, 600234.

Kumar, A., Singh, R., Singh, R., Senger, R.R. 2015. Molecular approaches for detection of plant pathogen. *Biotech Today an International Journal of Biological Sciences*. 5(2), 14–19.

REFERENCES

Abdullah, A.S., Turo, C., Moffat, C.S., Lopez-Ruiz, F.J., Gibberd, M.R., Hamblin, J., 2018. Real-time PCR for diagnosing and quantifying co-infection by two globally distributed fungal pathogens of wheat. *Frontiers in Plant Science*. 9, 1086.

Alemu, K., 2014. Real-time PCR and its application in plant disease diagnostics. *Advances in Life Science and Technology*. 27, 39–49.

Aslam, S., Tahir, A., Aslam, M.F., Alarm, M.W., Shedayi, A.A., Sadia., 2017. Recent advances in molecular techniques for the identification of phytopathogenic fungi – a mini review. *Journal of Plant Interactions* 12 (1), 493–504.

Badali, H., Nabili, M., 2012. Molecular tools in medical mycology; where we are! *Jundishapur Journal of Microbiology*. 6(1), 1–3.

Balodi, R., Bisht, S., Ghatak, A., Rao, K.H., 2017. Plant disease diagnosis: technological advancements and challenges. *Indian Phytopathology*. 70(3), 275–281.

Bhat, R.G., and Browne, G.T., 2010. Specific detection of Phytophthora cactorum in diseased strawberry plants using nested polymerase chain reaction. *Plant Pathology*. 59, 121–129.

Bilska, K., Kulik, T., Ostrowska-Kołodziejczak, A., Busko, M., Pasquali, M., Beyer, M., 2018. Development of a highly sensitive FcMito qPCR assay for the quantification of the toxigenic fungal plant pathogen *Fusarium culmorum*. *Toxins*. 10 (5), 211.

Borman, A.M., Linton, C.J., Miles, S.J., Johnson, E.M., 2008. Molecular identification of pathogenic fungi. *Journal of Antimicrobial Chemotherapy*. 61(Suppl 1), 7–12.

Brewer, M. T., Turner, A. N., Brannen, P. M., Cline, W. O., Richardson, E. A., 2014. *Exobasidium maculosum*, a new species causing leaf and fruit spots on blueberry in the southeastern USA and its relationship with other *Exobasidium* spp. parasitic to blueberry and cranberry. *Mycologia*. 106(3), 415–423.

Capote, N., Mara, A., Aguado, A., Snchez-Torres, P., 2012. Molecular tools for detection of plant pathogenic fungi and fungicide resistance, in *Plant Pathology*. Ed. C. J. R. Cumagun, RI, USA: InTech, 151–202.

Carnielli-Queiroz, L., Fernandes, P. M. B., Fernandes, A. A. R., Ventura, J. A., 2019. A rapid and reliable method for molecular detection of *Fusarium guttiforme*, the Etiological Agent of Pineapple Fusariosis. https://doi.org/10.1590/1678-4324-2019180591.

Chalupova, J., Raus, M., Sedlarova, M., Sebela, M. 2014. Identification of fungal microorganisms by MALDI-TOF mass spectrometry. *Biotechnology Advances*. 32, (1), 230–241.

Chang, C. C., Chen, C. C., Wei, S. C., Lu, H. H., Liang, Y. H., Lin, C. W., 2012. Diagnostic devices for isothermal nucleic acid amplification. *Sensors (Basel Switzerland)*. 12 (6), 8319–8337.

Chennappa, G., Adkar-Purushothama, C.R., Umdale S., Tamilvendan, K., Sreenivasa, M.Y. 2014. Impact of Pesticides on PGPR Activity of Azotobacter sp. Isolated from Pesticide Flooded Paddy Soils. *Greener Journal of Agricultural Sciences*. 4 (4), 117–129.

Cho, H., Hong, S. W., Kim, H. J., Kwak, Y. S. 2016. Development of a multiplex PCR method to detect fungal pathogens for quarantine on exported cacti. *Plant Pathololgy*. 3291, 53–57.

Conti, M., Cinget, B., Vivancos, J., Oudemans, P., Bélanger, R. R., 2019. A molecular assay allows the simultaneous detection of 12 fungi causing fruit rot in Cranberry. *Plant Diseases*. 103 (11), 2843–2850.

Dass, R.S., Sreenivasa, M.Y., Janardhana, G.R. 2007. High incidence of *Fusarium verticillioides* in animal and poultry feed mixtures produced in Karnataka, India. *Plant Pathology Journal* 6(2), 174–178.

Davari, M., van Diepeningen, A.D., Babai-Ahari, A., Arzanlou, M., Najafzadeh, M. J., van der Lee, T. A., 2012. Rapid identification of *Fusarium graminearum* species using Rolling circle amplification (RCA). *Journal of Microbiological Methods* 89 (1), 63–70.

Deepa, N., Adkar-Purushothama, C.R., Sreenivasa, M.Y., 2016b. "Multiplex PCR for the early detection of fumonisin producing *Fusarium verticillioides*". *Food Bioscience*. 13, 84–88.

Deepa, N., Adkar-Purushothama, C.R., Sreenivasa, M.Y. 2016c. "Nested PCR method for early detection of fumonisin producing *Fusarium verticillioides* in Pure Cultures, Cereal Samples and Plant parts". *Food Biotechnology*. 30(1), 18–29.

Deepa, N., Adkar-Purushothama, C.R., Sreenivasa, M.Y., 2021. Chapter 6: "Molecular technologies for the early detection of fungal phytopathogens associated with cereal crops". *Microbial genomic technologies to combat the problems of farming lands. Microbes in land use change management*. Elsevier. 137–154.

Deepa, N., Nagaraja, H., Sreenivasa, M.Y., 2016a. "Prevalence of fumonisin producing *Fusarium verticillioides* associated with cereals grown in Karnataka (India)". *Food Science and Human Wellness*. 5(3), 156–158.

Deepa, N., Rakesh, S., Sreenivasa, M.Y., 2018. "Morphological, pathological and mycotoxicological variations among *Fusarium verticillioides* isolated from cereals". *3 Biotech*. 8, 105.

Deepa, N., Sreenivasa, M.Y. 2017a. Fumonisins: A Review on its Global Occurrence, Epidemiology, Toxicity and Detection. *Journal of Veterinary Medicine and Research*. 4(6), 1093.

Deepa, N., Sreenivasa, M.Y. 2017b. *Fusarium verticillioides*, a Globally Important Pathogen of Agriculture and Livestock: A Review. *Journal of Veterinary Medicine and Research*. 4(4), 1084.

Deepa, N., Sreenivasa, M.Y., 2019a. "Molecular methods and key genes targeted for the detection of fumonisin producing *Fusarium verticillioides* – An updated review". *Food Bioscience*. 32: 100473.

Deepa, N., Sreenivasa, M.Y. 2019b. Biocontrol Strategies for Effective Management of Phytopathogenic Fungi Associated with Cereals, In: Singh JS., Singh DP. (eds) *New and Future Developments in Microbial Biotechnology and Bioengineering*, Elsevier, pp, 177–189, ISBN 9780444641915.

Deepa, N., Sreenivasa, M.Y., 2023. *Biodegradation of Fusarial mycotoxins associated with food and feeds*. Mycotoxinsite.com. Webpage.

Depotter, J. R. L., Rodriguez-Moreno, L., Thomma, B. P. H. A., Wood, T. A., 2017. The Emerging British *Verticillium longisporum* Population Consists of Aggressive Brassica Pathogens. *Phytopathology*. 107(11), 1399–1405.

Doyle, V. P., Tonry, H. T., Amsden, B., Beale, J., Dixon, E., Li, H., et al. (2019). First report of *Cercospora* cf. *flagellaris* on industrial hemp (*Cannabis saliva*) in Kentucky. *Plant Diseases*. 103, 1784.

Elverson, T. R., Kontz, B. J., Markell, S. G., Harveson, R. M., Mathew, F. M., 2020. Quantitative PCR Assays Developed for *Diaporthe helianthi* and *Diaporthe gulyae* for Phomopsis Stem Canker Diagnosis and Germplasm Screening in Sunflower (*Helianthus annuus*). Plant Diseases. 104(3), 793–800.

Faria, C. B., Abe, C. A., da Silva, C. N., Tessmann, D. J., Barbosa-Tessmann, I. P., 2012. New PCR assays for the identification of *Fusarium verticillioides*, *Fusarium subglutinans*, and other species of the *Gibberella fujikuroi* complex. International Journal of Molecular Sciences. 13(1), 115–132.

Gao, H., Zhang, K., Teng, X., Li, J., 2019. Rolling circling amplification for a single cell analysis and *in situ* sequencing. Trends in Analytical *Chemistry*. 121, 115700.

Gómez-Alpizar, L., Saalau, E., Picado, I., Tambong, J.T., Saborio, F. 2011. A PCR-RFLP assay for identification and detection of *Pythium myriotylum*, causal agent of the cocoyam root rot disease. *Letters in Applied Microbiology*. 52, 185–192.

Griffiths, M.W., 2014. Nucleic acid–based assays: Overview. In *Encyclopedia of Food Microbiology*. Eds. C. A. Batt and M. L. Tortorello, Oxford: Academic Press, 990–998.

Hariharan, G., Prasannath, K., 2021. Recent advances in molecular diagnostics of fungal plant pathogens: A mini review. *Frontiers in Cellular Infectious Microbiology*. 10, 600234.

He, Y. H., Isono, S., Shibuya, M., Tsuji, M., Adkar Purushothama, C. R., Tanaka, K., 2012. Oligo-DNA custom macroarray for monitoring major pathogenic and non-pathogenic fungi and bacteria in the phyllosphere of apple trees. *PLoS One*. 7 (3), e34249.

Henson, J.M., French, R. 1993. The polymerase chain reaction and plant disease diagnosis. *Annual Review of Phytopathology*. 31 (1), 81–109.

Heo, S., Kim, H. R., Lee, H. J., 2019. Development of a quantitative real-time nucleic acid sequence based amplification (NASBA) assay for early detection of apple scar skin viroid. *Plant Pathology Journal*. 35 (2), 164–171.

Jain, A., Sarsaiya, S., Wu, Q., Lu, Y., Shi, J., 2019. A review of plant leaf fungal diseases and its environment speciation. *Bioengineered*. 10 (1), 409–424.

Janardhana, G. R., Dass, R.S., Sreenivasa, M.Y. 2009. Fumonisins – A new class of Fusarial toxins. In *Frontiers in Fungal Ecology, Diversity and Metabolites*. Ed: K. A. Sridhar et al. I. K. International Publishing House Pvt. Ltd. New Delhi.

Johnson, K. L., Zheng, D., Kaewnum, S., Reid, C. L., Burr, T., 2013. Development of a magnetic capture hybridization real-time PCR assay for detection of tumorigenic A*grobacterium vitis* in grapevines. *Phytopathology*. 103, 633–640.

Kim, S.W., Li, Z., Moore, P.S., Monaghan, A.P., Chang, Y., Nichols, M., John, B., 2010. A sensitive non-radioactive northern blot method to detect small RNAs. *Nucleic Acids Research*. 38, 98–110.

Klemsdal, S.S., and Elen, O. 2006. Development of a highly sensitive nested PCR method using a single closed tube for detection of *Fusarium culmorum* in cereal samples. *Letters in Applied Microbiology*. 42, 544–548.

Kumar, R., Gupta, A., Srivastava, S., Devi, G., Singh, V. K., Goswami, S. K., 2020. Diagnosis and Detection of Seed-Borne Fungal Phytopathogens. In *Seed-Borne Diseases of Agricultural Crops: Detection, Diagnosis & Management*. Eds. R. Kumar and A. Gupta, Singapore: Springer, 107–142.

Lin, H., Jiang, X., Yi, J., Wang, X., Zuo, R., Jiang, Z., 2018. Molecular identification of *Neofabraea* species associated with bull's-eye rot on apple using rolling-circle amplification of partial EF-1a sequence. *Canadian Journal of Microbiology*. 64(1), 57–68.

Ling, K. S., Wechter, W. P., Somai, B. M., Walcott, R. R., and Keinath, A. P., 2010. An improved real-time PCR system for broad-spectrum detection of *Didymella bryoniae*, the causal agent of gummy stem blight of cucurbits. *Seed Science Technology*. 38(3), 692–703.

Mancini, V., Murolo, S., and Romanazzi, G., 2016. Diagnostic methods for detecting fungal pathogens on vegetable seeds. *Plant Pathology*. 65, 691–703.

Matsuda, Y., Sameshima, T., Moriura, N., Inoue, K., Nonomura, T., Kakutani, K., Nishimura, H., Kusakari, S., Takamatsu, S., Toyoda, H., 2005. Identification of individual powdery mildew fungi infecting leaves and direct detection of gene expression by single conidium polymerase chain reaction. *Phytopathology*. 95, 1137–1143.

Meghvansi, M. K., Khan, M. H., Gupta, R., Veer, V., 2013. Identification of a new species of *Cercospora* causing leaf spot disease in *Capsicum assamicum* in northeastern India. *Research in Microbiology*. 164(9), 894–902.

Michalecka, M., Bryk, H., Poniatowska, A., Puławska, J., 2016. Identification of *Neofabraea* species causing bull's eye rot of apple in Poland and their direct detection in apple fruit using multiplex PCR. *Plant Pathology*. 65, 643–654.

Mirmajlessi, S. M., Destefanis, M., Gottsberger, R. A., Mänd, M., Loit, E., 2015. PCR-based specific techniques used for detecting the most important pathogens on strawberry: A systematic review. *Systemic Review*. 4, 9.

Moeller, E.M., Chelkowski, J., Geiger, H.H. 1999. Species-specific PCR assays for the fungal pathogens *Fusarium moniliforme* and *Fusarium subglutinans* and their application to diagnose maize ear rot. *Journal of Phytopathology*. 147, 497–508.

Nagaraja, H., Chennappa, G., Poorna Chandra Rao, K., Mahadevprasad, G., Sreenivasa, M.Y. 2016. Diversity of toxic and phytopathogenic *Fusarium* species occurring on cereals grown in Karnataka state, India. *3Biotech*. 6, 57.

Narayanasamy, P., 2011. Diagnosis of fungal diseases of plants. In *Microbial plant pathogens-detection and disease diagnosis*. Dordrecht: Springer, 273– 284.

Ndayihanzamaso, P., Karangwa, P., Mostert, D., Mahuku, G., Blomme, G., Beed, F., 2020. The development of a multiplex PCR assay for the detection of *Fusarium oxysporum* f. sp. *cubense* lineage VI strains in East and Central Africa. *European Journal of plant Pathology*. 158, 495–509.

Nilsson, R.H., Abarenkov, K., Larsson, K.H., Kõljalg, U., 2011. Molecular identification of fungi: rationale, philosophical concerns, and the UNITE database. *Open Applied Infectious Journal*. 5, 81–86.

Patino, B., Mirete, S., Vazquez, C., Jimenez, M., Rodriguez, M.T., Gonzalez-Jaen, M.T. 2006. Characterization of *F. verticillioides* strains by PCR-RFLP analysis of the intergenic spacer region of the rDNA. *Journal of the Science of Food and Agriculture*. 86, 429–435.

Pecchia, S., Caggiano, B., Lio, D., Cafà, G., Le Floch, G., Baroncelli, R., 2019. Molecular detection of the seed-borne *pathogen Colletotrichum lupini* targeting the hyper-variable IGS region of the ribosomal cluster. *Plants*. 8 (7), 222.

Pepin, N., Punja, Z. K., Joly, D. L., 2018. Occurrence of powdery mildew caused by *Golovinomyces cichoracearum sensulato* on Cannabis sativa in Canada. *Plant Diseases*. 102, 2644.

Prasad, M.N.N., Shankara Bhat, S., Sreenivasa, M.Y. 2010. Antifungal activity of essential oils against *Phomopsis azadirachtae*- the causative agent of die-back disease of neem. *Journal of Agricultural Technology*. 6(1), 127–133.

Prasannath, K., Galea, V. J., and Akinsanmi, O. A., 2020. Characterisation of leaf spots caused by *Neopestalotiopsis clavispora* and *Colletotrichum siamense* in Macadamia in Australia. *European Journal of Plant Pathology*. 156, 1219–1225.

Qiao, T. M., Zhang, J., Li, S. J., Han, S., Zhu, T. H., 2016). Development of nested PCR, multiplex PCR, and loop-mediated isothermal amplification assays for rapid detection of *Cylindrocladium scoparium* on Eucalyptus. *Plant Pathology Journal*. 32(5), 414–422.

Raj, M., Nath, V. S., Sankar, M. S., Pravi, V., Archana, P. V., Jeeva, M. L., 2013. A highly sensitive nested PCR method using a single closed tube for the detection of *Colletotrichum gloeosporioides* causing greater yam anthracnose. *Journal of Root Crops*. 39(2), 163–167.

Rytkonen, K.T., Williams, T.A., Renshaw, G.M., Primmer, C.R., Nikinmaa, M. 2011. Molecular evolution of the metazoan PHD-HIF oxygen-sensing system. *Molecular Biology and Evolution*. 28, 1913–1926.

Schaad, N. W., Cheong, S. S., Tamaki, S., Hatziloukas, E., Panopoulos, N. J., 1995. A combined biological and enzymatic amplification (BIO-PCR) technique to detect *Pseudomonas syringae* pv. *phaseolicola* in bean extracts. *Phytopathology*. 85, 243–248.

Schlacht U.F., Moller E.M., Geiger H.G. 1997. Genetic diversity of *Gibberella fujikuroi* isolates from different geographic origins. Proceedings of the 5th European *Fusarium* seminar. *Cereal Research and Communication*. 25, 557–600.

Sikdar, P., Okubara, P., Mazzola, M., Xiao, C.L, 2014. Development of PCR assays for diagnosis and detection of the pathogens *Phacidiopycnis washingtonensis* and *Sphaeropsis pyriputrescens* in apple fruit. *Plant Diseases*. 98 (2), 241–246.

Silva, J.J., Viaro, H.P., Ferranti, L.S., Oliveira, A.L.M., Ferreira, J.M., Ruas, C.F., Ono, E.Y.S., Fungaro, M.H.P., 2017. Genetic structure of *Fusarium verticillioides* populations and occurrence of fumonisins in maize grown in southern Brazil. *Crop Protection*. 99, 160–167.

Sreenivasa, M.Y., Dass, R.S., Charith Raj, A.P., Janardhana, G.R. 2011b. Mycological evaluation of Maize grains produced in Karnataka (India) for the post-harvest fungal contamination. *World Applied Sciences Journal*. 13(4), 688–692.

Sreenivasa, M.Y., Dass, R.S., Charith Raj, A.P., Prasad, M.N.N., Premila N. Achar, Janardhana, G.R. 2011a. Assessment of the growth inhibiting effect of some plant essential oils on different *Fusarium* species isolated from sorghum and maize grains. *Journal of Plant Diseases and Protection*. 118 (6), 208–213.

Sreenivasa, M.Y., Dass, R.S., Janardhana, G.R. 2010. Post harvest fungi associated with sorghum grains produced in Karnataka (India). *Journal of Plant Protection Research*. 50(3), 335–339.

Sun, G., Liu, J., Li, G., Zhang, X., Chen, T., Chen, J., 2015. Quick and accurate detection and quantification of *Magnaporthe oryzae* in rice using real-time quantitative polymerase chain reaction. *Plant Diseases*. 99, 219–224.

Tör, M., Woods-Tör, A., 2017. Fungal and oomycete diseases. In *Encyclopedia of applied plant sciences*. Eds. B. Thomas, B. G. Murray and D. J. Murphy. New York, USA: Academic Press, 77–82.

Velculescu, V.E., 1995. Serial analysis of gene expression. *Science*. 270, 484–487.

Visentin, I., Montis, V., Doll, K., Alabouvette, C., Tamietti, G., Karlovsky, P., Cardinale F., 2012. Transcription of genes in biosynthetic pathway for fumonisin mycotoxins is epigenetically and differentially regulated in the fungal maize pathogen *Fusarium verticillioides*. *Eukaryotic Cell*. 11, 252–259.

Wang, X., Tang, C., Chen, J., Buchenauer, H., Zhao, J., Han, Q., 2009. Detection of *Puccinia striiformis* in latently infected wheat leaves by nested polymerase chain reaction. *Journal of Phytopathology*. 157, 490–493.

Wenneker, M., Thomma, B. P. H. J., 2020. Latent postharvest pathogens of pome fruit and their management: from single measures to a systems intervention approach. *European Journal of Plant Pathology*. 156, 663–681.

Yang, X., Hameed, U., Zhang, A. F., Zang, H. Y., Gu, C. Y., Chen, Y., 2017. Development of a nested-PCR assay for the rapid detection of *Pilidiella granati* in pomegranate fruit. *Scientific Reports*. 7, 40954.

Yang, Y., Zuzak, K., Feng, F., 2016. An improved simple method for DNA extraction from fungal mycelia. *Canadian Journal of Plant Pathology*. 38, 476–482.

Zhao, X., Lin, C. W., Wang, J., Oh, D. H., 2014. Advances in rapid detection methods for foodborne pathogens. *Journal of Microbiology and Biotechnology*. 24, 297–312.

5 Loop-Mediated Isothermal Amplification (LAMP) for the Detection and Quantification of Fungal Phytopathogens

S. Wily and Sandra V. Gomez-Gutierrez
Purdue University

Stephen B. Goodwin
Crop Production and Pest Control Research Unit, USDA–Agricultural Research Service

Accurate identification of fungal pathogens that are causal agents of plant diseases is necessary to make appropriate disease management. Loop-mediated isothermal amplification (LAMP) is a sensitive and specific tool for the detection of fungal pathogens in plants. LAMP is an isothermal technique for the amplification of nucleic acids. Four to six primers are used to amplify the target sequence of DNA. The result is a variety of DNA products of diverse sizes with a characteristic stem-loop structure and multiple loops. The amplified products are easily detected by fluorescence measurement and changes in the turbidity or color of the reaction by adding intercalating dyes. LAMP has been used to detect fungal phytopathogens in soybean, rice, grapes, cotton, banana, corn, and wheat, showing high specificity and sensitivity. Also, several portable devices for real-time LAMP have been developed and used under field conditions. One caveat is that this technique requires an extensive optimization process to be effective under field conditions due to cross-contamination.

DOI: 10.1201/9781003399704-5

GRAPHICAL ABSTRACT

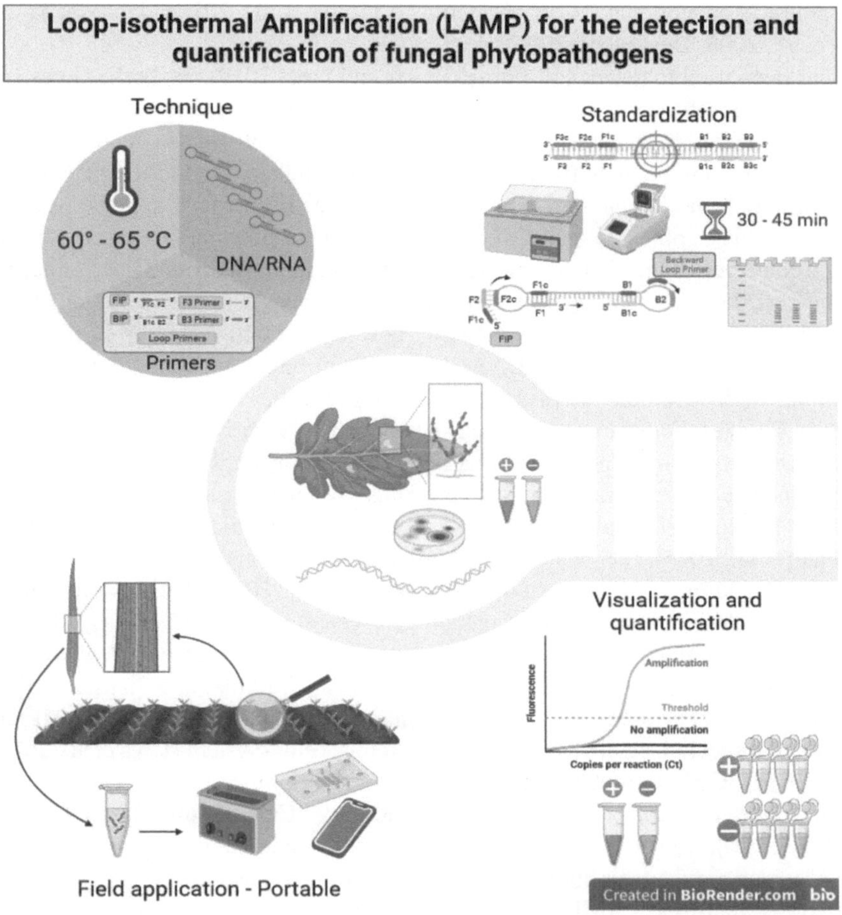

5.1 INTRODUCTION

Loop-mediated isothermal amplification (LAMP) was developed as an isothermal technique that performs the reaction in less time than other amplification methods and shows improved specificity and sensitivity in the results. Notomi et al. (2000) developed and optimized LAMP by using a special DNA polymerase and a set of four to six primers. The DNA polymerase has strand-displacement activity, and the primer sequences bind to eight unique regions in the target sequence. The amplification products of LAMP have increasing length as more cycles of the reaction occur. These products contain multiple loops and are usually called cauliflower structures (Wang et al., 2015). A ladder-like pattern is observed in the electrophoresis gel of the LAMP reaction solution, which demonstrates the formation of products of increasing length varying from 100 base pairs up until the wells in agarose gels (Kaur et al., 2020). LAMP was rapidly adopted as a diagnostic tool for plant monitoring as it proved suitable for diagnosing diseases in plants, seeds, spore traps, and soil. The isothermal nature of LAMP makes it an exceptionally convenient technique for detecting plant diseases, whether in the field or in minimally equipped laboratories. LAMP has high specificity and sensitivity due to the targeting of four to six regions by using two forward, two backward primers, and optionally two loop primers, which increase the speed of the reaction providing results in less time compared to conventional detection methods (Deepa et al., 2016a; Janardhana et al., 2009; Sreenivasa et al., 2008; Dass et al., 2007).

LAMP has been used to detect soil-borne (Panek and Frąc, 2019), air-borne (Lees et al., 2019), and seed-borne (Logeshwari et al., 2022) plant fungal diseases (Niessen, 2015). As a diagnostic method, LAMP can detect the pathogen in both hosts and alternative hosts by using pure culture or crude DNA extract from the plant tissue. For example, Choudhary et al. (2022) developed an assay for the detection of *Sarocladium oryzae* using pure cultures of the pathogen and in living infected tissue but also could be applied to seeds, weeds, and even environmental samples. His results showed that *S. oryzae* could be detected in two weed species that commonly surround rice fields. Furthermore, LAMP can detect the target pathogen in different infected host tissues and any other material that could carry the pathogen. For example, an assay developed by Si Ammour et al. (2020) allowed detection of *Botrytis cinerea* in crude DNA extracts from bunch trash and berries of grapevines. LAMP also has high specificity to distinguish target pathogens from other pathogenic species in environmental samples. For example, the assay developed by Villari et al. (2017) showed specificity for *Magnaporthe oryzae* in the presence of large quantities of background DNA.

Regarding its practical application, LAMP is a diagnostic technique suitable for laboratories with reduced access to expensive equipment that can be used for rapid diagnosis of plant diseases (Lu et al., 2015; Patel et al., 2015). For example, LAMP can be used in agricultural extension centers or to detect quarantine pathogens in border control (DeLude et al., 2022). This also includes its implementation in the field by extension plant pathologists (Patel et al., 2015). Assays for LAMP have been developed for use in laboratory conditions, for example using Genie II (Ayukawa et al., 2016) and III instruments (Achari et al., 2023), water bath (Ghosh, Tarafdar and Sharma, 2016; Hu et al., 2017), and heat block (Thiessen et al., 2016). In contrast, quantitative LAMP (qLAMP) assays have been carried out in specialized devices such as the StepOnePlus-System (Lees et al., 2019; Deepa & Sreenivasa, 2019a), CFX96 Touch Real-Time polymerase chain reaction (PCR) detection system (Villari et al., 2017; Deepa & Sreenivasa, 2019b), and the ICGENE mini portable instrument (Si Ammour et al., 2020). Most of the qLAMP assays involve using thermal cyclers or expensive real-time devices, which reduce their applicability for field applications.

Quantitative or real-time LAMP (qLAMP) is useful to detect quarantine pathogens and reduce the risk of contamination because it can be applied as a diagnostic at the point of care. As an example, qLAMP was used to detect the quarantine fungal pathogen *Ceratocystis platani*, which causes canker stain disease of plane tree (Aglietti et al., 2019). qLAMP uses labeled oligonucleotide probes and a quencher molecule to emit fluorescence through fluorescence resonance energy transfer (FRET). Another variation of LAMP that can be standardized for fungal pathogen detection is multiplex LAMP (mLAMP). This technique can differentiate various targets in a mixture reaction by using fluorescence-based methods. For example, mLAMP was used to detect and differentiate four species of the genus *Phytophthora* in Japan, an oomycete plant pathogen that causes tree mortality (Hieno et al., 2021). To date, we have not found a study that has applied mLAMP for detection of multiple plant fungal pathogens or lineages of the same plant pathogenic fungus. However, the need for expensive equipment, sophisticated reagents and a longer reaction time might increase the difficulty of applying mLAMP for regular field detection of fungal pathogens of plants. Therefore, further optimization is needed to standardize mLAMP for the detection of fungal pathogens in laboratory conditions and in the field (Liang et al., 2012; Feng et al., 2019).

Despite the limitations of LAMP, the technique has been promoted as a promising molecular assay for the detection of fungal plant pathogens with the potential for field applications (Duan et al. 2014a). Three major steps are required for the application of LAMP under field conditions: setting up the reaction mix, performing the amplification, and reporting the results. Setting up the reaction under field conditions can be conducted with premixed LAMP reagents so DNA can be added directly to the reaction tube (Patel et al., 2015).

5.2 AMPLIFICATION MECHANISM

LAMP reaction is performed using a water bath or heat block due to its isothermal nature. The amplification mechanism occurs at a constant temperature that ranges from 60 °C to 65 °C (Nagamine et al., 2002). The use of a special enzyme, called *Bst* polymerase, assists the DNA amplification process during the LAMP reaction. The *Bst* polymerase possesses auto-cycling strand-displacement activity, which eliminates the DNA denaturation stage (Notomi et al., 2000; Soroka et al., 2021). LAMP uses three sets of primers that consist of two inner and two outer primers, and an optional set of two loop primers. These six primers are complementary to eight different regions in the target DNA sequence (Notomi et al., 2000). The forward inner primer (FIP) and backward inner primer (BIP) initiate the amplification process, and during the later stages of amplification, they are required for self-priming. The forward (F3) and backward (B3) outer primers release the newly formed complementary strand during the non-cyclic stage of LAMP by annealing to the opposite strand (Parida et al., 2008; Gomez-Gutierrez and Goodwin, 2022). The optional loop forward (LF) and loop backward (LB) primers are complementary to the sequences that conform to the loops, which are formed during the cyclic amplification of LAMP. The use of LF and LB primers in the reaction increases the amplification products and reduces the time of the reaction (Nagamine et al., 2002). The most popular and specialized tool for designing the primer sets for LAMP is Primer Explorer V5 (Notomi et al., 2015).

The first step in the amplification is the synthesis of a complementary strand during the non-cyclic stage of LAMP. This initiates with the binding of the FIP to the complementary region in the target DNA (Figure 5.1. Step 1). The F3 primer releases the complementary strand formed by annealing the FIP through hybridization to its complementary sequence. This process forms a loop structure at one end (Figure 5.1. Steps 3 to 5). The recently released strand serves as a template for the BIP to anneal and initiate new DNA synthesis. Following BIP annealing, B3 also binds to the released strand causing strand-displacement DNA synthesis. This process releases the new complementary strand formed by the BIP (Figure 5.1., Steps 5 to 7). The new complementary single DNA produces a dumbbell structure, which is subsequently converted to a stem-loop DNA by self-primed DNA synthesis (Figure 5.1., Step 8). The stem loop is the starting material for the LAMP cycling stage. During this stage, if the loop primers are added to the reaction, they generate additional stem-loop structures that promote exponential amplification (Figure 5.1., Steps 9 to 12). In the elongation and recycling stages, the stem-loop structures are the templates for the annealing of BIP and the consequent strand-displacement reaction. Also, two double stem-loop DNA structures are generated. FIP and BIP will bind to these products in the elongation and recycling stages (Figure 5.1. Steps 9a, 11a, 13 to 16) to produce additional elongation structures (Figure 5.1. Steps 15 and 16). During every half cycle, the target DNA is amplified three times demonstrating the rapidity of LAMP (Notomi et al., 2000; Tomita et al., 2008; Bruce et al., 2015; Li and Macdonald, 2015; Gomez-Gutierrez and Goodwin, 2022). LAMP generates both elongated amplification products and smaller-length recycle amplicons, which produce the characteristic ladder-like pattern in gel visualization (Kaur et al., 2020).

5.3 VISUALIZATION OF AMPLIFICATION PRODUCTS

The visualization of LAMP products is possible using different techniques such as electrophoresis gel, turbidimetry (Garg et al., 2022), or visual observation of color change by using intercalating DNA dyes. The most common dyes used in assays for the identification of fungal phytopathogens are SYBR green I (Xu et al., 2021), HNB ((Duan et al. 2014a)), and calcein (Niessen, 2015), while turbidity is determined by Mg-pyrophosphate precipitation.

LAMP products also can be observed using lateral flow devices (Patel et al., 2015), portable foldable microdevices (Trinh and Lee, 2019; Prasannakumar et al., 2021), and most recently by chips. These chips are built with nanopore thin film able to immobilize primers, allowing to bind and

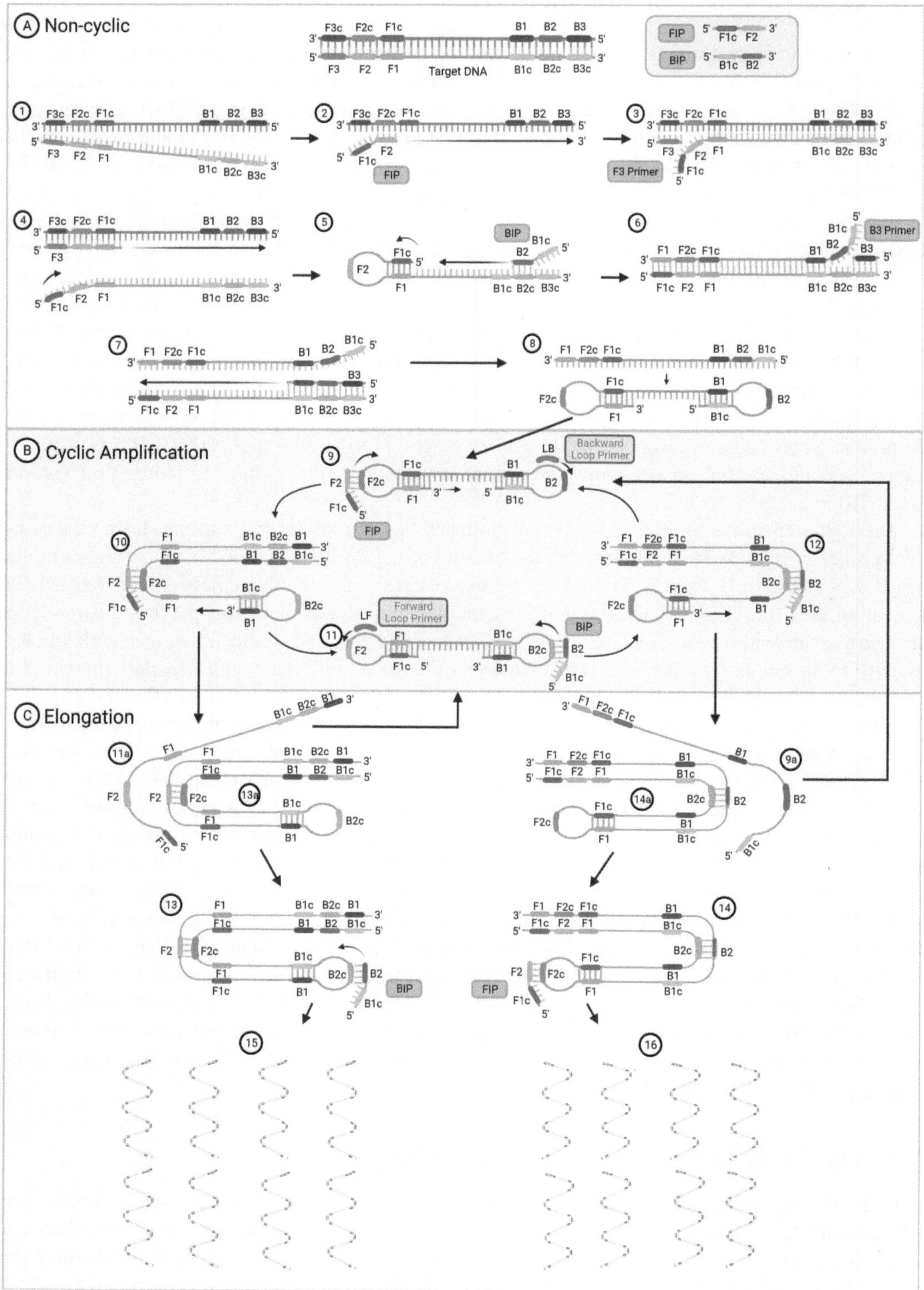

FIGURE 5.1 The loop-mediated isothermal amplification (LAMP) technique amplification process.

immobilize the LAMP product, which can be read with an optical spectrophotometer (Mao et al. 2023).

SYBR green is a cyanine dye that fluoresces when it is intercalated to double-stranded DNA or RNA. When this occurs, the color of the solution changes from orange to green (Dragan et al., 2012; Li et al., 2013). Calcein uses the pyrophosphates that are produced in high amounts during LAMP reactions to emit fluorescence in a positive reaction. The pyrophosphate is a subproduct of the synthesis of DNA by the polymerase and dNTPs. This subproduct binds to calcein, which was bound to manganese before the amplification (Tomita et al., 2008), producing fluorescence. Hydroxy naphthol blue (HNB) is a colorimetric indicator reagent for alkaline ions. The HNB color changes according to the pH of the reaction, with a pH in the range of 8.6–9.0 the reaction turns magenta or violet but is blue when the pH is lower. The change in the pH of the LAMP reaction is due to the reduction of the concentration of Mg2+, which is chelated by dNTPs during DNA extension. As a result, the reaction changes from violet to blue (Goto et al., 2009).

5.4 SENSITIVITY AND LIMIT OF DETECTION

The limit and time of detection (LoD) of LAMP reactions are determined by the source of DNA, the visualization method, and the colorimetric method. For example, the time of detection of *Phytophthora* spp. using DNA purified from plant tissue or soil is longer when compared to that of DNA extracted from pure cultures (Siegieda, Panek and Frąc, 2021). This is caused by the presence of inhibitors in DNA that is extracted from environmental samples. These inhibitors affect the rate of amplification. The LoD also may vary depending on whether it is calculated using genomic DNA or fragments of the target gene. Additionally, the LoD may fluctuate when results are observed on an electrophoresis gel as opposed to when different colorimetric dyes are used. For example, the sensitivity of calcein can be ten times lower compared to the sensitivity of SYBR green and HNB (Goto et al., 2009).

5.5 DNA SOURCES AND SAMPLE PREPARATION

Sources of DNA for the assays can be pure culture plates, infected plant tissue, soil, seeds, spore traps, and environmental samples, which can have large amounts of different DNAs. When LAMP is tested with environmental samples, it is recommended to calculate the values of true positive and true negative using a gold standard such as qPCR (Thiessen et al., 2016). LAMP assays have been combined with simple DNA extraction methods or using crude extractions from plant material (Kong et al., 2016; Si Ammour et al., 2020; Choudhary et al., 2022). Siegieda et al. (2021) showed that LAMP does not depend on the method of DNA isolation.

5.6 TARGET GENES FOR PRIMER DESIGN

To detect fungal plant pathogens, the most common target genes for primer design are the internal transcribed spacer (ITS) (Lees et al., 2019), β-tubulin ((Duan et al. 2014a)), EF-1alpha (Ghosh et al., 2015), Cytochrome b (Cytb) (Hu et al., 2017), and Actin (Choudhary et al., 2022). Sequences of particular regions in the genomes of the target pathogen, plus those of closely related pathogens, and any other organisms that may occur in the same samples can be aligned to identify a unique region in the pathogen of interest. This unique sequence region can be used to differentiate the target pathogen from other pathogens or different strains of the same pathogen. For example, during the primer design in an assay for the detection of *Alternaria solani*, 46 ITS sequences were compared to identify a specific region (Lees et al., 2019). Achari et al. (2023) also compared 356 genomes of 39 *formae speciales* of *Fusarium oxysporum* to identify a single orthogroup that is unique for *F. oxysporum* f.sp. *ciceris* as a molecular marker. Another approach of LAMP is the detection of point mutations in a pathogen population to distinguish different pathogen races (Ayukawa et al., 2016);

in this case, the primers need to be designed in the gene with the mutation of interest. In some assays, where mutations need to be avoided, Primer Explorer allows the design of mutation-sensitive or -insensitive primers (Mori and Notomi, 2009). *In silico* validation has been used to test the specificity of the designed primers against closely related organisms of the target pathogen or sequence (DeLude et al., 2022).

5.7 APPLICATIONS OF THE TECHNIQUE

LAMP has been used to detect fungal pathogens in artificially and naturally infected plant tissue and soil using crude DNA extracts without DNA purification steps. This is possible due to the tolerance of *Bst* polymerase to common inhibitors in unrefined DNA extracts such as phenols, sugars, and other compounds (Kaneko et al., 2007; Si Ammour et al., 2020). Some assays have been developed in conjunction with crude DNA extract procedures to allow for in-field detection of fungal plant pathogens.

This diagnostic method can detect latent invaders (Huang et al., 2011), which is the infection stage when the pathogen is in the host but is not causing symptoms that are visible to the naked eye (Luchi et al., 2016). This is possible due to its high sensitivity, which can detect down to 10 femtograms of starting template (Patel et al., 2015; Logeshwari et al., 2022). The high sensitivity of LAMP allows its use for the early detection of plant pathogens, preventing epidemics, or identifying and monitoring the introduction of new pathogens (Ghosh et al., 2015; Deepa et al., 2016b; Logeshwari et al., 2022).

LAMP has been used to detect pathogens at multiple levels including strains, races, special forms, and even the detection of strains with certain mutations, such as mutations that confer pathogen resistance to fungicides, which allows monitoring of fungicide resistance in a pathogen population. For example, Quinone outside inhibitor (QoIs) fungicides inhibit mitochondrial respiration by binding to the Qo site of the cytochrome b protein, but some fungal pathogens have developed resistance to these fungicides. Hu et al. (2017) developed two LAMP assays to monitor and evaluate the development of resistance to QoI fungicides in *Botrytis cinerea* by targeting different point mutations of the *cytochrome b* gene. Duan et al. (2014a) also developed an assay for the detection of the F167Y mutation of the β2-tubulin gene, which causes resistance against carbendazim in *Fusarium graminearum*.

The use of effector genes, virulence genes, and genes with mutations as the target region for amplification is a great option for designing primer sets to distinguish between different strains of the same species. For example, to distinguish *Fusarium oxysporum* f. sp. *ciceris* (Foc), which is a pathogenic form infecting chickpea (*Cicer arietinum*) from non-pathogenic strains, Achari et al. (2023) designed primers that target a unique putative effector gene.

5.8 ADVANTAGES OF THE TECHNIQUE

LAMP has several advantages compared to traditional methods for the detection of fungal pathogens and diagnosis of plant diseases. Traditional methods, which often are based on culturing the causal organism, require several steps such as preparation of the sample, single-hyphal isolation, incubation, and identification (Niessen, 2015). These methods are very slow and require thorough training for adequate isolation and microscopic identification, which relies on morphological characteristics or visual assessment and is prone to human error.

The constant amplification temperature and a reduced time for the reaction are the main advantages of LAMP. These characteristics allow the reaction to be performed without expensive equipment, which also generates results in a short time. LAMP is a very practical molecular technique for fast application due to its high specificity and sensitivity levels, which can surpass those of PCR techniques. These features are conferred by using six primers and a robust strand-displacement polymerase, which minimize inhibitor problems. The sensitivity of detection obtained with LAMP is key for the early detection of fungal pathogens in the field when no visible symptoms are apparent. This contributes to deciding on early disease management approaches, such as the elimination of infected plants or fungicide applications.

5.9 LIMITATIONS OF THE TECHNIQUE

Optimization of the LAMP assay could be a challenge due to cross-contamination, non-specific amplification, and self-dimerization that generates false positives. LAMP is characterized by its exponential efficiency in producing copies of the target gene. However, this great efficiency also increases the risk of cross-contamination by any aerosols generated during the amplification (Logeshwari et al., 2022). The contamination normally occurs in assays with colorimetric visualization where the dye is added at the end of the reaction. Cross-contamination occurs mainly during amplification and when the tube is open (Tomlinson, Barker and Boonham, 2007; Deepa et al., 2016c; Njiru et al., 2008). For this reason, it is recommended to use dyes that do not negatively affect the LAMP reaction and that can be added before amplification. This makes it possible to observe the results without opening the tubes to prevent carry-over contamination (Parida et al., 2008). Examples of these dyes are Calcein and HNB. Karthik et al. (2014) developed a tube assay based on agar dye capsules to prevent cross-contamination when SYBR green is used. Also, it is recommended to prepare the reaction in a different room than the one where the amplification is conducted (Goto et al., 2009). During the visualization step, it is recommended to work in a separate room from the one in which the amplification was performed. Also, the recommendation is to open the control tubes first and add the colorimetric dye before opening the tubes that contain DNA samples. These precautions contribute to reducing false positives if a colorimetric dye is added. Regular change of gloves and application of sterile pipetting techniques (Lau et al., 2010), using filter tips, individual tubes, and a biosafety cabinet (Mao et al., 2023) also can reduce cross-contamination. Inactivation of the polymerase by raising the temperature to 80 °C during the last 5–10 minutes of the reaction also could be beneficial.

Non-specific amplification affects both sensitivity (LoD) and specificity of the assay and generates false positives (Rolando et al., 2020). Non-specific amplification is caused by the complexity of the amplification structures generated during the reaction and the homo-dimerization/dimers of the inner primers. The formation of these structures is facilitated by the large number of primers used in

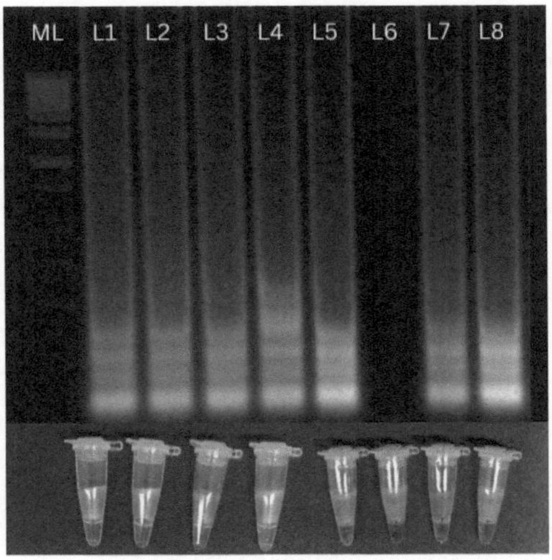

FIGURE 5.2 Different ladder-like patterns in electrophoresis obtained from negative controls and DNA template reactions with cross-contamination.

TABLE 5.1
Color of Positive and Negative Reactions Reported by Different Assays Using the Same Dye

Dye	Positive	Negative	Reference
SYBR Green I	Green	Orange	Huang et al., 2011
SYBR Green I	Brown	Fluorescent yellow	Fan et al., 2019
SYBR Green I	Yellow-green	Orange	Tian et al., 2017
SYBR Green I	Yellow	Orange	Siegieda et al., 2021
Warm Start Colorimetric LAMP 2X Master Mix (New England Biolabs)	Yellow	Red/pink	Choudhary et al., 2020
Quant-iT PicoGreen	Yellow	Orange	Tomlinson et al., 2007
Basic fuchsin	Purple	Colorless	Prasannakumar et al., 2021
HNB	Sky blue	Violet	Duan et al. 2014b; Ghosh et al., 2015; Hu et al., 2017; Logeshwari et al., 2022
HNB	Blue	Magenta to violet	Goto et al., 2009

the reaction, especially the inner primers, which are prone to the formation of stable hairpin structures due to their length (Meagher et al., 2018). This type of amplification can be identified when negative controls show a positive response or by mathematical models in conjunction with electrophoresis (Schneider et al., 2019). Target DNA amplification, non-specific amplification, and self-dimerization may show different ladder-like patterns in an electrophoresis gel, which makes it difficult to determine if primer dimers or contamination are occurring during the reaction. It is even possible to observe no amplification in a sample tube, while a ladder-like pattern is observed in control tubes due to cross-contamination with sample tubes that show correct DNA amplification (personal observation; Figure 5.2). Figure 5.2 indicates the electrophoresis gel of four replicates of a negative control reaction using nuclease-free water (L1–L4) and four replicates of an amplification reaction (L5–L8) targeting the intended gene. All reactions were carried out under the same conditions using WarmStart® Colorimetric LAMP 2X Master Mix. All reactions changed from a red color to a yellow color, except for a sample tube, after amplification. The control reactions show four bands in an electrophoresis gel with a similar thickness, except for one (L4) that shows a thicker band on the top. The sample reactions show only three bands in an electrophoresis gel, expecting one (L5) that shows an additional band. Reactions that show four bands in an electrophoresis gel could be a result of external contamination since the ladder-like pattern of the bands is not similar to the sample reaction. This illustrates the multiple challenges associated with standardization of the LAMP technique.

Another limitation of LAMP is that turbidity and colorimetric methods rely on the ability of the human eye to distinguish between positive and negative, which can lead to false positives and negatives (Patel et al., 2015). This is because the color intensity can vary between reactions. For example, when using SYBR green, positive reactions typically turn green, while negative reactions turn orange. However, there are instances where the reaction may turn yellow green. As a result, different assays report different color tonality for positive and negative reactions despite using the same dye (Table 5.1).

5.10 ASSAY OPTIMIZATION

For optimization of a LAMP assay, we suggest screening all the primers for self-dimerization and non-specific extension by amplifying the reactions with no DNA template, as this will reduce the chance of cross-contamination. Also test different combinations of temperature, time, and reagent concentrations. The critical components for the effectiveness of the reaction include $MgSO_4$, betaine, *Bst* polymerase, dNTPs, and dye concentration (Logeshwari et al., 2022). Magnesium is particularly important in the change of color and visualization of the LAMP products (Reuter et al., 2020; Zaczek-Moczydłowska et al., 2020). Primer concentration, especially the inner primers, also plays an important role in dimer formation as mentioned before. For colorimetric assays using SYBR green, it is necessary to test different concentrations of dye. High concentrations may cause the reaction to turn orange even in the presence of amplicons. Conversely, low concentrations may not change the color of the reactions (personal observation).

Once all the conditions are optimized, and no cross-contamination or self-dimerization is observed in the control tubes, the specificity of the primers can be tested against other fungal pathogens. This includes pathogens present in the target crop, alternative hosts, and weeds found in the field if the assay is used under field conditions with crude DNA extracts. Crude extracts may be contaminated with the DNA of the host and other pathogens and may contain different inhibitors.

Kits such as the Isothermal Mastermix and WarmStart® Colorimetric LAMP 2X Master Mix (NEB), Isothermal MasterMix (ISO-001, OptiGene, Horsham, UK), and Isothermal master mix (Nippon Gene, Tokyo, Japan) are commonly used to develop assays for the detection of fungal pathogens.

5.11 SUMMARY

LAMP is a molecular technique for DNA amplification. The reaction is performed at a constant temperature and uses three pairs of primers. Its isothermal nature, combined with the specificity and sensitivity of the reaction, represents an improvement over traditional methods such as PCR for detection of fungal pathogens in plants. The high sensitivity and specificity of LAMP results from the use of four to six primers for DNA amplification. These primers are designed to amplify a specific and unique region of the target organism. The reaction preparation and amplification are relatively simple and do not require expensive equipment. Also, there are several methods for the visualization of the amplification products, such as turbidimetry and colorimetric methods, electrophoresis gel, devices such as lateral flow (LFD), or more sophisticated equipment that can perform the amplification and produce amplification curves such as the Genie II instrument. LAMP is also considered an inexpensive diagnostic method since the amplification can be performed in relatively cheap, low-tech equipment such as a water bath or heat block. This allows its application in laboratories, agricultural centers, farms, and extension centers, which may not have access to sophisticated equipment such as a thermal cycler.

The application of LAMP has not been limited to the diagnosis of fungal diseases at the species level but also can be applied for the detection of specific strains or mutations such as those that confer resistance to fungicides. It is also used for the early and rapid detection of pathogens and as a tool for disease risk forecasting, allowing the detection of soil-, air-, and seed-borne fungal diseases. The detection can be conducted using infected live tissue such as leaves, fruits, seeds, branches, alternative hosts, and plant residues. LAMP can detect the target pathogen in environmental samples or tissue infected with multiple pathogens due to the specificity of the primers and the tolerance of the *Bts* polymerase to contaminants.

The main limitations of the assay are the cross-contamination between samples due to the high amplification rate and the formation of dimers due to the number and length of the primers. Dyes such as HNB, which is added before amplification, can reduce the risk of cross-contamination since the tubes are not open after amplification. Using different rooms for DNA extraction, reaction preparation, and amplification can also reduce the rate of cross-contamination. The formation of dimers can be avoided by proper primer design, in-silico validation, optimization of time, temperature, and reagent concentrations. Despite these limitations, LAMP is a practical diagnostic technique to detect fungal pathogens that affect plants and generate epidemics in crops of agricultural interest. LAMP has some advantages over conventional PCR, particularly when it is used under specific field conditions or in laboratories with limited equipment.

5.12 FUTURE PERSPECTIVES

LAMP assays have been combined with other technologies and performed in different practical equipment to conduct the amplification. These improvements need to be commercialized in user-friendly kits like PCR reagents to ensure easy application and avoid cross-contamination. Another future alternative for LAMP is the improvement of multiplex LAMP assays. Currently, mLAMP assays have not been standardized for detection of fungal pathogens in plants. However, the successful application of mLAMP to detect other organisms suggests that it will allow the identification of complex fungal diseases or the discrimination of several strains of fungi. This would be useful to detect multiple pathogens and save resources and time. For example, multiplex PCR has been used for the detection of quarantine pathogens on exported cacti, including *Bipolaris maydis*, *Phytophthora nicotianae*, *Phytophthora cactorum*, and *Fusarium oxysporum* (Cho et al., 2016). This alternative reduces the number of detection assays required to identify each pathogen and provides rapid and cost-effective results. Multiplex PCR also helps to identify different pathogens with similar control and fungicide requirements. For example, Chen (2015) generated a multiplex PCR for the detection of three pathogens in wheat. Development of effective protocols for multiplexing samples would greatly increase the utility of LAMP.

Regarding field applications, a portable device to conduct LAMP in a fast and efficient manner can serve to overcome the limitations of other techniques for the detection of fungal pathogens in plants. Portable fluorescent devices have been tested to conduct mLAMP, which can be an alternative for discriminating multiple pathogens in the same sample (Wang et al., 2015). However, the standardization of portable LAMP for field application requires the development of innovative devices and specialized reagents, which can increase the costs. On the other hand, if LAMP primers are correctly designed to differentiate fungal lineages based on polymorphisms, this will represent a huge advantage of LAMP to identify and manage epidemics in important crops. For example, LAMP was used to differentiate lineages of *Magnaporthe oryzae* that exclusively cause wheat blast epidemics, from other lineages that only cause opportunistic infections on wheat (Thierry et al., 2020).

STUDY QUESTIONS

1. Describe why false positives are commonly observed during the optimization of a LAMP assay and how this can be avoided.
2. Why is a single band not observed when the LAMP amplification products are visualized in an electrophoresis gel?
3. List the six primers used for LAMP amplification. Which ones increase the amplification rate?
4. Describe one application of LAMP for the diagnostic of fungal phytopathogens.
5. Mention one future improvement to make LAMP suitable for field detection of fungal pathogens.

ACKNOWLEDGMENT

This work was supported by the United States Department of Agriculture, Agricultural Research Service (USDA-ARS) research project 5020-21220-014-00D, and Purdue University.

FURTHER READING

Bolton, M.D., Thomma, B.P.H.J., 2012. *Plant Fungal Pathogens: Methods and Protocols*. Humana Press.
Gullino, M.L., Bonants, P.J.M., 2014. *Detection and Diagnostics of Plant Pathogens*. Springer. 139–150.
Lane, C.R., Beales, P.A., Hughes, K.J.D., 2023. *Fungal Plant Pathogens, 2nd Edition: Applied Techniques*. CABI. Pages 190–195.

REFERENCES

Achari, S.R., Mann, R.C., Sharma, M., Edwards, J., 2023. Diagnosis of *Fusarium oxysporum* f. sp. *ciceris* causing Fusarium wilt of chickpea using loop-mediated isothermal amplification (LAMP) and conventional end-point PCR. *Scientific Reports* 13, 2640. https://doi.org/10.1038/s41598-023-29730-6

Aglietti, C., Luchi, N., Pepori, A.L., Bartolini, P., Pecori, F. Raio, A., Capretti, P. & Santini, A. 2019. Real-time loop-mediated isothermal amplification: an early-warning tool for quarantine plant pathogen detection. *AMB Express* 9, 50. https://doi.org/10.1186/s13568-019-0774-9

Ayukawa, Y., Komatsu, K., Kashiwa, T., Akai, K., Yamada, M., Teraoka, T., Arie, T., 2016. Detection and differentiation of *Fusarium oxysporum* f. sp. *lycopersici* race 1 using loop-mediated isothermal amplification with three primer sets. *Letters in Applied Microbiology* 63, 202–209. https://doi.org/10.1111/lam.12597

Bruce, K.L., Leterme, S.C., Ellis, A.V., Lenehan, C.E., 2015. Approaches for the detection of harmful algal blooms using oligonucleotide interactions. *Analytical and Bioanalytical Chemistry* 407, 95–116. https://doi.org/10.1007/s00216-014-8193-x

Chen, S., Cao, Y., Li, T., Wu, X. 2015. Simultaneous detection of three wheat pathogenic fungal species by multiplex PCR. *Phytoparasitica* 43, 449–460. https://doi.org/10.1007/s12600-014-0442-1

Cho H.J., Hong S.W., Kim H.J., Kwak Y.S. 2016. Development of a Multiplex PCR Method to Detect Fungal Pathogens for Quarantine on Exported Cacti. *The Plant Pathology Journal*, 32, 53–57. https://doi.org/10.5423/PPJ.NT.09.2015.0184. Epub Feb 1. PMID: 26889115; PMCID: PMC4755675.

Choudhary, P., Goswami, S.K., Chakdar, H., Verma, S., Thapa, S., Srivastava, A.K., Saxena, A.K., 2022. Colorimetric loop-mediated isothermal amplification assay for detection and ecological monitoring of *Sarocladium oryzae*, an important seed-borne pathogen of rice. *Frontiers in Plant Science* 13, 936766. https://doi.org/10.3389/fpls.2022.936766

Choudhary, P., Rai, P., Yadav, J., Verma, S., Chakdar, H., Goswami, S.K., Srivastava, A.K., Kashyap, P.L., Saxena, A.K., 2020. A rapid colorimetric LAMP assay for detection of *Rhizoctonia solani* AG-1 IA causing sheath blight of rice. *Scientific Reports* 10, 22022. https://doi.org/10.1038/s41598-020-79117-0

Dass, R.S., Sreenivasa, M.Y., Janardhana, G.R. 2007. High incidence of *Fusarium verticillioides* in animal and poultry feed mixtures produced in Karnataka, India. *Plant Pathology Journal*. 6(2), 174–178.

Deepa, N., Charith Raj, A. P., Sreenivasa, M.Y. 2016b. Nested PCR method for the early detection of fumonisin producing *Fusarium verticillioides* in pure cultures, cereal samples and plant parts. *Food Biotechnology*. 30(1), 18–29.

Deepa, N., Charith Raj, A.P., Sreenivasa, M.Y. 2016c. Multiplex PCR method for the early detection of fumonisin producing *Fusarium verticillioides*. *Food Bioscience*. 13, 84–88.

Deepa, N., Nagaraja, H., Sreenivasa, M.Y. 2016a. Prevalence of fumonisin producing Fusarium verticillioides associated with cereals grown in Karnataka (India). *Food Science Human Wellness*. 5, 156–162. https://doi.org/10.1016/j.fshw.2016.07.001.

Deepa, N., and Sreenivasa M.Y. 2019a. Molecular methods and key genes targeted for the detection of fumonisin producing *Fusarium verticillioides* - An updated review. *Food Bioscience*. 32, 100473.

Deepa, N., Sreenivasa, M.Y. 2019b. Biocontrol Strategies for Effective Management of Phytopathogenic Fungi Associated with Cereals, In: Singh J.S., Singh D.P. (eds) *New and Future Developments in Microbial Biotechnology and Bioengineering*, Elsevier, pp, 177–189, ISBN 9780444641915.

DeLude, A., Wells, R., Boomla, S., Chuang, S.-C., Urena, F., Shipman, A., Rubas, N., Kuehu, D.L., Bickerton, B., Peterson, T., Dobhal, S., Arizala, D., Klair, D., Ochoa-Corona, F., Ali, M.E., Odani, J., Bingham, J.-P., Jenkins, D.M., Fletcher, J., Stack, J.P., Alvarez, A.M., Arif, M., 2022. Loop-mediated isothermal amplification (LAMP) assay for specific and rapid detection of *Dickeya fangzhongdai* targeting a unique genomic region. *Scientific Reports* 12, 19193. https://doi.org/10.1038/s41598-022-22023-4

Dragan A.I., Pavlovic, R., McGivney, J.B., Casas-Finet, J.R., Bishop, E.S., Strouse, R.J., Schenerman, M.A., Geddes, C.D., 2012. SYBR Green I: Fluorescence Properties and Interaction with DNA. *Journal of Fluorescence* 22, 1189–1199. https://doi.org/10.1007/s10895-012-1059-8

Duan, Y., Zhang, X., Ge, C., Wang, Y., Cao, J., Jia, X., Wang, J., Zhou, M., 2014a. Development and application of loop-mediated isothermal amplification for detection of the F167Y mutation of carbendazim-resistant isolates in *Fusarium graminearum*. *Scientific Reports* 4, 7094. https://doi.org/10.1038/srep07094

Duan, Y.B., Ge, C.Y., Zhang, X.K., Wang, J.X., Zhou, M.G. 2014b. Development and evaluation of a novel and rapid detection assay for *Botrytis cinerea* based on Loop-Mediated Isothermal Amplification. *PLoS ONE* 9(10): e111094. https://doi.org/10.1371/journal.pone.0111094

Fan, F., Hahn, M., Li, G.-Q., Lin, Y., Luo, C.-X., 2019. Rapid detection of benzimidazole resistance in *Botrytis cinerea* by loop-mediated isothermal amplification. *Phytopathology Research* 1, 10. https://doi.org/10.1186/s42483-019-0016-8

Feng, W., Hieno, A., Kusunoki, M., Suga, H., Kageyama, K., 2019. LAMP detection of four plant-pathogenic Oomycetes and its application in lettuce fields. *Plant Disease* 103, 298–307. https://doi.org/10.1094/PDIS-05-18-0858-RE

Garg, N., Ahmad, F.J., Kar, S., 2022. Recent advances in loop-mediated isothermal amplification (LAMP) for rapid and efficient detection of pathogens. *Current Research in Microbial Sciences* 3, 100120. https://doi.org/10.1016/j.crmicr.2022.100120

Ghosh, R., Nagavardhini, A., Sengupta, A., Sharma, M., 2015. Development of Loop-Mediated Isothermal Amplification (LAMP) assay for rapid detection of *Fusarium oxysporum* f. sp. *ciceris* - wilt pathogen of chickpea. *BMC Research Notes* 8, 40. https://doi.org/10.1186/s13104-015-0997-z

Ghosh, R., Tarafdar, A., Sharma, M., 2016. Rapid detection of *Fusarium oxysporum* f. sp. *ciceris* from disease infested chickpea fields by loop-mediated isothermal amplification. *Indian Phytopathology* 69, 47–50.

Gomez-Gutierrez, S.V., Goodwin, S.B., 2022. Loop-Mediated Isothermal Amplification for detection of plant pathogens in wheat (*Triticum aestivum*). *Frontiers in Plant Science* 13, 857673. https://doi.org/10.3389/fpls.2022.857673

Goto, M., Honda, E., Ogura, A., Nomoto, A., Hanaki, K.-I., 2009. Colorimetric detection of loop-mediated isothermal amplification reaction by using hydroxy naphthol blue. *BioTechniques* 46, 167–172. https://doi.org/10.2144/000113072

Hieno A, Li M, Otsubo K, Suga H, Kageyama, K. 2021. Multiplex LAMP detection of the genus *Phytophthora* and four *Phytophthora* species *P. ramorum*, *P. lateralis*, *P. kernoviae*, and *P. nicotianae*, with a plant internal control *Microbes Environ*, 36(2), ME21019. https://doi.org/10.1264/jsme2.ME21019. PMID: 34108359; PMCID: PMC8209452.

Hu, X.R., Dai, D.J., Wang, H.D., Zhang, C.Q., 2017. Rapid on-site evaluation of the development of resistance to quinone outside inhibitors in *Botrytis cinerea*. *Scientific Reports* 7, 13861. https://doi.org/10.1038/s41598-017-13317-z

Huang, C., Sun, Z., Yan, J., Luo, Y., Wang, H., Ma, Z., 2011. Rapid and precise detection of latent infections of wheat stripe rust in wheat leaves using Loop-Mediated Isothermal Amplification. *Journal of Phytopathology* 159, 582–584. https://doi.org/10.1111/j.1439-0434.2011.01806.x

Janardhana, G. R., Dass, R.S., Sreenivasa, M.Y. 2009. Fumonisins – A new class of Fusarial toxins. In *Frontiers in Fungal Ecology, Diversity and Metabolites*. Ed: K. A. Sridhar et al. I. K. International Publishing House Pvt. Ltd. New Delhi.

Kaneko, H., Kawana, T., Fukushima, E., Suzutani, T., 2007. Tolerance of loop-mediated isothermal amplification to a culture medium and biological substances. *Journal of Biochemical and Biophysical Methods* 70, 499–501. https://doi.org/10.1016/j.jbbm.2006.08.008

Karthik, K., Rathore, R., Thomas, P., Arun, T.R., Viswas, K.N., Dhama, K., Agarwal, R.K., 2014. New closed tube loop mediated isothermal amplification assay for prevention of product cross-contamination. *MethodsX* 1, 137–143. https://doi.org/10.1016/j.mex.2014.08.009

Kaur, N., Thota, N., Toley, B.J., 2020. A stoichiometric and pseudo kinetic model of loop mediated isothermal amplification. *Computational and Structural Biotechnology Journal* 18, 2336–2346. https://doi.org/10.1016/j.csbj.2020.08.020

Kong, X., Qin, W., Huang, X., Kong, F., Schoen, C.D., Feng, J., Wang, Z., Zhang, H., 2016. Development and application of loop-mediated isothermal amplification (LAMP) for detection of *Plasmopara viticola*. *Scientific Reports* 6, 28935. https://doi.org/10.1038/srep28935

Lau, Y.L., Meganathan, P., Sonaimuthu, P., Thiruvengadam, G., Nissapatorn, V., Chen, Y., 2010. Specific, sensitive, and rapid diagnosis of active toxoplasmosis by a Loop-Mediated Isothermal Amplification method using blood samples from patients. *Journal of Clinical Microbiology* 48, 3698–3702. https://doi.org/10.1128/JCM.00462-10

Lees, A.K., Roberts, D.M., Lynott, J., Sullivan, L., Brierley, J.L., 2019. Real-Time PCR and LAMP assays for the detection of spores of *Alternaria solani* and sporangia of *Phytophthora infestans* to inform disease risk forecasting. *Plant Disease* 103, 3172–3180. https://doi.org/10.1094/PDIS-04-19-0765-RE

Li, B., Du, J., Lan, C., Liu, P., Weng, Q., Chen, Q., 2013. Development of a loop-mediated isothermal amplification assay for rapid and sensitive detection of *Fusarium oxysporum* f. sp. *cubense* race 4. *European Journal of Plant Pathology* 135, 903–911. https://doi.org/10.1007/s10658-012-0136-9

Li, J., Macdonald, J., 2015. Advances in isothermal amplification: Novel strategies inspired by biological processes. *Biosensors and Bioelectronics* 64, 196–211. https://doi.org/10.1016/j.bios.2014.08.069

Liang, C., Chu, Y., Cheng, S., Wu, H., Kajiyama, T., Kambara, H., Zhou, G., 2012. Multiplex loop-mediated isothermal amplification detection by sequence-based barcodes coupled with nicking endonuclease-mediated pyrosequencing. *Analytical Chemistry* 84, 3758–3763. https://doi.org/10.1021/ac3003825

Logeshwari, R., Gopalakrishnan, C., Kamalakannan, A., Ramalingam, J., Saraswathi, R., 2022. A colorimetric hydroxy naphthol blue based loop-mediated isothermal amplification detection assay targeting the β-tubulin locus of *Sarocladium oryzae* infecting rice seed. *Frontiers in Plant Science* 13, 1077328. https://doi.org/10.3389/fpls.2022.1077328

Lu, C., Song, B., Zhang, H., Wang, Y., Zheng, X., 2015. Rapid diagnosis of soybean seedling blight caused by *Rhizoctonia solani* and soybean charcoal rot caused by *Macrophomina phaseolina* using LAMP assays. *Phytopathology* 105, 1612–1617. https://doi.org/10.1094/PHYTO-01-15-0023-R

Luchi, N., Capretti, P., Pazzagli, M., Pinzani, P., 2016. Powerful qPCR assays for the early detection of latent invaders: interdisciplinary approaches in clinical cancer research and plant pathology. *Applied Microbiology and Biotechnology* 100, 5189–5204. https://doi.org/10.1007/s00253-016-7541-5

Mao, S., Zhao, J., Ding, X., Vuong, V.A., Song, J., Que, L., 2023. Integrated sensing chip for ultrasensitive label-free detection of the products of Loop-Mediated Isothermal Amplification. *ACS Sensors* 8, 2255–2262. https://doi.org/10.1021/acssensors.3c00227

Meagher, R.J., Priye, A., Light, Y.K., Huang, C., Wang, E., 2018. Impact of primer dimers and self-amplifying hairpins on reverse transcription loop-mediated isothermal amplification detection of viral RNA. *Analyst* 143, 1924–1933. https://doi.org/10.1039/C7AN01897E

Mori, Y., Notomi, T., 2009. Loop-mediated isothermal amplification (LAMP): a rapid, accurate, and cost-effective diagnostic method for infectious diseases. *Journal of Infection and Chemotherapy* 15, 62–69. https://doi.org/10.1007/s10156-009-0669-9

Nagamine, K., Hase, T., Notomi, T., 2002. Accelerated reaction by loop-mediated isothermal amplification using loop primers. *Molecular and Cellular Probes* 16, 223–229. https://doi.org/10.1006/mcpr.2002.0415

Niessen, L., 2015. Current state and future perspectives of loop-mediated isothermal amplification (LAMP)-based diagnosis of filamentous fungi and yeasts. *Applied Microbiology and Biotechnology* 99, 553–574. https://doi.org/10.1007/s00253-014-6196-3

Njiru, Z.K., Mikosza, A.S.J., Armstrong, T., Enyaru, J.C., Ndung'u, J.M., Thompson, A.R.C., 2008. Loop-Mediated Isothermal Amplification (LAMP) method for rapid detection of *Trypanosoma brucei rhodesiense*. *PLoS Neglected Tropical Diseases* 2, e147. https://doi.org/10.1371/journal.pntd.0000147

Notomi, T., Mori, Y., Tomita, N., Kanda, H., 2015. Loop-mediated isothermal amplification (LAMP): principle, features, and future prospects. *Journal of Microbiology* 53, 1–5. https://doi.org/10.1007/s12275-015-4656-9

Notomi, T., Okayama, H., Masubuchi, H., Yonekawa, T., Watanabe, K., Amino, N., Hase, T., 2000. Loop-mediated isothermal amplification of DNA. *Nucleic Acids Research* 28, 63. https://doi.org/10.1093/nar/28.12.e63

Panek, J., Frąc, M., 2019. Loop-mediated isothermal amplification (LAMP) approach for detection of heat-resistant *Talaromyces flavus* species. *Scientific Reports* 9, 5846. https://doi.org/10.1038/s41598-019-42275-x

Parida, M., Sannarangaiah, S., Dash, P.K., Rao, P.V.L., Morita, K., 2008. Loop mediated isothermal amplification (LAMP): a new generation of innovative gene amplification technique; perspectives in clinical diagnosis of infectious diseases. *Reviews in Medical Virology* 18, 407–421. https://doi.org/10.1002/rmv.593

Patel, J.S., Brennan, M.S., Khan, A., Ali, G.S., 2015. Implementation of loop-mediated isothermal amplification methods in lateral flow devices for the detection of *Rhizoctonia solani*. *Canadian Journal of Plant Pathology* 37, 118–129. https://doi.org/10.1080/07060661.2014.996610

Prasannakumar, M.K., Parivallal, P.B., Pramesh, D., Mahesh, H.B., Raj, E., 2021. LAMP-based foldable microdevice platform for the rapid detection of *Magnaporthe oryzae* and *Sarocladium oryzae* in rice seed. *Scientific Reports* 11, 178. https://doi.org/10.1038/s41598-020-80644-z

Reuter, C., Slesiona, N., Hentschel, S., Aehlig, O., Breitenstein, A., Csáki, A., Henkel, T., Fritzsche, W., 2020. Loop-mediated amplification as promising on-site detection approach for *Legionella pneumophila* and *Legionella* spp. *Applied Microbiology and Biotechnology* 104, 405–415. https://doi.org/10.1007/s00253-019-10286-3

Rolando, J.C., Jue, E., Barlow, J.T., Ismagilov, R.F., 2020. Real-time kinetics and high-resolution melt curves in single-molecule digital LAMP to differentiate and study specific and non-specific amplification. *Nucleic Acids Research* 48, e42–e42. https://doi.org/10.1093/nar/gkaa099

Schneider, L., Blakely, H., Tripathi, A., 2019. Mathematical model to reduce loop mediated isothermal amplification (LAMP) false-positive diagnosis. *Electrophoresis* 40, 2706–2717. https://doi.org/10.1002/elps.201900167

Si Ammour, M., Castaldo, E., Fedele, G., Rossi, V., 2020. Use of LAMP for assessing *Botrytis cinerea* colonization of bunch trash and latent infection of berries in grapevines. Plants 9, 1538. https://doi.org/10.3390/plants9111538

Siegieda, D.G., Panek, J., Frąc, M., 2021. "Shining a LAMP" (Loop-Mediated Isothermal Amplification) on the molecular detection of phytopathogens *Phytophthora* spp. and *Phytophthora cactorum* in strawberry fields. *Pathogens* 10, 1453. https://doi.org/10.3390/pathogens10111453

Soroka, M., Wasowicz, B., Rymaszewska, A., 2021. Loop-Mediated Isothermal Amplification (LAMP): The better sibling of PCR? *Cells* 10, 1931. https://doi.org/10.3390/cells10081931

Sreenivasa, M.Y., Dass, R.S., Charith Raj, A.P., Janardhana, G.R. 2008. PCR-based detection of genus *Fusarium* and Fumonisin-producing isolates from freshly harvested Sorghum grains grown in Karnataka, India. *Journal of Food Safety*. 28, 236–237.

Thierry, M., Chatet, A., Fournier, E., Tharreau, D., Ioos, R., 2020. A PCR, qPCR, and LAMP toolkit for the detection of the wheat blast pathogen in seeds. *Plants* 9. https://doi.org/10.3390/plants9020277

Thiessen, L.D., Keune, J.A., Neill, T.M., Turechek, W.W., Grove, G.G., Mahaffee, W.F., 2016. Development of a grower-conducted inoculum detection assay for management of grape powdery mildew. *Plant Pathology* 65, 238–249. https://doi.org/10.1111/ppa.12421

Tian, Q., Lu, C., Wang, S., Xiong, Q., Zhang, H., Wang, Y., Zheng, X., 2017. Rapid diagnosis of soybean anthracnose caused by *Colletotrichum truncatum* using a loop-mediated isothermal amplification (LAMP) assay. *European Journal of Plant Pathology* 148, 785–793. https://doi.org/10.1007/s10658-016-1132-2

Tomita, N., Mori, Y., Kanda, H., Notomi, T., 2008. Loop-mediated isothermal amplification (LAMP) of gene sequences and simple visual detection of products. *Nature Protocols* 3, 877–882. https://doi.org/10.1038/nprot.2008.57

Tomlinson, J.A., Barker, I., Boonham, N., 2007. Faster, simpler, more-specific methods for improved molecular detection of *Phytophthora ramorum* in the field. *Applied and Environmental Microbiology* 73, 4040–4047. https://doi.org/10.1128/AEM.00161-07

Trinh, T.N.D., Lee, N.Y., 2019. A foldable isothermal amplification microdevice for fuchsin-based colorimetric detection of multiple foodborne pathogens. *Lab Chip* 19, 1397–1405. https://doi.org/10.1039/C8LC01389F

Villari, C., Mahaffee, W.F., Mitchell, T.K., Pedley, K.F., Pieck, M.L., Hand, F.P., 2017. Early detection of airborne inoculum of *Magnaporthe oryzae* in turfgrass fields using a quantitative LAMP assay. *Plant Disease* 101, 170–177. https://doi.org/10.1094/PDIS-06-16-0834-RE

Wang, Y., Wang, Y., Lan, R., Xu, H., Ma, A., Li, D., Dai, H., Yuan, X., Xu, J., Ye, C., 2015. Multiple endonuclease restriction real-time loop-mediated isothermal amplification: A novel analytically rapid, sensitive, multiplex loop-mediated isothermal amplification detection technique. *Journal of Molecular Diagnostics* 17, 392–401. https://doi.org/10.1016/j.jmol2015.03.002

Xu, L., Wang, Y., Zhu, S., Li, J., Chang, Y., Huang, L. 2021. Development and application of a LAMP assay for the Detection of the latent apple tree pathogen *Valsa mali*. *Plant Disease 4*, 1065–1071. https://doi.org/10.1094/PDIS-07-20-1449-RE

Zaczek-Moczydłowska, M.A., Mohamed-Smith, L., Toldrà, A., Hooper, C., Campàs, M., Furones, M.D., Bean, T.P., Campbell, K., 2020. A single-tube HNB-based Loop-Mediated Isothermal Amplification for the robust detection of the *Ostreid herpesvirus 1*. *International Journal of Molecular Science* 21, 6605. https://doi.org/10.3390/ijms21186605

6 Oligonucleotide Array Approaches for the Detection of Plant Pathogenic Fungi

Vignesh Dhandapani
Université de Sherbrooke

Teruo Sano
Hirosaki University

Charith Raj Adkar Purushothama
Université de Sherbrooke

Molecular biology techniques are the most powerful approach for the detection of fungal plant pathogens. Polymerase chain reaction (PCR) is one of the most widely used technology for fungal disease identification. However, it is restricted to the detection of one or two pathogens at a time and requires skilled personnel and a laboratory to detect. Over the past few decades, DNA oligonucleotide array techniques such as macroarray and microarray have played a key role in deciphering the underlying mechanism of gene regulation, defense genes, to understand the signal transduction processes, and so on due to their robustness and multiplexing capacity. So far, only a few studies have employed oligonucleotide techniques to screen fungal plant pathogens although their potential for use is immense. In this chapter, an overview of the application of these techniques in plant-fungal phytopathogen diagnosis, methodology, and the prospective of this technology is discussed.

DOI: 10.1201/9781003399704-6

GRAPHICAL ABSTRACT

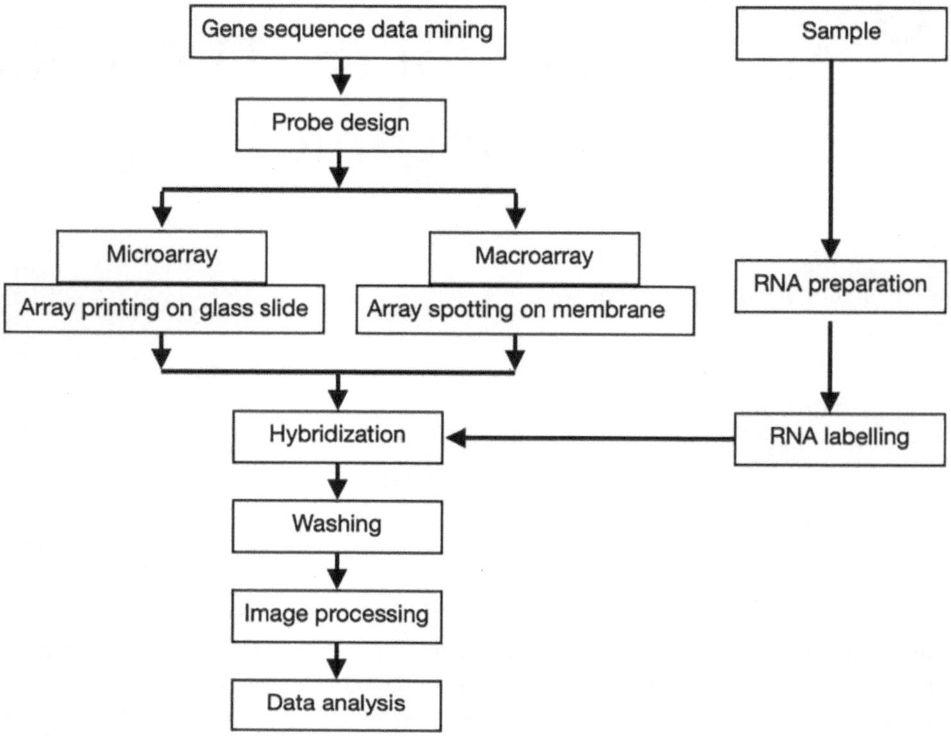

6.1 INTRODUCTION

More than 19,000 plant pathogenic fungi are known to cause diseases in crop plants, worldwide (Jain et al., 2019). This fungus can grow on both live plants and plant produce. They may remain dormant or prepare formation of organs such as sexual and asexual spores on these plant materials until the conditions are suitable for their proliferation. Fungal spores spread through biotic and abiotic factors and infect the entire crop or field (Pal and McSpadden Gardener, 2006). Some of the diseases caused by fungus include anthracnose, rust, leaf spot, wilt, blight, scab, mildew, dieback, damping-off, and root rot. The fungal diseases affect both quality and quantity of the crop (Jain et al., 2019; Deepa et al., 2019a; 2019b). Hence, the timely identification of the fungal pathogen and treatment is an effective management strategy to control further spread of the pathogen and yield loss (Sreenivasa et a*l*., 2008; Dass et al., 2007). Though visible signs of pathogens such as sporulation, sclerotia, or mycelia present on the plant enable the identification of the disease, often this is too late as the fungus spores start spreading (Janardhana et al., 2009). Additionally, it can be difficult to identify all the fungus species when multiple pathogens are present in a single sample (Zhang, McCarthy and Smart, 2008). Therefore, it is important to accurately identify the pathogen in early stage of infection for effective disease management.

Molecular techniques are widely used for fungal phytopathogens due to their sensitivity, rapidity, and reproducibility (Deepa et al., 2016a). Though polymerase chain reaction (PCR) is widely adopted in plant pathology laboratories, its application is restricted to screening a large number of pathogens. In this scenario, DNA oligonucleotide array (commonly known as microarray) is advantageous. DNA oligonucleotide arrays were originally developed for the analysis of whole genome gene expression in a single experiment (Trevino, Falciani and Barrera-Saldaña, 2007; Deepa et al., 2016b, 2016c). Considering their potential for applications in the field of microbiology, the technology was modified to detect and quantify microbes in each sample. Depending on the available probes, a single DNA oligonucleotide array assay enables identification and quantification of several thousand microbial strains (Bodrossy and Sessitsch, 2004). The DNA oligonucleotide array is a modified dot blot assay used to detect the expression of tens to thousands of genes at the same time. Based on the kind of surface material used for immobilizing the array, DNA oligonucleotide arrays can be broadly categorized into macroarray and microarray. In the case of microarray, the microscope slides are printed with thousands of tiny spots in defined positions, whereas in macroarray, spots in defined positions are immobilized on a nylon membrane. Each of these spots contains a known DNA sequence. The DNA molecule that is immobilized onto the slide/membrane acts as a probe to detect the target gene or gene of interest called "detector oligonucleotides" or a "probe" (Lievens and Thomma, 2005). Such spotted probes are hybridized with prelabeled "targets". The targets are prepared by labeling the nucleic acids extracted from the specimen with either chemical modification or with a fluorescent dye or radioactive isotopes (Agindotan and Perry, 2007). To date, only a few macroarrays and microarrays have been developed for the detection and quantification of fungal plant pathogens.

6.2 PRINCIPLE OF DNA OLIGONUCLEOTIDE ARRAY

The principle behind the microarray and macroarray is hybridization between two DNA/DNA or DNA/RNA strands when there is a complementarity. The hybridization is due to the property of the nucleic acids to form hydrogen bonds between the complementary nucleotide base pairs (A-G; C-T). The presence of a high number of complementary base pairs in a nucleotide sequence means tighter non-covalent bonding between the two strands. After washing off the non-specifically bonded sequences and less sequence complementarity bonded sequences, only strongly paired strands will remain hybridized (Govindarajan *et al.*, 2012).

In macroarray and microarray, the probes that are specific to the target of interest are immobilized on the glass slide or nylon membrane, respectively. During the hybridization, the prelabeled (fluorescent or radioactive isotope) "target" sequence obtained from the sample diffuses passively across the surface of the glass or membrane. During this step, the "target" sequences complementary to a probe will anneal and form a duplex. The non-specific binding is removed by washing, and then the hybridized targets are detected using an appropriate signal reader. The total strength of the signal for a given target (spot) depends upon the amount of target sample binding to the probes present on that spot. This signal can be used to quantify the target amount in the given sample (Ye *et al.*, 2001; Hofman, 2005).

6.3 DNA MACROARRAY AND ITS APPLICATION

Membrane-based DNA macroarrays require only a pin tool for array production (Zhang, McCarthy and Smart, 2008). Macroarrays are developed by spotting the detector oligonucleotides onto nylon membranes of relatively large dimensions. Although, the macroarray has lower throughput than the microarray, a 96-well microtiter plate-size membrane is able to contain over one thousand different detector oligonucleotides (Zhang, McCarthy and Smart, 2008). Further, these macroarray membranes can be washed and reused several times. Due to its low cost of production, and high throughput, macroarray has been employed by many researchers in plant disease diagnosis (Fessehaie, De Boer and Lévesque, 2003; Lievens *et al.*, 2003; Agindotan and Perry, 2007; He *et al.*, 2012, 2021).

DNA macroarray is widely used by researchers to detect fungal phytopathogens specific to crop plants. For example, in 2003 Lievens *et al.* (2003) developed a microarray to detect tomato vascular wilt pathogens, specifically, Fusarium wilt, caused by *Fusarium oxysporum* f. sp. *lycopersici*, and Verticillium wilt, caused by either *Verticillium albo-atrum* or *Verticillium dahliae*. These wilt pathogens cause devastating diseases of tomato (*Lycopersicon lycopersicum*) worldwide. Zhang *et al.* (2008) developed a macroarray system to detect 25 fungal and oomycetes pathogens of *Solanaceous* crops. In this system, 105 probes of 17–27-nucleotide long specific to 25 pathogens were designed

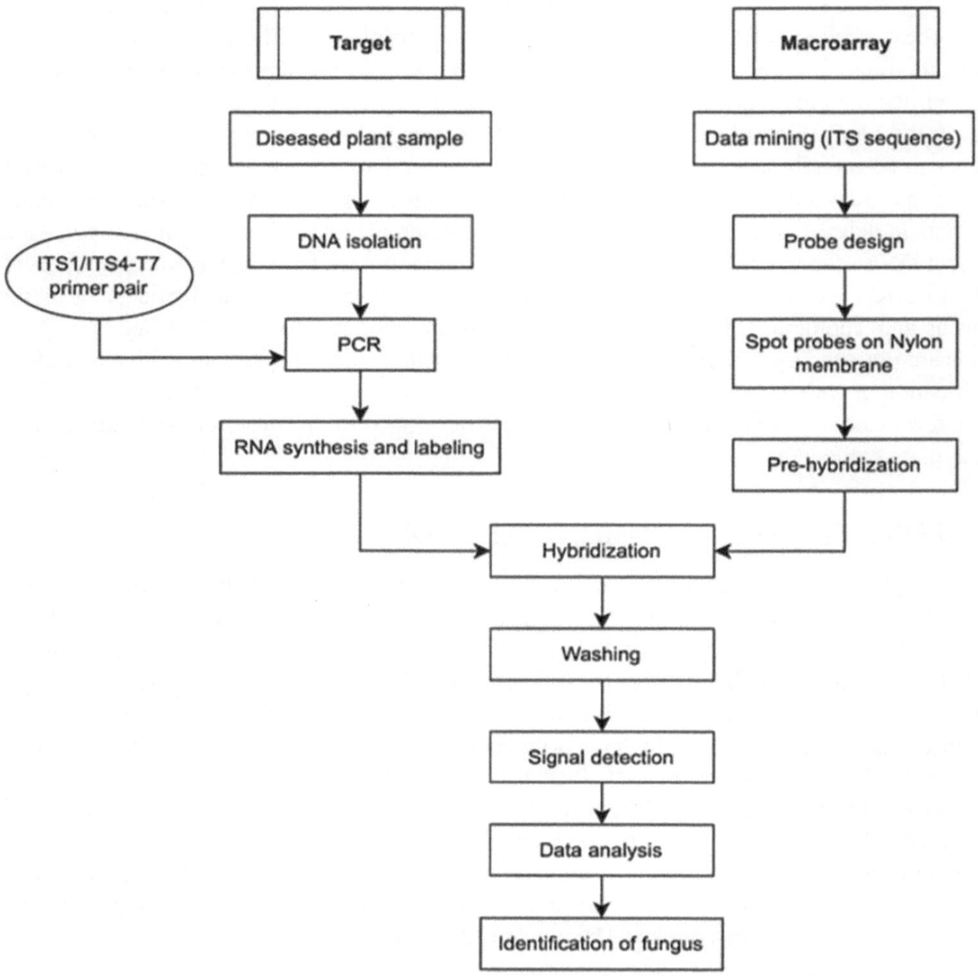

FIGURE 6.1 Flow chart describing steps involved in in-house macroarray assay for the detection of fungus in the laboratory.

and spotted on a nylon membrane. The 25 target pathogens were, *Alternaria alternata*, *A. solani*, *Botrytis cinerea*, *Cladosporium fulvum*, *Colletotrichum coccodes*, *C. gloeosporioides*, *Corynespora cassiicola*, *F. oxysporum*, *F. sambucinum*, *F. solani*, *Oidium neolycopersici*, *Sclerotinia sclerotiorum*, *Septoria lycopersici*, *Stemphylium solani*, *Verticillium albo-atrum*, *V. dahliae*, *Rhizoctonia solani*, *Phytophthora capsica*, *P. erythroseptica*, *P. infestans*, *P. nicotianae*, *Pythium aphanidermatum*, *P. irregulare*, *P. ultimum*, *Spongospora subterranea* f. sp. *Subterranean*. This array was validated by testing 46 infected field samples, and several nontarget species. Similarly, He *et al.* (2012) developed a macroarray technique to monitor pathogenic and non-pathogenic fungal species in the apple phyllosphere. In this work, macroarray was developed by immobilizing the fungal species-specific 40-nucleotide (nt) long oligo-DNA probes selected from the fungal *rDNA-ITS* (Internal Transcribed Spacer) region on the nylon membrane. For the detection of fungal species present in the phyllosphere of apple trees, the digoxigenin (DIG)-labeled cRNA obtained from the Internal Transcribed Spacer ITS region was used as a target. The probes designed here were able to detect the fungus that causes scab (*Venturia inaequalis*), Alternaria blotch (*Alternaria mali*), and Marssonina blotch (*Diplocarpon coronariae* [formerly *Marssonina coronaria*] teleomorph: *Diplocarpon mali*) in apple trees (He *et al.*, 2012) (Figure 6.1).

6.4 ADVANTAGES AND LIMITATIONS OF THE MACROARRAY

Macroarray is a modified dot blot assay that can be easily developed and employed in the laboratory for routine fungal pathogen screening. Due to its relatively low cost and high throughput (up to hundred targets), macroarray has been employed by many researchers for plant disease diagnosis in recent years (Zhang, McCarthy and Smart, 2008). As macroarray allows the screening of several targets at the same time, it is a more reliable and effective method to screen multiple fungal plants from any given sample. However, this array does not allow pathogen quantification and fails to define whether the detected DNA is from a live or dead organism (Úrbez-Torres *et al.*, 2015). As macroarray is designed based on the known probe sequence, it cannot be used to detect a pathogen whose probe is not spotted on the nylon membrane. Further, there is a chance of missing low abundance fungal pathogens due to the less hybridization sensitivity (Gammill and Lee, 2008). However, in the case of fungal pathogen detection, an initial PCR is added prior to the target labeling. This greatly enhances the sensitivity of microarrays (Table 6.1).

6.5 DNA MICROARRAY AND ITS APPLICATION

DNA microarray is also known as DNA chip, gene chip, or biochip. As described earlier, in DNA microarray, the microscopic spots consist of thousands of specific DNA sequences (probes) immobilized on the glass slide in a defined position (Bilska *et al.*, 2018). These probes are used to hybridize a target sample that is prelabeled with a chemiluminescence, fluorophore, or silver (Ha *et al.*, 2009). The slides are analyzed in a microarray plate reader as outlined in Figure 6.2. The advancement of DNA microarray technology has led to high throughput and multiple detection of various phytopathogens, including viruses, viroids, bacteria, and fungi (Hariharan and Prasannath, 2021). A variant of microarray called, ArrayTube, was developed using marker genes internal transcribed spacer (ITS), translational elongation factor 1 alpha (TEF-1α), and 16S ribosomal DNA (16S rDNA) with 42 probes. It allowed the detection of various sugar beet root rot pathogens that are prevalent in the field (e.g., *Aphanomyces cochlioides*), storage pathogens (e.g., *Botrytis cinerea*), and ubiquitous spoilage fungi (e.g., *Penicillium expansum*). Further, these assays proved to differentiate Fusarium species based on the single nucleotide polymorphism (SNP) (Liebe *et al.*, 2016). Similarly, PCR primers and florescent probes targeting six fungal potato pathogens (*Rhizoctonia solani*, *Spongospora subterranea*, *Alternaria solani*, *A. alternata*, *Fusarium* sp., and *Colletotrichum coccodes*) were developed in a 48-well microarrays (micromatrices) for qPCR (Nikitin *et al.*, 2018). To date, several microarray-based diagnostic assays to detect fungal pathogens have been cleared by the Food and Drug Administration (FDA), USA for analyzing blood samples (Beyda, Alam and Garey, 2013; Mylonakis *et al.*, 2015). This indicates

TABLE 6.1
Advantages and Disadvantages of DNA Macroarray and Microarray

Characteristics	Macroarray	Microarray
Probe density	Low	High
Cost	Low	High
Pathogen quantification	No	Yes
Needs specialized equipment	No	Yes
Reusable	Yes	No
Sensitivity	High (due to presence of higher amounts of sequence)	Low (could be enhanced)
High throughput	Yes (up to hundred targets)	Yes (up to thousand targets)
Unknown pathogen identification	No	No

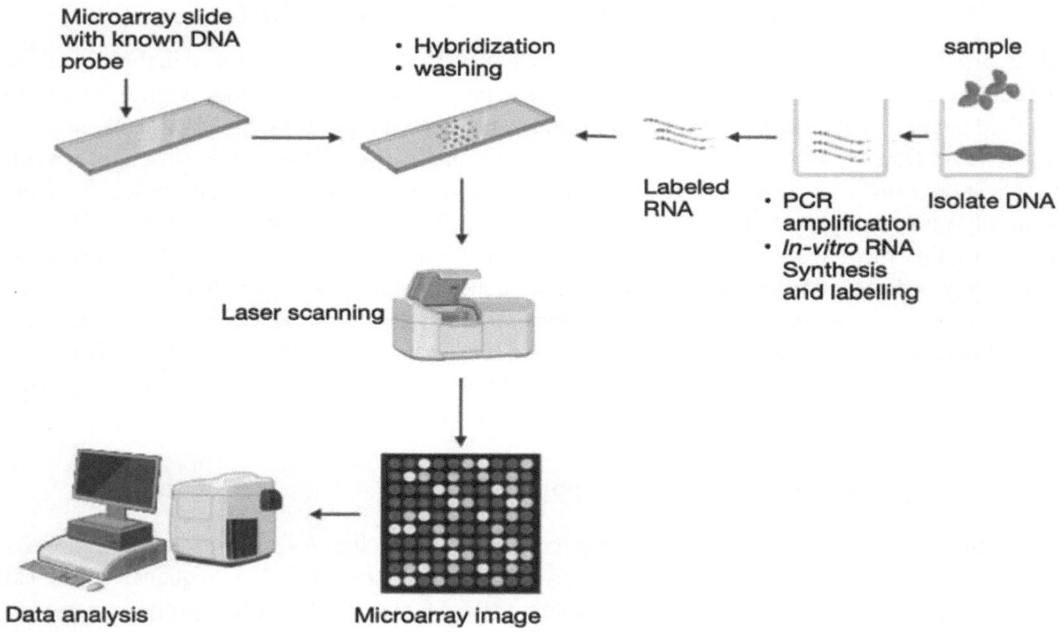

FIGURE 6.2 Flow chart outlining the steps involved in microarray assay.

the potential applications of this technology for monitoring quarantine pathogens at the port of entry.

6.6 ADVANTAGES AND LIMITATIONS OF THE MICROARRAY

Microarray enables higher probe density compared to macroarray. This allows for spotting several thousand probes on a single slide. That said, one microarray is capable of detecting thousands of different pathogens in a single experiment (Bodrossy and Sessitsch, 2004). However, microarray slides are not as reusable as macroarray membranes. Cost associated with the production of microarray slides is high. Further, microarray requires special equipment for analyzing the slides, thus increasing the overall cost of running a microarray assay. Though microarray can be used for quantification, lower sensitivity compared to PCR technologies limits its application where pathogen is in low abundance. However, microarray is more sensitive than Western blot and ELISA. Additionally, microarray can only detect the sequences that the array was designed to identify (Bumgarner, 2013). That said, it cannot be used to screen a new pathogen or a pathogen whose sequence is not present in the array (Table 6.1).

6.7 DEVELOPING AN IN-HOUSE MACROARRAY IN THE LABORATORY

This method is modified from He *et al.*, (2012).

1. **Selection of array sequences or probe**
   - On the basis of the nucleotide sequences available in the public databank, select 40-nt long oligo-DNA sequences of fungal rDNA-ITS specific for each fungal species that are of interest.
2. **Macroarray preparation**
   - Dissolve the commercially prepared oligo-DNA probes at a concentration of 1 mg/mL in a solution containing 50% (v/v) formamide, 35% (v/v) formaldehyde, and 1X saline-sodium citrate (SSC) buffer.

- Denature at 65°C for 15 min.
- Dilute with 4 volumes of 20XSSC buffer.
- Spot 1μL each of the probes on a positively charged nylon membrane.
- Immobilize by cross-linking the membranes using ultraviolet (UV) (120,000 μJ/cm^2) in a Stratagene Stratalinker UV2400 crosslinker.

3. **Preparation of DIG-labeled RNA targets for macroarray analysis**:
   - Extract the total nucleic acid from the sample.
   - PCR amplifies the fungus using ITS-specific primers and ITS1 (5'-TCCGTAGGTGAACCTGCGG-3') and ITS4-T7 (5'-AGAG-TAATACGACTCACTATAGGG-TCCTCCGCTTATTGATATGC-3'; underlined nucleotide indicates T7 promoter sequence).
   - Extract amplicons with equal volumes of phenol:chloroform (1:1).
   - Precipitate using ethanol and dissolve in 50μL of distilled water.
   - Perform *in vitro* transcription using T7 polymerase in the presence of DIG RNA-labeling mixture (Roche Diagnostics).
4. **Hybridization**
   - Place the array membrane in a glass hybridization bottle.
   - Prehybridize in 5-mL hybridization buffer containing 5X SSC buffer, 1% Denhardt solution, 1% sodium dodecyl sulfate (SDS), and 25 mg of yeast tRNA (Roche Diagnostics) at 58°C for 90 min.
   - Heat to denature DIG-labeled cRNA targets at 95°C for 10 min.
   - Add 5 μL of the DIG-labeled cRNA to the hybridization solution.
   - Hybridize at 58°C for 18 hours.
   - Wash membrane twice for 15 min with 70 mL of wash buffer 2X saline-sodium citrate (SSC)) at room temperature.
   - Wash membrane twice for 15 min with 70 mL of 2X SSC (0.5% SDS) at 58°C.
   - Place the membrane in 5-mL blocking solution containing 1% blocking reagent in 0.1 M maleic acid (0.15 M NaCl, pH 7.5).
   - Incubate at room temperature for 30 min.
   - Add 2 μL of anti-DIG-alkaline phosphatase (AP) (Fab fragment; Roche Diagnostics) and incubate for 30 min.
   - Wash membrane twice for 15 min each in 70 mL of washing buffer (0.1 M maleic acid (0.15 M NaCl, pH 7.5)) containing 0.3% Tween 20 (v/v).
   - Immerse membrane for 5 min in a 50-mL solution containing 0.1 M Tris-HCl (0.1 M NaCl and 0.05 M $MgCl_2$, pH 9.5).
   - Incubate the membranes with CSPD star (ready-to-use) (Roche Diagnostics) for 30 min to 2 hours in ChemiDoc XRS (Bio-Rad Laboratories) to detect chemiluminescent signals.
5. **Quantification**.
   - Quantify the signal intensity using Quantity One (Bio-Rad) or any other appropriate tool.

6.8 SUMMARY

Oligonucleotide array technologies such as microarray and macroarray offer a rapid, reliable, and cost-effective method for screening multiple pathogens from any given sample. Due to their high-throughput capacity to screen a large number of fungal pathogens in a single assay, it has a huge advantage over the PCR techniques. Due to the lower sensitivity of the technique, often low abundant pathogens will go undetected. However, using a locked nucleic acid monomer in the DNA probe increases the binding strength, thus improving the sensitivity of the array (Petersen and Wengel, 2003). On the other hand, the development of surface plasmon resonance imaging allows label-free detection and also enables the re-use of the array, thus helping to reduce the cost associated with array development (Nelson *et al.*, 2002). Currently, oligonucleotide array is widely used by researchers to screen fungal pathogens in various crop plants such as tomato, potato, and sugar beet. Further research is required to improve its field application, which includes but is not limited to reducing the steps involved in the assay, sample preparation, and associated costs.

6.9 FUTURE PERSPECTIVE

The current chapter has summarized the application of DNA oligonucleotide array technology to detect various fungal pathogens. Due to its enormous scope in screening multiple pathogens in an assay, it will directly impact the agriculture community once a cost and user-friendly device is available for public use. Though the DNA oligonucleotide array technology has many advantages over other diagnostic techniques, the difficulty and length of time needed in creating and testing the fungal pathogen is the key hurdle. In this scenario, adding a biosensor device capable of hybridizing and monitoring hybridization in real time could greatly decrease the number of steps involved in the microarray technique and the amount of time required to perform the assay. Developing biosensors based on nanotechnology could result in enhanced sensitivity and specificity. Designing a miniaturized system for point-on-care diagnosis could be a considerable improvement in microarray development. Future research directed toward the development of rapid detection mechanisms following macroarray and microarray hybridization could expand the range of applications.

STUDY QUESTIONS

1. What is the difference between macroarray and microarray technologies?
2. What is the principle of oligonucleotide array technologies?
3. How can an oligonucleotide array be designed in the laboratory to detect fungi infecting tomato plants?
4. What are the disadvantages of oligonucleotide array assay?

FURTHER READING

Chapter Nineteen – Oligonucleotide Microarray: Applications for Lyssavirus Speciation (https://doi.org/10.1016/B978-0-12-800014-4.00019-6) (Dastjerdi, Fooks and Johnson, 2014)

Overview of DNA Microarrays: Types, Applications, and Their Future (https://doi.org/10.1002/0471142727.mb2201s101) (Bumgarner, 2013)

Recent Advances in Molecular Diagnostics of Fungal Plant Pathogens: A Mini Review (https://doi.org/10.3389/fcimb.2020.600234) (Hariharan and Prasannath, 2021)

REFERENCES

Agindotan, B. and Perry, K. L. (2007) 'Macroarray, Detection of Plant RNA Viruses Using Randomly Primed and Amplified Complementary DNAs from Infected Plants', *Phytopathology*®, 97(1), pp. 119–127. doi: 10.1094/PHYTO-97-0119.

Beyda, N. D., Alam, M. J. and Garey, K. W. (2013) 'Comparison of the T2Dx instrument with T2Candida assay and automated blood culture in the detection of Candida species using seeded blood samples', *Diagnostic Microbiology and Infectious Disease*, 77(4), pp. 324–326. doi: 10.1016/j.diagmicrobio.2013.07.007.

Bilska, K. et al. (2018) 'Development of a Highly Sensitive FcMito qPCR Assay for the Quantification of the Toxigenic Fungal Plant Pathogen Fusarium culmorum', *Toxins*, 10(5), p. 211. doi: 10.3390/toxins 10050211.

Bodrossy, L. and Sessitsch, A. (2004) 'Oligonucleotide microarrays in microbial diagnostics.', *Current Opinion in Microbiology*, 7(3), pp. 245–254. doi: 10.1016/j.mib.2004.04.005.

Bumgarner, R. (2013) 'Overview of DNA Microarrays: Types, Applications, and Their Future', *Current Protocols in Molecular Biology*, 101(1). doi: 10.1002/0471142727.mb2201s101.

Dass, R.S., Sreenivasa, M.Y., Janardhana, G.R. 2007. High incidence of *Fusarium verticillioides* in animal and poultry feed mixtures produced in Karnataka, India. *Plant Pathology Journal*. 6(2), pp. 174–178.

Dastjerdi, A., Fooks, A. R. and Johnson, N. (2014) 'Oligonucleotide Microarray', in *Current Laboratory Techniques in Rabies Diagnosis, Research and Prevention*. Elsevier, pp. 193–203. doi: 10.1016/B978-0-12-800014-4.00019-6.

Deepa, N., Charith Raj, A. P., Sreenivasa, M.Y. 2016b. Nested PCR method for the early detection of fumonisin producing *Fusarium verticillioides* in pure cultures, cereal samples and plant parts. *Food Biotechnology*. 30(1), pp. 18–29.

Deepa, N., Charith Raj, A.P., Sreenivasa, M.Y. 2016c. Multiplex PCR method for the early detection of fumonisin producing *Fusarium verticillioides*. *Food Bioscience*. 13, pp. 84–88.

Deepa, N., Nagaraja, H., Sreenivasa, M.Y. 2016a. Prevalence of fumonisin producing Fusarium verticillioides associated with cereals grown in Karnataka (India). *Food Science Human Wellness*. 5, pp. 156–162. doi: 10.1016/j.fshw.2016.07.001.

Deepa, N and M Y Sreenivasa. 2019a. Molecular methods and key genes targeted for the detection of fumonisin producing *Fusarium verticillioides* - An updated review. *Food Bioscience*. 32, p. 100473.

Deepa N, Sreenivasa, M.Y. 2019b. Biocontrol Strategies for Effective Management of Phytopathogenic Fungi Associated with Cereals, In: Singh JS., Singh DP. (eds) *New and Future Developments in Microbial Biotechnology and Bioengineering*, Elsevier, pp, 177–189, ISBN 9780444641915.

Fessehaie, A., De Boer, S. H. and Lévesque, C. A. (2003) 'An oligonucleotide array for the identification and differentiation of bacteria pathogenic on potato.' *Phytopathology*, 93(3), pp. 262–269. doi: 10.1094/PHYTO.2003.93.3.262.

Gammill, L. S. and Lee, V. M. (2008) 'Chapter 16 Gene Discovery: Macroarrays and Microarrays', pp. 297–312. doi: 10.1016/S0091-679X(08)00216-1.

Govindarajan, R. et al. (2012) 'Microarray and its applications.', *Journal of pharmacy & bioallied sciences*, 4(Suppl 2), pp. S310–S312. doi: 10.4103/0975-7406.100283.

Ha, Y. et al. (2009) 'Simultaneous Detection of Acidovorax avenae subsp. citrulli and Didymella bryoniae in Cucurbit Seedlots Using Magnetic Capture Hybridization and Real-Time Polymerase Chain Reaction', *Phytopathology®*, 99(6), pp. 666–678. doi: 10.1094/PHYTO-99-6-0666.

Hariharan, G. and Prasannath, K. (2021) 'Recent Advances in Molecular Diagnostics of Fungal Plant Pathogens: A Mini Review', *Frontiers in Cellular and Infection Microbiology*, 10. doi: 10.3389/fcimb.2020.600234.

He, Y. H. et al. (2012) 'Oligo-DNA custom macroarray for monitoring major pathogenic and non-pathogenic fungi and bacteria in the phyllosphere of apple trees', *PLoS One*. doi: 10.1371/journal.pone.0034249.

He, Y.-H. et al. (2021) 'Microbial Diversity in the Phyllosphere and Rhizosphere of an Apple Orchard Managed under Prolonged "Natural Farming" Practices,' *Microorganisms*, 9(10). doi: 10.3390/microorganisms 9102056.

Hofman, P. (2005) 'DNA Microarrays', *Nephron Physiology*, 99(3), pp. 85–89. doi: 10.1159/000083764.

Jain, A. et al. (2019) 'A review of plant leaf fungal diseases and its environment speciation', *Bioengineered*, 10(1), pp. 409–424. doi: 10.1080/21655979.2019.1649520.

Janardhana, G. R., Dass, R.S., Sreenivasa, M.Y. 2009. Fumonisins – A new class of Fusarial toxins. *In Frontiers in Fungal Ecology, Diversity and Metabolites*. Ed: K A Sridhar et al. I. K. International Publishing House Pvt. Ltd. New Delhi.

Liebe, S. et al. (2016) 'Development of a DNA Microarray-Based Assay for the Detection of Sugar Beet Root Rot Pathogens', *Phytopathology®*, 106(1), pp. 76–86. doi: 10.1094/PHYTO-07-15-0171-R.

Lievens, B. and Thomma, B. P. H. J. (2005) 'Recent Developments in Pathogen Detection Arrays: Implications for Fungal Plant Pathogens and Use in Practice', *Phytopathology®*, 95(12), pp. 1374–1380. doi: 10.1094/PHYTO-95-1374.

Lievens, B. et al. (2003) 'Design and development of a DNA array for rapid detection and identification of multiple tomato vascular wilt pathogens', *FEMS Microbiology Letters*, 223(1), pp. 113–122. doi: 10.1016/S0378-1097(03)00352-5.

Mylonakis, E. et al. (2015) 'T2 Magnetic Resonance Assay for the Rapid Diagnosis of Candidemia in Whole Blood: A Clinical Trial', *Clinical Infectious Diseases*, 60(6), pp. 892–899. doi: 10.1093/cid/ciu959.

Nelson, B. P. et al. (2002) 'Label-free detection of 16S ribosomal RNA hybridization on reusable DNA arrays using surface plasmon resonance imaging', *Environmental Microbiology*, 4(11), pp. 735–743. doi: 10.1046/j.1462-2920.2002.00350.x.

Nikitin, M. et al. (2018) 'Preserved Microarrays for Simultaneous Detection and Identification of Six Fungal Potato Pathogens with the Use of Real-Time PCR in Matrix Format', *Biosensors*, 8(4), p. 129. doi: 10.3390/bios8040129.

Pal, K. K. and McSpadden Gardener, B. (2006) 'Biological Control of Plant Pathogens', *The Plant Health Instructor*. doi: 10.1094/PHI-A-2006-1117-02.

Petersen, M. and Wengel, J. (2003) 'LNA: a versatile tool for therapeutics and genomics', *Trends in Biotechnology*, 21(2), pp. 74–81. doi: 10.1016/S0167-7799(02)00038-0.

Sreenivasa, M.Y., Dass, R.S., Charith Raj, A.P., Janardhana, G.R. 2008. PCR-based detection of genus *Fusarium* and Fumonisin-producing isolates from freshly harvested Sorghum grains grown in Karnataka *India,' Journal of Food Safety*. 28, pp. 236–237.

Trevino, V., Falciani, F. and Barrera-Saldaña, H. A. (2007) 'DNA microarrays: a powerful genomic tool for biomedical and clinical research.' *Molecular medicine (Cambridge, Mass.)*, 13(9–10), pp. 527–541. doi: 10.2119/2006-00107.Trevino.

Úrbez-Torres, J. R. et al. (2015) 'Development of a DNA Macroarray for the Detection and Identification of Fungal Pathogens Causing Decline of Young Grapevines', *Phytopathology®*, 105(10), pp. 1373–1388. doi: 10.1094/PHYTO-03-15-0069-R.

Ye, R. W. et al. (2001) 'Applications of DNA microarrays in microbial systems', *Journal of Microbiological Methods*, 47(3), pp. 257–272. doi: 10.1016/S0167-7012(01)00308-6.

Zhang, N., McCarthy, M. L. and Smart, C. D. (2008) 'A Macroarray System for the Detection of Fungal and Oomycete Pathogens of Solanaceous Crops', *Plant Disease*, 92(6), pp. 953–960. doi: 10.1094/PDIS-92-6-0953.

7 An Overview of Application and Advancements of the Lateral Flow Assay Technique in Effective Detection of Fungal Phytopathogens

P. V. Bhavani
Pentavalent Bio Sciences Private Limited

In the current era, culture-based conventional method of identification of food pathogens is not highly preferred as it is time consuming and requires to be performed by a skilled microbiologist in a laboratory. In recent decades, evident progress has been achieved in the adoption of molecular techniques as a replacement for culture-based techniques. In this background, lateral flow assay (LFA) is one such molecular technique that is proven to be a reliable approach for the detection of microorganisms. The technique is based on the interaction of natural and/or synthetic biomolecules such as antigen-antibody or DNA-DNA or probe DNA-DNA. This assay is easy to perform, rapid, and, more notably, is a point-of-care diagnostic tool. Hence, it has been successfully developed and commercially used in several diagnostic fields. The performance ease of a given LFA is also attributed to its characteristic feature that it does not require a skilled person or a laboratory to perform the assay. In this chapter, the recent developments and applications of LFA, specifically in the agriculture field for the detection of fungal pathogens that infect plants and plant products (phytopathogens), are explained and discussed at length with suitable examples.

DOI: 10.1201/9781003399704-7

GRAPHICAL ABSTRACT

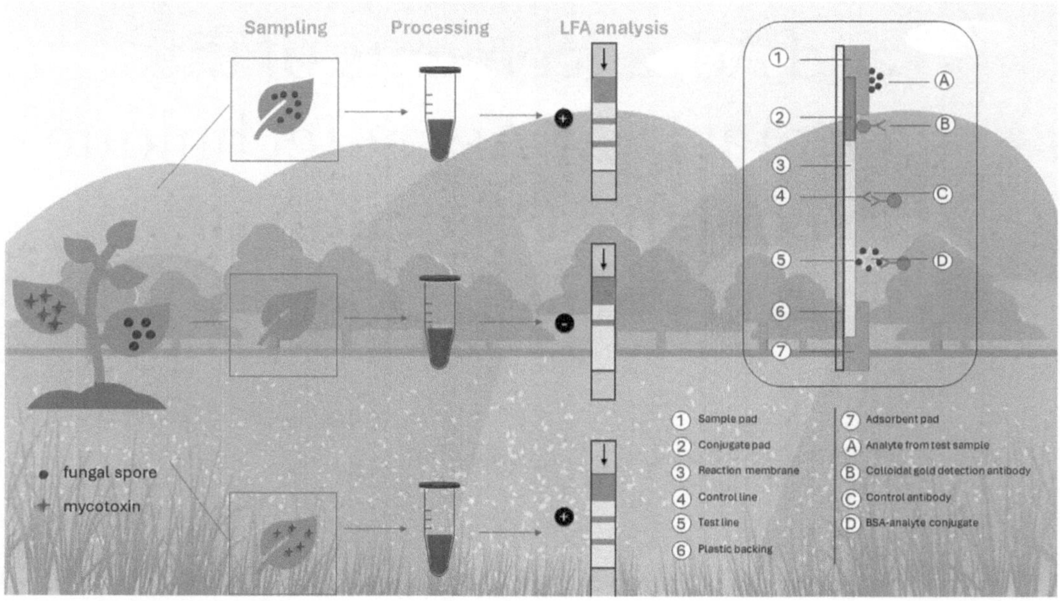

7.1 INTRODUCTION

With the growing population, global demand for the production of food crops and vegetables is also increasing. To meet this growing requirement, the maintenance of balance of food production is essential. Achieving this is a challenge because the encounter of diseases, weeds and insects in agriculture leads to crop losses. Diseases in the crops due to the infection by phytopathogens (bacteria, fungi, viruses, parasites, protozoans and nematodes) are a severe threat to the agricultural yield and economy (Urusov et al., 2018; Sreenivasa et al., 2011a; 2008).

There are several old and emerging plant diseases of which around 70–80% are transmitted by fungal phytopathogens like species of *Aspergillus, Fusarium, Penicillium, Verticillium, Alternaria* and *Ascochyta* (Nagaraja et al., 2016; Deepa et al., 2016; Sreenivasa et al., 2011b). For efficient management of plant diseases, accurate yet simple rapid diagnosis of fungal phytopathogens at the species level is necessary (Deepa and Sreenivasa, 2019b; Prasad et al., 2010; Chennappa et al., 2014). The conventional approach to a disease diagnosis is essentially designed to align with the criteria of Koch's postulates (Manawasinghe et al., 2021). The conventional diagnostic techniques are highly accurate, but the output is prolonged and not suitable for large-scale use, and thus undermines the timely disease management process (Janardhana et al., 2009; Dass et al., 2007).

Molecular and serological diagnostic techniques have been demonstrated as the reliable alternative approach, because of the research advancements (Deepa and Sreenivasa, 2019a). These developments have resulted in accurate and rapid disease identification in plants, even before the onset of symptoms. Several quantitative and qualitative techniques have been developed for fungal phytopathogens for use in efficient and timely disease management practices (Gautam and Kumar, 2020; Chakraborty and Chakraborty, 2021; Sreenivasa et al., 2010, 2011b). Also, precise, and frequent field inspections using rapid diagnostic techniques enable early detection and prevention of disease outbreak, both at regional and global levels (Deepa et al., 2016b, 2016c). Lateral flow assay (LFA) is one such rapid user-friendly immunodiagnostic technology that enables accurate and rapid disease diagnosis. The test result is highly reliable since the assay has an internal test control. Additional benefit of the LFA approach is the non-dependency of the test in a laboratory facility. The results can be obtained at the agricultural site itself as the test can be performed by non-skilled/non-trained personnel, making the disease diagnosis decentralized and on-field/point-of-care (POC) (Market, 2013). The assay has been widely used in several diagnostic fields including cancer markers, hormones, microbes, mycotoxins and pesticides.

The technique of LFA is based on the interaction of natural and/or synthetic biomolecules such as antigen-antibody or secondary antibody approach or probe DNA-DNA. In the majority of the LFAs, antigen–antibody interaction is exploited, wherein the antigen component of the assay is of fungal origin while the antibody (polyclonal or monoclonal) is the antigen-specific component produced *in vitro*. An antibody can be developed in the laboratory using a mouse model and hybridoma technology. Thus, the disease diagnosis using LFA is based on the principle of quantitative detection of the specific binding of the plant pathogen antigen with the raised antibody. These biochemical interactions are read using a variety of biorecognition elements such as nanoparticles (gold, carbon), latex beads, enzymes, quantum dots and others (Bahadır and Sezgintürk, 2016). In this chapter, the basic working principle of LFA, and recent improvization of the assay with growing technology are discussed. Also, different LFAs developed for use in agriculture with specific applications in the detection of fungal phytopathogens are depicted. The chapter also covers the major advantages and limitations of LFA. The future possible perspectives in this area are also briefed.

7.2 WORKING PRINCIPLE OF A STANDARD LATERAL FLOW ASSAY SYSTEM

In principle, an LFA works on a combination of biochemical interaction (using a range of biomolecules and reagents) and chromatography. The biochemical interaction might be between antigen and antibody, or between a nucleic acid and nucleic acid, or nucleic acid and a given target analyte. The chromatography system is used to enable the interaction of the biomolecules. Finally, the successful interaction is visually confirmed with the help of elements that recognize and calorimetrically indicate the same. These elements are thus generally termed "biorecognition elements" that are embedded in an LFA unit.

A typical LFA has four critical components, namely (1) a sample pad, (2) a conjugate pad, (3) a reaction membrane, and (4) an adsorbent pad (Figure 7.1). First, a sample pad is used as a compartment to apply the test sample, in which the target analyte might be present or absent. Second, a conjugate pad is used to immobilize the biorecognition element (antigen or antibody or DNA probe based on the assay format) in the form of a particulate conjugate which usually is combined with a labelled tag (e.g., colloidal gold, fluorescent latex particle, and paramagnetic monodisperse latex particle). Third, that is, a reaction membrane is the actual platform meant for the interaction between the biomolecules (antigen and antibody, nucleic acid and nucleic acid, or nucleic acid and a target analyte). In this reaction membrane, two lines are present, a control line (which acts as the internal positive control) and a test line (where the actual interaction of the biomolecules occurs). Fourth,

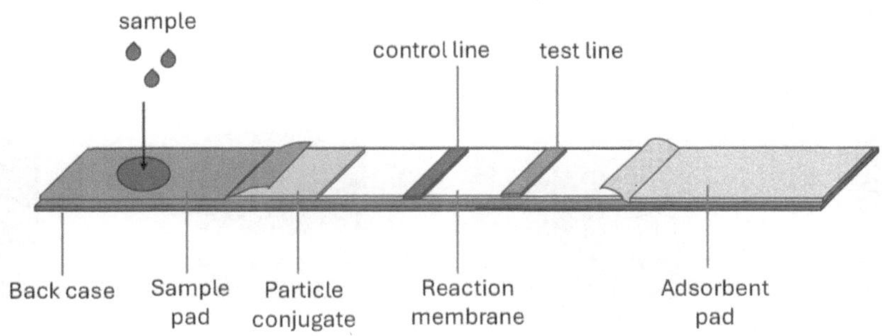

FIGURE 7.1 Schematic representation of a typical lateral flow assay device.

and finally, an adsorbent pad is meant to adsorb and reserve the waste/non-interactive components present in the test sample. Upon organized assembly of these four parts (overlapping on one another and then mounting the assembly on a backing card using any suitable pressure-sensitive adhesive), a simple and miniaturized bioreactor is formed that can be used in the detection of a given analyte.

When an analytical test is run to detect the presence or absence of a specific target compound/organism in a sample, the sample is pre-treated according to the assay, and then a few drops of it are applied onto the sample pad. This treated sample chromatographically migrates from the sample pad to the conjugate pad. Once the sample reaches this conjugate pad, here it hydrates the dried conjugate to gain mobility. Concomitantly, the analyte that is present in the sample binds with the conjugate to form analyte-conjugate hybrids. Further, these hybrids migrate by the test line on the reaction membrane. Excess non-reactive agents in the sample migrate past the test and control lines to finally get entrapped in the adsorbent pad. At this point, the results visually develop on the test line and the control line, which shall be seen by the naked eyes by an individual. Key point is that the control line which acts as the internal control always develops the visual result in the control line. This indicates that the LFA unit is accurate and thus the result obtained in the test line is also accurate. All these steps from sample addition to the formation of visual lines in control and test lines take as less as 5 minutes only. Thus, LFA is a highly rapid, easy-to-perform, cost-effective assay that can be commissioned in agriculture fields.

7.3 APPLICATIONS OF LATERAL FLOW ASSAY IN PLANT HEALTH MANAGEMENT

An effective framework in plant health management has a direct impact on plant productivity, food security and the agriculture ecosystem. Timely diagnosis of infection in plants plays an importance in disease prevention by awarding the spread of pathogens and in turn enabling appropriate management of disease outbreaks. A molecular technique such as LFA is tailor-made for this objective as it detects infection in real-time, rapidly and on-field. A lateral flow assay can be employed in plant health management as the result obtained from the assay can be undoubtedly applied in the prevention of pathogen spread in non-infected areas, early surveillance of infection, early detection of latent fungal pathogens before the onset of symptoms and accurate pathogen detection in containment and disease management (Figure 7.2). Several immunochromatographic assays on the LFA platform have been developed for use in plant health management by rapid detection of phytopathogens, including bacteria, fungi, viruses, related toxins, pesticides, and drug residues. This chapter, however, mainly covers the developments and advancements in the application of LFA in the accurate detection of fungal phytopathogens and their toxins (mycotoxins).

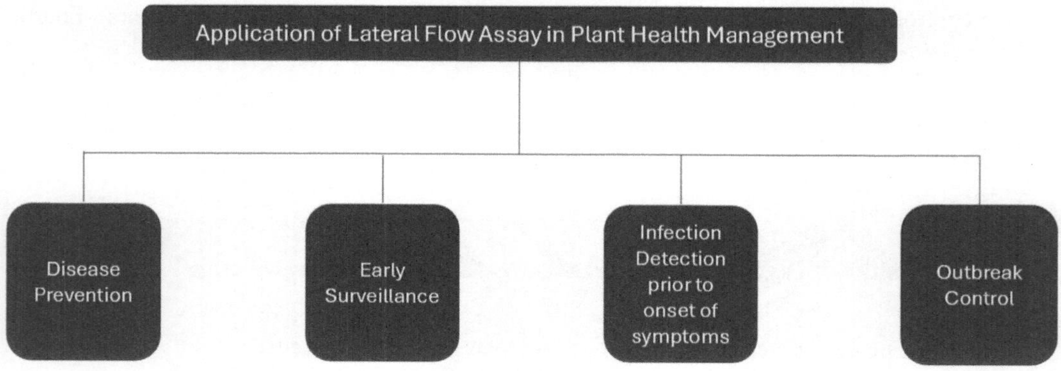

FIGURE 7.2 Application of lateral flow assay technique for plant pathogen detection in different stages of plant health management.

7.4 LATERAL FLOW ASSAY IN DETECTION OF FUNGAL PHYTOPATHOGENS

Fungal phytopathogens are known to secrete secondary toxic metabolites, termed mycotoxins, on the infected plant regions. The presence of these mycotoxins is considered evidence of fungus infestation in the agricultural field. Therefore, these mycotoxins are typically employed as promising diagnostic biomarkers on the LFA platform. These small toxic secondary metabolites produced by fungi are usually found in agricultural food produce and have direct and deleterious effects on consumer's health. A summary of studies on the development of LFA for the detection of these mycotoxins is enlisted in Table 7.1. Mycotoxins are basically encountered in various fungus-infested foodstuffs such as cereals. These secondary metabolites are proven to be hazardous to both humans and animals upon consumption beyond regulatory limits. Many of them are carcinogens too. Though several LFA formats have been successfully developed worldwide for simple and rapid detection of these toxins in agricultural food products, there is still a need for assays with higher sensitivity in terms of limit of detection (LOD) (Krska and Molinelli, 2009).

7.4.1 Detection of Fungal Phytopathogens Using Mycotoxins as Diagnostic Markers

Fusarium species such as *F. acuminatum, F. cerealis, F. culmorum, F. cerealis, F. crookwellense, F. oxysporum, F. semitectum* and *F. sporotrichioides* (Ropejko and Twarużek, 2021; Deepa and Sreenivasa, 2017a; 2017b) are facultative parasites of agricultural importance. Upon successful infection, these pathogens secrete a mycotoxin called zearalenone (ZEN). The presence of ZEN is witnessed in the *Fusarium*-infested crops, including maize, oats, wheat, sorghum, barley, rice, and rye (Bennett and Klich, 2003). ZEN is a non-steroidal fungal toxin with a resemblance to oestrogen. It is also detected in processed products like milk, bread, chocolate and flour (Rogowska et al., 2019). For specific detection of these *Fusarium* species on the LFA platform, ZEN is targeted as the diagnostic marker. Kolosova and co-workers (2007) employed the antibody-antigen interaction principle between a commercial IgG monoclonal antibody (mAb) that is generated using bovine serum albumin (BSA)-conjugated ZEN and ZEN mycotoxin itself. This commercial mAb was conjugated with 40 nm colloidal-gold nanoparticles resulting in the development of colloidal-gold-labelled mAb with an optical density of 10.2. Using this mAb as the capture antibody, an immuno-dipstick-based LFA was developed and the presence of ZEN in wheat samples was detected within 10 min. Here, the assay detection limit for ZEN in contaminated sample was found to be 100 μg/kg. In the same study, LFA for the detection of yet another mycotoxin Deoxynivanelol (DON) was also developed with a similar design using another commercial anti-DON mAb. The detection limit of this DON detection LFA was found to be 1500 μg/kg of contaminated wheat. DON produced by *Fusarium* species, namely *F. graminearum* and *F. culmorum*, is a type B trichothecene (Krska and Josephs, 2001; Deepa et al., 2021). This mycotoxin has strong cytotoxic and immunosuppressive activities, leading to diarrhoea and emesis in subjects consuming DON-contaminated food products (Kolosova et al., 2007). Several other groups have used the same antigen-antibody interaction principle employing different mAbs or polyclonal antibodies (pAbs) conjugated using different particles such as amorphous carbon nanoparticles, and quantum dots. All the attempts demonstrated different yet satisfactory detection limits of ZEN and DON in contaminated samples tested ranging from cereals and pig feeds (Kolosova et al., 2008; Xu et al., 2010; Foubert et al., 2017; Zhang et al., 2017; Li et al., 2018). Often, these LFAs are developed in a multiplexed format, wherein on a single immuno-dip stick more than one mycotoxin is detected (Kolosova et al., 2007; Song et al., 2014).

Aspergillus flavus and *A. parasiticus* are facultative parasites reported to infect a range of plants and plant products. These two species produce a group of mycotoxins named aflatoxins (AFT), which is of agricultural importance. Basically, there are four major types of AFT namely, B_1, B_2, G_1 and G_2. Of these four, B_1 and G_1 are carcinogenic and B_1 is the most prevalent (Ngom et al., 2010). The presence of these mycotoxins in agricultural food produce indicates fungal invasion in the

TABLE 7.1
Overview of Studies on Lateral Flow Assay Formats and Tests Reported for the Detection of Mycotoxins from Contaminated Agricultural Samples

Sl. No.	Mycotoxins	Fungal Phytopathogen	Assay format	Labels	Samples	Detection Sensitivity	Detection Time	Reference
1	Zearalenone	*Fusarium* sp.	Immunochromatography	Colloidal gold	Wheat	100 µg/kg	10 min	Kolosova et al., 2007
			Immuno-dipstick	Amorphous carbon nanoparticles	Maize	0.4 µg/L	NR[a]	Zhang et al., 2017
			Luminescence	Quantum dots	Cereals	10 µg/L	NR[a]	Li et al., 2018
			Luminescence	Quantum dots	Wheat	80 µg/kg	NR[a]	Foubert et al., 2017
2	Deoxynivanelol	*Fusarium* sp.	Immunochromatography	Colloidal gold	Wheat	1500 µg/kg	10 min	Kolosova et al., 2007
			Immunochromatography	Colloidal gold	Wheat, maize, pig feed	50 ng/mL	10 min	Kolosova et al., 2008; Xu et al., 2010
			Immuno-dipstick	Amorphous carbon nanoparticles	Maize	0.5 µg/L	NR[a]	Zhang et al., 2017
			Luminescence	Quantum dots	Wheat	1000 µg/kg	NR[a]	Foubert et al., 2017
			Immuno-dipstick	Gold nano particles	Corn extracts	2 ng/mL	15 min	Urusov et al., 2018
3	Aflatoxin B_1	*A. flavus, A. parasiticus*	Immunochromatography	Colloidal gold	Rice, corn, wheat flour	2.5 ng/mL	10 min	Xiulan et al., 2006
4	Aflatoxin B_2	*A. flavus, A. parasiticus*	Immuno-dipstick	Magnetic nanogold microspheres	Peanuts, hazelnuts, pistacia, almonds	0.9 ng/mL	15 min	Tang et al., 2009
5	Fumonisin B_1, B_2, B_3	*Fusarium* sp.	Competitive immunochromatography	Colloidal gold	Maize	199 µg/kg	4 min	Molinelli et al., 2009
6	T-2 toxin (Trichothecene)	*Fusarium* sp.	Immuno-dipstick	Amorphous carbon nanoparticles	Maize	0.4 µg/L	NR[a]	Zhang et al., 2017
			Luminescence	quantum dots	Blood	80 µg/kg	NR[a]	Foubert et al., 2017

[a] NR – Not reported

agricultural field and their detection is a priority. An LFA based on immunochromatography was developed by Xiulan and co-workers (2006) for the detection of AFT B_1 using anti-B_1-BSA specific colloidal-gold-labelled mAb as capture antibody. The assay detection limit was found to be within the range of 2–50 μg/kg of contaminated samples such as rice, corn, and wheat. In another study, Tang and co-workers (2009) were successful in developing an LFA for the detection of AFT B_2. Here, the biorecognition element consisted of magnetic nanogold microspheres (MnGMs) with nano-Fe_2O_3 particles as core and gold nanoparticles conjugated with monoclonal anti-AFT B_2 antibodies. The reaction membrane, like any other LFA, consisted of one each of control and test lines. The control line was spotted with goat anti-mouse IgG and the test line with AFT B_2–BSA conjugates. The detection limit with MnGMs was found to be highly sensitive (0.9 ng/mL) when compared to the conventional gold nanoparticle-based strip (3.1 ng/mL). This study is not only about the development of a mycotoxin detection LFA but also about an illustrative and orientational work that suggests that the detection sensitivity for a given antigen can be increased by using different biorecognition elements and/or antibodies. This basic knowledge of trying different reaction agents such as biorecognition elements, antibodies, reaction membranes and revealing agents all enables in improvement of detection sensitivity of fungal phytopathogens. *Fusarium* species produce another group of toxins called fumonisins. There are more than 25 structural analogues of fumonisins and among them, fumonisin B_1, B_2 and B_3 are the most prevalent analogues found in agricultural commodities (Rheeder et al., 2002). Fumonisins resemble sphingosine present in cell membranes and thus disrupt the sphingosine metabolism in the body, causing pulmonary oedema. These mycotoxins are also carcinogenic. On the LFA platform, a highly sensitive and robust fumonisin detection assay was developed by Molinelli and co-workers (2009). Briefly, the assay could detect as little as 4000 μg of total B fumonisins in 1 kg of contaminant maize within 4 min.

Aspergillus species including *A. ochraceus*, *A. glaucus*, *A. sydowii*, *A. repens, A. albertensis*, *A. auricomus*, *A. clavatus*, *A. fumigatus*, *A. foetidus*, *A. vadensis*, *A. tubingensis*, *A. versicolor* and several others are highly effective fungal phytopathogens. These species exhibit their pathogenicity by producing a group of mycotoxins, named ochratoxins. Ochratoxins mainly include ochratoxin A (OTA) and its methyl and ethyl ester (ochratoxin C), ochratoxin B (OTB), and its methyl and ethyl ester (ochratoxin α). Among all, OTA is the most toxic and prevalent ochratoxin and is very stable. It is resistant to acidity and high temperatures. Hence, it is very difficult to remove or destroy this toxin from contaminated foodstuffs. OTA is produced by species of *Penicillium*, *Petromyces*, and *Neopetromyces* (Sanchis Almenar et al., 2013). OTA is commonly found in cereal grains, peanuts, sorghum, spices, dried fruits, olives, beer, cheese, green coffee beans, etc. The toxic effects of OTA are mainly on kidneys leading to lesions, necrosis and karyomegaly. OTA acute poisoning is known to affect various organs, including brain, heart, spleen, liver, kidneys, and reproductive system. For the detection of OTA, co-workers (2012) developed a one-step LFA that could semi-quantitatively detect OTA in wines and grape musts. The detection limit was reported to be 1 μg/L, and the assay was simple and rapid enabling accurate detection of OTA in 5 min. In an attempt to enhance the detection sensitivity of a typical gold-based competitive lateral flow immunoassay, Anfossi and co-workers (2013) deposited silver on gold nanoparticles and used in the assay. As a result, the antibody and OTA interaction signal was 10-fold enhanced compared to a typical gold-based LFA and reduced the quantity of the specific and competitive antibodies required for the reaction. In another study, Fadlalla and co-workers (2020) used a mAb that could detect both OTA and OTB to develop two formats of LFA – (Anfossi et al., 2013) using colloidal-gold nanoparticles (CGNs) and (Bahadır & Sezgintürk, 2016) silver nanoflower gold strips (AuNFs). This study also revealed the effect of using different biorecognition elements on detection sensitivity. The limit of detection of AuNFs- and CGNs-based LFAs for both OTA and OTB were 1 and 5 μg/mL, respectively. Recently, a rapid, accurate and sensitive LFA on magnetic biolabels for quantitative detection of OTA in contaminated agricultural commodities was reported. The LOD of this assay was found to be 11 pg/mL, almost a 1000-fold increase in sensitivity, compared to a simple non-linear magnetized lateral flow assay.

Fusarium species including *Fusarium culmorum, F. equiseti, F. graminearum, F. langsethiae, F. poae* and *F. sporotrichioides* are highly contagious fungal phytopathogens. These species exhibit their pathogenicity by producing a group of mycotoxins, named trichothecenes (tetracyclic sesquiterpenoids). DON also belongs to this group. Other important mycotoxins in this group are HT-2 toxin (HT-2) and T-2 toxin (T-2). T-2 toxin (T-2) is the most toxic type A trichothecene mycotoxin contaminating cereals, and it causes human immunotoxicity and hematotoxicity (McCormick et al., 2011). Several lateral flow assays have been developed for the detection of T-2 (Zhang et al., 2015; Qie et al., 2019). For instance, Zhang and co-workers (2015) developed two sensitive T-2 specific LFAs using fluorescent microspheres and AuNPs as biorecognition elements with LOD of 100 and 400 μg/kg in rice samples, respectively. Qie and co-workers (2019) reported an innovative anti-ovalbumin antibody-based LFA for the detection of T-2 toxin. Here, both T-2 toxin in the test sample and ovalbumin-modified T-2 toxin were made to competitively bind on the anti-T-2 toxin mAb conjugated with CdSe/Zns quantum dot beads and thus formed immuno-complexes were made to fluoresce at 610 nm. This assay demonstrated a highly sensitive LOD of 10 fg/mL. In a recent study, even more sensitive and novel "turn-off" mode LFA was developed for quantitative detection of T-2. Here, a range of biorecognition elements such as AuNPs, amorphous carbon nanoparticles (ACNPs), quantum dots nanospheres (QDs), and time-resolved fluorescent microspheres (TRFMs) were tried. Here, AuNPs and ACNPs were employed as adsorbers, and QDs and TRFMs as fluorescents. Upon optimization, the LOD for these formats was found to be within a range of 0.22 to 0.50 ng/mL, and ACNPs-TRFMs combination showed the highest sensitivity (Zhang et al., 2022).

Fungal phytopathogenic species including *Aspergillus, Byssochlamys*, and *Penicillium* are known to produce patulin (PAT), a highly toxic mycotoxin. Typically, it is detected in fruits, vegetables and their products. The presence of PAT in agricultural food produce is considered an indication of fungal contamination and thus unfit for consumption. Intake of PAT-contaminated agricultural food products is known to induce nausea, agitation, oedema, convulsions, intestinal haemorrhage, genotoxic, neurotoxic and teratogenic effects. A few assays based on lateral flow immunoassay have been successfully devised and tested for rapid and simple detection of PAT during routine screening of food products. One such promising product has been reported to be developed using PAT-specific aptamer. In this aptamer-based LFA, Streptavidin-coated gold nanoparticles were used for the detection. Detection sensitivity of this assay in the test sample (apple juice) was in a range of 7.07 – 359.5 ng/mL (Tang et al., 2022). An interestingly novel approach was incorporated in the development of an LFA using recombinase polymerase amplification (RPA) reaction, instead of antigen-antibody reaction (de Almeida, 2021). Basically, RPA is an isothermal amplification technique with unique advantages in operation, sensitivity, performance at a constant and low temperature, and cost effectiveness. In this study, the Isoepoxydon dehydrogenase (*Idh*) gene of *P. expansum* was amplified in RPA reactions and the resultant amplicons were detected on the lateral flow assay platform. The total reaction time to obtain results is within 5 to 15 min.

7.4.2 Commercial LFA Kits Available for Detection of Mycotoxins

Several LFA kits are available in the commercial market for specific and accurate detection of mycotoxins. A list of some of those commercial kits and additional details are presented in Table 7.2. The ease of use in the case of LFA tests makes it possible to analyse fungal contamination in several samples in a very short duration, and thus they are very beneficial for screening plant diseases and to analyse the toxin levels semi-quantitatively.

7.4.3 Detection of Fungal Phytopathogens Using Fungal Tissue Types as Diagnostic Markers

Though majorly mycotoxins have been the primary detection/diagnostic biomarkers in the case of fungal phytopathogens, in very few cases, even the fungal tissue including mycelial extracts and

TABLE 7.2
Commercially Available Lateral Flow Assay Kits for Detection of Mycotoxins

Mycotoxin	Product Name	Manufacturer	Limit of Detection
Deoxynivalenol	Reveal SQ for Deoxynivalenol	Neogen Corporation (Lansing, MI)	<0.5 ppm
Deoxynivalenol	Rida®Quick Deoxynivalenol	R-Biopharm AG (USA)	1.25 ppm
Deoxynivalenol	DONCheck	Vicam (Watertown, MA)	1.0 ppm
Deoxynivalenol	DON-V	Vicam (Watertown, MA)	0.2-5.0 ppm
Deoxynivalenol	AgraStrip® Deoxynivalenol	Romer Laboratories (Union, MO)	0.07–6.0 ppm
Fumonisin	Reveal SQ for Fumonisin	Neogen Corporation (Lansing, MI)	<1.0 ppm
Fumonisin	Rida®Quick Fumonisin	R-Biopharm AG (USA)	0.8 ppm
Fumonisin	AgraStrip® Fumonisin	Romer Laboratories (Union, MO)	0.17–8.0 ppm
Fumonisin	ROSA FAST5 Fumonisin	Charm Sciences Inc (USA)	5–25 ppm
Fumonisin	ROSA FAST5-X Fumonisin	Charm Sciences Inc (USA)	0 to 25 ppm
Fumonisin	ROSA WET-S5 Fumonisin	Charm Sciences Inc (USA)	> 40 ppm
Aflatoxins B_1, B_2, G_1 and G_2	AgraStrip® Aflatoxin A	Romer Laboratories (Union, MO)	10 ppb
Total aflatoxin	QuickTox for QuickScan	Envirologix Inc. (Portland, Maine)	2.5 ppb
Total aflatoxin (AFB_1 and AFB_2)	QuickTox™ Kit for Aflatoxin	Envirologix Inc. (Portland, Maine)	20 ppb
Total aflatoxins (AFB_1, AFB_2, AFG_1, AFG_2)	Reveal® for Aflatoxin	Neogen Corporation (Lansing, MI)	20 ppb
Aflatoxin AFM_1	Reveal® for Aflatoxin M_1	Neogen Corporation (Lansing, MI)	500 ppt
Total aflatoxins (AFB_1, AFB_2, AFG_1, AFG_2)	Reveal®Q+	Neogen Corporation (Lansing, MI)	2–150 ppb
Total aflatoxins (AFB_1, AFB_2, AFG_1, AFG_2)	Reveal® for Aflatoxin SQ	Neogen Corporation (Lansing, MI)	<10 ppb
Total aflatoxins (AFB_1, AFB_2, AFG_1, AFG_2)	Reveal® for Aflatoxin SQ Peanut	Neogen Corporation (Lansing, MI)	<10 ppb
Aflatoxins	RidaQuick® Aflatoxin	R-Biopharm AG (USA)	≥ 4 ppb
Aflatoxins	RIDA®QUICK Aflatoxin RQS	R-Biopharm AG (USA)	4 ppb
Aflatoxins	ROSA® Aflatoxin	Charm Sciences Inc (USA)	0–150 ppb
Aflatoxins	ROSA®Fast Aflatoxin	Charm Sciences Inc (USA)	0–150 ppb
Aflatoxins	ROSA®Aflatoxin P/N	Charm Sciences Inc (USA)	10–20 ppb
Aflatoxins	ROSA®BEST Aflatoxin P/N	Charm Sciences Inc (USA)	10–20 ppb
Aflatoxin AFM_1	Charm SL Aflatoxin M_1	Charm Sciences Inc (USA)	0–750 ppt
Aflatoxin AFM_1	Charm II Aflatoxin M_1	Charm Sciences Inc (USA)	0.025 ppb or 0.25 ppb
Zearalenone	QuickTox for QuickScan Zearalenone	Envirologix Inc. (Portland, Maine)	520 ppb
Zearalenone	TotalTox™ Zearalenone	Envirologix Inc. (Portland, Maine)	250-2000 ppb
Ochratoxin	ROSA OCHRAQ-G	Charm Sciences Inc (USA)	5–30 ppb
Ochratoxin	QuickTox for QuickScan Ochratoxin	Envirologix Inc. (Portland, Maine)	1.5–100 ppb
T-2/HT-2 toxin	ROSA T-2/HT-2	Charm Sciences Inc (USA)	25–200 ppb
T-2/HT-2 toxin	QuickTox for QuickScan T-2/HT-2 Flex	Envirologix Inc. (Portland, Maine)	25–600 ppb

spores are used as detection targets on the LFA platform. To achieve this, ideally polyclonal antibodies against the fungal tissue are generated in laboratory rabbits. Alternatively, monoclonal antibodies are also generated using the conventional hybridoma technology. For instance, dipstick assays for specific detection of *Penicillium islandicum* (Dewey et al., 1990). In some cases, IgY antibodies are generated by immunizing chickens with fungal tissue samples and isolated from their egg yolk. For example, IgY antibodies were generated against *Colletotrichum falcatum* and *Fusarium subglutinans* infecting sugarcane, and they were found to be equally effective as the polyclonal antibodies generated in rabbits (Vöhringer and Sander, 2001). Phenomenal success has been long achieved in developing immuno-dipstick-based lateral flow assays for efficient detection of fungus-like phytopathogens (oomycetes). Monoclonal antibodies specific to the zoospores of *Phytophthora cinnamomi* were used to develop such an assay that could detect as less as 40 zoospores from 1 millilitre of the processed infected plant sample within 45 min (Cahill and Hardham, 1994). Pettitt and co-workers (2002) were successful in developing a dipstick assay to detect zoospores of *Pythium* spp. and *Phytophthora* spp. at a satisfactory sensitivity comparable to that of the conventional bait methods.

7.5 ADVANTAGES OF LATERAL FLOW ASSAY

Lateral flow assay is the only available format of detection tool that addresses most of the criteria necessary for decentralized diagnostic applications. Some of those diagnostic criteria are the assay is one-step, rapid, simple, versatile, and longer shelf-life without the need for cold storage. All these criteria, in fact, fall under the **ASSURED** (**a**ffordable, **s**ensitive, **s**pecific, **u**ser friendly, **r**apid and **r**obust, **e**quipment-free and **d**eliverable) criteria of a diagnostic assay, as per World Health Organization recommendation. These unique set of diagnostic characteristics will enable LFA assay to be a point-of-care and rapid testing tool for several decades to come.

7.6 LIMITATIONS OF LATERAL FLOW ASSAY

It is important to note that the majority of the LFA assays for the detection of fungal phytopathogens are still used at the research laboratory stage and have not been implemented for on-field use by the concerned authorities for large-scale use. This could be due to the current limitations for its use in routine application on a large scale of samples: that it is designed as an individual test rather than high-throughput screening. Though at laboratory tests, the detection sensitivity is found satisfactory, on-field detection sensitivity is reported to be relatively low and this also could be a reason for finding it unsuitable for detecting latent infections.

7.7 SUMMARY

Fungal phytopathogens are comprehensively investigated as they have a critical and negative effect on agricultural yield and food security. Contamination of crops by fungi that occurs at the pre-harvest stage, and foodstuffs and feeds at the post-harvest stage often leads to mycotoxin production in these agricultural commodities. All these toxins have ill effects on the consumer's health and some of the toxins are extremely potent carcinogens. This potential of fungal phytopathogens and their mycotoxins to cause deleterious harm to the consumer's health has led to the development of many detection methods. Lateral flow assay has demonstrated huge advantages and potential in this field. In this chapter, the construction of a typical LFA strip, the components used to build an LFA strip, and the working principle of the assay are elaboratively explained. Further, various LFAs that have been reported for use are the detection of fungal phytopathogens and their mycotoxins have been discussed in terms of their technical improvization and their effect on detection sensitivity and overall performance of the detection tool. The commercial success of the technique is reflected by an extensive list of LFA kits that are available for specific detection of mycotoxins from agricultural commodities. Advantages and limitations associated with the technique and future implications have been mentioned.

7.8 FUTURE PROSPECTIVE

A commendable number of studies have been reported in developing mycotoxin/fungi detection using the LFA platform majorly focusing on a wide range of agricultural commodities, different biorecognition elements, improved sensitivity, and lesser time. However, only a few products are available for commercial use and a huge scope is foreseen in this dimension. To make LFAs a commercial success, several advancements in result analysis and interpretation are being introduced. For instance, visual reading of the results using assay readers, handheld reflectometry, scanners, fluorescence visualizers and other imaging systems. Readers with numerical output of the result by sensing the colour by its specific wavelength and converting it to its corresponding specific detection concentration and further using Bluetooth functionality for data export for further analysis are also developed. A few key components need improvization to address the effective point-of-care application of LFAs. Some of them specify as collection and handling of sample, concentration and pre-treatment of sample, LFA cassette design and result readout. Key considerations in the selection of mAb or pAb, mode of selection of specific antibodies, method of antibody conjugation, and immobilization of conjugated antibodies are highly recommended to improvize the overall performance of a given LFA kit.

STUDY QUESTION

1. Is lateral flow assay useful in the detection of fungal phytopathogens in the field of agriculture?
2. Is a limit of detection of a given lateral flow assay satisfactory for on-field detection of fungal infection, both pre-harvest and post-harvest?
3. Is there more to be done and could be done to improvize the performance and data interpretation in the case of lateral flow assay?
4. Is there a scope for further research and development in the case of lateral flow assay for its broader application?

FURTHER READING

Tang, X., Zhang, Q., Isabel Pividori, M., Zhang, Z., Marty, J.L., Catanante, G., 2022. A sensitive aptasensor using Biotin-Streptavidin system for patulin detection in apple juice. *Biosensors*, 12(2), 59.

REFERENCES

Anfossi, L., Di Nardo, F., Giovannoli, C., Passini, C., Baggiani, C., 2013. Increased sensitivity of lateral flow immunoassay for ochratoxin A through silver enhancement. *Anal and Bioanal Chem*, 405, 9859–9867.

Bahadır, E.B., Sezgintürk, M.K., 2016. Lateral flow assays: Principles, designs and labels. *Trends Analyt Chem*, 82, 286–306.

Bennett, J.W., Klich. M., 2003. Mycotoxins. *Clin Microbiol Rev*, 16(3), 497–516.

Cahill, D.M., Hardham, A.R., 1994. A dipstick immunoassay for the specific detection of *Phytophthora cinnamomi* in soils. *Phytopathol*, 84(11), 1284–1292.

Chakraborty, B. N., Chakraborty, U., 2021. Molecular detection of fungal pathogens and induction of phytoimmunity using bioinoculants. *Indian Phytopathol* 74, 307–322.

Chennappa, G., Adkar-Purushothama, C.R., Umdale S., Tamilvendan, K., Sreenivasa, M.Y. 2014. Impact of Pesticides on PGPR Activity of Azotobacter sp. Isolated from Pesticide Flooded Paddy Soils. *Greener J Agri. Sci.* 4 (4), 117–129.

de Almeida, E.I.A.M., 2021. Development of fast methods for the detection of patulin producing fungi (Doctoral dissertation, Universidade do Minho (Portugal)).

Dass, R.S., Sreenivasa, M.Y., Janardhana, G.R. 2007. High incidence of *Fusarium verticillioides* in animal and poultry feed mixtures produced in Karnataka, India. *Plant Pathol. J.* 6(2), 174–178.

Dewey, F.M., MacDonald, M.M., Phillips, S.I., Priestley, R.A., 1990. Development of monoclonal-antibody-ELISA and-DIP-STICK immunoassays for *Penicillium islandicum* in rice grains. *Microbiol*, 136(4), 753–760.

Deepa, N., Charith Raj, A.P., Sreenivasa, M.Y. 2016c. Multiplex PCR method for the early detection of fumonisin producing *Fusarium verticillioides*. *Food Biosci.* 13, 84–88.

Deepa, N and M Y Sreenivasa. 2019a. Molecular methods and key genes targeted for the detection of fumonisin producing *Fusarium verticillioides* – An updated review. *Food Biosci.* 32, 100473.

Deepa, N., Charith Raj, A. P., Sreenivasa, M.Y. 2016b. Nested PCR method for the early detection of fumonisin producing *Fusarium verticillioides* in pure cultures, cereal samples and plant parts. *Food Biotechnol.* 30(1), 18–29.

Deepa N., and Sreenivasa, M.Y. 2019b. Biocontrol Strategies for Effective Management of Phytopathogenic Fungi Associated with Cereals, In: Singh J.S., Singh D.P. (eds) *New and Future Developments in Microbial Biotechnology and Bioengineering*, Elsevier, pp, 177–189, ISBN 9780444641915.

Deepa, N., and Sreenivasa, M.Y. 2017a. Fumonisins: A review on its global occurrence, epidemiology, toxicity and detection. *J Vet Med Res.* 4(6), 1093.

Deepa, N., and Sreenivasa, M.Y. 2017b. *Fusarium verticillioides*, a Globally Important Pathogen of Agriculture and Livestock: A Review. *J Vet Med Res.* 4(4), 1084.

Deepa, N., Nagaraja, H., Sreenivasa, M.Y. 2016. Prevalence of fumonisin producing Fusarium verticillioides associated with cereals grown in Karnataka (India). *Food Sci Human Wellness.* 5, 156–162. http://doi.org/10.1016/j.fshw.2016.07.001

Deepa, N., Achar, P.N., Sreenivasa, M.Y. 2021. Current perspectives of biocontrol agents for management of *Fusarium verticillioides* and Its Fumonisin in cereals—A Review. *J Fungi.* 7, 776.

Fadlalla, M.H., Ling, S., Wang, R., Li, X., Yuan, J., Xiao, S., Wang, K., Tang, S., Elsir, H., Wang, S., 2020. Development of ELISA and lateral flow immunoassays for ochratoxins (OTA and OTB) detection based on monoclonal antibody. *Front Cell Infect Microbiol*, 10, 80.

Foubert, A., Beloglazova, N.V., De Saeger, S., 2017. Comparative study of colloidal gold and quantum dots as labels for multiplex screening tests for multi-mycotoxin detection. *Analytica Chimica Acta*, 955, 48–57.

Gautam, A. K., Kumar, S., 2020. Techniques for the detection, identification, and diagnosis of agricultural pathogens and diseases. In *Natural remedies for pest, disease and weed control.* Academic Press, 135–142.

Janardhana, G. R., Dass, R.S., Sreenivasa, M.Y. 2009. Fumonisins – A new class of Fusarial toxins. In *Frontiers in Fungal Ecologuy, Diversity and Metabolites.* Ed: K.A. Sridhar et al., I.K. International Publishing House Pvt. Ltd. New Delhi.

Kolosova, A.Y., De Saeger, S., Sibanda, L., Verheijen, R., Van Peteghem, C., 2007. Development of a colloidal gold-based lateral-flow immunoassay for the rapid simultaneous detection of zearalenone and deoxynivalenol. *Anal and Bioanal Chem*, 389, 2103–2107.

Kolosova, A.Y., Sibanda, L., Dumoulin, F., Lewis, J., Duveiller, E., Van Peteghem, C. and De Saeger, S., 2008. Lateral-flow colloidal gold-based immunoassay for the rapid detection of deoxynivalenol with two indicator ranges. *Anal Chim Acta*, 616(2), 235–244.

Krska, R., Josephs, R., 2001. The state-of-the-art in the analysis of estrogenic mycotoxins in cereals. *Fresenius J Anal Chem*, 369,.469–476.

Krska, R., Molinelli, A., 2009. Rapid test strips for analysis of mycotoxins in food and feed. *Anal and Bioanal Chem*, 393, 67–71.

Li, S.J., Sheng, W., Wen, W., Gu, Y., Wang, J.P., Wang, S., 2018. Three kinds of lateral flow immunochromatographic assays based on the use of nanoparticle labels for fluorometric determination of zearalenone. *Microchimica Acta*, 185, 1–8.

Manawasinghe, I.S., Phillips, A.J., Xu, J., Balasuriya, A., Hyde, K.D., Stępień, Ł., Harischandra, D.L., Karunarathna, A., Yan, J., Weerasinghe, J., Luo, M., 2021. Defining a species in fungal plant pathology: beyond the species level. *Fungal Diversity*, 109(1), 267–282.

Market, F.C., 2013. Lateral flow immunoassay systems: evolution from the current state of the art to the next generation of highly sensitive, quantitative rapid assays. *Immunoass Handb*, 89, 89–107.

McCormick, S.P., Stanley, A.M., Stover, N.A., Alexander, N.J., 2011. Trichothecenes: from simple to complex mycotoxins. *Toxins*, 3(7), 802–814.

Molinelli, A., Grossalber, K., Krska, R., 2009. A rapid lateral flow test for the determination of total type B fumonisins in maize. *Anal and Bioanal Chem*, 395, 1309–1316.

Nagaraja, H., Chennappa, G., Poorna Chandra Rao, K., Mahadevprasad, G., Sreenivasa, M.Y. 2016. Diversity of toxic and phytopathogenic *Fusarium* species occurring on cereals grown in Karnataka state, India. *3Biotech.* 6, 57

Ngom, B., Guo, Y., Wang, X., Bi, D., 2010. Development and application of lateral flow test strip technology for detection of infectious agents and chemical contaminants: a review. *Anal and Bioanal Chem*, 397, 1113–1135.

Pettitt, T.R., Wakeham, A.J., Wainwright, M.F., White, J.G., 2002. Comparison of serological, culture, and bait methods for detection of *Pythium* and *Phytophthora* zoospores in water. *Plant Pathol*, 51(6), 720–727.

Qie, Z., Yan, W., Gao, Z., Meng, W., Xiao, R. Wang, S., 2019. Ovalbumin antibody-based fluorometric immunochromatographic lateral flow assay using CdSe/ZnS quantum dot beads as label for determination of T-2 toxin. *Microchimica Acta*, 186, 1–8.

Prasad M.N.N., Shankara Bhat, S., Sreenivasa, M.Y. 2010. Antifungal activity of essential oils against *Phomopsis azadirachtae* – the causative agent of die-back disease of neem. *J Agri Technol.* 6 (1), 127–133.

Rheeder, J.P., Marasas, W.F., Vismer, H.F., 2002. Production of fumonisin analogs by *Fusarium* species. *Appl Environ Microbiol*, 68(5), pp.2101–2105.

Rogowska, A., Pomastowski, P., Sagandykova, G., Buszewski, B., 2019. Zearalenone and its metabolites: Effect on human health, metabolism and neutralisation methods. *Toxicon*, 162, 46–56.

Ropejko, K., Twarużek, M., 2021. 'Zearalenone and Its Metabolites-General 534 Overview, Occurrence, and Toxicity,' *Toxins*, 13(1), 35.

Sanchis Almenar, V., Ramos Girona, A.J., Marin Sillué, S., 2013. Other moulds and mycotoxins. *Guide Foodborne Pathogens*, 14, 284–302.

Song, S., Liu, N., Zhao, Z., Njumbe Ediage, E., Wu, S., Sun, C., De Saeger, S., Wu, A., 2014. Multiplex lateral flow immunoassay for mycotoxin determination. *Anal Chem*, 86(10), 4995–5001.

Sreenivasa, M.Y., Dass, R.S., Charith Raj, A.P., Janardhana, G.R. 2008. PCR-based detection of genus *Fusarium* and Fumonisin-producing isolates from freshly harvested Sorghum grains grown in Karnataka, *India J Food Saf.* 28, 236–237.

Sreenivasa, M.Y., Dass, R.S., Janardhana, G.R. 2010. Post harvest fungi associated with sorghum grains produced in Karnataka (India). *J Plant Protection Res.* 50(3), 335–339.

Sreenivasa, M.Y., Dass, R.S., Charith Raj, A.P., Prasad, M.N.N., Achar, P. N., Janardhana, G.R. 2011a. Assessment of the growth inhibiting effect of some plant essential oils on different *Fusarium* species isolated from sorghum and maize grains. *J Plant Dis Prot.* 118 (6), 208–213.

Sreenivasa, M.Y., Dass, R.S., Charith Raj, A.P., Janardhana, G.R. 2011b. Mycological evaluation of Maize grains produced in Karnataka (India) for the post-harvest fungal contamination. *World Appl Sci J.* 13(4), 688–692.

Tang, D., Sauceda, J.C., Lin, Z., Ott, S., Basova, E., Goryacheva, I., Biselli, S., Lin, J., Niessner, R. and Knopp, D., 2009. Magnetic nanogold microspheres-based lateral-flow immunodipstick for rapid detection of aflatoxin B2 in food. *Biosens Bioelectron*, 25(2), 514–518.

Tang, X., Zhang, Q., Isabel Pividori, M., Zhang, Z., Marty, J.L., Catanante, G., 2022. A sensitive aptasensor using Biotin-Streptavidin system for patulin detection in apple juice. *Biosensors*, 12(2), 59.

Urusov, A.E., Gubaidullina, M.K., Petrakova, A.V., Zherdev, A.V., Dzantiev, B.B., 2018. A new kind of highly sensitive competitive lateral flow immunoassay displaying direct analyte-signal dependence. Application to the determination of the mycotoxin deoxynivalenol. *Microchimica Acta*, 185, 1–7.

Vöhringer, C., Sander, E., 2001. Comparison of antibodies in chicken egg yolk (IgY) and rabbit (IgG) for quantitative strain detection of *Colletotrichum falcatum* and *Fusarium subglutinans* with ELISA/Vergleich von Antikörpern aus Hühnereigelb (IgY) und Kaninchen (IgG) für den quantitativen Nachweis von Colletotrichum falcatum und Fusarium subglutinans Stämmen im ELISA. *J Plant Dis Prot*, 45, 39–48.

Xiulan, S., Xiaolian, Z., Jian, T., Xiaohong, G., Jun, Z., Chu, F.S., 2006. Development of an immunochromatographic assay for detection of aflatoxin B1 in foods. *Food Control*, 17(4), 256–262.

Xu, Y., Huang, Z.B., He, Q.H., Deng, S.Z., Li, L.S., Li, Y.P., 2010. Development of an immunochromatographic strip test for the rapid detection of deoxynivalenol in wheat and maize. *Food Chem*, 119(2), 834–839.

Zhang, X., Ding, M., Mao, Y., Huang, X., Xie, X., Song, L., Qiao, M., Zhang, J., Wang, T., Zhu, H., Li, Z., 2022. A comparative study of "turn-off" mode and "turn-on" mode lateral flow immunoassay for T-2 toxin detection. *Sens Actuators B Chem*, 359, 131545.

Zhang, X., Wu, C., Wen, K., Jiang, H., Shen, J., Zhang, S., Wang, Z., 2015. Comparison of fluorescent microspheres and colloidal gold as labels in lateral flow immunochromatographic assays for the detection of T-2 toxin. *Molecules*, 21(1), 27.

Zhang, X., Yu, X., Wen, K., Li, C., Mujtaba Mari, G., Jiang, H., Shi, W., Shen, J., Wang, Z., 2017. Multiplex lateral flow immunoassays based on amorphous carbon nanoparticles for detecting three fusarium mycotoxins in maize. *J Agric Food Chem*, 65(36), 8063–8071.

8 DNA Barcoding for Identification of Fungal Pathogens

S. Mahadevakumar
Andaman and Nicobar Regional Centre

K. Ajithkumar and A. S. Savitha
University of Agricultural Sciences

*K. R. Sridhar**
Mangalore University

In recent years, there have been numerous new diseases reported on various commercial crops. Earlier taxonomists recognized host and morphological features for the identification of phytopathogens. However, with the advent of molecular tools, DNA barcoding has made significant changes to the facets of fungal taxonomy and the molecular detection of associated pathogens. Several common diseases are mainly diagnosed by morphology and later by ITS barcoding. However, due to advancements in sequencing platforms, more genes are exploited and utilized for molecular detection and systematic appraisal of pathogens. Therefore, the precise recognition of plant pathogenic fungi up to the species level by multi-locus phylogenetic assessment is crucial. In this chapter, recent developments in utilizing multi-locus sequence strategy to detect fungal phytopathogens are addressed explicitly for recognized cryptic species. The most important cryptic species or species complexes recognized in recent times belong to the genera *Colletotrichum*, *Diaporthe*, *Didymella*, *Fusarium* and *Pestalotiopsis*-like organisms. Identification of these is obscured by sequence analysis of ITS and LSU. However, taxonomic ambiguity was successfully resolved using specific additional gene sequencing and concatenated phylogenetic analysis. This chapter addresses the necessity of multi-locus barcoding of phytopathogens like *Colletotrichum*, *Diaporthe*, and *Didymella* taxa to understand the possibility of different genes in resolving species-level identity in each of the recognized genera. A few successful attempts from our research group are illustrated, where species were identified or delimited based on multi-gene sequence analysis. Future prospects for the convenience of multi-locus sequence barcoding for molecular diagnosis along with other sequence platforms are in competition, which emphasizes better understanding.

* All authors contributed to the design and structure of this chapter. All authors verified and approved the final version.

DOI: 10.1201/9781003399704-8

GRAPHICAL ABSTRACT

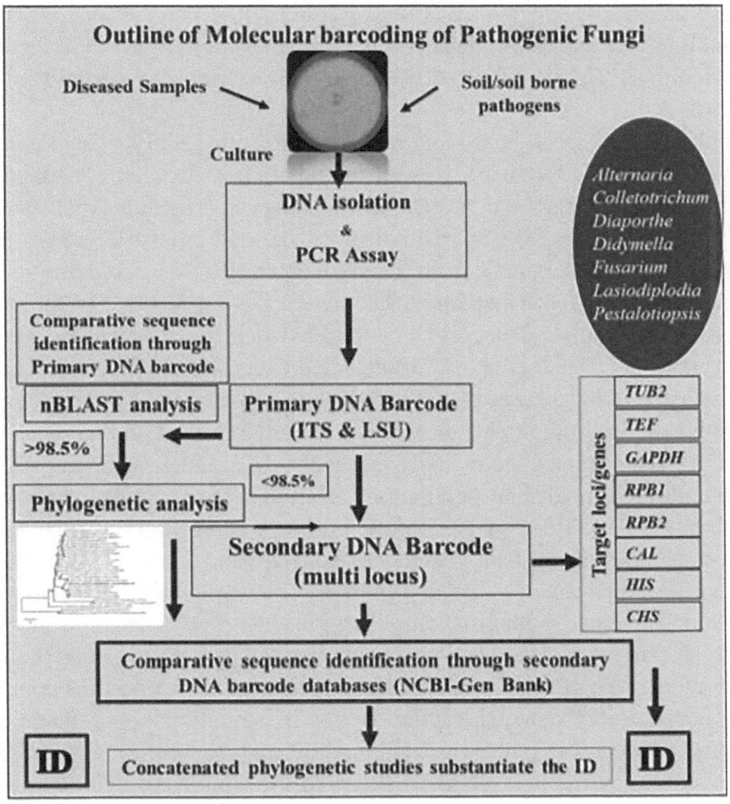

8.1 INTRODUCTION

The causes of numerous harmful plant diseases and commercially significant crops that cause significant damage globally are pathogenic fungi. They can be limited to one or a few host species, or they can infect a large variety of plant species. While a few of them are obligatory parasites that need a living host in order to develop and proliferate, the majority are saprophytes that can survive in soil, water and the atmosphere without the help of a host. Pathogenic fungal isolates can be distinguished from one another based on their aggressiveness (pathotypes or races), capacity to fuse genetically distinct strains to generate stable vegetative heterokaryons physical traits and host range (*formae speciales*) (Capote et al. 2012). The first step towards controlling a plant disease or initiating preventive or therapeutic measures is the proper detection and identification of the pathogen (Mahadevakumar and Sridhar 2020). It is clear that using molecular techniques will help get over these restrictions and increase accuracy and dependability. Numerous plant pathogenic fungi have been detected, identified and quantified using a wide range of molecular techniques (Deepa et al., 2016a, 2016b).

The method of identifying the pathogens responsible for infection and disease development is one of the basic necessities of diagnosis. The pathogens were previously recognized only by morphological characterization (Sreenivasa et al., 2008). The conventional techniques for identifying plant pathogens required visual inspection of several afflicted areas of the host to discover symptoms The development of microscopy and optics made it simple for taxonomists to assign taxa to the various infections that were discovered. Innovative concepts were being explored to replace or supplement conventional taxonomic and morphological identification of pathogens with digital information. Taxonomists have now achieved higher taxonomic precision by eliminating the non-reproducibility associated with phenotypic identification through molecular identification based on several genes and genomic segments (Janardhana et al., 2009; Dass et al., 2007). A considerable degree of repeatability and resolution can be achieved when using certain of these genes and genomic areas as "taxon identifiers". These brief (~600 bp) genomic sections were employed as identification DNA barcodes.

Beginning with animals, DNA barcoding has now spread to nearly every kind of life, albeit to varying degrees (Hebert et al. 2003a, b). This technique has been extensively adapted for the identification of animals and plants to comprehend diversity and facilitate conservation measures. Researchers in different sectors, like animal nutrition, the food industry, forensic science, biotechnology and illness diagnostics, benefit from the application of DNA barcoding technology (Chase et al., 2005; Ortega et al., 2018; Deepa and Sreenivasa, 2019a; 2019b). In the recent past, DNA barcoding has proven to be a very successful method for identifying latent diseases or cases where plants show no symptoms. Hsieh et al. (2020) detected *Cercospora*, *Colletotrichum* and *Fusarium* in *Platostoma palustre* (Chinese masona) by combining histone H3, internal transcribed spacer (ITS) and partial β-tubulin. In the fields of ecology, the environment and medical exploration, barcodes have been utilized in metagenomic approaches that have aided in the knowledge of both culturable and non-culturable microbial populations (Harrison et al., 2020; Hoshino and Inagakim, 2017; Shahraki et al., 2019). Studies on population biology and biosurveillance benefited from the analysis of environmental samples using barcodes (Konwarh and Sharma 2020). The ability to diagnose diseases has been transformed by methods based on barcoding genes, such as polymerase chain reaction (PCR), loop-mediated isothermal amplification and others.

Despite having enormous potential from the beginning, DNA barcoding had drawbacks that were overlooked in research. One of them was considering the mean instead of the congeneric distances and the smallest interspecificity in quantifying the gap in barcoding, which made it problematic to assess the variation between the type strain and unknown strain matrices enhanced by taxon sampling (Meier et al. 2008). A crucial aspect of DNA is that no single gene could be an ideal barcode, meaning a gene that is always the same within a species but varies between species or that no such genome area has been discovered as of yet. Disease surveillance and forecasting are being used to try and understand the pathosystem. Global biosecurity rules are being modified to track and prevent the dissemination of alien pathogenic hosts, primarily ornamentals, through international trade (Luchi et al. 2020). Plant pathologists need quick and inexpensive protocols for this. Future research on molecular methods, particularly DNA barcoding, may be able to resolve this issue. Dual DNA barcoding is one technique that is gaining popularity, particularly with regard to fungal diseases (Hoang et al. 2019). In this chapter, the molecular barcoding of plant pathogenic fungi and their implications for diagnosis, taxonomy and management are discussed.

8.2 DNA BARCODING – AN OVERVIEW

The fundamental concept behind DNA barcoding is the identification of a specimen using a short DNA sequence. According to Kress and Erickson (2008), a gene or genomic region must meet certain requirements in order to be used as a barcode: (i) universal and possess sufficient phylogenetic information; (ii) necessary sequence variability; (iii) adequate conservation of sequence in the regions surrounding the variable sequences; (iv) not be too lengthy to amplify, sequence or analyse. Finding a genetic region that meets all of the aforementioned requirements is likely the most difficult task in barcoding. In actuality, these typically meet the needs of the barcode applications rather than meeting every need. For instance, Valentini et al. (2009) noted that while taxonomists may be further interested in phylogenetic information, sequence variability is important for discrimination and universality. Ecologists may place more emphasis on characteristics linked to conserved sequences, such as having a short sequence and length universal primers. However, the quest for the perfect barcode continues, and at least for plants and animals, we currently have highly reliable barcodes.

Research has demonstrated that DNA barcoding is useful for more than just identifying species; it may also be used to define boundaries between species, find new species and delineate existing ones. When a lack of such a DNA area with low barcode gaps can be initially identified in plants, the strategy becomes problematic (Spooner 2009; Whitworth et al. 2007). According to findings published by Hollingsworth et al. (2009a, 2009b, 2011), about 90% of the plants could also be barcoded. It was suggested that the widespread barcodes for plants (*rbc*L and *mat*K) have a requirement that other segments be tested and incorporated (Hollingsworth, 2011). The addition of nrDNA ITS to this 2-loci combination increased species discrimination from 50–62% to 77–82% (Hollingsworth 2011; Hollingsworth et al., 2011). In addition to the numerous attempts to barcode plants and animals, there were also attempts to barcode fungi. As the field of molecular identification of fungi gained momentum, several papers from throughout the world demonstrated its potential (Okubara et al., 2008; Schaad and Frederick, 2002; Steenkamp et al., 1999; Thorsten-Lumbsch and Leavitt, 2011). Schoch et al. (2012) conducted a thorough study of the nuclear ribosomal ITS region to conclude that it can be considered a universal DNA barcode, which was a significant development in the field of the fungal kingdom. These reports made possible the establishment of barcoding techniques for fungal diseases.

Applying the mean rather than the lowest interspecific distances causes misidentification by exaggerating the size of the "Barcoding Gap" (Meier et al. 2008). For congeneric species, DNA barcoding papers and BOLD continue to publish the mean rather than the shortest interspecific distances. According to Kashyap et al. (2017), identifying intra- or interspecific variation is necessary for DNA-based species variation. It may be challenging to distinguish newly divergent species or new species that have emerged as a result of hybridization, mutation and selection pressure because the ranges of numerous types of differences are unknown and may vary among taxa (Kashyap et al. 2017). An additional noteworthy obstacle is that, although the raw chromatograms could be examined during review, this rarely happens, which results in low-quality sequence data being deposited in GenBank.

A study by van Velzen et al. (2012) demonstrated similarity-based (NN and BLAST) and identification techniques (BLOG and DNA-BAR) to significantly outpace tree-based techniques (NJ and PAR). Van Velzen et al. (2012) applied simulated DNA barcode details to recently deviating species. Thus, character-based distance metrics are more relevant compared to distance-based studies, which weaken the entire DNA barcode perception.

8.3 DNA BARCODING IN DIAGNOSTICS AND BIOSURVEILLANCE

Because molecular signatures are accurate, repeatable and capable of early detection, they offer advantages over other conventional methods for the identification and monitoring of diseases. It is evident that developing effective control measures and monitoring harmful diseases to prevent their spread to new areas is greatly aided by the use of DNA barcodes in diagnostics and biosurveillance. The successful diagnosis and identification of bacterial, viral and fungal infections along with their hosts has been achieved through the use of DNA barcoding.

The technique of identifying species of fungi by amplification, sequencing and comparison of particular DNA sequences with sequences stored in a DNA barcode database is known as fungal DNA barcoding. A DNA barcode is a distinctive 400–800 nucleotide pattern in DNA sequence that can be easily extracted from thousands of specimens or cultures and unmistakably examined by computer programs to determine the species (Crous et al. 2015).

8.3.1 Barcoding and Diagnosis of Plant Pathogens

In order to restrict the dissemination of disease and support efficient management methods, it is imperative to monitor plant health and discover plant pathogens early on. The microbiome, a collection of various microbial populations, influences disease outcomes in addition to the primary pathogen (Hu et al. 2018). Plant pathogens that are primarily of concern include bacteria, fungi, invasive plants, nematodes, oomycetes, phytoplasmas and viruses. Although molecular diagnostics has evolved significantly in the past few years, there are still numerous obstacles in the way of its advancement and practical use to help plant disease diagnosis (Li et al. 2019).

8.3.2 Barcoding and Diagnosis of Fungi and Oomycetes

Following the discovery of *rbc*L and *mat*K genes as plant identification barcodes (Hollingsworth et al. 2011), it was suggested that the Fungal kingdom was the only major eukaryotic group still lacking a formal barcode. The ITS region was proposed by Schoch et al. (2012) as a potential fungal DNA barcode marker. The preserved 5.8S region is flanked by two parts of the ITS region, ITS1 and ITS2. Three characteristics of ITS make it a good barcode: (1) modest size ranging from 500 to 800 bp, (2) universality, and (3) a conserved section with a hypervariable region. For Basidiomycetes, the ITS1 barcode typically yields a higher resolution than that of Ascomycetes; however, for ITS2, the opposite is true (Banchi et al. 2018; Buyck et al. 2018). Ascomycota and Glomeromycota have shorter ITSs than Basidiomycota, Zygomycota and Chytridiomycota. According to some studies, ITS2 alone may be adequate to investigate operational taxonomic richness in fungi if the entire ITS is unavailable (Yang et al. 2018). However, Wang et al. (2015) argued that ITS1 would be a more suitable barcode for eukaryotes. Schoch et al. (2012) found that PCR amplification of ribosomal RNA genes was only observed in 65% of the early diverging lineages, suggesting that further work is necessary.

8.4 CRYPTIC FUNGI AND NEED FOR MULTI-LOCUS BARCODING

The remarkable yet subtle, until recently unnoticed variety of the plant-associated Glomeromycota, Sebacinales and fungus-like Peronosporomycetes would remain almost unknown without the application of molecular phylogeny (Stockinger et al., 2010; Weiss et al., 2004). In forest soils or deep-sea environmental sampling has shown a profusion of unidentified phylogenetic lineages and suggested that the fungal species that are currently known may represent only a small fraction of the total fungal population and variety, which is a crucial aspect of identification. According to Leal et al. (2019), there are cryptic species of *Coniferiporia sulphurascens* in both the Eurasian and North American (NA) populations. This species is the causative agent of laminated root rot in conifers. Together with six other loci (actin, laccase-like multicopper oxidase, ITS, RNA polymerase II LSU and TEF 1-α), they examined both small and large mitochondrial RNA. The fact that no allele at any of the loci under study was shared between the Eurasian and North American populations of *C. sulphurascens* was one of the major findings. In forests, the soil environment is substantially more complicated for soilborne diseases. These infections occur underground, last for extended periods of time and manifest much later than the above-ground symptoms (Grünwald et al., 2019; Luchi et al., 2020).

Ultimately, a major barrier to its implementation is the ability to distinguish between different fungal species based on a common gene and identification. Quarantine practices routinely employ a number of molecular techniques for general diagnostic procedures. For instance, for the widespread international quarantine pathogen *Tilletia indica*, three major molecular techniques are employed to confirm the present surveillance procedures of the EPPO: (i) conventional PCR, (ii) real-time PCR and (iii) restriction enzyme assessment of the ITS 1 region (EPPO 2015).

8.4.1 *Colletotrichum*

The genus *Colletotrichum* (sexual morph: *Glomerella*) belongs to the family Glomerellaceae and its type species is *C. lineola* Corda. It is identified as one of the top 10 fungal pathogens on the globe and is known to induce anthracnose on numerous economically significant crop plants (Dean et al.

2012; Mahadevakumar and Sridhar 2021). It lowers the commercial value of the fruit business and results in large yield losses (Dowling et al., 2020; Peres et al., 2005). It is clear that Colletotrichum species are widely distributed and omnipresent. Anthracnose of chilli, tomato, beans, maize, flower crops, horticulture crops and decorations were among the several Colletotrichum species that were isolated from diseased plants (Mahadevakumar et al., 2019; Chung et al., 2020; Wang et al., 2021). Furthermore, *Colletotrichum* species were infrequently discovered to be saprobes but were most frequently reported to be endophytes (Osono et al., 2009; Thongkantha et al., 2008). Additionally, some species were isolated concurrently in the capacities of saprobe, pathogen and endophyte (Riolo et al., 2021). The species *Colletotrichum* has been distinguished by its appressoria, asci, ascomata, ascospores, conidia, colony, conidiomata, conidiophores, illness symptoms and setae (Damm et al., 2012, 2019; Talhinhas and Baroncelli 2021). These characteristics, however, don't provide a lot of information for differentiating *Colletotrichum* species. It is therefore likely that several names are substitutes (Soares et al., 2021).

There are currently 1040 entries in the Index Fungorum and 1129 species altogether registered in MycoBank; many of these entries are synonyms, orthographic variants and illegitimate entries (accessed on January 25, 2024). On the other hand, information found in the species Fungorum indicates that there are 545 species of *Colletotrichum* in existence. Nevertheless, using a mix of morphological and molecular analyses, only about 280 were verified (Bhunjun et al. 2019; Hyde et al. 2020a; Liu et al. 2022; Zapata and Palfner 2022; Zheng et al. 2022). The official fungal barcoding marker has been suggested to be the ITS regions (Schoch et al., 2012). A contemporary taxonomy framework for *Colletotrichum* was presented by Liu et al. (2022) using phylogenetic analysis at the genome-scale and multi-locus levels (ACT, CHS, GAPDH, TUB2 and TEF1). In a recent investigation, Qiao et al. (2021) demonstrated that *C. menglaense* is perfectly grouped in the *C. gloeosporioides* species complex and therefore, explored multi-locus phylogeny to discriminate closely linked species. For example, the ITS-based phylogenetic analysis revealed that *C. menglaense* could not be separated from *C. queenslandicum*, *C. salasolae* and *C. siamense*. The ITS, ACT, CHS and GAPDH sequences were used to determine the phylogenetic location of *C. menglaense*.

Researchers in the past utilized morphology-based or single gene-based techniques for identification and classification of the genus *Colletotrichum*, which failed to resolve the phylogenetic relationships as the number of species increased and the cryptic nature was evident in many of the species complex groups. Species delineation is difficult due to a lack of distinctive phenotypic variation. These issues were resolved by recent technological advancements, especially the robust molecular tools and techniques that provided a solid platform to explore additional marker genes. Multi-locus sequence analysis resolved the identities of many cryptic Colletotrichum species. Understanding biodiversity, conservation, evolution, host-parasite interaction and managing plant pathogens to develop quarantine protocols depends on the accurate identification of *Colletotrichum* species (Bhunjun et al., 2021a; Jayawardena et al., 2016a, 2016b).

8.4.2 *Diaporthe*

Nitschke has introduced the genus *Diaporthe* (synonym, *Phomopsis*), which is represented by *Diaporthe eres*, which belongs to the family Diaporthaceae and order Diaporthales (Senanayake et al. 2017). Numerous reports have identified *Diaporthe* species as saprobes, endophytes and pathogens on a variety of host types. These species are known to cause a wide range of diseases that affect economically significant plants and crops, including branch canker on camphor trees (*Cinnamomum camphora*), dieback of forest trees, leaf spots of common hop, leaf spots of *Ixora* spp., leaf spots of tea, melanose of *Citrus* spp., pear shoot canker of sunflower, stem canker of sunflower, stem-end rot of *Citrus* spp. and trunk diseases of grapevines. As of January 2024, the MycoBank contains about 1339 names under the genus *Diaporthe* and 1055 under Phomopsis; the Species Fungorum contains 908 *Diaporthe* species and 888 entries under Phomopsis. The Index Fungorum contains 1266 entries under *Diaporthe* and 985 entries for Phomopsis. Traditionally, phenotypic characteristics of *Diaporthe* species, such as colony form, ascomata and conidiomata appearance, ascospore and conidiospore variation and host associations, were used to identify them. However, the discovery of several mysterious species as a result of advances in DNA sequencing has altered our knowledge of *Diaporthe*'s natural classification. Because *Diaporthe* strains exhibit diversity under varying environmental conditions, traditional morphological features and host relationships have proven insufficient for distinguishing between species in recent genetic phylogeny-based research.

The phylogenetic recognition of *Diaporthe* species was explicitly re-evaluated by multi-locus phylogeny using a pooled data matrix of β-tubulin (TUB), calmodulin (CAL), partial sequences from the translation elongation factor 1-α (EF 1-α) as well as rDNA ITS molecular markers (Udayanga et al., 2011). Though earlier workers' output contributed to ITS alone and/or a few genes, only in recent research developments has a combination of markers been employed to differentiate the species and examine many of the previously recognized and described species to provide a multi-locus backbone tree of known Diaporthe. Therefore, the natural categorization of *Diaporthe* species is most typically determined by multi-locus phylogeny depending on beta-tubulin 2 (tub2), calmodulin (CAL), DNA sequences of ITS (internal transcribed spacer) and translation elongation factor 1-α (tef1-α).

8.4.3 *Didymella*

In 1880, Saccardo introduced *Didymella* under the type genus of the family Didymellaceae, with *Didymella exigua* as a type species. The main sexual morphological features of *Didymella* are solitary, pseudothecial ostiolate ascomata possessing multi-layered, pseudoparenchymatous walls, cylindrical or saccate to clavate eight-spored and bitunicate asci possessing uniseptate or multiseptate hyaline or brownish ascospores. The primary characteristics of *Didymella*'s asexual morphs include solitary or confluent, poroid or ostiolate, conidiomata pycnidial with multi-layered, conidiomatal walls pseudoparenchymatous, phialidic and conidiogenous cells hyaline producing smooth conidia, which vary in shape, aseptate and occasionally hyaline or pigmented. Pure cultures of chlamydospores frequently contain unicellular organisms. As of right now, *Didymella* has 597 records in the Index Fungorum and 438 entries in the Species Fungorum. The majority of Didymellaceae species are plant-associated fungi, and some are saprobes that are typically found on woody and herbaceous plants, however, several of them are known to cause diseases.

Despite its age and abundance of species, *Didymella*'s initial taxonomic classifications remain ambiguous. Following its first description in the Mycosphaerellaceae family, the genus was subsequently assigned to the Phaeosphaeriaceae, Pleosporaceae and Venturiaceae genera *incertae sedis*. On indication from phylogenetic studies of combined LSU and SSU sequences, a new family, Didymellaceae, with a type genus *Didymella* to accommodate *Ascochyta*, *Didymella*, *Phoma* and other linked *Phoma*-like species is created (De Gruyter et al. 2012). According to Aveskamp et al. (2010), *Didymella* seems to be polyphyletic, with some of its members lumped in with *Peyronellaea, Macroventuria, Microsphaeropsis* and *Leptosphaerulina* species. They also proposed that a taxonomy revision for *Didymella* is urgently needed. Using a morpho-molecular method, Chen et al. (2015b) provided additional clarification on the generic delimitation in Didymellaceae, which limited *Didymella* to a monophyletic cluster that included 37 species. Subsequently, 49 more species have been added using phylogenetic and morphological analysis. It has always been challenging to correctly identify species in this family, mostly based on host plant relationships and morphology (Aveskamp et al. 2010; Chen et al. 2015a). Nevertheless, a relatively sturdy phylogenetic backbone for taxonomic determination is provided by a strong tree based on ITS regions and intervening 5.8S nrDNA (ITS), partial 28S large subunit nrDNA (LSU) sequences, partial regions of β-tubulin (tub2) and RNA polymerase II second largest subunit (rpb2) genes (Chen et al., 2015a). A thorough explanation was given by Chen et al. (2017), based on 108 isolates of Didymellaceae from 40 plant species belonging to 27 families in China and a few other nations. Among them, 68 isolates, or 32 new taxa, are characterized based on multi-locus phylogenetics and morphological traits in a polyphasic manner. Although morphological traits play a crucial role in the identification of fungi, morphological trait-based research is not exhaustive. The molecular phylogenetic assessment has facilitated the elucidation of the evolutionary connections amongst the members of Dothideomycetes, and multi-locus analyses have further established the species boundaries for *Didymella* due to the increased availability of sequences for Dothideomycetes species. Nevertheless, there are no widely recognized criteria for choosing barcodes for phylogenetic analysis in studies on *Didymella*.

Plant disease diagnosis has evolved as a result of the quick rise in the use of molecular tools, sophisticated sequencing platforms and diagnostic techniques. Molecular diagnosis, or sequence-based identification, has emerged as the most popular and successful technique as a result of the advancements in PCR and Sanger sequencing platforms. Because of this, the traits that were once used for identifying and classifying fungi have now evolved, beginning with morphology and development and progressing to DNA sequences and genomics research. Because phenotypic traits often undergo convergent evolution, it has been shown that molecular phylogeny performs fundamentally better in fungal identification than traditional approaches. The diagnosis of fungi and their systematics has been drastically transformed by phylogenetic study, and it constantly advances our knowledge of fungal evolution. According to the ITS-rDNA (SSU) phylogeny, fungi were categorized as a monophyletic group in 2000 and categorized into nine subgroups (Tehler et al. 2000). In 2007, the phylogeny of the fungal kingdom was restructured using multi-locus DNA sequences. When molecular phylogenetic analyses of fungi first began, single-gene sequences were the only ones employed to develop the phylogeny (Liu et al., 1999; Tehler et al., 2000). The following years saw the first reports of multi-locus phylogenies in the literature; these phylogenies usually contain less than six loci (Blackwell et al., 2006; Spatafora et al., 2006; Zhang et al., 2006). Subsequently, relatively superior resolution was achieved based on single-gene trees. Assembling the Fungal Tree of Life (AFTOL) projected the LSU and SSU of rDNA to evaluate the fungal phylogeny (http://aftol.org/about.php), which formed the source for the classification of fungi (Hibbett et al. 2007; James et al. 2006). This included a number of genes coding proteins such as ATP6, mtSSU, RPB1, RPB2 and TEF1. Multi-locus phylogeny has also been shown in a large number of recently published fungal monographs of *Alternaria*, *Cladosporium*, *Phoma*, Massarineae and Sordariomycetes (Maharachchikumbura et al. 2016).

8.5 REFERENCE SEQUENCES AND ONLINE RESOURCES

Numerous open databases are available for usage related to DNA barcoding sequence repositories, curated reference sequence files and identification resources (Table 8.1). Genus- or species-specific elements are made possible by focused-group databases for particular uses connected to diagnostic and research requirements. Numerous sequencing resources are available through the NCBI GenBank database. Nonetheless, there are certain online resources that provide reference sequence data that may be utilized and examined with any sequence produced by a research project.

8.6 SUMMARY

Identification of plant pathogens was primarily based on morpho-cultural characterization. However, as the number of pathogens and species is huge, there were cryptic species established, which necessitated identification through advanced molecular tools. If one or a small number of barcoding loci are able to accomplish this, additional sequencing is not necessary. More data are necessary when complete species discrimination is not possible, but it could be complemented with other useful applications (population genetics and phylogenetics) or when the barcode loci are not suitable for recovering from damaged fungal DNAs. In the end, this extra information may possibly extend to whole genomes. As costs continue to decline, this is becoming more and more possible. Therefore,

TABLE 8.1
List of Databases Available for DNA Barcodes for Identification of Fungal Pathogens

Database	Description	Web link
1000 Fungal Genome Project (1KFG)	It is a comprehensive collection of fungal genomes sequenced	http://1000.fungalgenomes.org/home/
Barcode of Life Datasystem (BOLD)	A web platform that provides an integrated environment for the assembly and use of DNA barcode data	http://www.barcodinglife.org
Canadian Centre for DNA Barcoding	Birthplace of DNA Barcoding and offers unparalleled access to species identifications for over 180,000 animals, plants and molds	http://www.dnabarcoding.ca
CBS-KNAW: Centraal Bureau voor Schimmel cultures BioloMICS Databases	Database based on the curated sequence of ITS	http://www.cbs.knaw.nl/Collections/
Consortium for the Barcode of Life (CBOL)	CBOL promotes barcoding through working groups, networks, workshops, conferences, outreach, and training	http://www.barcodeoflife.org
FungiDB	It integrates whole genome sequence and annotation and also includes experimental and environmental isolate sequence data	http://fungidb.org/fungidb/
Index Fungorum	Database dedicated for nomenclature of fungi	http://www.indexfungorum.org/
International Commission on the Taxonomy of Fungi (ICTF)	Dedicated database to improve information related to fungal taxonomy	http://www.fungaltaxonomy.org/
ITS database for ectomycorrhizal fungi (UNITE)	An rDNA sequence database to provide a stable platform for sequence identification of ectomycorrhizal and basidiomycetes. It is equipped with a BLAST interface for rapid similarity searches against the records of the database	http://unite.ut.ee
Mushroom Observer	It contains information of more than one million mushroom species	http://mushroomobserver.org/
Q-bank	Comprehensive database on quarantine plant pest and diseases	http://www.q-bank.eu/
Quarantine Barcode of Life (QBOL)	Developing DNA barcoding to identify quarantine organisms in support of plant health	http://www.qbol.org
RefSeq Target Loci (RTL)	Database based on the curated sequences of ITS, 18S, and 28S	http://www.ncbi.nlm.nih.gov/refseq/

DNA barcoding is a game changer in systematics as well as plant disease diagnosis. However, understanding technological advancements and their applications requires proper training and skill sets.

8.7 FUTURE PERSPECTIVE

Though there is an evolution in the technology used in identification, DNA barcoding stands out as unique and resolves most of the ambiguity without leaving scope for further confusion. The ultimate objective of DNA barcoding is to establish a phylogenetic order for all species on the globe. The accessibility of inexpensive sequencing per sample is crucial, given the task at hand, in order to meet the objectives.

STUDY QUESTIONS

1. Add a note on Biosurveillance of DNA barcoding.
2. What are cryptic fungi?
3. DNA barcoding and its types.

CONFLICT OF INTEREST

Authors declare no conflict of interest.

ACKNOWLEDGEMENT

We acknowledge everyone for the support given for the completion of the chapter.

FURTHER READING

Bhunjun, C.S., Phukhamsakda, C., Jayawardena, R.S., Jeewon, R., Promputtha, I., Hyde, K.D., 2021b. Investigating species boundaries in *Colletotrichum*. *Fungal Diversity* 107, 107–127. https://doi.org/10.1007/s13225-021-00471-z

Hsieh, C.W., Chuang, Y.Y., Lee, M.Z., Kirschner, R., 2020, First Inventory of fungi in symptomless and symptomatic Chinese Mesona indicates phytopathological threat. *Plant Disease*, 104(9), 2391–2397https://doi.org/10.1007/s13225-018-0397-5

REFERENCES

Aveskamp, M.M., de-Gruyter, J., Woudenberg, J.H.C., Verkley, G.J.M., Crous, P.W., 2010. Highlights of the Didymellaceae: A polyphasic approach to characterise *Phoma* and related pleosporalean genera. *Studies in Mycology*, 65, 1–60.

Banchi, E., Ametrano, C.G., Stanković, D., Verardo, P., Moretti, O. et al., 2018. DNA metabarcoding uncovers fungal diversity of mixed airborne samples in Italy. *PLoS One* 13(3), e0194489. https://doi.org/10.1371/journal.pone.0194489

Bhunjun, C.S., Jayawardena, R.S., Wei, D.P., Huanraluek, N., Abeywickrama, P.D. et al., 2019. Multigene phylogenetic characterisation of *Colletotrichum artocarpicola* sp. nov. from *Artocarpus heterophyllus* in northern Thailand. *Phytotaxa*, 418, 273–286. https://doi.org/10.11646/phytotaxa.418.3.3

Bhunjun, C.S., Phillips, A.J.L., Jayawardena R.S., Promputtha, I., Hyde, K.D. 2021a. Importance of molecular data to identify fungal plant pathogens and guidelines for pathogenicity testing based on Koch's Postulates. *Pathogens*, 10(9), 1096.

Bhunjun, C.S., Phukhamsakda, C., Jayawardena, R.S., Jeewon, R., Promputtha, I., Hyde, K.D., 2021b. Investigating species boundaries in *Colletotrichum*. *Fungal Diversity* 107, 107–127.https://doi.org/10.1007/s13225-021-00471-z

Blackwell, M., Hibbett, D.S., Taylor, J.W., Spatafora, J.W. 2006. Research Coordination Networks: A phylogeny for kingdom Fungi (Deep Hypha). *Mycologia*, 98, 829–837.https://doi.org/10.1080/15572536.2006.11832613

Buyck, B., Zoller, S., Hofstetter, V., 2018. Walking the thin line… ten years later: the dilemma of above- versus below-ground features to support phylogenies in the Russulaceae (Basidiomycota). *Fungal Diversity*, 89(1), 267–292.

Capote, N., Pastrana, A.M., Aguado, A., Sánchez-Torres, P., 2012. Molecular tools for detection of plant pathogenic fungi and fungicide resistance. In: *Plant Pathology*, Cumagun, C.J. (ed.)., In Tech. https://doi.org/978-953-51-0489-6(http://www.intechopen.com/books/plantpathology/molecular-tools-for-detection-of--plant-pathogenic-fungi-and-fungicide-resistance accessed on January 10, 2024).

Chase, M.W., Salamin, N., Wilkinson, M., Dunwell, J.M., Kesanakurthi, R.P. et al., 2005. Land plants and DNA barcodes: Short-term and long-term goals. *Philosophical Transactions of the Royal Society B*, 360(1462), 1889–1895. https://doi.org/10.1098/rstb.2005.1720

Chen, Q., Hou, L.-W., Duan, W.-J., Crous, P. W., Cai, L. 2017. Didymellaceae revisited. *Studies in Mycology*, 87, 105–159. https://doi.org/10.1016/j.simyco.2017.06.002

Chen, Q., Jiang, J.R., Zhang, G.Z., Cai, G.Z., Croys, P.W. 2015a. Resolving the *Phoma* enigma. *Studies in Mycology*, 82, 137–217. https://doi.org/10.1016/j.simyco.2015.10.003.

Chen, Q., Zhang, K., Zhang, G., Cai, L., 2015b. A polyphasic approach to characterise two novel species of *Phoma* (Didymellaceae) from China. *Phytotaxa*, 197, 267–281. https://doi.org/10.11646/phytotaxa.197.4.4

Chung, P.C., Wu, H.Y., Huang, Y.W., Ariyawansa, H.A., Hu, H.-P. et al., 2020. Diversity and pathogenicity of *Colletotrichum* species causing strawberry anthracnose in Taiwan and description of a new species, *Colletotrichum miaoliense* sp. nov. *Scientific Reports*, 10, 14664. https://doi.org/10.1038/s41598-020-70878-2

Crous, P.W., Hawksworth, D.L., Wingfield, M.J., 2015. Identifying and naming plant-pathogenic fungi: Past, present, and future. *Annual Reviews of Phytopathology*, 53, 247–267. https://doi.org/10.1146/annurev-phyto-080614-120245

Damm, U., Cannon, P.F., Woudenberg, J.H.C., Crous P.W., 2012. The *Colletotrichum acutatum* species complex. *Studies in Mycology*, 73, 37–113. https://doi.org/10.3114/sim0010

Damm, U., Sato, T., Alizadeh, A., Groenewald, J.Z., Crous, P.W., 2019. The *Colletotrichum dracaenophilum, C. magnum* and *C. orchidearum* species complexes. *Studies in Mycology*, 92, 1–46. https://doi.org/10.1016/j.simyco.2018.04.001

Dass, R.S., Sreenivasa, M.Y., Janardhana, G.R. 2007. High incidence of *Fusarium verticillioides* in animal and poultry feed mixtures produced in Karnataka, India. *Plant Pathology Journal*. 6(2), 174–178.

De Gruyter, J., Woudenberg, J.H.C., Aveskamp, M.M., 2012. Redisposition of *Phoma*-like anamorphs in Pleosporales. *Studies in Mycology*, 75, 1–36.

Dean, R., Van Kan, J.A.L., Pretorius, Z.A., Hammond-Kosack, K.E., Pietro, A.D. et al., 2012. The Top 10 fungal pathogens in molecular plant pathology. *Molecular Plant Pathology*, 13, 414–430. https://doi.org/10.1111/j.1364-3703.2011.00783.x

Deepa, N., Charith Raj, A. P., Sreenivasa, M.Y. 2016a. Nested PCR method for the early detection of fumonisin producing *Fusarium verticillioides* in pure cultures, cereal samples and plant parts. *Food Biotechnology*. 30(1), 18–29.

Deepa, N., Charith Raj, A.P., Sreenivasa, M.Y. 2016b. Multiplex PCR method for the early detection of fumonisin producing *Fusarium verticillioides*. *Food Bioscience*. 13, 84–88.

Deepa, N and M Y Sreenivasa. 2019a. Molecular methods and key genes targeted for the detection of fumonisin producing *Fusarium verticillioides* – An updated review. *Food Bioscience*. 32, 100473.

Deepa N and M.Y. Sreenivasa. 2019b. Biocontrol Strategies for Effective Management of Phytopathogenic Fungi Associated with Cereals, In: Singh J.S., Singh D.P. (eds) *New and Future Developments in Microbial Biotechnology and Bioengineering*, Elsevier, pp, 177–189, ISBN 9780444641915.

Dowling, M., Peres, N., Villani, S. and Schnabel, G., 2020. Managing *Colletotrichum* on Fruit Crops: A "Complex" Challenge. Plant Disease, 104(9), 2301–2316. https://doi.org/10.1094/PDIS-11-19-2378-FE

EPPO, 2015. EPPO A1 and A2 lists of pests recommended for regulation as quarantine pests.

Grünwald, N.J., LeBoldus, J.M., Hamelin, R.C., 2019, Ecology and evolution of the sudden oak death pathogen *Phytophthora ramorum. Annual Reviews of Phytopathology*, 57, 301–321. https://doi.org/10.1146/annurev-phyto-082718-100117

Harrison, J., Calder, W.J., Shuman, B.N. Buerkle, C.A., 2020, The quest for absolute abundance: the use of internal standards for DNA barcoding in microbial ecology. *Molecular Ecology Resources*, 21(1), 30–43. https://doi.org/10.1111/1755-0998.13247

Hebert, P.D.N., Cywinska, A., Ball, S.L., DeWaard, J.R. (2003a). Biological identifications through DNA barcodes. *Proceedings of the Royal Society B: Biological Sciences*, 270(1512), 313–321.

Hebert, P.D.N., Ratnasingham, S., DeWaard, J.R., 2003b, Barcoding animal life: cytochrome c oxidase subunit 1 divergences among closely related species. *Proceedings of the Royal Society B: Biological Sciences*, 270(supplement 1), S96–S99. https://doi.org/10.1098/rsbl.2003.0025\

Hibbett, D.S., Binder, M., Bischoff, J.F., Blackwell, M., Cannon, P.F. et al., 2007, A higher-level phylogenetic classification of the Fungi. *Mycological Research*, 111, 509–547. https://doi.org/10.1016/j.mycres.2007.03.004

Hoang, M.T.V., Irinyi, L., Chen, S.C., Sorrell, T.C., Meyer, W., 2019. Dual DNA barcoding for the molecular identification of the agents of invasive fungal infections. *Frontiers in Microbiology*, 10, 1647. https://doi.org/10.3389/fmicb.2019.01647

Hollingsworth, M.L., Andra, C.A., Forrest, L.L., Richardson, J., Pennington, R.T. et al., 2009a. Selecting barcoding loci for plants: Evaluation of seven candidate loci with species-level sampling in three divergent groups of land plants. *Molecular Ecology Resources*, 9(2), 439–457. https://doi.org/10.1111/j.1755-0998.2008.02439.x

Hollingsworth, P.M., 2011. Refining the DNA barcode for land plants. *Proceedings of the National Academy of Sciences, USA*, 108(49), 19451–19452. https://doi.org/10.1073/pnas.1116812108

Hollingsworth, P.M., Forrest, L.L., Spouge, J.L., Hajibabaei, M., Ratnasingham, S. et al., 2009b. A DNA barcode for land plants. *Proceedings of the National Academy of Sciences, USA*, 106(31), 12794–12797.

Hollingsworth, P.M., Graham, S.W., Little, D.P., 2011. Choosing and using a plant DNA barcode. *PLoS One*, 6(5), e19254. https://doi.org/10.1371/journal.pone.0019254

Hoshino, T., Inagakim, F., 2017. Correction: Application of stochastic labeling with random-sequence barcodes for simultaneous quantification and sequencing of environmental 16S rRNA genes. *PLoS One*, 12(1), e0169431. https://doi.org/10.1371/journal.pone.0173546

Hsieh, C.W., Chuang, Y.Y., Lee, M.Z., Kirschner, R., 2020, First Inventory of fungi in symptomless and symptomatic Chinese Mesona indicates phytopathological threat. *Plant Disease*, 104(9), 2391–2397. https://doi.org/10.1094/PDIS-03-20-0475-RE; https://doi.org/10.1007/s13225-018-0397-5

Hu, Y., Green, G., Milgate, A., Stone, E., Rathjen, J., Schwessinger, B., 2018. Pathogen detection and microbiome analysis of infected wheat using a portable DNA sequencer. *Phytobiomes Journal*, 3(2), 92–101. https://doi.org/10.1094/PBIOMES-01-19-0004-R

Hyde, K.D., Jeewon, R., Chen, Y.J., Bhunjun, C.S., Calabon, M.S. et al., 2020a. The numbers of fungi: is the descriptive curve flattening? *Fungal Diversity*, 103, 219–271. https://doi.org/10.1007/s13225-020-00458-2

Hyde, K.D., Norphanphoun, C., Maharachchikumbura, S.S.N., Bhat, D.J., Jone, E.B.G. et al. 2020b. Refined families of Sordariomycetes. *Mycosphere*, 11, 305–1059. https://doi.org/10.5943/mycosphere/11/1/7

James, T.Y., Kauff, F., Schoch, C.L., Matheny, P.B., Hofstetter, V. et al., 2006. Reconstructing the early evolution of Fungi using a six-gene phylogeny. *Nature*, 443, 818–822. https://doi.org/10.1038/nature05110

Janardhana, G. R., Dass, R.S., Sreenivasa, M.Y. 2009. Fumonisins – A new class of Fusarial toxins. In *Frontiers in Fungal Ecology, Diversity and Metabolites*. Ed: K. A. Sridhar et al. I. K. International Publishing House Pvt. Ltd. New Delhi.

Jayawardena, R.S., Hyde, K.D., Damm, U., Cai, L., Liu, M. et al., 2016a. Notes on currently accepted species of *Colletotrichum*. *Mycosphere*, 7, 1192–1260. https://doi.org/10.5943/mycosphere/si/2c/9

Jayawardena, R.S., Hyde, K.D., Jeewon, R., Li, X.H., Yan, J.Y., 2016b. Mycosphere essay 6: Why it is important to correctly name Colletotrichum species? *Mycosphere*, 7, 1076–1092. https://doi.org/10.5943/mycosphere/si/2c/1

Kashyap, P.L., Rai, P., Kumar, S., Chakdar, H., Srivastava, A.K., 2017. DNA barcoding for diagnosis and monitoring of fungal plant pathogens. In: *Molecular markers in mycology*. Fungal Biology, Springer, Cham, 87–122. https://doi.org/10.1007/978-3-319-34106-4_5

Konwarh, R., Sharma, P.L., 2020. Nanosensor platforms for surveillance of plant pathogens and phytometabolites/analytes vis-à-vis plant health status. In: *Nanomaterials for Agriculture and Forestry Applications*, Husen, A., Jawaid, M. (eds.). Elsevier, pp. 357–385. https://doi.org/10.1016/B978-0-12-817852-2.00014-7

Kress, W.J., Erickson, D.L., 2008. DNA barcodes: Genes, genomics, and bioinformatics. *Proceedings of the National Academy of Sciences, USA*, 105(8), 2761–2762. https://doi.org/10.1073/pnas.0800476105

Leal, I., Bergeron, M.J., Feau, N., Tsui, C.K.M., Foord, B. et al., 2019. Cryptic speciation in western North America and Eastern Eurasia of the pathogens responsible for laminated root rot. *Phytopathology*, 109(3), 456–468. https://doi.org/10.1094/PHYTO-12-17-0399-R

Li, Z., Paul, R., Tis B.T., Saville, A.C., Hansel, J.C. et al., 2019. Non-invasive plant disease diagnostics enabled by smartphone-based fingerprinting of leaf volatiles. *Nature Plants*, 5(8), 856–866. https://doi.org/10.1038/s41477-019-0476-y

Liu, F., Ma, Z.Y., Hou, L.W., Diao, Y.Z., Wu, W.P. et al., 2022, Updating species diversity of Colletotrichum, with a phylogenomic overview. *Studies in Mycology*, 101, 1–56. https://doi.org/10.3114/sim.2022.101.01

Liu, Y.J., Whelen, S., Hall, B.D., 1999. Phylogenetic relationships among ascomycetes: Evidence from an RNA polymerase II subunit. *Molecular Biology and Evolution*, 16, 1799–1808. https://doi.org/10.1093/oxfordjournals.molbev.a026092

Luchi, N., Ioos, R., Santini, A., 2020. Fast and reliable molecular methods to detect fungal pathogens in woody plants. *Applied Microbiology and Biotechnology*, 104(6), 2453–2468. https://doi.org/10.1007/s00253-020-10395-4

Mahadevakumar, S., Chandana, C., Janardhana, G. R., 2019, Association of *Colletotrichum truncatum* with anthracnose disease of Tuberose (*Polianthes tuberosa*) in India. *Crop Protection*, 118, 1–5.

Mahadevakumar, S., Sridhar, K.R., 2020. Diagnosis of *Pythium* by Classical and Molecular Approaches. In: *The genus Pythium – Diagnosis, Diseases and Management*. Rai, M., Abd-Esalam, K.A., Ingle, A.P. (eds.). CRC Press, 200–224. https://doi.org/10.1201/9780429296406

Mahadevakumar, S., Sridhar, K.R., 2021. Diversity of Plant Pathogenic Fungi in Agricultural Crops. In: *Plant, Soil and Microbes in Tropical Ecosystems*. Dubey, S.K., Verma, S.K. (eds.). Springer Nature, 101–149. https://doi.org/10.1007/978-981-16-3364-5_6

Maharachchikumbura, S.S.N., Hyde, K.D., Jones, E.B.G., McKenzie, E.H.C., Bhat, J.D. et al. 2016. Families of sordariomycetes. *Fungal Diversity*, 79, 1–317. https://doi.org/10.1007/s13225-016-0369-6

Meier, R., Zhang, G., Ali, F., 2008. The use of mean instead of smallest interspecific distances exaggerates the size of the "barcoding gap" and leads to misidentification. *Systematic Biology*, 57(5), 809–813. https://doi.org/10.1080/10635150802406343

Okubara, P.A., Schroeder, K.L., Paulitz, T.C., 2008. Identification and quantification of Rhizoctonia solani and *R. oryzae* using real-time polymerase chain reaction. *Phytopathology*, 98(7), 837–847. https://doi.org/10.1094/PHYTO-98-7-0837

Ortega, S.F., Tomlinson, J., Hodgetts, J., Spadaro, D., Gullino, M.L., Boonham, N., 2018. Development of loop-mediated isothermal amplification assays for the detection of seedborne fungal pathogens *Fusarium fujikuroi* and *Magnaporthe oryzae* in Rice Seed. *Plant Disease*, 102(8), 1549–1558. https://doi.org/10.1094/PDIS-08-17-1307-RE

Osono, T., Ishii, Y., Takeda, H., Seramethakun, T., Khamyong, S. et al. 2009. Fungal succession and lignin decomposition on *Shorea robusta* leaves in a tropical seasonal forest in northern Thailand. *Fungal Diversity*, 36, 101–119.

Peres, N.A., Timmer, L.W., Adaskaveg, J.E., Correll, J.C., 2005. Lifestyles of *Colletotrichum acutatum*. *Plant Disease*, 89, 784–796. https://doi.org/10.1094/PD-89-0784

Qiao, M., Li, J., Fang, L.L., Li, J.Y., Yu, Z.F., 2021. Morphology, Phylogeny and Pathogenicity of *Colletotrichum menglaense* sp. nov., Isolated from air in China. *Pathogens*, 10(10), 1243. https://doi.org/10.3390/pathogens10101243

Riolo, M., Aloi, F., Pane, A., Cara, M., Cacciola, S.O., 2021. Twig and shoot dieback of Citrus, a new disease caused by Colletotrichum Species. *Cells*, 10(2), 449. https://doi.org/10.3390/cells10020449.

Schaad, N.W., Frederick, R.D., 2002. Real-time PCR and its application for rapid plant disease diagnostics. *Canadian Journal of Plant Pathology*, 24(3), 250–258. https://doi.org/10.1080/07060660209507006

Schoch, C.L., Seifert, K.A., Huhndorf, S., Robert, V., Spouge, J.L. et al., 2012. Nuclear ribosomal internal transcribed spacer (ITS) region as a universal DNA barcode marker for Fungi. *Proceedings of the National Academy of Sciences, USA*, 109(16), 6241–6246.

Senanayake, I.C., Crous, P.W., Groenewald, J.Z., Maharachchikumbura, S.S., Jeewon, R. et al., 2017. Families of diaporthales based on morphological and phylogenetic evidence. *Studies in Mycology*, 86, 217–296. https://doi.org/10.1016/j.simyco.2017.07.003

Shahraki, A.H., Chaganti, S.R., Heath, D., 2019. Assessing high-throughput environmental DNA extraction methods for meta-barcode characterization of aquatic microbial communities. *Journal of Water and Health*, 17(1), 37–49.

Soares, V.F., Velho, A. C., Carachenski, A., Astolfi, P., Stadnik, M.J., 2021. First Report of *Colletotrichum karstii* causing Anthracnose on Strawberry in Brazil. *Plant Disease*, 105,10, 3295. https://doi.org/10.1094/PDIS-03-21-0518-PDN

Spatafora, J.W., Sung, G.H., Johnson, D., Hesse, C., O'Rourke, B. et al., 2006. A five-gene phylogeny of Pezizomycotina. *Mycologia*, 98, 1018–1028. https://doi.org/10.1080/15572536.2006.11832630

Spooner, D.M., 2009. DNA barcoding will frequently fail in complicated groups: an example in wild potatoes. *American Journal of Botany*, 96(6), 1177–1189.

Sreenivasa, M.Y., Dass, R.S., Charith Raj, A.P., Janardhana, G.R. 2008. PCR-based detection of genus *Fusarium* and Fumonisin-producing isolates from freshly harvested Sorghum grains grown in Karnataka, India. *Journal of Food Safety*. 28, 236–237.

Steenkamp, E.T., Wingfield, B.D., Coutinho, T.A., Wingfield, M.J., Marasas, W.F.O., 1999. Differentiation of *Fusarium subglutinans f. sp.* pini by histone gene sequence data. *Applied and Environmental Microbiology*, 65(8), 3401–3406. https://doi.org/10.1086/501484

Stockinger, H., Krüger, M., Schüßler, A., 2010. DNA barcoding of Arbuscular mycorrhizal fungi. *New Phytologist*, 187(2), 461–474. https://doi.org/10.1111/j.1469-8137.2010.03262.x

Yang, R.H., Su, J.H., Shang, J.J., Wu, Y.Y., Li, Y. et al., 2018. Evaluation of the ribosomal DNA internal transcribed spacer (ITS), specifically ITS1 and ITS2, for the analysis of fungal diversity by deep sequencing. *PLoS ONE*, 13(10), e0206428. https://doi.org/10.1371/journal.pone.0206428

Talhinhas, P., Baroncelli, R., 2021. Colletotrichum species and complexes: geographic distribution, host range and conservation status. *Fungal Diversity*, 111, 109–198. https://doi.org/10.1007/s13225-021-00491-9

Tehler, A., Farris, J.S., Lipscomb, D.L., Kallersjo, M., 2000. Phylogenetic analyses of fungi based on large rDNA data sets. *Mycologia*, 92, 459–474.

Thongkantha, S., Lumyong, S., McKenzie, E.H.C., Hyde, K.D., 2008. Fungal saprobes and pathogens occurrence on tissues of Dracaenaloureiri and Pandanus spp. *Fungal Diversity*, 30,149–179.

Thorsten-Lumbsch, H., Leavitt, S.D., 2011. Goodbye morphology? A paradigm shift in the delimitation of species in lichenized fungi. *Fungal Diversity*, 50(1), 59. https://doi.org/10.1007/s13225-011-0123-z

Udayanga, D., Liu, X, McKenzie E.H., Chukeatirote, E., Bahkali, A.H.A. et al., 2011. The genus Phomopsis: biology, applications, species concepts and names of common pathogens. *Fungal Diversity*, 50, 189–225. https://doi.org/10.1007/s13225-011-0126-9

Valentini, A., Pompanon, F., Taberlet, P., 2009. DNA barcoding for ecologists. *Trends in Ecology and Evolution*, 24(2), 110–117. https://doi.org/10.1016/j.tree.2008.09.011

Van Velzen, V.R., Weitschek, E., Felici, G., Bakker, F.T., 2012. DNA barcoding of recently diverged species: relative performance of matching methods. *PLoS One*, 7(1), e30490. https://doi.org/10.1371/journal.pone.0030490

Wang, X., Liu, C., Huang, L., Bengtsson-Palme, J., Chen, H. et al. 2015. ITS 1: a DNA barcode better than ITS 2 in eukaryotes? *Molecular Ecology Resources*, 15, 573–586. https://doi.org/10.1111/1755-0998.12325

Wang, Y.H., Ban, S., Wang, W.G., Li, Y., Wang, K., et al., 2021. *Pleurocordyceps* gen. nov. for a clade of fungi previously included in *Polycephalomyces* based on molecular phylogeny and morphology. *Journal of Systematics and Evolution*, 59(5): 1065–1080. https://doi.org/10.1111/jse.12705\

Weiss, M., Selosse, M.A., Rexer, K.H., Urban, A., Oberwinkler, F. 2004. Sebacinales: A hitherto overlooked cosm of heterobasidiomycetes with a broad mycorrhizal potential. *Mycological Research*, 108(9), 1003–1010. https://doi.org/10.1017/S0953756204000772

Whitworth, T.L., Dawson, R.D., Magalon, H., Baudry, E., 2007. DNA barcoding cannot reliably identify species of the blowfly genus *Protocalliphora* (Diptera: Calliphoridae). *Philosophical Transactions of the Royal Society B*, 274(1619), 1731–1739.

Zapata, M., Palfner, G., 2022. Colletotrichum (Ascomycota, Glomerellaceae) associated with native woody plants in natural ecosystems: What do we know? *Forest Pathology*, 52, e12748. https://doi.org/10.1111/efp.12748

Zhang, N., Castlebury, L.A., Miller, A.N., Huhndorf, S.M., Schoch, C.L., et al. 2006. An overview of the systematics of the Sordariomycetes based on a four-gene phylogeny. *Mycologia*, 98, 1076–1087. https://doi.org/10.3852/mycologia.98.6.1076\

Zheng, H., Yu, Z.F., Jiang, X.W., Fang, L., Qiao, M. 2022. Endophytic Colletotrichum Species from aquatic plants in southwest China. *Journal of Fungi*, 8, 87. https://doi.org/10.3390/jof8010087

9 Next-Generation Sequencing, a New Opportunity for Discovering Fungal Phytopathogens

M. Sarvajith
King Abdullah University of Science and Technology

Plant pathogenic fungi cause massive losses in quantity and quality of crop output. This is a serious global issue for the agriculture sector in terms of the economy. To ensure effective disease control, it is crucial to have accurate detection and identification methods for plant-infecting fungi. This helps to promote better disease management and prevent further spread of the infection. PCR-based diagnostics using traditional and variant assays, and hybridization techniques, are well-established. These techniques, however, necessitate prior knowledge of pathogen sequence. With advances in DNA/RNA-based Next-generation Sequencing (NGS), it is now possible to identify novel and emerging infections by sequencing fungal genomes across many platforms with no prior knowledge of the pathogen sequence. While it improves the coverage and resolution of the dynamic behavior of the gene. It can also provide insights into alterations at a single base location in nucleic acids, allowing for deeper analysis into emergent pathogen populations, genome biology, and the evolution of species. In this chapter, a detailed account of the evolution of sequencing technology and available platforms is discussed. Further a framework for using NGS to identify fungal phytopathogens, the benefits and limitations of the sequencing method for early, and real-time, identification is emphasized.

 DOI: 10.1201/9781003399704-9

GRAPHICAL ABSTRACT

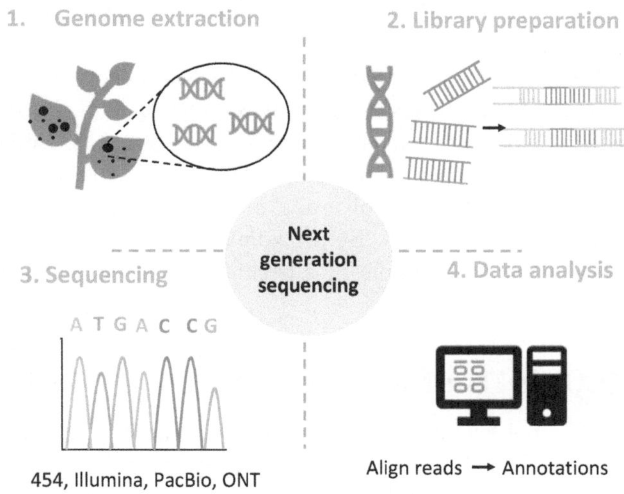

9.1 INTRODUCTION

Fungal pathogens have a significant impact on crop yields and represent a major challenge in agriculture. The effects of fungal diseases on crop production can be substantial and result in reduced yields, loss of quality, increased production costs, crop vulnerability, development of pathogen resistance, and food concerns among others. Efforts to mitigate the impact of fungal pathogens on crop yields include the development of resistant crop varieties, the use of integrated disease management strategies, early detection and monitoring, and research into sustainable and eco-friendly approaches to disease control. Effective disease management is essential to ensure food production remains sufficient to meet the growing global demand for agricultural products (Rani *et al.*, 2021; Sreenivasa *et al.*, 2011b; Beccari *et al.*, 2022; Deepa & Sreenivasa 2019b; Prasad *et al.*, 2010).

Traditionally, the phytopathogens were identified through culture-dependent methods such as isolation, and identification based on morphology and growth characteristics. While these methods were useful, they require significant time, human resource and expertise and often can be inconclusive. With the advent of techniques such as quantitative polymerase chain reaction (q-PCR); Loop-Mediated Isothermal Amplification (LAMP); micro, macro-array, and florescent silica nanoparticles combined antibodies, etc. – it is possible to identify pathogens much faster than traditional techniques such as those discussed in previous chapters. These techniques are often categorized as "targeted methods", with their major limitation being the ability to detect only a few targeted pathogens (Jongman *et al.*, 2020; Deepa & Sreenivasa 2019a; Sreenivasa *et al.*, 2011a, 2011b). Unlike targeted methods, sequencing-based approaches, that is, next-generation sequencing (NGS) is a powerful and comprehensive approach that allows the identification and characterization of nearly all fungal pathogens in a single analysis (Landegren *et al.*, 1988; Sreenivasa *et al.*, 2010; Diskin *et al.*, 2017). Therefore, NGS-based approach (metagenomics) is called a non-targeted method. It is highly suitable for detecting pathogens in plant tissue, seeds, and food products. It is a revolutionary technology for the field of plant pathology and pathogen surveillance by providing high-throughput and highly sensitive methods (Taylor *et al.*, 2014; Kazeeroni & Al-Sadi, 2016).

The high-throughput amplicon sequencing or simply high-throughput sequencing (HTS) methodologies can be traced back to their initial development for bacterial communities, primarily focused on the 16S region of the smaller ribosomal subunit. The hypervariable regions of 16S rRNA are highly conserved and provide a robust foundation for bacterial community analysis. Subsequent advancements extended the HTS to include diverse microbial communities, including fungi. Although HTS is a powerful tool, especially in surveillance, its usage is often overlooked and limited in the field of mycopathology. Unlike bacterial 16S, fungal DNA amplification focuses on the internal transcribed region (ITS) spacer (Schoch *et al.*, 2012). ITS is situated between 18S and 26S/28S of ribosomal RNA and is considered a universal DNA barcode for fungi. In general, ITS comprises two segments (ITS1 and ITS2) separated by a conserved 5.8S region. Most of the studies design primers strategically in the conserved region between 5.8S and 18S or 28S. This unique configuration facilitates a broader phylogenetic coverage (Dass *et al.*, 2007; Janardhana *et al.*, 2009; Bakker, 2018). This chapter first introduces historical account of improvements and available platforms in sequencing technology, now HTS. By providing a framework for using NGS to identify fungal phytopathogens with specific examples, the benefits and limitations of the sequencing platforms are discussed.

9.2 EARLY MOLECULAR TECHNIQUES (PRE-NGS AND NGS ERA)

Before the advent of NGS, amplification-based techniques such as PCR (polymerase chain reaction) (Punja and Rodriguez 2018), RFLP (restriction fragment length polymorphism) (Duttweiler *et al.*, 2008), immunological assays like ELISA (Enzyme-linked immunosorbent assay) (Musgrave 1984), as well as nucleic acid-based techniques such as RT-PCR (Reverse transcription-PCR) (Brown *et al.*, 2011; Deepa *et al.*, 2016a, 2016b, 2016c), were widely employed to detect and identify fungal phytopathogens. The former methods offer a fast and cost-effective means of detection, however, lacks sensitivity (Grünwald et al., 2012). In contrast, the latter methods, like RT-PCR and end-point PCR, provide the required sensitivity for pathogen surveillance applications but are constrained in their capacity to identify a broad spectrum of pathogens (Postnikova *et al.*, 2008; Sreenivasa *et al.*, 2008). Moreover, these techniques necessitate prior knowledge of the specific proteins or genomic sequences of the targeted pathogens, rendering the detection of uncharacterized plant pathogens challenging (Johnston-Monje and Lopez Mejia 2020; Deepa and Sreenivasa, 2017a; 2017b). In addition, these techniques are highly specific and can detect pathogens even when they are low in number often, these are used for identifying known fungal pathogens in plant tissues. Therefore, these methods are limited to specific targets and lack the capacity for comprehensive pathogen profiling.

9.3 EMERGENCE OF DNA SEQUENCING

The emergence of DNA sequencing technologies, particularly the Sanger sequencing technique, facilitated scientists to decode the genetic makeup of different organisms, including fungal phytopathogens (Sanger, Nicklen, and Coulson 1977). In the same year, another technology based on chemical degradation was also reported (Maxam and Gilbert 1977). Through sequencing targeted genes, such as the ITS region, it is now possible to achieve more precise identification of fungal species. While Sanger sequencing was historically the go-to method for DNA sequencing, it came with significant drawbacks such as being time-intensive, costly, and having limited throughput (Suzuki. 2020). This made it less feasible for large-scale sequencing of fungal genomes and detection of phytopathogens.

9.3.1 First-Generation Sequencing (Sanger Sequencing)

As the first-generation sequencing technology, Sanger sequencing emerged in the 1970s. It was instrumental in elucidating the genomes of various phytopathogenic fungi, providing a foundation for understanding their genetic makeup. Sanger sequencing involves enzymatic synthesis of target DNA with chain termination (Sanger, Nicklen, and Coulson 1977). The first automatic sequencers were developed by Applied Biosystems Instruments (ABI) (França et al., 2002). The sequencing process employs a dye-termination method, allowing rapid sequencing of a single reaction through distinct light wavelengths for each dideoxynucleotide triphosphate (ddNTP). Sanger sequencing was used to characterize *Puccinia emaculata*, rust fungi, using ITS. The authors developed diagnostic markers to evaluate the variability among different cultivars (Uppalapati et al., 2013). Other studies aimed to design and map the resistance using SNP (single nucleotide polymorphisms) in powdery mildew disease in wheat (Yao *et al.*, 2022) or identify endophytic fungi in *Livistona chinensis* (Guo *et al.*, 2000). Although Sanger sequencing has the advantage of obtaining large genome fragments with less data volume, making genome assembly more cost-effective than other sequencing technologies, it has notable limitations such as speed restrictions, potential cloning bias, volume limitations, and repetitive regions (Tables 9.1 and 9.2). These limitations can lead to incomplete genome sequencing, which affects the accuracy of the results (Tedersoo *et al.*, 2010).

9.3.2 Second-Generation Sequencing (NGS)

The introduction and advancements in sequencing technologies during the mid-2000s marked a major milestone. Platforms such as Illumina and 454 pyrosequencing were introduced to provide high-throughput sequencing, which enables the analysis of fungal communities in environmental samples, the detection of specific pathogens, and the assessment of genetic diversity within pathogen populations (Tedersoo *et al.*, 2010; Nicolaisen *et al.*, 2014). The current NGS platforms have emerged as a prevalent technology in modern sequencing, offering substantial advancements over traditional Sanger sequencing (Pop and Salzberg 2008; Magi *et al.*, 2010). NGS encompasses several platforms with reads (up to a million), and read length (20–1000 nt) (Tucker *et al.*, 2009) (Figure 9.1).

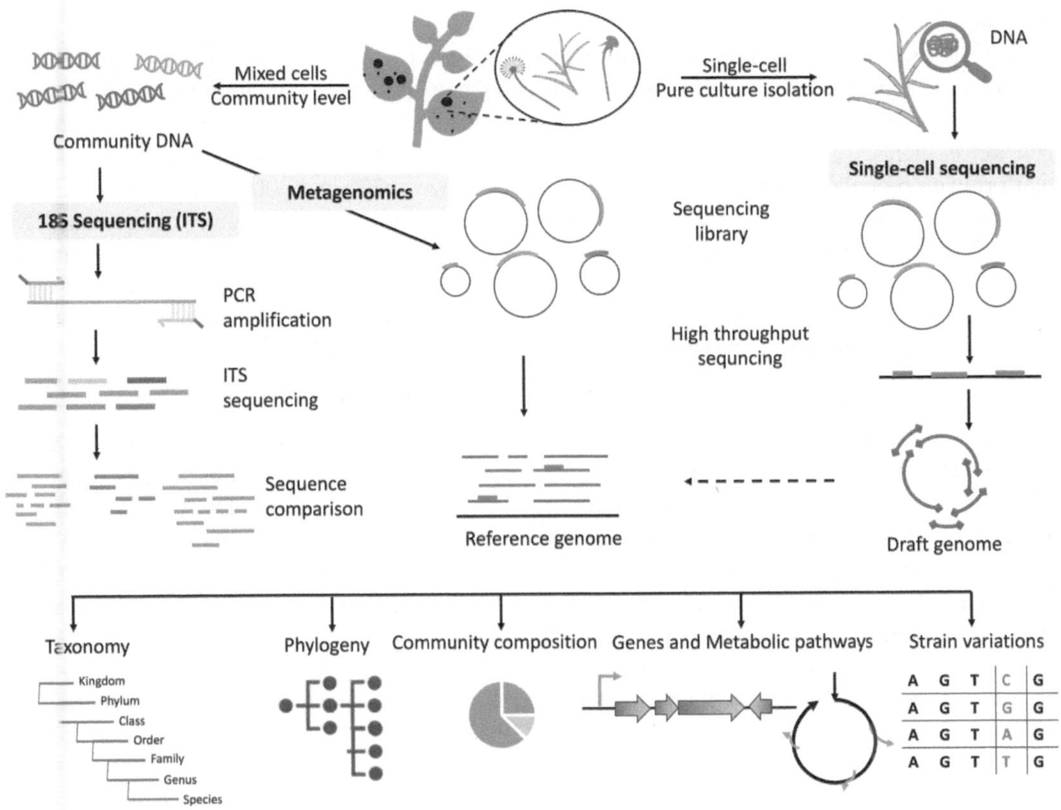

FIGURE 9.1 Schematic representation of workflow of next-generation sequencing technologies.

9.3.3 Examples of NGS Sequencing Platforms

9.3.3.1 454 Pyrosequencing

Roche Applied Science developed 454 pyrosequencing (Margulies *et al.*, 2005). In this method, the genomic DNA is randomly fragmented, and adapters are attached. The adapters containing DNA fragments are bound to beads before emulsion PCR. The emulsion PCR generates millions of DNA copies, which are denatured to ssDNA. The ssDNA is loaded into sequencing plate wells. During sequencing, a pyrophosphate is released as a result of the synthesis of new DNA, which is then detected. 454 platform generates long reads compared to Illumina and has an advantage in *de novo* sequencing and assembly of complex genomes. However, it is less commonly used today due to its low throughput and high cost per base. 454 can generate between 80 and 120 Mb in 4 h for 200–300 bp reads (Barba *et al.*, 2014).

Pyrosequencing platforms can capture a comprehensive range of fungal diversity, even in challenging environments. By employing 454 pyrosequencing platforms Taylor *et al.* (2014) identified over 200 fungal taxa in *Vitis vinifera*, surpassing the culture-dependent methods by 10-fold. By using the 26S rRNA gene instead of the commonly used ITS region, it was possible to overcome the issue of length polymorphism and obtain more accurate results. Similarly, Kazeeroni and Al-Sadi (2016) investigated the effect of fungicide on the fungal diversity in date palms and acid limes within the semi-oasis farms of the Arabian Peninsula. Their study revealed that the use of fungicides significantly impacted the diversity, and abundance of fungal species. Specifically, the decrease in fungal diversity led to the dominance of Ascomycota, and Microsporidia among others. The use of fungicide may have been effective in controlling the fungal disease, but it also led to unintended consequences on the overall fungal ecosystem within these agricultural systems. Thus, pyrosequencing is a powerful tool for assessing diversity and implementing interventions with ecological implications and framing practices.

9.3.3.2 Ion Torrent

It is a Thermo Fisher Scientific-developed platform based on ion-semiconductor sequencing. It employs a chip featuring an array of microwells, each housing a bead containing multiple identical fragments (emulsion PCR). As the new nucleotides are added during DNA polymerization, a hydrogen ion is released causing a measurable change in the pH (Merriman *et al.*, 2012). Ion torrent platform can produce 10 Gb throughput with lengths between 200 and 600 bp. Kemler *et al.*, 2013 obtained more than 2.3 million reads when sequencing the ITS1 nuclear-encoded ribosomal RNA from *Eucalyptus grandis*. While Leboldus *et al.* 2015 utilized to characterize genotypes of *Pyrenophora teres f. maculate* and *Sphaerulina musiva*. Despite its capacity to generate millions of sequences reads, it does not simplify the complexity of the fungal genome and underestimates its diversity due to unknown taxonomic affiliations.

9.3.3.3 Sequencing by Oligonucleotide Ligation and Detection (SOLiD)

ABI developed SOLiD in 2008, and it was commercialized by Life Technologies. It utilizes a distinctive DNA sequencing approach based on DNA ligase to identify mismatches during sequencing (Ho *et al.*, 2011). The process involves emulsion PCR by adding adapters to DNA. Followed by an 8-mer (having fluorescent labeling at the one end) is added through sequential ligation of DNA fragments that emit color. Each binding is associated with a color from a fluorescent molecule. Although only four fluorophores are used for 16 combinations (2+2 nucleotide), the color-space technique enables the identification of the position of each nucleotide. In this platform, each DNA fragment hybridizes with one of the 16 fluorophore-labeled probes having two known bases at positions n and n+1. Subsequently, the probe is cleaved, releasing a fluorophore. This leaves a free 5' phosphate end with two known bases followed by degenerate bases. In the subsequent extension, 10 rounds of hybridization, ligation, and cleavage continue. After which, the cycle is again initiated but at the position n+2. SOLiD platform achieves maximum accuracy (>99.99%) as each nucleotide is sequenced twice. Consequently, the technology minimizes the risk of misinterpretation by relying on two consecutive colors for each nucleotide (Voelkerding *et al.*, 2009).

9.3.3.4 Solexa/Illumina Sequencing

It is an advanced and widely used high-throughput DNA sequencing platform. This method is based on the amplification of the DNA on a solid surface with the stepwise addition of fluorescently labeled nucleotides. Each incorporation event is captured and recorded, producing massive amounts of short reads. Illumina platforms are known for their accuracy and cost-effectiveness, making them suitable for a broad range of applications, including whole-genome sequencing, RNA-Seq, and targeted sequencing. Illumina MiSeq platform is widely used to investigate the microbial community changes of bacteria using V4 of 16S and ITS1 loci genes associated with stem-end rot and postharvest diseases in mangoes (Diskin *et al.*, 2017). The authors identified changes in microbial communities at different stages of harvest, cold storage, and throughout shelf life. For instance, Sporidiobolaceae dominated at harvest but transiently decreased due to cold storage and disappeared entirely after shelf-life storage. Conversely, bacteria belonging to Chitinophagaceae, undetected at harvest was the second-most abundant bacteria during shelf-life storage. The shift was corroborated by the increased abundance of pathogenic fungi and stem-end rot pathogens (Diskin *et al.*, 2017). Similarly, Abdelfattah *et al.* (2016) used the MiSeq platform to investigate the spatial and compositional variations in fungal communities of apples grown organic and conventional method. Ascomycota and Basidiomycota were observed to be the predominant fungi respectively with unidentified fungi. This incomplete characterization indicates that the fungal biomes are not fully characterized, possibly due to limited genetic variations within the analyzed gene regions. Nevertheless, Illumina MiSeq was successfully utilized in deciphering complex relationships and compositional variations within microbial populations offering new insights for disease management.

9.3.4 Third-Generation Sequencing (TGS)

Recent developments in NGS, particularly with the emergence of Pacific Biosciences (PacBio) and Oxford Nanopore Technologies (ONT), commonly referred to as TGS, have enhanced the detection and characterization genomes. TGS offers long reads, real-time sequencing, and the ability to directly sequence, including RNA. PacBio utilizes SMRT (single molecule real-time) technology, which eliminates the requirement of PCR for library preparation (Eid *et al.*, 2009). It utilizes closed and circular single-stranded DNA template that undergoes automatic replication. DNA molecules are bound to specialized DNA mode waveguides (ZMWs). ZMWs are highly sensitive. It captures the lowest amount of light emitted as a result of the addition of a single fluorescently labeled nucleotide. During sequencing, the light emitted by fluorescently labeled nucleotides is recorded in real time (each nucleotide has different fluorophores). The SMRT by PacBio is based on sequencing by synthesis. In contrast, the ONT involves the insertion of nanopores into an electrically resistant membrane. By applying a potential across the membrane, current flow is generated exclusively to the nanopore. Any disruption in this current during the passage of nucleotides through its membrane indicates a specific single molecule. The DNA library preparation for ONT involves a designed hairpin structure to ligate double DNA strands, enabling the reading of both strands continuously. ONT's technology yields ultra-long reads (ULRs), with some exceeding 300 kb and approaching 1 million bp. Additionally, ONT's pocket-sized sequencers are portable without requiring an elaborate laboratory setup, facilitating cost-effective transportation for field applications. Notably, PacBio and ONT's TGS features enable real-time long-read sequencing with minimal alignment and mapping errors during library construction (Table 9.1).

By utilizing the ONT platform, several plant pathogens mostly bacteria were successfully identified (Chalupowicz *et al.*, 2019). However, only a few fungal taxa could be identified, and most of the tomato and cucumber leaf spots tested for phytopathogens remained unknown. The major limitations in identifying the phytopathogens were due to poor DNA quality, diluted pathogen DNA, and contamination of endophytic, epiphytic, and host genomes (Chalupowicz *et al.*, 2019). Moreover, the RefSeq database (https://www.ncbi.nlm.nih.gov/refseq/targetedloci/) has a notable disparity in genome representation. For instance, only 251 fungal genomes (complete and draft) are recorded

TABLE 9.1
Comparison of Different Sequencing Platforms

	First generation	Second generation	Third generation
Sequencing approach	Chain termination end-labelled DNA fragments	Sequencing by synthesis (Illumina), sequencing by ligation (Roche 454)	Single-molecule real-time (SMRT)
Read length	1000 bp	50–300 bp	>1000 bp
Throughput	Low	High	High
Resolution	Averaged across many copies of DNA molecule	Averaged across many copies of DNA molecule	Single DNA molecule
Error rates	low	High	High
Cost	High cost per base Low cost per run	Low cost per base high cost per run	Low cost per base high cost per run
Time	Hours	Days	Day
Platforms	Sanger sequencing	Roche 454, Illumina, Ion torrent, SOLiD	Oxford Nanopore Technology (ONT), PacBio, Helioscope
Applications	Genome sequencing, specific gene sequence	Whole-genome sequencing, RNA sequencing, DNA methylation analysis, population and functional genomics, development of markers, metagenomics	De novo genome assembly,structural variants, genome architecture, gene diversity, metagenomics

against 7509 viruses and 37,399 bacterial genomes. The representation of the fungal genome is lower than the Archaeal at 533 genomes (Lu and Salzberg 2018). Much of the fungal genome coverage lacks association with plant pathogens highlighting the need for further study in fungal sequence-based approaches in plant pathology. Nevertheless, the characteristic features of TGS are valuable for studying *de novo* genome assembly, structural variant detection, and understanding complex genomic regions of fungal pathogens.

A study by Loit *et al.* (2019) compared the performance of PacBio and ONT platforms for the identification of fungal pathogens and oomycetes in conifer needles, potato leaves, and tubers. The study used ITS4 in combination with ITS1F and ITS100. Both PacBio and ONT platforms could handle >7000 bases amplicons. The PacBio (Sequel) demonstrated consistent results with morphology-based and species-specific PCR-based technology. However, it could not distinguish closely related species (e.g. *D. pini* and *D. septosporum*). In contrast, ONT (MinION) showed higher chimeric sequences, that were not identified by UCHIME software, requiring manual screening. Both of these platforms presented similar results but had drawbacks (Table 9.2). The authors concluded Sequel to better suited for metabarcoding up to long sequences (4.5 kb), while MinION to better suited for rapid identification of dominant species, mainly due to higher error rates (1–3%) and biases associated with longer reads (Loit et al., 2019).

9.3.5 Advantages of TGS

TGS methods represent the latest advances in DNA sequencing technology, offering several advantages over earlier generations. TGS methods have overcome many limitations of traditional Sanger sequencing and second-generation sequencing (NGS) technologies.

I. *Long reads*: TGS generates markedly longer reads than NGS. Long reads are especially advantageous when studying fungal genomes, which can be complex and contain repetitive elements. For example, De Miccolis Angelini *et al.* (2019) obtained about 44 Mb length using Illumina and PacBio platforms. About 13,000 proteins were annotated in *Monilinia*

TABLE 9.2
Comprehensive Details on Advantages and Drawbacks of Sequencing Platforms

	Platform	Principle	Advantages	Drawbacks
Illumina	MiSeq	Sequencing by synthesis	Short run time High coverage Low cost	Short read length
	HiSeq		High throughput Low cost	Short read length Long run time
Roche	454 GS FLX+	Pyrosequencing	High throughput High coverage Long reads	High error rate in homopolymeric regions High cost
Helicos	Heliscope	Single molecule sequencing	No template bias	Short read length Expensive instrument
ABI Life technologies	5500 SOLiD	Sequencing by ligation	High throughput Low reagent cost	Long run time Short reads
	PGM	Proton detection	Short run time Low cost	High reagent cost High error rate
PacBio	PacBio	SMRT	Short run time Very long reads Low read cost	High error rate
ONT	MinION	Nanopore	Longest read length Portable and low cost	High cost per Mb No standard protocol

Source: (adapted from Tyagi and Bhide 2020)

fructicola using >12,000 assembled genes. Similarly, Loit *et al.* (2019) obtained 4.5 kb using MinIOn as detailed above.

II. *Reduced assembly challenges*: The long reads generated by TGS methods simplify the genome assembly process, and can more easily bridge repetitive regions and gaps in the genome. This is crucial for accurately characterizing fungal genomes, which often contain repetitive DNA elements, and transposal elements (Rao *et al.*, 2018).

III. *Characterization of epigenetic modifications*: TGS technologies can provide information about epigenetic modifications, such as DNA methylation, in fungal phytopathogens. Epigenetic regulation plays a significant role in the pathogenicity and virulence of these pathogens, and TGS can shed light on these mechanisms.

IV. *Real-time sequencing*: ONT sequencing, in particular, offers the unique feature of real-time sequencing where sequencing can be monitored as it happens, allowing for rapid detection and identification of fungal phytopathogens in the field or even in a clinical setting.

V. *Rapid turnaround*: TGS platforms like ONT are known for their rapid turnaround times, which can be essential in situations where quick identification of phytopathogens is critical for disease management.

VI. *Reduced DNA preparation*: TGS methods often require less complex DNA preparation compared to NGS, reducing the chances of introducing bias or artifacts during library preparation. This can be especially advantageous when working with limited or degraded DNA samples.

VII. *Cost-effective*: While TGS technologies were initially more expensive than NGS, the costs have been decreasing steadily. The advantages of long reads and reduced assembly challenges outweigh the higher per-base sequencing costs.

VIII. *Direct RNA sequencing*: TGS methods can sequence RNA molecules directly without the need for reverse transcription and amplification, enabling the study of fungal gene expression and transcriptomics with high accuracy.

IX. *Application in pathogen evolution*: TGS can provide insights into the evolution and adaptation of fungal phytopathogens, as it can accurately characterize structural and genetic variations. For example, Fournier *et al.* (2010) identified 15 distinct single nucleotide polymorphisms (SNP profiles) in *Pandora neoaphidis*.
X. *Practical field applications*: The real-time sequencing capabilities of TGS platforms make them suitable for on-site diagnostics and monitoring of fungal phytopathogens, which can be crucial for early detection and response in agricultural settings.

9.3.6 Limitations of TGS

While TGS technologies have numerous advantages, there are also some challenges and disadvantages associated with their use for fungal phytopathogen detection:

I. *Error rates*: TGS in its earlier iterations were prone to higher error rates compared to NGS sequencing platforms. However, there have been several advancements to improve accuracy, and the error rates can still be a concern, particularly in regions with complex genomic structures or in the presence of homopolymers.
II. *Cost*: TGS is still more expensive than NGS, which may hinder its adoption in large-scale projects.
III. *Sample requirement*: TGS technologies often require high-quality and high-molecular-weight DNA, which can be challenging to obtain from certain sample types, especially when dealing with environmental samples or degraded DNA.
IV. *Data analysis*: TGS data can be computationally intensive and may require specialized bioinformatics expertise. The long reads generated by TGS technologies may present challenges in terms of alignment, assembly, and downstream analysis, requiring robust bioinformatics pipelines.

9.3.7 Sequencing Approach

9.3.7.1 Metagenome Sequencing

Metagenomics involves the direct sequencing of genetic material extracted from environmental samples. Metagenomics provides a holistic view of the microbial community of an ecosystem, enabling the identification and characterization of fungal species associated with plants. Metagenomic sequencing circumvents the need for pathogen-specific probes or primers, requiring minimal prior knowledge about the potential identity of the pathogen. Metagenomics offers several advantages for broad-spectrum analysis of complex microbial communities, but it also has its drawbacks. In traditional sequencing methods, the thorough characterization of individual organisms was achievable. In contrast, metagenomics typically generates short DNA fragments, making it challenging to characterize a single microorganism. In the metagenome of multiple or mixed organisms, genomic information often remains undiscovered, rendering the assembly of an entire genome for a specific microbe impossible. Despite these limitations, metagenomics provides a unique opportunity to obtain genome information of all organisms, whether they are known or unknown/unidentified organisms (Nagaraja *et al.*, 2016).

Metagenomic approaches employing NGS have already found application in clinical studies. It is now gaining popularity in plant disease diagnosis (Gu *et al.*, 2019; Piombo *et al.*, 2021). For faster diagnostics, the MinION nanopore sequencer, SMRT platform, emerges as particularly promising. Distinguishing itself from other NGS platforms, the MinION offers longer reads, enhancing genome assembly and detection precision. Its unique advantage lies in providing initial results within minutes of initiating a sequencing run, and its portability enables its use in diverse settings, including remote locations and even space. In the context of plant disease diagnostics, the MinION's metagenomic sequencing capabilities, utilizing DNA or RNA directly extracted from plants, facilitate swift pathogen detection and identification in virtually any laboratory setting, including field conditions (Chalupowicz *et al.*, 2019). Despite these capabilities, the MinION has predominantly been applied for identifying plant pathogenic viruses (Bronzato et al., 2018) and bacteria (Mechan et al., 2020). Limited studies have explored the use of MinION (ONT) (Hu *et al.*, 2019; Loit *et al.*, 2019). This is mainly due to the challenges arising from the inadequate information available in databases and the technical complexities associated with extracting fungal DNA from plant tissues. Nevertheless, it is a powerful tool undergoing development. Liacsa *et al.* (2019) using metagenomics identified 546 and 154 bacterial taxa and 147 and 122 fungal taxa, respectively, in the soil of *Tabebuia chrysantha* and *T. billbergii*. Eukaria was the most prominent domain with Glomeromycota and Ascomycota covering 95% total fungal genome in *T. chrysantha* while *T. billbergii* has Ascomycota covering 98% total genome.

9.3.7.2 Whole-Genome Sequencing (WGS)

WGS involves sequencing the entire genome of specific fungi, providing a detailed and comprehensive analysis of its genetic contents. WGS is used to explore the genomic diversity within a particular fungal species, identify virulence factors, study genetic variations associated with fungicide resistance, and understand the overall genomic architecture of individual pathogens. The main advantages of WGS are the in-depth genetic details of a specific fungal pathogen and high-resolution genomic data for detailed comparative analyses. Besides its advantages, the need for pure cultures is challenging for some fungi, and the time and resources required for extensive genomic analysis. Moreover, large datasets generated from WGS demand robust computational infrastructure for analysis and interpretation. For example, with the WGS of *Mycosphaerella graminicola*, a causal fungal agent of Septoria tritici blotch in wheat, it was possible to understand the genetic basis of fungicide resistance, identify the genes associated with virulence factors, and even evolutionary dynamics, genome changes associated with different strains (Goodwin *et al.*, 2011; Stukenbrock et al., 2010). Similarly, several studies have utilized WGS to study genomic and transcriptomic features of *Magnaporthe oryzae*, the causative agent of rice blast disease (Zhang *et al.*, 2018); *Colletotrichum* species responsible for anthracnose diseases (O'Connell *et al.*, 2012; Gan *et al.*, 2013); *Botrytis cinerea* and *Fusarium graminearum*, causative agent of gray mold and wheat head blight, respectively (Amselem *et al.*, 2011; Caseys *et al.*, 2021; Alouane *et al.*, 2021), etc. While studying boxwood, WGS identified various bacterial, fungal, or oomycete boxwood pathogens, including *Calonectria pseudonaviculata* and *Calonectria henricotiae* (Yang *et al.*, 2022).

9.3.7.3 Hybrid Sequencing

Hybrid sequencing combines data from multiple sequencing technologies to leverage the strengths of each platform. By utilizing short reads from Illumina sequencing and long reads from PacBio or ONT, a hybrid approach can offer a more comprehensive assembly of the fungal genome. Short reads provide high accuracy and are useful for genome polishing and error correction, while long reads resolve repetitive regions, structural variations, etc., enabling the reconstruction of more complete and continuous genomics. This approach is particularly useful for *de novo* genome assembly. In addition to this, it can mitigate the challenges associated with incomplete reference databases and improve the precision of species identification. For the first time, *Berkeleyomyces rouxiae*, a black root rot fungus that infects *Cannabis sativa* was sequenced using hybrid genome assembly (Dumigan *et al.*, 2023). About 28.2 Mb genome size containing more than 400 contigs was obtained. After genome annotation, ~ 7000 protein-coding genes and >59,000 functional annotations were described suggesting the successful integration of two techniques for sequence-based diagnostic development (Dumigan *et al.*, 2023).

9.4 GENERAL STAGES OF NGS

9.4.1 Template Preparation

Preparing genomic DNA for use in mate-pair and fragment templates usually involves DNA fragmentation (Metzker, 2005). DNA fragments are generated via physical shearing and enzymatic shearing. In NGS, DNA templates are immobilized on solid surface support. Often, immobilization may be one of three ways: (a) immobilization by covalent attachment of primer molecules, that is, a template created by random fragmentation and addition of common adaptors, and then hybridized by immobilized primers. (b) Through single-strand priming and extension, single molecule templates are covalently bound to the matrix, and (c) attachment of primed templates to DNA polymerase-bound solid matrix (Metzker, 2005). This spatial immobilization of template sites facilitates massive parallel reactions.

9.4.2 Sequencing

The sequencing method varies depending on the platform. For instance, Illumina uses sequencing-by-synthesis, a four-color Cyclic Reversible Termination (CRT) cycle, whereas Helicos BioSciences employs one-color CRT cycles. CRT is a sequencing technology that involves a cyclic process consisting of nucleotide incorporation, fluorescence imaging, and cleavage (Metzker 2005). CRT employs reversible terminators, which are modified nucleotides that can be added to the growing DNA chain during sequencing. The "reversible" aspect allows for the removal of the terminator group after imaging, enabling the next cycle of nucleotide addition. The process occurs in cycles where a single nucleotide is added to the growing DNA chain, followed by fluorescence imaging for identification. One characteristic feature is the way DNA synthesis is controlled to terminate after each addition of nucleotides, ensuring sequence accuracy.

In the case of sequencing by ligation, a probe labeled with a fluorescent dye selectively hybridizes the complementary sequence next to the primed template (Landegren *et al.*, 1988). Subsequently, the fluorescent dye-labeled probe is ligated to the primer by the DNA ligase enzyme. Unbound probes are then removed through washing, and fluorescence imaging is conducted to ascertain the ligated probe. This cycle can be repeated in two ways: by removing the fluorescent dye and regenerating the 5' phosphate group for ligation or by the addition of a new primer. The key feature of SOLiD sequencing is its use of a color code, or "color space", to represent DNA sequences (Valouev *et al.*, 2008). Each base is represented by a pair of colors, and the sequence is determined by the series of color changes during the ligation steps. The ligation and detection steps are repeated through multiple cycles to determine the sequence of the DNA fragment.

Pyrosequencing is a bioluminescent technique where the release of phosphate molecules through enzymatic reaction is measured indirectly as light. Unlike other methods, pyrosequencing manipulates DNA polymerase through single, limited dNTP additions. To construct the DNA sequence, the sequence and light intensity are recorded as flowgrams (Ronaghi *et al.*, 1998).

9.4.3 Genome Assembly

Generally, the genome is aligned with the reference sequence or *de novo*. *de novo* assembly utilizes algorithms to construct contigs or scaffolds. These contigs are a set of DNA sequences that, when overlapped, provide a continuous sequence of fungal genomes. However, this approach may lead to gaps and ambiguities in repetitive sequence regions. This can be overcome by additional sequencing or using other complementary technologies (Pop and Salzberg 2008; Reinhardt *et al.*, 2009).

9.4.4 Annotations

Annotating a genome includes recognizing and labeling different functional components of genomes like genes, regulatory regions, and other features. For fungal phytopathogen genomes, annotation plays a crucial role in understanding the genetic basis of pathogenicity and host interactions. For example, in the case of *Magnaporthe oryzae*, the rice blast disease pathogen, the gene AVR-PITA was identified for its role in pathogenicity (Khang *et al.*, 2008). Further, gene clusters that produce secondary metabolites, such as those involved in the production of mycotoxins, non-coding

RNA, repeat elements, etc., can also be assigned. The assignment of ecological functions to fungal operational taxonomic units (OTUs) is performed utilizing FUNGuild (a python-based tool). The UNITE database (https://unite.ut.ee/) uses the Species Hypothesis (SH) to identify OTUs (Nilsson *et al.*, 2014). All the genome data is submitted in NCBI (National Centre for Biotechnology Information) database and is available freely.

9.4.5 Validation

It is a critical step in ensuring the accuracy and reliability of a genome assembly derived from NGS data. Various methods can be employed to validate the assembly, including experimental and computational approaches. Some of the methods include (a) read mapping and coverage analysis: uniform and expected coverage support the reliability of the assembly. Aberrant coverage patterns may indicate potential misassemblies or regions of low confidence. (b) orthologous gene comparison: presence and conservation of orthologous known genes in the assembled genome. (c) comparative genomics: consistency in gene content, synteny, and overall genomic organization with related species. (d) functional validation: experimental confirmation of the predicted function to experimental evidence. By combining multiple validation approaches, genome assembly can be strengthened.

9.5 APPLICATIONS OF NGS

I. Detection of fungal phytopathogens: Fungal phytopathogens can be detected based on sequence data that matches known fungal genomes or by identifying specific fungal genes associated with pathogenicity or virulence. Sequence reads that correspond to fungal pathogens are extracted from the metagenomic data.
II. Quantification and relative abundance: The relative abundance of fungal phytopathogens in the community to assess the severity of the infection.
III. Identification of novel pathogens: The presence of novel or less-studied fungal species or strains, potentially leading to the discovery of previously unknown phytopathogens.
IV. Functional genomics: The transcriptomic and genomic studies can be used to understand the genes and mechanisms involved in fungal pathogenicity, host-pathogen interactions, and the development of resistance in plants.
V. Diagnostics and surveillance: NGS can be used for diagnostic purposes for rapid and accurate identification. Specifically, RNA-Seq-based NGS can be used as a rapid detection method. It is also a valuable tool for surveillance programs to monitor the spread of fungal diseases over time (Chalupowicz et al., 2019).
VI. Quarantine measures: Strict quarantine measures can be imposed to avoid the spread of disease of exotic fungal pathogens in regions where they are not already established.
VII. Designing biological control agents: Biological control methods involve the use of beneficial microorganisms, such as mycoparasites and antagonistic bacteria, to suppress fungal pathogens. Biopesticides and biofungicides containing these beneficial organisms can be used as an eco-friendly alternative to chemical control.
VIII. Development of molecular diagnostics: The sequences provide information to design primers required for other molecular diagnostics techniques such as PCR.

9.6 SUMMARY

NGS and TGS technologies have emerged as powerful tools for the detection and characterization of fungal phytopathogens. NGS, with its high-throughput capabilities, allows analysis of multiple samples simultaneously, providing a comprehensive view of fungal communities in various ecosystems. It has been instrumental in identifying putative plant pathogenic fungi and understanding their ecological functions. TGS technologies like PacBio and ONT offer long-read capabilities, contributing to more accurate genome assemblies and aiding in the identification of virulence factors. These advancements in sequencing technologies have facilitated the development of metagenomic approaches for detecting fungal microbial communities from environmental samples. The application of these technologies not only enhances our understanding of fungal phytopathogens but also opens avenues for real-time and field-based detection. Future studies must include refining sensitivity and resolution, integrating with other omics data, and leveraging global surveillance networks for monitoring pathogen spread. Overall, NGS and TGS play a pivotal role in advancing our knowledge of fungal phytopathogens and hold great promise for improving disease management strategies in agriculture.

9.7 FUTURE PERSPECTIVE

The use of NGS technology for the detection of fungal phytopathogens is in the very early phase. Primarily due to a lack of fungal database. Unlike well-studied bacteria and viruses, fungal genomics still lacks research. The ongoing efforts focus on improving the sensitivity and resolution of sequencing technologies. In particular, the emphasis is on refining library preparation methods and optimizing sequencing chemistry. TGS technologies, such as PacBio and ONT, offer long-read capabilities, enabling more comprehensive insights into the genomic architecture of fungal phytopathogens. Efforts are made for the integration of metagenomic approaches to detect complex fungal communities from the environment. Additionally, the development of portable, real-time sequencing devices for on-site detection in the field holds promise for rapid and early disease diagnosis in agricultural settings. The use of genomics with other omics data, machine learning applications, and functional genomics approaches would provide a holistic understanding of pathogenicity mechanisms. This will greatly assist in informed targeted disease management. Collaborative global surveillance networks and sharing genomic data help monitor pathogens and outbreaks (if any).

STUDY QUESTIONS

1. How does the application of NGS technology impact the accuracy and resolution of fungal phytopathogen detection compared to traditional methods?
2. What novel insights into the diversity and dynamics of fungal phytopathogen populations can be gained through metagenomic approaches?
3. What are the long-read technologies? How do they assist in disease management?
4. Write the general methodology of the ONT-sequencing platform.
5. Can these technologies be successfully employed for on-site monitoring and early diagnosis of plant diseases, and what are the practical challenges associated with field applications?

FURTHER READING

Goodwin, S., McPherson, J.D. and McCombie, W.R., 2016. Coming of age: ten years of next-generation sequencing technologies. *Nature Reviews Genetics*, 17(6), 333–351.

Lindahl, B.D., Nilsson, R.H., Tedersoo, L., Abarenkov, K., Carlsen, T., Kjøller, R., Kõljalg, U., Pennanen, T., Rosendahl, S., Stenlid, J. and Kauserud, H., 2013. Fungal community analysis by high-throughput sequencing of amplified markers–a user's guide. *New Phytologist*, 199(1), 288–299.

Nilsson, R.H., Tedersoo, L., Ryberg, M., Kristiansson, E., Hartmann, M., Unterseher, M., Porter, T.M., Bengtsson-Palme, J., Walker, D.M., De Sousa, F. and Gamper, H.A., 2015. A comprehensive, automatically updated fungal ITS sequence dataset for reference-based chimera control in environmental sequencing efforts. *Microbes and Environments*, 30(2), 145–150.

Yang, S., Johnson, M.A., Hansen, M.A., Bush, E., Li, S. and Vinatzer, B.A., 2022. Metagenomic sequencing for detection and identification of the boxwood blight pathogen *Calonectria pseudonaviculata*. *Scientific Reports*, 12(1), 1399.

REFERENCES

Abdelfattah, A., Wisniewski, M., Li Destri Nicosia, M.G., Cacciola, S.O., Schena, L., 2016. Metagenomic analysis of fungal diversity on strawberry plants and the effect of management practices on the fungal community structure of aerial organs. *PloS One*, 11(8), 0160470.

Alouane, T., Rimbert, H., Bormann, J., González-Montiel, G.A., Loesgen, S., Schäfer, W., Freitag, M., Langin, T., Bonhomme, L., 2021. Comparative genomics of eight Fusarium graminearum strains with contrasting aggressiveness reveals an expanded open pangenome and extended effector content signatures. *International Journal of Molecular Sciences*, 22(12), 6257.

Amselem, J., Cuomo, C.A., van Kan, J.A., Viaud, M., Benito, E.P., Couloux, A., Coutinho, P.M., de Vries, R.P., Dyer, P.S., Fillinger, S., Fournier, E., 2011. Genomic analysis of the necrotrophic fungal pathogens *Sclerotinia sclerotiorum* and *Botrytis cinerea*. *PLoS genetics*, 7(8), 1002230.

Bakker M.G., 2018. A fungal mock community control for amplicon sequencing experiments. *Molecular Ecology Resources*, 18(3), 541–556.

Barba, M., Czosnek, H., Hadidi, A., 2014. Historical perspective, development and applications of next-generation sequencing in plant virology. *Viruses*, 6(1), 106–136.

Beccari, G., Hao, G., Liu, H., 2022. Fusarium pathogenesis: Infection mechanisms and disease progression in host plants. *Frontiers in Plant Science*, 13, 1020404.

Bronzato, B. A., Sherman, D., Stone, A., Gopakumar, A., Wilson, V., Schneider, W., King, J., 2018. Nanopore sequencing as a surveillance tool for plant pathogens in plant and insect tissues. *Plant Disease*, 102(8), 1648–1652.

Brown, N.A., Bass, C., Baldwin, T.K., Chen, H., Massot, F., Carion, P.W., Urban, M., Van De Meene, A.M., Hammond-Kosack, K.E., 2011. Characterisation of the Fusarium graminearum-wheat floral interaction. *Journal of Pathogens*, 2011, 626345.

Caseys C., Shi, G., Soltis, N., Gwinner, R., Corwin, J., Atwell, S., Kliebenstein, D.J., 2021. Quantitative interactions: the disease outcome of *Botrytis cinerea* across the plant kingdom. *G3*, 11(8), jkab175.

Chalupowicz, L., Dombrovsky, A., Gaba, V., Luria, N., Reuven, M., Beerman, A., Lachman, O., Dror, O., Nissan, G., Manulis-Sasson, S., 2019. Diagnosis of plant diseases using the Nanopore sequencing platform. *Plant Pathology*, 68(2), 229–238.

Dass, R.S., Sreenivasa M.Y., Janardhana G.R. 2007. High incidence of *Fusarium verticillioides* in animal and poultry feed mixtures produced in Karnataka, India. *Plant Pathology Journal*, 6(2), 174– 178.

De Miccolis Angelini, R.M., Romanazzi, G., Pollastro, S., Rotolo, C., Faretra, F., Landi, L., 2019. New high-quality draft genome of the brown rot fungal pathogen *Monilinia fructicola*. *Genome Biology and Evolution*, 11(10), 2850–2855.

Deepa N, Charith Raj A P and Sreenivasa M Y. 2016b. Nested PCR method for the early detection of fumonisin producing *Fusarium verticillioides* in pure cultures, cereal samples and plant parts. *Food Biotechnology*. 30(1), 18–29.

Deepa N, Charith Raj A P, and Sreenivasa M Y. 2016c. Multiplex PCR method for the early detection of fumonisin producing *Fusarium verticillioides*. *Food Bioscience*. 13, 84–88.

Deepa N., Nagaraja H., Sreenivasa M. Y. 2016a. Prevalence of fumonisin producing Fusarium verticillioides associated with cereals grown in Karnataka (India). *Food Science Human Wellness*, 5, 156–162. https://doi.org/10.1016/j.fshw.2016.07.001.

Deepa N., Sreenivasa M. Y. 2017b. *Fusarium verticillioides*, a Globally Important Pathogen of Agriculture and Livestock: A Review. *Journal of Veterinary Medicine and Research* 4(4), 1084.

Deepa N, Sreenivasa M Y. 2017a. Fumonisins: A Review on its Global Occurrence, Epidemiology, Toxicity and Detection. *Journal of Veterinary Medicine and Research* 4(6), 1093.

Deepa N, Sreenivasa M Y. 2019a. Molecular methods and key genes targeted for the detection of fumonisin producing *Fusarium verticillioides* – An updated review. *Food Bioscience*. 32, 100473.

Deepa N, M.Y. Sreenivasa. 2019b. Biocontrol Strategies for Effective Management of Phytopathogenic Fungi Associated with Cereals, In: Singh JS., Singh DP. (eds) *New and Future Developments in Microbial Biotechnology and Bioengineering*, Elsevier, pp, 177–189, ISBN 9780444641915.

Diskin, S., Feygenberg, O., Maurer, D., Droby, S., Prusky, D., Alkan, N., 2017. Microbiome alterations are correlated with occurrence of postharvest stem-end rot in mango fruit. *Phytobiomes*, 1(3), 117–127.

Dumigan, C.R., Maddock, S., Bray-Stone, D., Deyholos, M.K., 2023. Hybrid Genome Assembly of *Berkeleyomyces rouxiae*, an emerging cannabis fungal pathogen causing black root rot in an aeroponic facility. *Plant Disease*, 107(9), 2679–2686.

Duttweiler, K.B., Sun, G.Y., Batzer, J.C., Harrington, T.C., Gleason, M.L., 2008. An RFLP-based technique for identifying fungi in the sooty blotch and flyspeck complex on apple. *Plant Disease*, 92(5), 794–799.

Eid, J., Fehr, A., Gray, J., Luong, K., Lyle, J., Otto, G., Peluso, P., Rank, D., Baybayan, P., Bettman, B., Bibillo, A., 2009. Real-time DNA sequencing from single polymerase molecules. *Science*, 323(5910), 133–138.

Fournier, A., Widmer, F., Enkerli, J., 2010. Assessing winter-survival of *Pandora neoaphidis* in soil with bioassays and molecular approaches. *Biological Control*, 54(2), 126–134.

França, L.T., Carrilho, E., Kist, T.B., 2002. A review of DNA sequencing techniques. *Quarterly Reviews of Biophysics*, 35(2), 169–200.

Gan, P., Ikeda, K., Irieda, H., Narusaka, M., O'Connell, R.J., Narusaka, Y., Takano, Y., Kubo, Y., Shirasu, K., 2013. Comparative genomic and transcriptomic analyses reveal the hemibiotrophic stage shift of Colletotrichum fungi. *New Phytologist*, 197(4), 1236–1249.

Goodwin, S.B., Ben M'Barek, S., Dhillon, B., Wittenberg, A.H., Crane, C.F., Hane, J.K., Foster, A.J., Van der Lee, T.A., Grimwood, J., Aerts, A., Antoniw, J., 2011. Finished genome of the fungal wheat pathogen *Mycosphaerella graminicola* reveals dispensome structure, chromosome plasticity, and stealth pathogenesis. *PLoS Genetics*, 7(6), 1002070.

Grünwald, N.J., Werres, S., Goss, E.M., Taylor, C.R., Fieland, V.J., 2012. *Phytophthora obscura* sp. nov., a new species of the novel Phytophthora subclade 8d. *Plant Pathology*, 61(3), 610–622.

Gu, W., Miller, S., Chiu, C.Y., 2019. Clinical metagenomic next-generation sequencing for pathogen detection. *Annual Review of Pathology: Mechanisms of Disease*, 14, 319–338.

Guo, L.D., Hyde, K.D. and Liew, E.C.Y., 2000. Identification of endophytic fungi from *Livistona chinensis* based on morphology and rDNA sequences. *The New Phytologist*, 147(3), 617–630.

Ho, S.Y., Lanfear, R., Phillips, M.J., Barnes, I., Thomas, J.A., Kolokotronis, S.O., Shapiro, B., 2011. Bayesian estimation of substitution rates from ancient DNA sequences with low information content. *Systematic Biology*, 60(3), 366–375.

Hu, Y., Green, G.S., Milgate, A.W., Stone, E.A., Rathjen, J.P., Schwessinger, B., 2019. Pathogen detection and microbiome analysis of infected wheat using a portable DNA sequencer. *Phytobiomes Journal*, 3(2), 92–101.

Janardhana, G. R., Dass R. S. and Sreenivasa M. Y. 2009. Fumonisins – A new class of Fusarial toxins. In *Frontiers in Fungal Ecology, Diversity and Metabolites*. Ed: K A Sridhar et al. New Delhi: I. K. International Publishing House Pvt. Ltd.

Johnston-Monje, D., Lopez Mejia, J., 2020. Botanical microbiomes on the cheap: Inexpensive molecular fingerprinting methods to study plant-associated communities of bacteria and fungi. *Applications in Plant Sciences*, 8(4), 11334.

Jongman, M., Carmichael, P.C., Bill, M., 2020. Technological advances in phytopathogen detection and metagenome profiling techniques. *Current Microbiology*, 77, 675–681.

Kazeeroni, E.A., Al-Sadi, A.M., 2016. 454-pyrosequencing reveals variable fungal diversity across farming systems. *Frontiers in Plant Science*, 7, 314.

Kemler, M., Garnas, J., Wingfield, M.J., Gryzenhout, M., Pillay, K.A., Slippers, B., 2013. Ion Torrent PGM as tool for fungal community analysis: a case study of endophytes in Eucalyptus grandis reveals high taxonomic diversity. *PLoS One*, 8(12), 81718.

Khang, C.H., Park, S.Y., Lee, Y.H., Valent, B., Kang, S., 2008. Genome organization and evolution of the AVR-Pita avirulence gene family in the *Magnaporthe grisea* species complex. *Molecular plant-microbe interactions*, 21(5), 658–670.

Landegren, U., Kaiser, R., Sanders, J., Hood, L., 1988. A ligase-mediated gene detection technique. *Science*, 241(4869), 1077–1080.

Leboldus, J.M., Kinzer, K., Richards, J., Ya, Z., Yan, C., Friesen, T.L., Brueggeman, R., 2015. Genotype-by-sequencing of the plant-pathogenic fungi *P yrenophora* teres and *S phaerulina musiva* utilizing I on Torrent sequence technology. *Molecular Plant Pathology*, 16(6), 623–632.

Liacsa, L.X., Solis-Castro, R.L., Mialhe, E., García-Seminario, R., 2019. Metagenomic analysis of the bacterial and fungal community associated to the rhizosphere of *Tabebuia chrysantha* and *T. billbergii*. *Current Microbiology*, 76(9), 1073–1080.

Loit, K., Adamson, K., Bahram, M., Puusepp, R., Anslan, S., Kiiker, R., Drenkhan, R., Tedersoo, L., 2019. Relative performance of MinION (Oxford Nanopore Technologies) versus Sequel (Pacific Biosciences) third-generation sequencing instruments in identification of agricultural and forest fungal pathogens. *Applied and Environmental Microbiology*, 85(21), e01368–19.

Lu, J., Salzberg, S.L., 2018. Removing contaminants from databases of draft genomes. *PLoS Computational Biology*, 14(6), 1006277.

Magi, A., Benelli, M., Gozzini, A., Girolami, F., Torricelli, F., Brandi, M.L., 2010. Bioinformatics for next generation sequencing data. *Genes*, 1(2), 294–307.

Margulies, M., Egholm, M., Altman, W.E., Attiya, S., Bader, J.S., Bemben, L.A., Berka, J., Braverman, M.S., Chen, Y.J., Chen, Z., Dewell, S.B., 2005. Genome sequencing in microfabricated high-density picolitre reactors. *Nature*, 437(7057), 376–380.

Maxam, A. M., Gilbert, W., 1977. A new method for sequencing DNA. *Proceedings of the National Academy of Sciences USA* 74(2), 560–564.

Mechan, M.E., Sharma, P., Aguilera Flores, M., Yang, S., Pollok, J., Tian, L., Huang, C., Rideout, S., Heath, L.S., Li, S., Vinatzer, B.A., 2020. Strain-level identification of bacterial tomato pathogens directly from metagenomic sequences. *Phytopathology*, 110(4), 768–779.

Merriman, B., Team, I.T., Rothberg, J.M., 2012. Progress in ion torrent semiconductor chip based sequencing. *Electrophoresis*, 33(23), 3397–3417.

Metzker, M.L., 2005. Emerging technologies in DNA sequencing. *Genome Research*, 15(12), 1767–1776.

Metzker, M.L., 2010. Sequencing technologies-the next generation. *Nature Reviews Genetics*, 11(1), 31–46.

Musgrave, D.R., 1984. Detection of an endophytic fungus of *Lolium perenne* using enzyme-linked immunosorbent assay (ELISA). *New Zealand Journal of Agricultural Research, 27*(2), 283–288.

Nagaraja, H., Chennappa, G., Poorna Chandra Rao, K., Mahadevprasad, G., M Y Sreenivasa. 2016. Diversity of toxic and phytopathogenic *Fusarium* species occurring on cereals grown in Karnataka state, India. *3Biotech*. 6, 57.

Nicolaisen, M., Justesen, A.F., Knorr, K., Wang, J., Pinnschmidt, H.O., 2014. Fungal communities in wheat grain show significant co-existence patterns among species. *Fungal Ecology*, 11, 145–153.

Nilsson, R.H., Hyde, K.D., Pawłowska, J., Ryberg, M., Tedersoo, L., Aas, A.B., Alias, S.A., Alves, A., Anderson, C.L., Antonelli, A., Arnold, A.E., 2014. Improving ITS sequence data for identification of plant pathogenic fungi. *Fungal Diversity*, 67, 11–19.

O'Connell, R.J., Thon, M.R., Hacquard, S., Amyotte, S.G., Kleemann, J., Torres, M.F., Damm, U., Buiate, E.A., Epstein, L., Alkan, N., Altmüller, J., 2012. Lifestyle transitions in plant pathogenic Colletotrichum fungi deciphered by genome and transcriptome analyses. *Nature Genetics*, 44(9), 1060–1065.

Piombo, E., Abdelfattah, A., Droby, S., Wisniewski, M., Spadaro, D., Schena, L., 2021. Metagenomics approaches for the detection and surveillance of emerging and recurrent plant pathogens. *Microorganisms*, 9(1), 188.

Pop, M., Salzberg, S.L., 2008. Bioinformatics challenges of new sequencing technology. *Trends in Genetics*, 24(3), 142–149.

Postnikova, E., Baldwin, C., Whitehouse, C.A., Sechler, A., Schaad, N.W., Sampath, R., Harpin, V., Li, F., Melton, R., Blyn, L., Drader, J., 2008. Identification of bacterial plant pathogens using multilocus polymerase chain reaction/electrospray ionization-mass spectrometry. *Phytopathology*, 98(11), 1156–1164.

Prasad, M.N.N., Shankara Bhat S., Sreenivasa M.Y. 2010. Antifungal activity of essential oils against *Phomopsis azadirachtae*- the causative agent of die-back disease of neem. *Journal of Agricultural Technology*, 6 (1), 127–133.

Punja, Z.K., Rodriguez, G., 2018. Fusarium and Pythium species infecting roots of hydroponically grown marijuana (*Cannabis sativa* L.) plants. *Canadian Journal of Plant Pathology*, 40(4), 498–513.

Rani, L., Thapa, K., Kanojia, N., Sharma, N., Singh, S., Grewal, A.S., Srivastav, A.L., Kaushal, J., 2021. An extensive review on the consequences of chemical pesticides on human health and environment. *Journal of Cleaner Production*, 283, 124657.

Rao, S., Sharda, S., Oddi, V., Nandineni, M.R., 2018. The landscape of repetitive elements in the refined genome of chilli anthracnose fungus *colletotrichum truncatum*. *Frontiers in Microbiology*, 9, 2367.

Reinhardt, J.A., Baltrus, D.A., Nishimura, M.T., Jeck, W.R., Jones, C.D., Dangl, J.L., 2009. De novo assembly using low-coverage short read sequence data from the rice pathogen *Pseudomonas syringae* pv. oryzae. *Genome Research*, 19(2), 294–305.

Ronaghi, M., Uhlén, M., Nyrén, P., 1998. A sequencing method based on real-time pyrophosphate. *Science*, 281(5375), 363–365.

Sanger, F., Nicklen, S., Coulson, A.R., 1977. DNA sequencing with chain-terminating inhibitors. *Proceedings of the National Academy of Sciences*, 74(12), 5463–5467.

Schoch C.L., Seifert, K.A., Huhndorf, S., Robert, V., Spouge, J.L., Levesque, C.A., Chen, W., Fungal Barcoding Consortium, Fungal Barcoding Consortium Author List, Bolchacova, E., Voigt, K., 2012. Nuclear ribosomal internal transcribed spacer (ITS) region as a universal DNA barcode marker for Fungi. *Proceedings of the National Academy of Sciences*, 109(16), 6241–6246.

Sreenivasa, M.Y., Dass, R.S., Charith Raj, A. P., Janardhana, G. R.. 2008. PCR-based detection of genus *Fusarium* and Fumonisin-producing isolates from freshly harvested Sorghum grains grown in Karnataka, India. *Journal of Food Safety*, 28, 236–237.

Sreenivasa, M.Y., Dass, R. S., Charith Raj, A. P., Janardhana, G. R.. 2011b. Mycological evaluation of Maize grains produced in Karnataka (India) for the post-harvest fungal contamination. *World Applied Sciences Journal*, 13(4), 688 – 692.

Sreenivasa, M.Y., Dass, R.S., Janardhana G. R. 2010. Post harvest fungi associated with sorghum grains produced in Karnataka (India). *Journal of Plant Protection Research*, 50(3), 335–339.

Sreenivasa, M.Y., Dass, Regina S., Charith Raj, A.P., Nagendra Prasad, Mysore N., Achar, Premila N., Janardhana, Gotravalli R., 2011a. Assessment of the growth inhibiting effect of some plant essential oils on different *Fusarium* species isolated from sorghum and maize grains. *Journal of Plant diseases and Protection*, 118 (6), 208–213.

Stukenbrock, E.H., Jørgensen, F.G., Zala, M., Hansen, T.T., McDonald, B.A., Schierup, M.H., 2010. Whole-genome and chromosome evolution associated with host adaptation and speciation of the wheat pathogen *Mycosphaerella graminicola*. *PLoS Genetics*, 6(12), 1001189.

Suzuki, Y., 2020. Advent of a new sequencing era: Long-read and on-site sequencing. *Journal of Human Genetics*, 65(1), 1.

Taylor, M.W., Tsai, P., Anfang, N., Ross, H.A., Goddard, M.R., 2014. Pyrosequencing reveals regional differences in fruit-associated fungal communities. *Environmental microbiology*, 16(9), 2848–2858.

Tedersoo, L., Nilsson, R.H., Abarenkov, K., Jairus, T., Sadam, A., Saar, I., Bahram, M., Bechem, E., Chuyong, G., Kõljalg, U., 2010. 454 Pyrosequencing and Sanger sequencing of tropical mycorrhizal fungi provide similar results but reveal substantial methodological biases. *New Phytologist*, 188(1), 291–301.

Tucker, T., Marra, M., Friedman, J.M., 2009. Massively parallel sequencing: the next big thing in genetic medicine. *The American Journal of Human Genetics*, 85(2), 142–154.

Tyagi, P., Bhide, M., 2020. History of DNA sequencing. *Folia Veterinaria*, 64(2), 66–73.

Uppalapati, S.R., Serba, D.D., Ishiga, Y., Szabo, L.J., Mittal, S., Bhandari, H.S., Bouton, J.H., Mysore, K.S., Saha, M.C., 2013. Characterization of the rust fungus, Puccinia emaculata, and evaluation of genetic variability for rust resistance in switchgrass populations. *Bioenergy Research*, 6, 458–468.

Valouev, A., Ichikawa, J., Tonthat, T., Stuart, J., Ranade, S., Peckham, H., Zeng, K., Malek, J.A., Costa, G., McKernan, K., Sidow, A., 2008. A high-resolution, nucleosome position map of *C. elegans* reveals a lack of universal sequence-dictated positioning. *Genome Research*, 18(7), 1051–1063.

Voelkerding, K.V., Dames, S.A., Durtschi, J.D., 2009. Next-generation sequencing: from basic research to diagnostics. *Clinical Chemistry*, 55(4), 641–658.

Yang, S., Johnson, M.A., Hansen, M.A., Bush, E., Li, S., Vinatzer, B.A., 2022. Metagenomic sequencing for detection and identification of the boxwood blight pathogen *Calonectria pseudonaviculata*. *Scientific reports*, 12(1), 1399.

Yao, D., Ijaz, W., Liu, Y., Hu, J., Peng, W., Zhang, B., Wen, X., Wang, J., Qiu, D., Li, H., Xiao, S., 2022. Identification of a Pm4 allele as a powdery mildew resistance gene in wheat line xiaomaomai. *International Journal of Molecular Sciences*, 23(3), 1194.

Zhang, N., Cai, G., Price, D.C., Crouch, J.A., Gladieux, P., Hillman, B., Khang, C.H., LeBrun, M.H., Lee, Y.H., Luo, J., Qiu, H., 2018. Genome wide analysis of the transition to pathogenic lifestyles in Magnaporthales fungi. *Scientific Reports*, 8(1), 5862.

10 In-situ Hybridization for the Detection of Fungal Phytopathogens

B. Shruthi, G. Adithi, and Sreenivasa Marikunte Yanjarappa
University of Mysore

Currently, numerous fungal pathogens are predicted. However, very few fungi have been reported. Several fungal species cause diseases both to humans and plants. These diseases lead to huge economic losses to the agricultural sector globally as well as have adverse impacts on human health. Consequently, detecting the pathogen in the early stage is very essential to reduce the fungal growth and enable effective management strategies. Primary methods that are used to detect fungi mostly count on microscopic and morphological methods, which require immense time, manpower and classical taxonomy knowledge. To overcome such difficulties, molecular techniques came into existence. Various molecular approaches are progressively becoming important and beneficial approaches in every phase to diagnose fungal pathogens. The foremost goal of all diagnostic approaches is to identify the organism rapidly, accurately and also reliably. At present, there are numerous effective molecular approaches which have transformed the research in fungal detection such as probe development, PCR and DNA barcoding strategies, RNA sequence-based next-generation sequencing, etc., out of which, in-situ hybridization is one of the efficient techniques which may contribute to the rapid and accurate identification of fungal pathogen. In this chapter, the detection mechanism, applications and advantages of in-situ hybridization techniques to detect fungal pathogens are discussed.

DOI: 10.1201/9781003399704-10

GRAPHICAL ABSTRACT

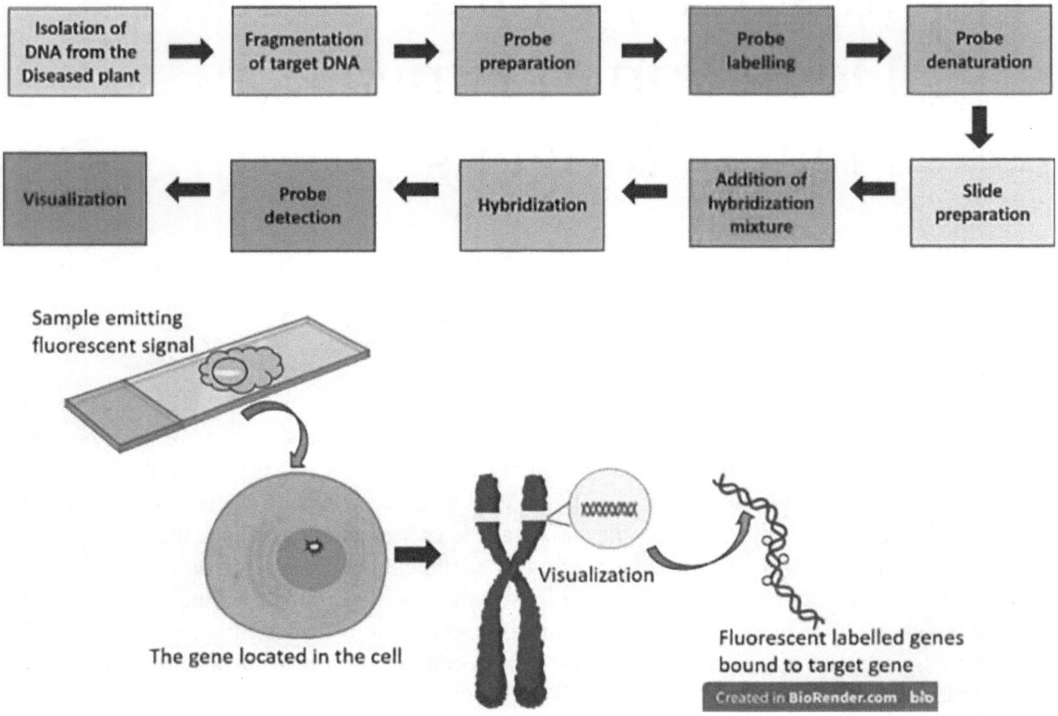

10.1 INTRODUCTION

Plant diseases and infection play a vital role in the destruction of natural resources in agriculture, which are the leading cause of a drop in the foremost annual food production in the world (Tyśkiewicz et al., 2022). The main causes of plant diseases are soil-borne pathogens, out of which fungal pathogens are the main culprits affecting the plants to a major extent, resulting in the eradication of one-third of crops every year (Almeida et al., 2019; Sreenivasa et al., 2010). From the beginning of the 21st century, diseases which are caused by microbial pathogens have led to the loss of around 16% of crop yields, about 70–80% of loss is caused by fungal pathogens. These diseases and infections are expected to increase due to global changes occurring because of pathogens and diseases exhibiting pathogenic behaviour (Rodrigo et al., 2022). Fungi cause the greatest impact on losses of crop production which includes extensive foliage and losses of fruits and vegetables during post-harvest. The fungal diseases comprise leaf spots, powdery mildew, rust, blight, root and crown rots, damping off, smut, anthracnose and vascular wilts. Some of the infective fungi comprise *Pythium*, *Phytophthora*, *Fusarium* and *Rhizoctonia* spp, which are the main reasons for seedling damp-off, root and crown rot in most of the vegetables and ornamental plants. *Aspergillus*, *Fusarium*, *Penicillium* and *Alternaria* are also the major causes of plant diseases, which also affect food commodities by producing mycotoxins (Mdee et al., 2009). Fungal diseases cause significant losses to agronomic and food industries globally. *P. digitatum*, *P. expansum* causes spoilage of oranges, and Aflatoxins B1 and B2 and Fumi toxins produced from *A. flavus* and *A. fumigatus* result in harmful effects on animals and human health (Mahlo et al., 2016).

Phytopathogens not only affect the survival and growth of plants but also result in great economic losses; transmission of plant pathogens can communicate horizontally through vectors like soil, water and insects. It also transmits vertically by mother plants (Chen and Nan, 2015). Some phytopathogenic diseases are not caused by a single pathogen, but they are the result of the interaction of multiple pathogens (Peng et al., 2021). The fungal phytopathogens are the most harmful agents in plants, which increase economic losses and reduce food availability affecting the increasing population worldwide (Savary et al., 2019; Prasad et al., 2010; Chennappa et al., 2014). There are a few global issues faced due to these phytopathogenic fungi, which include, in the 19th century, Irish potato famine, which was instigated by fungal-like oomycete, directed to devastation by causing demises, migration, economic and political calamity in Ireland (Mokyr and Gráda, 2002). These phytopathogenic fungi are also responsible for the baring of landscapes instigated by Dutch elm blight and chestnut blight, and they completely ruined 30% of food crops in 2012 (Vicente et al., 2021). Currently, there is about 80% of crops globally affected by phytopathogens, which are also because of conditions such as climate change and increase in temperature and humidity (Zabiák et al., 2020). The regular applications of pesticides and agrochemicals are effective agents in preventing the most pathogenic phytopathogens, but they are expensive and very harmful and damaging to the environment and ecosystem (Cornelissen and Melchers, 1993). Phytopathogens are a threat to ecosystem stability and food security, so it's vital to develop an operative context for primary detection and internment to prevail over the biological incursion of phytopathogenic fungi (Luchi et al., 2020). To develop such context, early recognition needs correct species identification and appropriate identification of disease and causative organisms to prevent pathogens from entering new environmental areas. The emergence of diseases can be because of the generation of new virulent genotypes over mutation or recombination among native genotypes. Hence, it is required to identify and detect phytopathogens beyond the species level (Manawasinghe et al., 2021).

10.2 DIFFERENT METHODS TO IDENTIFY FUNGAL PHYTOPATHOGENS

Fungal ailments in commercially imperious plants cause a substantial drop in both excellence and profit, frequently leading to the damage of a whole plant. In order to abate the losses, it is vital to detect and recognize the pathogens at an initial stage. Early detection and precise identification of fungal pathogens can prevent the spread of infection (Ray et al., 2017).

Conventional methods for the detection and identification of phytopathogen, causing diseases in floras, primarily depend on morphologic, microbiological and biochemical credentials. This method includes the identification of pathogens by the interpretation of visual signs, culturing and plating methods (Ray et al., 2017). Microscopic observation includes analysis of pathogens based on their morphologic features such as spore morphology, sporulation forms, production and features of sporulating structures producing asexual and sexual spore forms, which are used for taxonomical cataloguing of fungi (Narayanasamy, 2010). Isozyme examination is used to distinguish, differentiate and recognize morphologically alike or closely associated species. (Ray et al., 2017). Antibody-based techniques are used to detect or enumerate the binding of the investigative antibody with the target antigen. There are numerous ways to detect antibody/antigen binding, but often these include linking the antibody to an enzyme that can be cast off to produce a colour change when a substrate is added (Ward et al., 2004). The immunologic-based detection comprises ELISA (enzyme-linked immunosorbent assay), RIA (radio-immunosorbent assay), western blot analysis, dot immunobinding assay (DIBA), dipstick immunoassay, tissue blot immunoassay (TBIA) and immunosorbent electron microscopy (Ray et al., 2017).

The failure to effectively identify phytopathogens from culture-based morphological methods has led to the expansion of culture-independent molecular methods (Lievens and Thomma, 2005). Molecular-based identification techniques include polymerase chain reaction (PCR-based) technologies (Deepa et al., 2016b; Deepa et al., 2016c; Deepa and Sreenivasa, 2019a; 2019b), DNA probing technology, microarray technology, hybridization-based techniques and isothermal amplifications. Current developments in the field of molecular biology have formed opportunities for the expansion of innovative diagnostic approaches, which are fetching inimitability not only in the detection and typing of diseases (Knez et al., 2013).

10.3 *IN-SITU* HYBRIDIZATION AND DETECTION MECHANISM

In-situ hybridization is a technique applied to focus and spot specific DNA and RNA sequences in the cells, conserved tissue sections or whole tissue through hybridizing the complementary strand of nucleotide probe to the exact sequence. These hybrids formed and labelled with chromogen or radioactive can be visualized by autoradiography (Jensen, 2014). In-situ hybridization is a very sensitive technique which allows the recognition and localization of specific DNA or RNA molecules. This procedure was first defined by (Gall and Pardue) in 1969 (Nouri-Aria, 2008). It provides vital information about the localization of gene expression in heterogeneous tissue. This technique is delicate and detects the amount of mRNA contained in the complete cell (Wilcox, 1993). There were many variations in in-situ hybridization methods developed in the past decades. The method has undergone many disparities and modifications to become a significant tool in elementary scientific research and clinical diagnosis (Jin and Lloyd, 1997). This hybridizing technique can be imagined with two sorts of probes, which are fluorescent (fluorescent in-situ hybridization; FISH) or chromogen based (chromogenic in-situ hybridization; CISH), which work on a similar principle (Veselinyová et al., 2021). Fluorescent in-situ hybridization (FISH) was developed as a substitute for the older method of in-situ hybridization. The first application of fluorescent in-situ hybridization was reported in 1980 in which RNA was labelled with a fluorophore and used as a probe for definite DNA sequence (Bauman et al., 1981). It was used for the visualization of nucleic acids and permitted substantial developments in resolution, safety and speed. It was also developed for concurrent recognition of multiple targets, quantifiable analysis and live-cell imagination (Levsky and Singer, 2003).

Chromogenic in-situ hybridization (CISH) is a comparatively innovative method to determine gene amplification via peroxidase-based chromogenic reaction, which could be observed using a conventional bright field microscope. It governs the definite degree of gene amplification (Hanna and Kwok, 2006).

The detection mechanism of in-situ hybridization majorly depends on the probes used; the use of probes includes certain limitations and conditions. Basically, probes are required for the detection of appropriate specific labelled sequences of interest. Preservation of morphological details in the tissue is required to identify the position of the labelled probe. The annealing and parting of two hybridized strands depends on certain aspects, which include the nature of probes and aimed molecules, temperature, pH, salt concentration, the melting temperature of hybrids and configuration of hybridization and washing solutions (Qian et al., 2004). A large variability of probes is used for ISH (in-situ hybridization) based on their applications. For DNA detection, double-stranded DNA probes are commonly used (Temsamani and Agrawal, 1996), while for RNA recognition, RNA probes (riboprobes) are regularly used since RNA-RNA hybridization is more constant compared to DNA-RNA hybridization (Brown, 1998). And also, oligonucleotides are used for the recognition of high profusion RNA like hormone mRNAs, which have high specificity, and specificity of ISH is important, so careful selection of oligonucleotide is required (Mitsuhas, 1996).

10.3.1 Tissue Preparation

The preparation of tissue for ISH is very important, which includes cellular and tissue materials. ISH has been useful for cell samples, which include smears, cytospins and cell pellets. In addition, tissue sections include frozen, paraffin, semithin and ultrathin plastic segments. Both integral cells and new frozen tissues are idyllic for ISH due to less injured nucleotide sequences, formalin static archival tissues are used for ISH for tissues stored for numerous years (Qian et al., 2004).

10.3.2 Pretreatment of Cell and Tissue

Sequences of pretreatment phases before hybridization increase the proficiency and reduce nonspecific background staining; few methods are regularly used in the ISH procedures, which have protease treatment (Proteinase K), which is measured as an important stage to raise the obtainability of marked nucleic acid (Oliver et al., 1997). The concentration of proteinase K and length of the sequences depend on the tissue kind and extent of fixation, extended incubation and under assimilation, which leads to inactivation of probes (Qian et al., 2004).

10.3.3 Types of Probes Used for In-situ Hybridization

The contest in emerging various ISH approaches is to expand the sensitivity of probes used to picture the target sequence. Different types of probes are used in this technique to detect the specimen (Tables 10.1 and 10.2)

10.3.3.1 Double-stranded DNA Probes

These are obtained by amplification and cloning of the exact sequence of DNA or cDNA, which are obtained by reverse transcription of mRNA, by means of vectors such as bacteriophages, bacteria, plasmids and cosmids. These probes have a high level of specificity and sensitivity because of their length.

10.3.3.2 Riboprobes or Complementary RNA Probes

In-situ hybridization with the use of RNA riboprobes is very popular and sensitive, which requires exact conditions compared to DNA probes. Riboprobes are composed by an in-vitro transcription with the use of DNA as a template. These riboprobes were used for the detection of viruses (Brown, 1998).

10.3.3.3 Oligonucleotide Probes

These probes contain single-stranded DNA molecules, which are of the ranges of 20 to 50 bases; these can be made by automated DNA synthesizer. These oligonucleotides produce excellent

TABLE 10.1
Different Radioactive Isotopes and Non-Radioactive Isotopes Used in In-situ Hybridization

Sl. No	Radioactive isotopes	Non-Radioactive Isotopes
1	^{3}H - Tritium	Digoxigenin
2	^{35}S - Sulphur	Biotin
3	^{125}I - Iodine	Fluorescein
4	^{32}P - Phosphorus	Alkaline phosphatase
5	^{33}P - Phosphorus	Bromodeoxyuridine (BrdU)

TABLE 10.2
Advantages and Disadvantages of Probes

Type of probe	Advantages	Disadvantages
Double stranded DNA (dsDNA) probe	Stable, easier to obtain, available	Self-hybridize, less sensitive, required denaturation before hybridization
Single stranded DNA (ssDNA) probe	Stable, easier, specific, resistant to RNases, good tissue penetration, without self-hybridize	Time consuming, expensive
RNA probes (Riboprobes)	Thermal stability, specific, better tissue penetration,	Sensitive to RNases
Synthetic oligonucleotides	Economical, stable, available, more specific, resistant to RNases, better reproducibility	These are not sensitive, mainly used for hormones, careful selection of nucleotide is required.

signals, these probes will be labelled with their tail and these probes are generally not so sensitive compared to cDNA and RNA probes. Hence, these probes are generally used in gene expressions such as hormone mRNAs (Jin and Lloyd, 1997).

10.3.4 Labelling and Signal Detection

Labelling and signal detection include two steps: traditional step of using radioisotope labelling, which includes tritium, sulfur-35, iodine-125 and isotopes of phosphorous, which are visualized using autoradiography, and non-isotope labelling method including biotin, digoxigenin, fluorescein, alkaline phosphatase or bromodeoxyuridine (BrdU). These can be envisioned by histochemistry or immunohistochemistry (IHC) systems (Jin and Lloyd, 1997) (Table 10.1). Radioisotopes are more sensitive compared to nonradioactive labelling such as poor stability; it requires longer processing time, more rigorous waste disposal and hours together of washing procedures (Looi et al., 1992). Nonradioactive method of labelling and detection is a direct method and much simpler and quicker compared to radioactive in-situ hybridization method (Kiyama et al., 1990). The detection process is presented in Figure 10.1.

10.4 DETECTION OF FUNGI BY IN-SITU HYBRIDIZATION

Primary identification of soil-borne diseases is important for the estimation of disease growth and effective disease handling (Milner et al., 2019). Hydrophobins are exceptional mycological extracellular proteins that yield amphipathic films at edges, arbitrate interaction with hydrophobic faces and are noted as vital in phytopathogenicity. The infective ascomycete Claviceps *purpurea*, instigating ergot disease in cereals and grass and ergotism in cattle, a genetic factor concealing an unusual kind of hydrophobin, has been identified, which seemed to be induced in the time of alkaloid production axenic

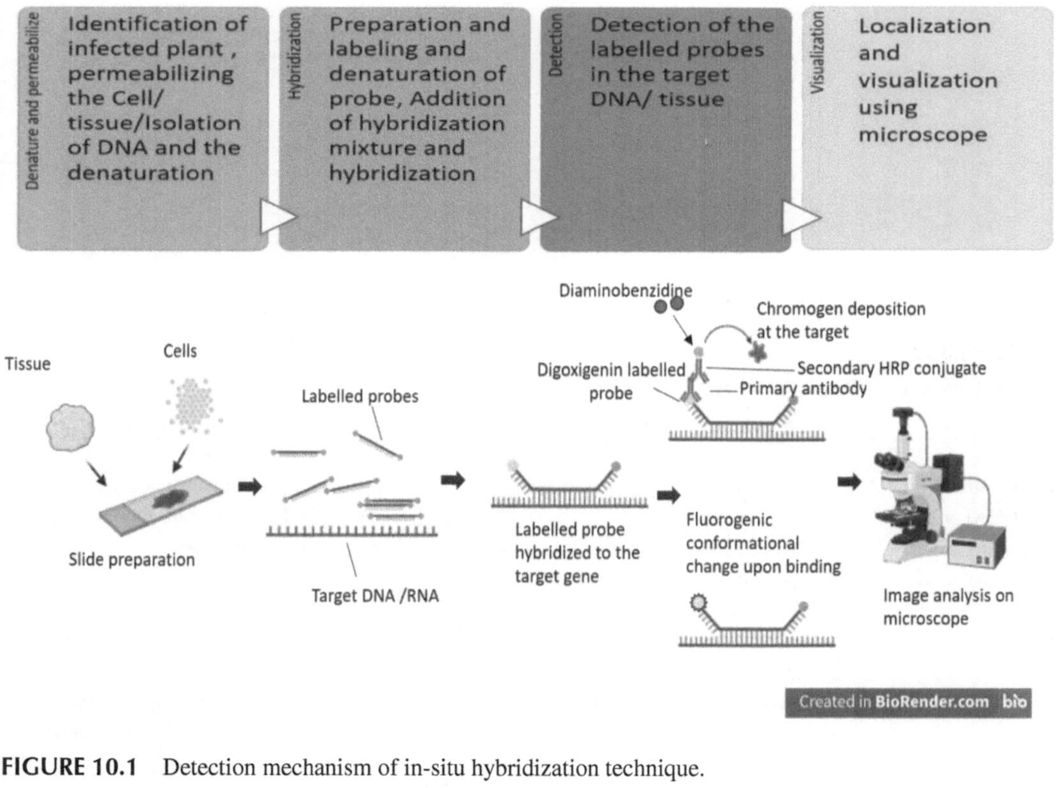

FIGURE 10.1 Detection mechanism of in-situ hybridization technique.

culture of an ergotalkaloid synthesising strain of *Claviceps* (Tenberge et al., 1998). To explicate occurrence and purpose of this hydrophobin at the time of infection of rye, the nonradioactive in-situ hybridization method was effectively acclimatized to the fungi and improved in the infective contact system. Semithin cryosections were demonstrated to be appropriate for microscopic gene expression investigation with immune-mediated alkaline phosphatase staining for the identification of digoxigenin-tagged cRNA probes. Precise hybridization of the equipped antisense riboprobe to hydrophobin mRNA was oriented in the nonradioactive northern blots (Tenberge et al., 1998). *Alternaria alternata*, which causes black spot disease in Japanese pear, is one of the most injurious sicknesses in Japanese pear farming (Yamamoto et al., 2016). In the study, the sites of lack of spot disease resistance/susceptibility associated genome portions were pondered by FISH with BAC clone (BAC-FISH) on Japanese pear (*Pyrus pyrifolia* (*Burm.* f.) *Nakai*) chromosomes. The BAC clone, together with black spot infection-linked gene, was tagged as a probe for the FISH analysis. Black spot disease-associated genomic sites were identified in telomeric regions of two average-sized chromosomes. These two regions and the six telomeric 18S-5.8S-25S rDNA regions were located on dissimilar chromosomes as determined from the outcomes of multi-colour FISH. The efficiency of the physical plotting of beneficial genes on the pear genetic material attained by the BAC-FISH technique was explicitly proved (Yamamoto et al., 2016).

To identify and enumerate soil inhabitants of *Sclerotium rolfsii*, the fundamental factor of southern blight of tomatoes, Q-PCR was cast off to measure the amount of soil overall moulds and particular *S. rolfsii* DNA. Suggestively, maximum number of overall fungi and S. *rolfsii* DNA were identified in contaminated than in the uncontaminated soil in glasshouse inoculation experiment. FISH was effectively established to identify the profusion of *S. rolfsii* by whole-cell hybridization and was later put on for visible identification of the abundance of *S. rolfsii* in soil samples of glasshouse inoculation trial. Innate profusion of *S. rolfsii* in the soil with a DNA quantity of 0.06 pg μl^{-1}, which was correspondent to 8 pg g^{-1} soil, was effectively identified. The hybridization signals exposed in the soil intended for *S. rolfsii* had a substantial association with the quantity of infested DNA concentrations of *S. rolfsii* ((Milner et al., 2019).

Fusarium species are one of the major fungi which infect a large commodity of crops (Sreenivasa et al., 2008; Janardhana et al., 2009; Deepa and Sreenivasa. 2019b;). *Fusarium* produces a mycotoxin called fumonisin, which causes a range of diseases in animals and is also a carcinogenic compound (Deepa and Sreenivasa. 2017a; 2017b; Deepa et al., 2016a; Deepa et al., 2021; Nagaraja et al., 2016; Sreenivasa et al., 2011a, 2011b). Even though precise and hasty diagnosis of *Fusarium* infections is now obligatory, morphologic resemblances amongst fungi make it problematic to histologically distinguish *Fusarium* spp. from different fungi. Okubo et al. (2013) presented the ISH method as an appreciated implement to detect *Fusarium* spp. and accentuated the practicality of the method. *Fusarium oxysporum* f. sp. *lycopersici* was detected by ISH. The detection was done using mRNA in a filamentous phytopathogenic fungus by pricking microinjection of a photobiotin-labelled antisense RNA Probe (Toyoda et al., 1992). *Fusarium* is one of the significant phytopathogens, but now this fungus has newly occurred as an opportunistic pathogen of human beings. The histologic difference of *Fusarium* from *Aspergillus* and *Scedosporium* is chiefly challenging because these fungi may instigate similar clinical structures and display filament-like growth in host tissues. Therefore, there is a necessity to set up hasty and consistent techniques, which are appropriate to diagnose pathogens. The purpose of the authors in the examination was to assess and set up ISH using peptide nucleic acid (PNA) probes directing the 28S rRNA to recognize *Fusarium* species in tissue sections. As a result, hefty positive signals were detected within fungi existing in tissues. Additionally, this probe responded sturdily with both *F. solani* and *Fusarium oxysporum* (Shinozaki et al., 2011).

The authors also identified cytoplasmic rRNAs of disease structures of phytopathogenic fungus, *Erysiphe graminis* f. sp. *hordei*, which causes chalky mildew disease in barley. This identification is carried out by precise ISH method on micro-injected photo biotin-tagged nucleic acid probes and concludes grouping of ISH with microinjection. This would make it conceivable to examine gene expression in a broad array of filamentous fungi counting the fungal phytopathogens, particularly in non-sectioned fungal cells (Toyoda et al., 1992).

Dematiaceous fungi are a varied group of "darkly" pigmented moulds, which comprise melanin in the cell walls and are normally seen in the soil globally (Montone et al., 2011). ISH for profuse fungi rRNA sequences might offer a mode for identifying *dematiaceous fungi*. In a study, a 24-base artificial biotin-tagged oligonucleotide probe pointing to rRNA sequences of variable dematiaceous fungi was established. This probe was verified and found optimistic for *Curvularia* species, *Alternaria* species, *Cladosporium* species, *Bipolaris* species, *Scedosporium prolificans*, *Scopulariopsis* species and dematiaceous species. ISH exhibited positive results for 24 specimens out of 24 of 29 specimens (Montone et al., 2011).

Allergic fungal sinusitis (AFS) is a consequential form of sinonasal fungal infection that is often associated with *Aspergillus* or Dematiaceous fungi. An investigation was executed to determine the occurrence of *Aspergillus* or *Penicillium* in AFS by means of ISH for *Aspergillus* and *Penicillium* ribosomal RNA (rRNA). To notice the pigmented fungi, Fontana-Masson melanin stain (FMMS) was utilized (Perez-Jaffe et al., 1997). The in-situ hybridization (ISH) method has rarely been used for the localization of endophytic fungi in plant tissue. The ISH technique with digoxigenin-labelled oligonucleotide probes has effectively been used for the detection of *Aciculosporium* in bamboo tissues and fluorescent oligonucleotide probes have been used for the detection of *Aureobasidium pullulans* on apple leaves (Pirttilä, 2011).

10.5 OTHER APPLICATIONS OF IN-SITU HYBRIDIZATION

The in-situ hybridization method includes various clinical research and diagnostic applications such as gene expression studies by detecting mRNA, diagnosis of various viral infections, in the part of cell cycle and apoptosis (Figure 10.2), in studies such as cytogenetics in examination of fatal aberrations and cancers (McNicol and Farquharson, 1997).

10.5.1 Infectious Diseases

Detection of infectious agents such as viruses, bacteria, and fungi in tissue sections and the infectious diseases caused and their aetiology can be imagined by ISH methods (Jin and Lloyd, 1997). ISH method has the main application in recognition of human immunodeficiency virus (HIV), cytomegalovirus (CMV), human papillomavirus (HPV), herpes simplex virus (HSV), hepatitis B virus (HBV), Epstein-Barr virus (EBV) and JC viruses (Jin and Lloyd, 1997). Ribosomal RNA (rRNA) is the target which would be used to depict infectious agents for in-situ hybridization.

10.5.2 Gene Expression

In-situ hybridization has developed as a prevailing tool for the appraisal of gene expression in tumour cells at the cellular level. This hybridization analysis provides the advantage of detecting cells on the origin of their contents of exact mRNAs coding the product of interest. ISH affords the data of individual cells rather than multiple cells or total cellular population. In-situ hybridization provides the risky approach for investigation of heterogenicity in tumours that characteristically contain cells at diverse phases of neoplastic evolution. At different levels of variation and functional activity, ISH have proved as vital technique for examination of oncogene overexpression in the variability of diverse categories of tumours (Delellis, 1994).

10.5.3 Cytogenetics

In-situ hybridization has now developed practical techniques in molecular genetics and cytogenetics. Fluorescent in-situ hybridization (FISH) technique is furthermost frequently applied in this field. In-probes used in FISH are labelled with fluorescein or labelled with biotin or digoxigenin, and fluorochrome is used in additional signal detects (Gray et al., 1994). FISH has several

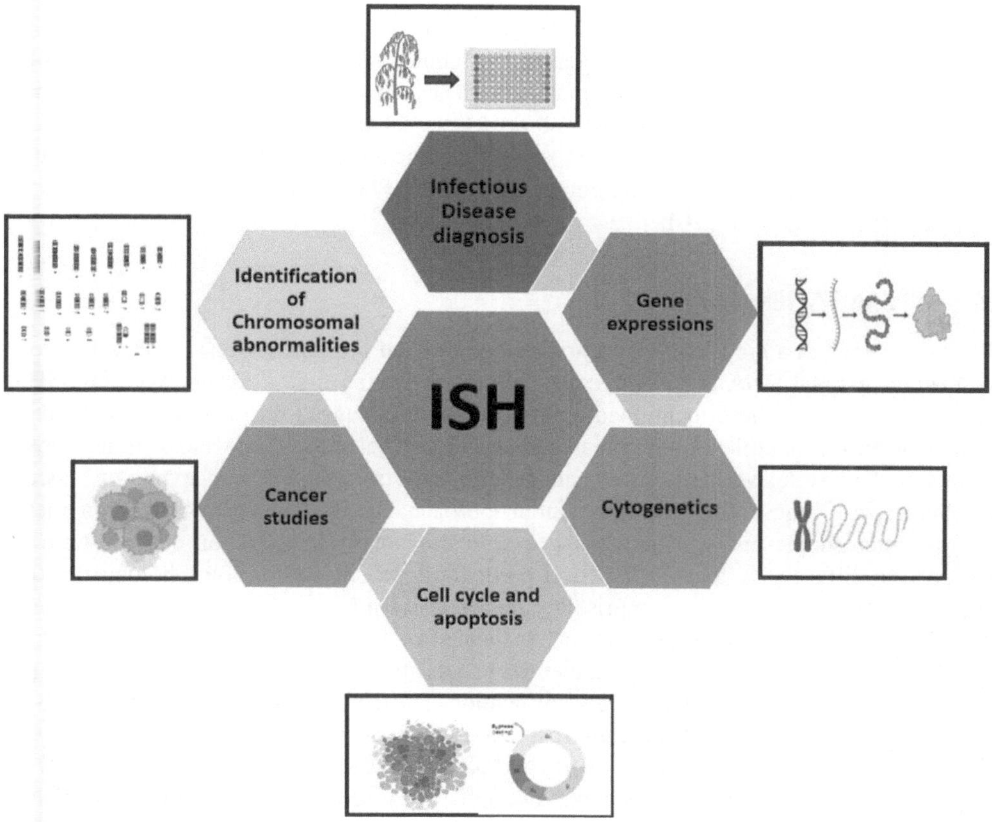

FIGURE 10.2 Applications of in-situ hybridization.

applications such as exposure of numerical and operational chromosomal irregularities, identifying marker chromosomes, nursing effects of therapy, recognition of early decline or slight residual disease, identifying the origin of bone marrow cells subsequently, stem cell transplantation, identifying lineages of neoplastic cells, inspection of the karyotypic outline of interphase cells, as well as non-dividing or terminally differentiated cells and finding of gene amplification and deletion (Gozzetti and Le Beuu, 2000).

10.5.4 Cell Cycle and Apoptosis

In-situ hybridization is used for the examination of the cell cycle, propagation and identification of cells transcribing mRNAs for histone proteins, which are amplified through S-phase. This technique is also helpful in detecting mRNA encrypting cell cycle-associated proteins such as cyclin D1, which shows altered expression in tumours (Mcnicol and Farquharson, 1997). In-situ detects the active sites of the transcription in this method. DNase is used to partially digest DNA to produce single-strand DNA breaks, which are filled with DNA polymerase. Those will be incorporated with radiolabelling. This judicious use of DNase doesn't damage the chromatin as it is completely coiled; it only labels the actively transcribing genes. It also helps to define the spatial distribution of apoptotic cells in the histological materials (Ansari et al., 1993).

The apoptosis or programmed cell death, which shows characteristics such as cell shrinkage, condensation of chromatin, apoptotic body formation and nuclear fragmentation of polymorphonuclear neutrophils (PMN), can be clearly shown through in-situ end labelling (ISEL). With this technique, apoptotic cells are detected in all tissue sections (Koulouri et al., 1999).

10.5.5 Non-coding RNAs

There are several non-coding RNAs present in the human genome. Most studies are microRNAs (miRNAs) and long non-coding RNAs (lincRNAs). These non-coding RNAs play a central role as tumour suppressor genes and oncogenes and are also vital for tumorigenesis. These non-coding RNA can be detected by Immunohistochemistry. The applications of ISH for oncogenic and tumour suppressor long non-coding RNAs have comprised studies of breast cancer, thyroid cancer and prostate cancer (Ying-Hsiaa et al., 2019).

10.6 ADVANTAGES

ISH has developed as a prevalent laboratory method utilized for gene expression investigation and to localize the explicit DNA or RNA in cells. Numerous diverse approaches to executing ISH have been defined, and also these mentioned approaches have advanced into significant appliances in fundamental basic scientific examinations and in clinical detection (Gall and Pardue, 1969). As such, this method exclusively offers to locate genetic material, which is overlapping on subcellular and cellular parts, consenting investigation unapproachable by further hybridization methods. Therefore, microscopic and/or macroscopic examination of the preparation can illustrate if the aimed gene is diffusely or focally dispersed particularly which portion, morphologically or biochemically distinct type of cells comprise it, and the subcellular localization of the genome (e.g., nuclear, cytoplasmic, nucleolar, perinuclear, or located in a specific region of a specific chromosome). Also, the benefits of this detection method in a morphologic context, ISH has become prevalent due to its eminent sensitivity in gene identification. The technique predominantly prevails when the targeted gene to be identified is focally scattered in a minute proportion (Moench, 1987). One of the aims of ISH is to locate the genes *in-situ* and to visualize the consequences inside cells at the time of conserving cell integrity. This method permits expressive structural and histologic elucidation of the localized product(s) in heterogeneous tissues (Ying-Hsiaa et al., 2019).

10.7 LIMITATIONS

There are numerous confines for ISH, especially for fungi in cytologic and tissue substratum. First, signal can be feeble if the levels of rRNA are limited; though, most of the fungal rRNA exist in high number of copies. By means of ineffectually developed probes, indication might be faint or imperfect (Montone and Guarner 2013). Also, another aspect of fungal ISH is dissimilarity in cell walls, creating enzymatic pretreatment and microwave pretreatment which are essential stages aimed at ideal signal development. In spite of these approaches, still several fungi find it hard to steadily generate Strong ISH signals. It is probable that ISH identifies only viable samples (Montone and Guarner 2013). In earlier work, Park et al. (2003) revealed that ISH for fungi is frequently weaker in entwined masses of fungi (fungus ball) and repeatedly displays enormous staining in the peripheral region of fungal balls and on the foremost edge of tissue invasion where the fungi are more expected to be viable. Considerable tissue necrosis is frequently associated with condensed ISH signals (Montone et al., 2011).

10.8 SUMMARY

Plants which comprise commercially vital crops are contaminated by a huge number of phytopathogens, which cause utmost harmful diseases that are accountable for significant loss of harvest globally. Identification and diagnosis of plant-infecting fungi are considered the crucial stages in emerging approaches for fungal management. Fungal plant disease detection depends on varied strategies extending from conventional taxonomy to progressive molecular approaches. All over the past 10 years, ISH has developed into a beneficial analytical approach in diagnostic pathology laboratories. This method is also referred to as hybridization histochemistry. It offers helpful evidence to detect and compute fungi. In this method, a single-stranded RNA probe is utilized, which is also referred to as riboprobe. This technique also provides numerous benefits over antibody-mediated detection, whenever the serum concentrations levels are maximum. The foremost benefit of this method is the maximum use of the short-supply tissue such as clinical biopsies and cultured cells. Hundreds of diverse hybridizations can be performed on the same tissue. ISH has some drawbacks. First, tagged radio probes are very expensive and harmful material. It must be handled, transported and disposed very carefully. Second, a disadvantage of using in-situ hybridization technique is the difficulty in identifying targets that have low DNA and RNA copies.

10.9 FUTURE PERSPECTIVES

During past decades, the expansion of molecular diagnostic assays intended for detecting plant pathogens has been laborious and required a high level of technical proficiency. In recent years, the sum of procedures economically established to notice fungal pathogens. This has enlarged awareness of innovative molecular diagnostic tools for identifying plant pathogens. The escalation of variability of new molecular techniques drastically altered methods in plant pathology. Isolation of fungi for diagnostic purposes is no longer obligatory, outcomes from molecular tools are attained in hours as an alternative to days, and finding procedures are more detailed and delicate. These exact and delicate techniques are predominantly of interest for susceptible long-living plants such as forest trees. Evolving technologies permit the profitable and high-throughput recognition and quantification of pathogens with rapidity, sensitivity, and comfort of use (Luchi et al., 2020). Further studies were seen advanced in in-situ hybridization by combining many ISH methods for simultaneous recognition of several target nucleic acids and by growing sensitivity of ISH indications by amplification of detecting systems or by amplification of DNA or RNA targets with in-situ polymerase chain reaction (in-situ PCR). Even if there are many advances and improvements in molecular techniques, the flawless molecular method is not yet accessible; all methods contain both advantages and restrictions.

STUDY QUESTIONS

1 What is fluorescent in-situ hybridization (FISH)?
2 Expand CSIH
3 What are oligonucleotide probes?
4 What are riboprobes?
5 What are the applications of in-situ hybridization?

ACKNOWLEDGEMENT

We thank the Department of Microbiology and Department of Biotechnology, Manasagangotri, and University of Mysore. G. Adithi is thankful to the Science and Engineering Research Board, Department of Science & Technology, Government of India, New Delhi, for the support (REF. CRG/2021/008325). B. Shruthi is grateful to the Indian Council for Medical Research (ICMR; New Delhi) for providing SRF fellowship (Myco/Fell/8/2022-ECD-II).

FURTHER READING

Aslam, S., Tahir, A., Aslam, M. F., Alam, M. W., Shedayi, A. A. & Sadia, S. (2017) Recent advances in molecular techniques for the identification of phytopathogenic fungi – a mini review. *Journal of Plant Interactions*, *12*(1), 493–504. https://doi.org/10.1080/17429145.2017.1397205

Chu, Y. H., Hardin, H., Zhang, R., Guo, Z., & Lloyd, R. V. (2019). In situ hybridization: Introduction to techniques, applications and pitfalls in the performance and interpretation of assays. In *Seminars in Diagnostic Pathology* (Vol. 36, Issue 5, pp. 336–341). W.B. Saunders. https://doi.org/10.1053/j.semdp.2019.06.004

Jensen, E. (2014). Technical review: In situ hybridization. *Anatomical Record*, *297*(8), 1349–1353. https://doi.org/10.1002/ar.22944

Qian, X., Jin, L., & Lloyd, R. V. (2004). In Situ Hybridization: Basic Approaches and Recent Development. *Journal of Histotechnology*, 27(1), 53–67. https://doi.org/10.1179/his.2004.27.1.53

REFERENCES

Almeida, F., Rodrigues, M. L., and Coelho, C. (2019). The Still Underestimated Problem of Fungal Diseases Worldwide. *Frontiers in Microbiology 10*, 214. https://doi.org10.3389/fmicb.2019.00214

Ansari, B., Coates, P. J., Greenstein, B. D., & Hall, P. A. (1993). In situ end-labelling detects DNA strand breaks in apoptosis and other physiological and pathological states. *Journal of Pathology*, *170*(1), 1–8.

Bauman, J., Duijn, V., Lubsen, N., Sondermeijer, P., Hennig, W., & Kubli, E. (1981). Rapid and High-Resolution Detection of in situ Hybridisation to Polytene Chromosomes Using Fluorochrome-labeled RNA. *Chromosoma (Berl.)*, *84*, 1–18.

Brown, C. (1998). In situ hybridization with riboprobes: an overview for veterinary pathologists. *Veterinary Pathology*, *35*(3), 159–167.

Chen, T., & Nan, Z. (2015). Effects of phytopathogens on plant community dynamics: A review. *Acta Ecologica Sinica*, *35*(6), 177–183. https://doi.org/10.1016/j.chnaes.2015.09.003

Chennappa G., Adkar-Purushothama C. R., Umdale S., Tamilvendan K., Sreenivasa M. Y., 2014. Impact of Pesticides on PGPR Activity of Azotobacter sp. Isolated from Pesticide Flooded Paddy Soils. *Greener Journal of Agricultural Sciences*, *4*(4), 117–129

Cornelissen, B. J. C., & Melchers, L. S. (1993). Strategies for control of fungal diseases with transgenic plants. *Plant Physiology*, *101*, 709.

Deepa N., Achar P. N., and Sreenivasa M. Y. 2021. Current perspectives of biocontrol agents for management of *Fusarium verticillioides* and Its Fumonisin in cereals—A Review. *Journal of Fungi 2021*(7), 776.

Deepa N., Charith Raj A. P. and Sreenivasa M. Y. 2016b. Nested PCR method for the early detection of fumonisin producing *Fusarium verticillioides* in pure cultures, cereal samples and plant parts. *Food Biotechnology 30*(1), 18–29.

Deepa N., Charith Raj A. P., and Sreenivasa M. Y. 2016c. Multiplex PCR method for the early detection of fumonisin producing *Fusarium verticillioides*. *Food Bioscience 13*, 84–88.

Deepa N., Nagaraja H., Sreenivasa M. Y. 2016a. Prevalence of fumonisin producing Fusarium verticillioides associated with cereals grown in Karnataka (India). *Food Science and Human Wellness*, 5, 156–162. http://dx.doi.org/10.1016/j.fshw.2016.07.001.

Deepa N. and Sreenivasa M. Y. 2017a. Fumonisins: A Review on its Global Occurrence, Epidemiology, Toxicity and Detection. *Journal of Veterinary Medicine and Research* 4(6): 1093.

Deepa N. and Sreenivasa M. Y. 2017b. *Fusarium verticillioides*, a Globally Important Pathogen of Agriculture and Livestock: A Review. *Journal of Veterinary Medicine and Research*, 4(4), 1084.

Deepa N. and Sreenivasa M. Y. 2019a. Molecular methods and key genes targeted for the detection of fumonisin producing *Fusarium verticillioides* – An updated review. *Food Bioscience*, 32, 100473

Deepa N. and Sreenivasa M. Y. 2019b. Biocontrol strategies for effective management of phytopathogenic fungi associated with cereals, In: Singh J. S., Singh D. P. (eds) *New and Future Developments in Microbial Biotechnology and Bioengineering*, Elsevier, pp, 177–189.

DeLellis, R. A. 1994. In situ hybridization techniques for the analysis of gene expression: applications in tumor pathology. *Human Pathology*, *25*(6): 580–585.

Gall, J. G. and Pardue, M. L., 1969. Formation and detection of RNA-DNA hybrid molecules in cytological preparations. *Proceedings of the National Academy of Sciences*, *63*(2), 378–383.

Gozzetti, A. and Le Beuu, M. M. (2000). *Fluorescence In Situ Hybridization: Uses and Limitations*.

Gray, L. E., Ostby, J. S. and Kelce, W. R. (1994). Developmental effects of an environmental antiandrogen: the fungicide vinclozolin alters the sex differentiation of the male rat. *Toxicology and Applied Pharmacology*, *129*(1), 46–52.

Hanna, W. M., and Kwok, K. (2006). Chromogenic in-situ hybridization: A viable alternative to fluorescence in-situ hybridization in the HER2 testing algorithm. *Modern Pathology*, *19*(4), 481–487. https://doi.org/10.1038/modpathol.3800555

Janardhana G. R., Dass R. S. and Sreenivasa M. Y. 2009. Fumonisins – A new class of Fusarial toxins. In *Frontiers in Fungal Ecology, Diversity and Metabolites*. Ed: K. A. Sridhar et al. I. K. International Publishing House Pvt. Ltd. New Delhi.

Jensen, E. (2014). Technical review: In situ hybridization. *Anatomical Record*, *297*(8), 1349–1353. https://doi.org/10.1002/ar.22944

Jin, L., & Lloyd, R. (1997). In Situ Hybridization: Methods and Applications. *Journal of Clinical Laboratory Analysis* 11(1), 2–9.

Kiyama, H., Emson, P. C., & Tohyama, M. (1990). Recent progress in the use of the technique of non-radioactive in situ hybridization histochemistry: new tools for molecular neurobiology. *Neuroscience Research*, 9(1), 1–29.

Knez, K., Spasic, D., Janssen, K. P. F., & Lammertyn, J. (2013). Emerging technologies for hybridization based single nucleotide polymorphism detection. *Analyst*, *139*(2), 353–370. https://doi.org/10.1039/c3an01436c

Koulouri, O., Lappin, D. F., Radvar, M., & Kinane, D. F. (1999). Cell division, synthetic capacity and apoptosis in periodontal lesions analysed by in situ hybridisation and immunohistochemistry. *Journal of Clinical Periodontology*, *26*(8), 552–559. https://doi.org/10.1034/j.1600-051X.1999.260810.x

Levsky, J. M., & Singer, R. H. (2003). Fluorescence in situ hybridization: Past, present and future. *Journal of Cell Science*, *116*(14), 2833–2838. https://doi.org/10.1242/jcs.00633

Lievens, B. and Thomma, B.P., 2005. Recent developments in pathogen detection arrays: implications for fungal plant pathogens and use in practice. *Phytopathology*, *95*(12), 1374–1380.

Looi, M. D., Path, M. R. C., and Cheah P. L., 1992. In situ hybridisation: principles and applications, *The Malaysian Journal of Pathology*, *14*(2): 69–76

Luchi, N., Ioos, R., & Santini, A. (2020). Fast and reliable molecular methods to detect fungal pathogens in woody plants. *Applied Microbiology and Biotechnology* *104*(6), 2453–2468. https://doi.org/10.1007/s00253-020-10395-4

Mahlo, S. M., Chauke, H. R., McGaw, L., & Eloff, J. (2016). Antioxidant and antifungal activity of selected medicinal plant extracts against phytopathogenic fungi. *African Journal of Traditional, Complementary, and Alternative Medicines*, *13*(4), 216–222. https://doi.org/10.21010/ajtcam.v13i4.28

Manawasinghe, I. S., Phillips, A. J. L., Xu, J., Balasuriya, A., Hyde, K. D., Stępień, Ł., Harischandra, D. L., Karunarathna, A., Yan, J., Weerasinghe, J., Luo, M., Dong, Z., & Cheewangkoon, R. (2021). Defining a species in fungal plant pathology: beyond the species level. *Fungal Diversity*, *109*(1), 267–282. https://doi.org/10.1007/s13225-021-00481-x

McNicol, A. M., & Farquharson, M. A. (1997). In situ hybridization and its diagnostic applications in pathology. *The Journal of Pathology: A Journal of the Pathological Society of Great Britain and Ireland*, *182*(3), 250–261.

Mdee, L. K., Masoko, P., & Eloff, J. N. (2009). The activity of extracts of seven common invasive plant species on fungal phytopathogens. *South African Journal of Botany*, *75*(2), 375–379. https://doi.org/10.1016/j.sajb.2009.02.003

Milner, H., Ji, P., Sabula, M., & Wu, T. (2019). Quantitative polymerase chain reaction (Q-PCR) and fluorescent in situ hybridization (FISH) detection of soilborne pathogen Sclerotium rolfsii. *Applied Soil Ecology*, *136*, 86–92. https://doi.org/10.1016/j.apsoil.2019.01.002

Mitsuhas, M. (1996). Technical Report: Part 1. Basic Requirements for Designing Optimal Oligonucleotide Probe Sequences. *Journal of Clinical Laboratory Analysis*. 10(5),277–284.

Moench, T.R. (1987). In situ hybridization. *Molecular and Cellular Probes*, 1, 195–205.

Mokyr, J. and Gráda, C.Ó., (2002). What do people die of during famines: the Great Irish Famine in comparative perspective. *European Review of Economic History*, *6*(3), 339–363.

Montone, K. T., & Guarner, J. (2013). In Situ Hybridization for rRNA Sequences in Anatomic Pathology Specimens, Applications for Fungal Pathogen Detection: A Review. http://www.anatomicpathology.

Montone, K. T., Livolsi, V. A., Lanza, D. C., Feldman, M. D., Kennedy, D. W., Palmer, J., Chiu, A. G., & Nachamkin, I. (2011). Rapid In-situ Hybridization for Dematiaceous Fungi Using a Broad-spectrum Oligonucleotide DNA Probe. www.molecularpathology.com

Nagaraja H., Chennappa G., Poorna Chandra Rao K., Mahadevprasad G., and M Y Sreenivasa. 2016. Diversity of toxic and phytopathogenic *Fusarium* species occurring on cereals grown in Karnataka state, India. *3Biotech*, 6, 57.

Narayanasamy, P., 2010. *Microbial Plant Pathogens-Detection and Disease Diagnosis: Viral and Viroid Pathogens*, (Vol. 3). Springer Science & Business Media.

Nouri-Aria, K. T. (2008). In situ hybridization. *Allergy Methods and Protocols*, 138, 331–347.

Okubo, Y., Shinozaki, M., Wakayama, M., Nakayama, H., Sasai, D., Ishiwatari, T., Nemoto, T., Naobumi, T., & Shibuya, K. (2013). Applied gene histopathology: Identification of Fusarium species in FFPE tissue sections by in situ hybridization. *Methods in Molecular Biology*, *968*, 141–147. https://doi.org/10.1007/978-1-62703-257-5_10

Oliver, K. R., Heavens, R. P., & Sirinathsinghji, D. J. S. (1997). Quantitative Comparison of Pretreatment Regimens Used to Sensitize In Situ Hybridization Using Oligonucleotide Probes on Paraffin-embedded Brain Tissue. *The Journal of Histochemistry & Cytochemistry*, *45*(12), 1707–1713.

Park, K., Kim, J., Lim, S., Han, S. and Lee, J.Y., 2003. Comparing fluorescence in situ hybridization and chromogenic in situ hybridization methods to determine the HER2/neu status in primary breast carcinoma using tissue microarray. *Modern Pathology*, *16*(9), 937–943.

Peng, Y., Li, S. J., Yan, J., Tang, Y., Cheng, J. P., Gao, A. J., Yao, X., Ruan, J. J., & Xu, B. L. (2021). Research Progress on Phytopathogenic Fungi and Their Role as Biocontrol Agents. In *Frontiers in Microbiology* (Vol. 12). Frontiers Media S.A. https://doi.org/10.3389/fmicb.2021.670135

Perez-Jaffe, L. A., Lanza, D. C., Loevner, L. A., Kennedy, D. W., Montone, K. T., & Surgery, D. N. (1997). *In Situ Hybridization for Aspergillus and PeniciZZium in Allergic Fungal Sinusitis: A Rapid Means of Speciating Fungal Pathogens in Tissues.*

Pirttilä, A. M. (2011). Colonization of Tree Shoots by Endophytic Fungi. https://www.researchgate.net/publication/234095361

Prasad, M.N.N., Shankara Bhat, S. and Sreenivasa, M.Y. 2010. Antifungal activity of essential oils against *Phomopsis azadirachtae* – the causative agent of die-back disease of neem. *Journal of Agricultural Technology*, 6(1): 127–133.

Qian, X., Jin, L., & Lloyd, R. V. (2004). In Situ Hybridization: Basic Approaches and Recent Development. *Journal of Histotechnology*, *27*(1), 53–67. https://doi.org/10.1179/his.2004.27.1.53

Ray, M., Ray, A., Dash, S., Mishra, A., Achary, K. G., Nayak, S., & Singh, S. (2017). Fungal disease detection in plants: Traditional assays, novel diagnostic techniques and biosensors. *Biosensors and Bioelectronics*, *87*, 708–723. https://doi.org/10.1016/j.bios.2016.09.032

Rodrigo, S., García-Latorre, C., & Santamaria, O. (2022). Metabolites produced by fungi against fungal phytopathogens: Review, implementation and perspectives. *Plants*, *11*(1). https://doi.org/10.3390/plants11010081

Savary, S., Willocquet, L., Pethybridge, S. J., Esker, P., McRoberts, N., & Nelson, A. (2019). The global burden of pathogens and pests on major food crops. *Nature Ecology and Evolution*, *3*(3), 430–439. https://doi.org/10.1038/s41559-018-0793-y

Shinozaki, M., Okubo, Y., Sasai, D., Nakayama, H., Murayama, S. Y., Ide, T., Wakayama, M., Hiruta, N., & Shibuya, K. (2011). Identification of Fusarium species in formalin-fixed and paraffin-embedded sections by in situ hybridization using peptide nucleic acid probes. *Journal of Clinical Microbiology*, *49*(3), 808–813. https://doi.org/10.1128/JCM.01149-10

Sreenivasa, M. Y., Dass, R. S., Charith Raj, A. P., Nagendra Prasad, M. N., Achar, P. N., & Janardhana, G. R., 2011a. Assessment of the growth inhibiting effect of some plant essential oils on different *Fusarium* species isolated from sorghum and maize grains. *Journal of Plant Diseases and Protection*, 118(6), 208–213.

Sreenivasa, M. Y., R. S. Dass, A. P. Charith Raj and G. R. Janardhana. 2008. PCR-based detection of genus *Fusarium* and Fumonisin-producing isolates from freshly harvested Sorghum grains grown in Karnataka, India. *Journal of Food Safety*, 28, 236–237.

Sreenivasa, M. Y., Dass, R. S., A. P. Charith Raj and G. R. Janardhana. 2011b. Mycological evaluation of Maize grains produced in Karnataka (India) for the post-harvest fungal contamination. *World Applied Sciences Journal*, 13(4), 688–692.

Sreenivasa, M.Y., Dass Regina Sharmila, and Janardhana G. R. 2010. Post harvest fungi associated with sorghum grains produced in Karnataka (India). *Journal of Plant Protection Research*, 50(3): 335–339.

Temsamani, J., & Agrawal, S. (1996). Enzymatic labeling of nucleic acids. *Molecular Biotechnology*, *5*, 223–232.

Tenberge, K. B., Stellamanns, P., Plenz, G., & Robenek, H. (1998). Nonradioactive in situ hybridization for detection of hydrophobin mRNA in the phytopathogenic fungus Claviceps purpurea during infection of rye. *European Journal of Cell Biology*, *75*(3), 265–272. https://doi.org/10.1016/s0171-9335(98)80122-2

Toyoda, H., Nishiguchi, T., Matsuda, Y., Nonomura, T., & Ouchi, S. (1992). In Situ Hybridization Detection of mRNA in a Filamentous Phytopathogenic Fungus, Fusarium oxysporum f. sp. lycopersici, by Pricking Microinjection of Photobiotin-labeled Antisense RNA Probe. *Bioscience, Biotechnology, and Biochemistry*, *56*(8), 1306–1307. https://doi.org/10.1271/bbb.56.1306

Tyśkiewicz, R., Nowak, A., Ozimek, E., & Jaroszuk-Sciseł, J. (2022). Trichoderma: The Current Status of Its Application in Agriculture for the Biocontrol of Fungal Phytopathogens and Stimulation of Plant Growth. *International Journal of Molecular Sciences*, *23*(4). https://doi.org/10.3390/ijms23042329

Veselinyová, D., Mašlanková, J., Kalinová, K., Mičková, H., Mareková, M. and Rabajdová, M., 2021. Selected in situ hybridization methods: principles and application. *Molecules*, *26*(13), 3874.

Vicente, T. F. L., Lemos, M. F. L., Félix, R., Valentão, P., & Félix, C. (2021). Marine macroalgae, a source of natural inhibitors of fungal phytopathogens. *Journal of Fungi*, *7*(12). https://doi.org/10.3390/jof7121006

Ward, E., Foster, S. J., Fraaije, B. A., & Mccartney, H. A. (2004). Plant pathogen diagnostics: immunological and nucleic acid-based approaches. *The Annals of Applied Biology*, 145(1), 1–16.

Wilcox, J. N. (1993). Fundamental principles of in situ hybridization. *Journal of Histochemistry & Cytochemistry*, 41, (12), 1725–1733.

Yamamoto, M., Terakami, S., Takada, N., & Yamamoto, T. (2016). Physical mapping of black spot disease resistance/susceptibility-related genome regions in Japanese pear (Pyrus pyrifolia) by BAC-FISH. *Breeding Science*, *66*(3), 444–449. https://doi.org/10.1270/jsbbs.15085

Ying-Hsiaa, C., Heathera, H., Ranrana, Z., Zhenyingb, G., Ricardo, L. (2019). In situ hybridization: Introduction to techniques, applications and pitfalls in the performance and interpretation of assays. https://doi.org/10.1053/j.semdp.2019.06.004

Zabiák, A., Károlyi, G. M., & Sándor, E. (2020). Investigation of host-specificity of phytopathogenic fungi isolated from woody plants. *Acta Agraria Debreceniensis*, *1*, 155–160. https://doi.org/10.34101/actaagrar/1/3721

11 Serological Assays for the Detection of Phytopathogenic Fungi

S. Dhanush and P. K. Maheshwar
University of Mysore

Fungi are cosmopolitan in nature, and most of them are toxigenic. Phytopathogenic fungal species can cause huge losses in quantity of crop yields. Annually, 25–50% of worldwide harvested crops are contaminated with mycotoxins produced by the members of phytopathogenic genera *Aspergillus*, *Fusarium*, *Penicillium*, *Claviceps* and *Alternaria*. Precise rapid detection and identification of these fungi are essential to facilitate the effective management of disease in agricultural produce. Researchers engaged in the field of mycology need to tackle the problem of identifying phytopathogenic fungi, as well as their associated mycotoxin. Phytopathogenic fungi are known to produce five distinct mycotoxins, viz., deoxynivalenol/Nivalenol, zearalenone, ochratoxin, fumonisins and aflatoxins. Contamination of food and feed with phytopathogenic fungi and mycotoxins constitutes a major issue for food safety, food security and public health. Therefore, detection of phytopathogenic fungi is fundamental to ensure food safety. Immunological methods enable quick and precise detection of mycotoxin. Enzyme-linked immunosorbent assay (ELISA), chemiluminescence-based microplate reader and lateral flow stripassay, combined with advanced immunoassay technologies such as fluorescence polarization immunoassay, lateral flow immunochromatographic assay, are essential for the safety of food and feed. Biotic stress caused by phytopathogenic fungi substantially contributes to overall crop yield losses. Hence, immunoassays are crucial for the detection of toxigenic fungi and their associated mycotoxin in the agriculture sector. The present chapter is mainly oriented towards the discussion of advantages, disadvantages and protocols involved in serological techniques that can be applied for the detection of fungal phytopathogens.

DOI: 10.1201/9781003399704-11

GRAPHICAL ABSTRACT

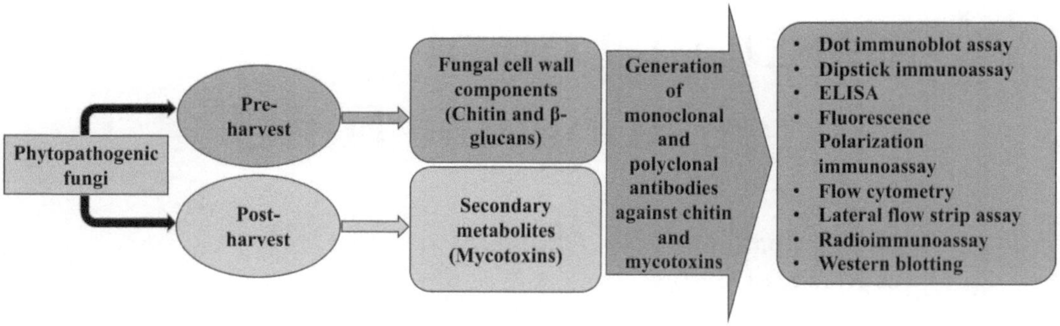

11.1 INTRODUCTION

The fungi are a group of eukaryotic achlorophyllous multicellular organisms; their distribution in nature is cosmopolitan. They are characterized by cell walls made up of a complex matrix of chitin and glycoproteins (*N*- or *O*- glycosylated with mannose side chains). The cell wall is an essential factor that aids in morphogenesis, growth and virulence (Geoghegan et al., 2017). They also have unique and unusual biochemical pathways, which produce low-molecular-weight bioactive compounds. A majority of these bioactives have potent physiological and pharmaceutical activities like penicillin, statins, cyclosporins and mycotoxins (Janardhana et al., 2009; Dass et al., 2007).

Mycotoxins are toxic, low-molecular-weight (300–700 Da) secondary fungal metabolites characterized by variety of heteroatom-containing functional groups (Marroquín-Cardona et al., 2014). Agricultural produce and foods infested with toxigenic fungi contain five distinct mycotoxins or mycotoxin groups: deoxynivalenol/Nivalenol, zearalenone, ochratoxin, fumonisins and aflatoxins (Tola and Kebede, 2016). A larger part of these toxigenic fungi is present in soil, on plants, agricultural produce, foods and spices as they need a specialized niche for their survival and spread (Horn, 2003; Logrieco et al., 2009; Aiko and Mehta, 2016; Sreenivasa et al., 2008; Deepa et al., 2016b, 2016c). Annually, 25–50% of worldwide agricultural produce pre-harvest and post-harvest are infested and contaminated by the members of genera *Aspergillus*, *Fusarium*, *Penicillium*, *Claviceps* and *Alternaria* (Alshannaq and Yu, 2017; Agriopoulou et al., 2020; Chiotta et al., 2020; Sreenivasa et al., 2011a, 2011b). By detecting a fungal infection, control measures can be implemented to minimize crop losses through the development of disease-resistant varieties and fungicides (Peng et al., 2021).

The detection of fungal phytopathogens can be performed pre-harvest, which helps in the early detection and controlling of diseases or post-harvest, to improve storage quality and conditions (Deepa and Sreenivasa, 2019a, 2019b). There are several methods for detecting phytopathogenic fungi, including immunological methods, molecular methods and morphological methods (Wakeham and White, 1996; Jellison and Jasalavich, 2000; Aslam et al., 2017).

11.2 SEROLOGICAL DETECTION OF PHYTOPATHOGENIC FUNGI

The spores of phytopathogenic fungi are found in soil; they actively inflict illness on plants, and their secretions persist in food products. Isolating fungal cellular products, such as cell wall and spore surfaces, soluble antigens, damaged cells, and extracellular antigens, is necessary for the serological detection of fungi (Nagaraja et al., 2016). The cell wall and spore surfaces are made up of chitin and *β*-glucans. Soluble antigens and broken cells are components apart from the cell and spore wall. Mainly proteins, enzymes and extracellular antigens are predominantly glycoproteins, mycotoxins and extracellular polysaccharides (Werres and Steffens, 1994).

Monoclonal and polyclonal antibodies are generated against the fungal antigens by injecting adjuvant combined antigens into animals like white rabbits, mice and mammalian cell cultures. The antibodies generated against the fungal components can be used to detect their presence using techniques like dipstick assay, dot blot assay, enzyme-linked immunosorbent assay (ELISA) and immunofluorescence. (Wakeham and White, 1996; Ward et al., 2004).

11.3 PRE-HARVEST DETECTION

The pre-harvest detection of fungi comprises of identification of spores in soil, cell wall and cellular components in infected plants and extracellular glycoproteins and polysaccharides.

11.3.1 Isolation of Fungal Spores from Soil

The rhizosphere soil and infected roots are collected and homogenized with a buffered saline solution. Density gradient ultracentrifugation with 30% sucrose density is performed to extract the spores to the supernatant. The supernatant is collected, passed through a filter with a 3 μm diameter and resuspended in a buffered saline. The buffered saline with the spores is then subjected to sonication to expose the wall polysaccharides and glycoproteins (Wakeham and White, 1996).

11.3.2 Isolation of Fungal Cellular Components from Infected Plants

The infected plant tissues are homogenized using a buffer solution with an appropriate pH; the homogenate is filtered and centrifuged to obtain a pellet. The pellet is resuspended in the buffer, and the proteins and enzymes are purified using ammonium sulphate precipitation and SDS-PAGE (Garg and Gupta, 2010).

11.3.3 Production of Antibodies against Fungal Antigenic Determinants

The production of antibodies is accomplished by using animals such as mice, rats and rabbits or using mammalian cell cultures. Rabbits and mice are injected with intramuscular injection of an adjuvant combined with antigen to elicit the immune response. Generally, 4–6 weeks following the initial immunization, the blood from animals containing the antisera will be collected. The collected blood sample will undergo clotting separating the serum and blood cells. The serum will be collected and purified by affinity chromatographic techniques (Hamaoka et al., 1971; Wakeham and White, 1996; Ward et al., 2004).

Raising antibodies in animals is a resource- and time-intensive process; in order to meet the rapid demand for antibodies mammalian cell culture is used. This process is easily scalable, robust and meets the quality and safety criteria (Figure 11.1). The production of antibodies in cell cultures requires an expression vector with efficient transcription of antibody genes and a cell line capable of efficiently translating antibody mRNAs, assembling and modifying the antibody at high rates with minimal accumulation of incorrectly processed polypeptides, and having sufficient secretory capacity for secreting the resulting assembled antibody. Commonly used cell lines are the Chinese Hamster Ovary (CHO) cell lines, murine lymphoid lines like NS0 and Sp2/0-Ag14 derived from plasmacytoma cells originating from BALB/c mice, hamster line BKH21, human PER.C6 cell line, etc. The antibodies obtained from the cell culture pass through a variety of filters to remove cellular impurities. The purification of antibodies will be done using a Protein A affinity column with ion exchange, hydrophobic interaction or size exclusion chromatography (Birch and Racher, 2006).

11.4 POST-HARVEST DETECTION METHODS

The major loss of harvested food products occurs due to three factors: biological (storage pests), microbial (bacteria and fungi) and chemical (residues of insecticides and pesticides). Fungi have

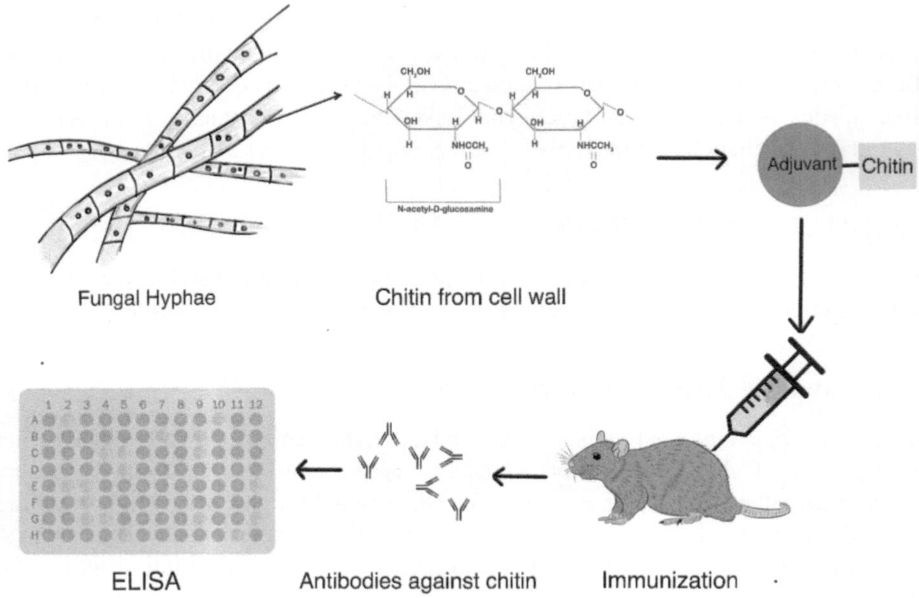

FIGURE 11.1 Overview of antibody production using mice.

TABLE 11.1
List of Serological Methods for Mycotoxin Detection

- Dot immunoblot assay
- Dip stick immunoassay
- Enzyme linked Immunosorbent Assay (ELISA)
- Fluorescence polarization immunoassay
- Flow cytometry
- Lateral flow strip assay
- Radio-immunoassay (RIA)
- Western Blotting

been designated to rank second only to insect pests as a cause of damage during the storage of grains (Bhat et al., 2010). The growth of fungi on agricultural products may or may not be conspicuous, but their metabolites such as mycotoxins persist. The post-harvest detection of phytopathogenic fungi involves the detection of fungal cell wall components and mycotoxins in food products (Deepa et al., 2016a, 2016b, 2016c). Serological methods like ELISA, flow-through immunoassay, lateral flow strip assay, radioimmunoassay, dot immunoblot assay, dipstick immunoassay, etc., serve as a rapid method for analysis of mycotoxins (Ray et al., 2017; Singh and Mehta, 2020) (Table 11.1).

11.4.1 Dot Immunoblots Assay or Dot Blot Assay

Dot immunoblot assays are carried out on a nitrocellulose membrane, which acts as the antigen carrier. Antigens are added to the membrane as dots and then a specific primary antibody is applied. The antigen-antibody reaction is visualized using a suitable detection system.

Chromogenic substrates such as 3,3,5,5-tetramethylbenzidine (TMB), 3,3′-diaminobenzidine tetrahydrochloride (DAB), 5-Bromo-4-Chloro-3-Indolyl Phosphate (BCIP), Nitro Blue Tetrazolium

(NBT), O-phenylenediamine dihydrochloride (OPD) etc., are used for visual and colorimetric detection of antigen-antibody detection. The dot blot assay can also be used to measure the living fungal material where the mycelium grows and serves as an antigen, this is performed by applying infected seeds directly onto the nitrocellulose membrane.

Dot bot assays permit rapid detection and analysis of antigens, DNA, proteins, and screening of monoclonal antibodies, and the samples do not require separation by electrophoresis which saves time. The disadvantages of this technique are as follows – it is a semi-quantitative test, and the samples used have to be purified to get specific results as impure samples show nonspecific binding (Werres and Steffens, 1994; Kandan et al., 2010; Shekhar et al., 2010; Ray et al., 2017).

11.4.2 Dipstick Immunoassay

In dipstick immunoassay, the antibodies are coated onto the nitrocellulose membrane. The membrane is then incubated in a sample containing mycotoxins and chromogenic substrates. The dipstick can be used in the field for real-time detection of pathogens. The dipstick is incubated with the sample and the labelled secondary antibody. At first, the mycotoxin binds to the primary antibody immobilized on the membrane, to which the labelled secondary antibody is bound. The secondary antibody is an enzyme-conjugated antibody, the enzyme is released upon binding and breaks down the chromogenic substrate, resulting in the change of colour of the sample. The colour change is measured colorimetrically or visually. The advantages of dipstick immunoassay are high specificity, cost-effectiveness and portability. The assay is not quantitative and may exhibit cross-reactivity (Werres and Steffens, 1994; Dewey et al., 1997; Lattanzio et al., 2012; Ray et al., 2017; Mishra et al., 2018).

11.4.3 Enzyme-linked Immunosorbent Assay

ELISA is a quantitative test based on the competitive interactions between antigens and antibodies labelled with enzyme conjugates. The mycotoxin serves as an antigen. ELISA is also called double sandwich technique. The antibodies generated against the antigen are adsorbed onto the surface of a polystyrene plate. The sample with antigen is added to the plate, the antigens will bind to the antibodies. Another antibody with an enzyme called a conjugated antibody is added to the plate; these antibodies will bind to the antigen. The chromogenic substrate added to the plate will be modified by the enzyme, and the intensity of the coloured reaction products formed will be proportional to the amount of antigen that is bound. Hence, by determining the colour intensity produced, antigen concentration can be calculated. It is a rapid, specific and simple method for the analysis of mycotoxins in food and its products. The disadvantages include cross-reactivity, requirement of a specific matrix and detection of only one mycotoxin at a time (Gaastra, 1984; Zheng et al., 2006; Alshannaq and Yu, 2017; Ray et al., 2017; Singh and Mehta, 2020). The limit of detection is 50ng, below which ELISA cannot detect the antigen/antibody.

11.4.4 Fluorescence Polarization Immunoassay

Fluorescence polarization immunoassay is based on a competitive binding reaction between mycotoxin and a mycotoxin-fluorescein tracer for a mycotoxin-specific antibody. The antibody can react with both mycotoxin and the tracer. The unbound fluorophore upon receiving plane polarized light undergoes excitation and emission resulting from its tumbling movement, the light emitted will be highly depolarized. When the fluorophore is bound to an antibody, it cannot get excited easily and will emit a light having less degree of depolarization. The depolarized light emitted by the fluorophore can be grouped into horizontal and vertical components using a polarizer and can be used to calculate the concentration of mycotoxin in the sample.

When the mycotoxin concentration is high, the binding of mycotoxin to the antibody will be maximum, which will result in a high concentration of unbound tracers, leading to the emission of highly depolarized light. Subsequently, when the concentration of mycotoxin is low, tracer molecules competitively bind to the antibody, thereby resulting in the emission of low depolarized light. The polarization value is inversely proportional to the concentration of mycotoxin.

This immunoassay offers two advantages over ELISA: that is, there is no involvement of an enzyme reaction – hence, no time is wasted on colour development – and it doesn't require a wash step to separate bound and unbound particles. The drawback of this technique includes detection of low-molecular-weight antigens, photobleaching of sample and phototoxicity (Phipps et al., 2008; Sharma et al., 2014; Singh and Mehta, 2020).

11.4.5 Flow Cytometry

Flow cytometry is a technology used to rapidly analyse single cells or particles in a fluid stream as they flow through single or multiple lasers. Characteristics like size, granularity and fluorescent features of the analyte are used in the differentiation and separation of components. When the particles pass through the light beam, they scatter the light; the scattered light is measured in forward and side directions to determine its size and granularity, respectively. The scattering of light is an attribute of its structural and morphological properties; the fluorescence emission is the measure of the number of probes bound to the analysing particle. Fluorescent proteins like green fluorescent protein (GFP), fluorescent dyes (propidium iodide) or fluorescently conjugated antibodies are used to label the particles. Flow cytometry has applications in many biological disciplines such as virology, cancer biology, molecular biology and the monitoring of infectious diseases (Adan et al., 2017; Ray et al., 2017; McKinnon, 2018).

11.4.6 Lateral Flow Strip Assay or Lateral Flow Test

It is an immunochromatographic test based antigen–antibody reactions for the quick analysis of analytes with high sensitivity and specificity. It consists of a strip containing a sample pad and a conjugate pad with an antibody conjugated to a gold particle, a membrane containing a second antibody and an absorbent pad. The sample is applied onto the sample pad, which will react with the anti-mycotoxin antibodies forming an immune complex, the immune complex migrates together and binds with the second antibody present in the membrane. The membrane contains a test zone where the mycotoxin-antibody and the second antibody complex will dry and absorb any free anti-mycotoxin antibody-gold particle, resulting in no colour change, indicating a positive sample. When the sample has no mycotoxins, the gold particles accumulate in the test zone and result in a colour change, indicating a negative result. The control zone always shows a colour change. This technique offers low-cost rapid results without the involvement of detection devices and has commercial applications in real-time mycotoxin detection (Zheng et al., 2006; Singh and Mehta, 2020).

11.4.7 Radioimmunoassay (RIA)

Radioimmunoassay uses a similar principle to ELISA but differs in the use of radioactive isotopes for labelling the antibodies. Instead of using a spectrophotometer to visualize the antigen-antibody reaction autoradiography is performed to measure the radiation emitted by the radioactive isotopes. This technique is rarely used as it is expensive but can help in the detection of antigens whose antibodies cannot be conjugated with an enzyme (Sharon et al., 1993; Werres and Steffens, 1994; Ray et al., 2017; Azim et al., 2018).

11.4.8 Western Blotting

Western blotting is an immunodetection technique that helps in the detection of antigens. The mycotoxins are separated using SDS-polyacrylamide gel electrophoresis. The separated mycotoxins are then transferred onto a nitrocellulose membrane. The nitrocellulose membrane is incubated in a solution of primary antibodies, followed by a buffer wash to remove unbound antibodies. The membrane is treated with a secondary antibody conjugated with an enzyme capable of breaking down a chromogenic substrate. Generally, the enzyme used is horseradish peroxidase, and the chromogenic substrate is H_2O_2-TMB. Western blotting is a laborious process and requires highly specific antibodies against an antigen (Werres and Steffens, 1994; Kurien and Scofield, 2006; Yang and Ma, 2009).

11.5 ADVANTAGES

The main advantage of serological detection methods is that they require very little automation in comparison to techniques like polymerase chain reaction (PCR). Serological techniques enable the user to test a sample on any platform, in the lab as well as on the field. These could be the best rapid platform tests available for plant pathology. Serological techniques are rapid, easily standardized for new specimens, and user-friendly.

11.6 DISADVANTAGES

Serological assays are extremely sensitive and therefore require a high-precision sample extraction. The reactions are influenced by enzyme inhibitors that may be present in biological samples. Nucleases and other antigen molecules in samples or in reagents may show false negative/positive results. Extremely high care needs to be taken in handling biological specimens, reagents, as well as reaction products.

11.7 APPLICATIONS

Serological techniques may be used to detect all possible diseased specimens of plant sap, root, shoot, etc. Even seed-borne fungi may be detected in a few minutes and therefore, can be used by agriculture-based researchers in field conditions. The technique may be used for qualitative detection of seed-born fungal pathogens in the seed certification agency as a part of qualitative assessment. The technique offers huge applications in the taxonomic characterization of fungal pathogens.

11.8 SUMMARY

Fungi are cosmopolitan and are capable of inducing a variety of plant diseases and spoilage of agricultural products. The early detection of fungi and their metabolic by-products helps in controlling the spread of diseases and spoilage of produce. The detection of these phytopathogenic fungi can be performed pre-harvest and post-harvest. The serological detection of pre-harvest fungi involves immunologically detecting the fungal cell wall components. In post-harvest detection, fungal secondary metabolites such as mycotoxins are detected. Techniques such as dipstick assays, dot immunoblot assays, ELISA, fluorescence polarization assay, flow cytometry, radioimmunoassay, western blotting, etc. Serological techniques provide easy and rapid detection of fungi. Some techniques are relatively cheap like dipstick assay, immune blot assay and lateral flow strip assay, while others like fluorescence polarization immunoassay, western blotting and ELISA are on the expensive side but offer highly specific results.

11.9 FUTURE PERSPECTIVES

Serological diagnostic methods are vital components of human society. They occupy top positions, especially in human disease diagnostics all over the globe. Thus, it has created a huge market for entrepreneurs in developing serological kits for almost every known microorganism affecting the human population. Nevertheless, the technology has yet to reach its full scale in the diagnosis of fungal pathogens. The serological techniques are widely used for the diagnosis of phytopathogenic viruses but seldom used for fungal diagnosis. Therefore, in the forthcoming years, the precision offered by serological techniques should be introduced more vigorously to mycopathology, especially for seed-borne fungal detection.

STUDY QUESTIONS

1. What is the application of green fluorescent protein?
2. Write the comparison between western blot and ELISA.
3. Discuss lateral flow test.
4. Describe radioimmunoassay.
5. What is the principle involved in flow cytometry?

ACKNOWLEDGEMENT

The authors would like to thank the University of Mysore, India.

FURTHER READING

Singh, J. and Mehta, A. (2020) 'Rapid and sensitive detection of mycotoxins by advanced and emerging analytical methods: A review', *Food Science and Nutrition*, 8(5), pp. 2183–2204. doi: 10.1002/fsn3.1474.

REFERENCES

Adan, A. et al. (2017) 'Flow cytometry: basic principles and applications', *Critical Reviews in Biotechnology*, 37(2), pp. 163–176. doi: 10.3109/07388551.2015.1128876

Agriopoulou, S., Stamatelopoulou, E. and Varzakas, T. (2020) 'Advances in analysis and detection of major mycotoxins in foods', *Foods*, 9(4), pp. 1–23. doi: 10.3390/foods9040518

Aiko, V. and Mehta, A. (2016) 'Prevalence of toxigenic fungi in common medicinal herbs and spices in India', *3 Biotech*, 6(2), pp. 1–10. doi: 10.1007/s13205-016-0476-9

Alshannaq, A. and Yu, J.H. (2017) 'Occurrence, toxicity, and analysis of major mycotoxins in food', *International Journal of Environmental Research and Public Health*, 14(6). doi: 10.3390/ijerph14060632

Aslam, S. *et al.* (2017) 'Recent advances in molecular techniques for the identification of phytopathogenic fungi – A mini review', *Journal of Plant Interactions*, 12(1), pp. 493–504. doi: 10.1080/17429145.2017.1397205

Azim, M.A.U. et al. (2018) 'Chemiluminescence immunoassay: Basic mechanism and applications', *Bangladesh Journal of Nuclear Medicine*, 18(2), pp. 171–178. doi: 10.3329/bjnm.v18i2.35240

Bhat, R., Rai, R.V. and Karim, A.A. (2010) 'Mycotoxins in food and feed: Present status and future concerns', *Comprehensive Reviews in Food Science and Food Safety*, 9(1), pp. 57–81. doi: 10.1111/j.1541-4337.2009.00094.x

Birch, J.R. and Racher, A.J. (2006) 'Antibody production', *Advanced Drug Delivery Reviews*, 58(5–6), pp. 671–685. doi: 10.1016/j.addr.2005.12.006

Chiotta, M.L. *et al.* (2020) 'Toxigenic fungal species and natural occurrence of mycotoxins in crops harvested in Argentina', *Revista Argentina de Microbiologia*, 52(4), pp. 339–347. doi: 10.1016/j.ram.2020.06.002

Dass, R.S., Sreenivasa, M.Y. and Janardhana, G.R. (2007) High incidence of *Fusarium verticillioides* in animal and poultry feed mixtures produced in Karnataka, India. *Plant Pathology Journal*, 6(2), pp. 174–178.

Deepa, N., Charith Raj, A.P., Sreenivasa, M.Y. (2016b) Nested PCR method for the early detection of fumonisin producing *Fusarium verticillioides* in pure cultures, cereal samples and plant parts. *Food Biotechnology*, 30(1), 18–29.

Deepa, N., Charith Raj, A.P., Sreenivasa, M.Y. (2016c) Multiplex PCR method for the early detection of fumonisin producing *Fusarium verticillioides*. *Food Bioscience*, 13, 84–88.

Deepa, N., Nagaraja, H., Sreenivasa M. Y. 2016a. 'Prevalence of fumonisin producing Fusarium verticillioides associated with cereals grown in Karnataka (India)', *Food Science and Human Wellness*, *5*(3), pp. 156–162.

Deepa, N. and Sreenivasa, M.Y. (2019a) Molecular methods and key genes targeted for the detection of fumonisin producing *Fusarium verticillioides* – An updated review. *Food Bioscience*, 32, 100473.

Deepa, N. and Sreenivasa, M.Y. (2019b) Biocontrol strategies for effective management of phytopathogenic fungi associated with cereals. In J.S. Singh and D.P. Singh (Eds.), *New and future developments in microbial biotechnology and bioengineering*, Elsevier, pp, 177–189, ISBN 9780444641915.

Dewey, F.M., Thornton, C.R. and Gilligan, C.A. (1997) 'Use of monoclonal antibodies to detect, quantify and visualize fungi in soils', *Advances in Botanical Research*, 24(C), pp. 275–308. doi: 10.1016/S0065-2296(08)60076-2

Gaastra, W. (1984) Enzyme-Linked Immunosorbant Assay (ELISA). In: Walker, J.M. (eds) Proteins. Methods in Molecular Biology™, vol 1. Humana Press. doi: 10.1385/0-89603-062-8:349

Garg, N. and Gupta, H. (2010) 'Isolation and purification of fungal pathogen (*Macrophomina phaseolina*) induced chitinase from moth beans (*Phaseolus aconitifolius*)', *Journal of Pharmacy and Bioallied Sciences*, 2(1), p. 38. doi: 10.4103/0975-7406.62708

Geoghegan, I., Steinberg, G. and Gurr, S. (2017) 'The role of the fungal cell wall in the infection of plants', *Trends in Microbiology*, 25(12), pp. 957–967. doi: 10.1016/j.tim.2017.05.015

Hamaoka, T. *et al.* (1971) 'Antibody production in mice', *British Society of Immunology*, 17(1), pp. 871–881.

Horn, B.W. (2003) 'Ecology and population biology of aflatoxigenic fungi in soil', *Journal of Toxicology* 22, pp. 351–379. doi: 10.1081/TXR-120024098

Janardhana, G.R., Dass, R.S. and Sreenivasa, M.Y. (2009) Fumonisins – A new class of Fusarial toxins. In K.A. Sridhar et al. (Eds), *Frontiers in fungal ecology, diversity and metabolites*. New Delhi: I. K. International Publishing House Pvt. Ltd.

Jellison, J. and Jasalavich, C. (2000) 'A review of selected methods for the detection of degradative fungi', *International Biodeterioration and Biodegradation*, 46(3), pp. 241–244. doi: 10.1016/S0964-8305(00)00075-5

Kandan, A. *et al.* (2010) 'Development and evaluation of an enzyme-linked immunosorbent assay (ELISA) and dot immunobinding assay (DIBA) for the detection of Ganoderma infecting palms', *Archives of Phytopathology and Plant Protection*, 43(15), pp. 1473–1484. doi: 10.1080/03235400802536915

Kurien, B.T. and Scofield, R.H. (2006) 'Western blotting', *Methods*, 38(4), pp. 283–293. doi: 10.1016/j.ymeth.2005.11.007

Lattanzio, V.M.T. *et al.* (2012) 'Multiplex dipstick immunoassay for semi-quantitative determination of Fusarium mycotoxins in cereals', *Analytica Chimica Acta*, 718, pp. 99–108. doi: 10.1016/j.aca.2011.12.060

Logrieco, A., Moretti, A. and Solfrizzo, M. (2009) 'Alternaria toxins and plant diseases: an overview of origin, occurrence and risks', 2(May), p. 3920. doi: 10.3920/WMJ2009.1145

Marroquín-Cardona, A.G. *et al.* (2014) 'Mycotoxins in a changing global environment - A review', *Food and Chemical Toxicology*, 69, pp. 220–230. doi: 10.1016/j.fct.2014.04.025

McKinnon, K.M. (2018) 'Flow cytometry: An overview', *Current Protocols in Immunology*, 2018(February), pp. 5.1.1–5.1.11. doi: 10.1002/cpim.40

Mishra P. *et al.* (2018) 'An immunochromatographic dipstick as an alternate for monitoring of heroin metabolites in urine samples', *RSC Advances*, 8(41), pp. 23163–23170. doi: 10.1039/C8RA02018C

Nagaraja, H., Chennappa, G., Poorna Chandra Rao, K., Mahadevprasad, G. and Sreenivasa, M.Y. (2016) Diversity of toxic and phytopathogenic *Fusarium* species occurring on cereals grown in Karnataka state, India. *3Biotech*, 6, 57.

Peng, Y. *et al.* (2021) 'Research progress on phytopathogenic fungi and their role as biocontrol agents', 12(May). doi: 10.3389/fmicb.2021.670135

Phipps, R.J. *et al.* (2008) Chapter 2 Current methods for the separation and analysis of cocaine analytes. *Handbook of Analytical Separations*. 2nd ed. Elsevier B.V. doi: 10.1016/S1567-7192(06)06002-5

Ray, M. et al. (2017) 'Fungal disease detection in plants: Traditional assays, novel diagnostic techniques and biosensors', *Biosensors and Bioelectronics*, 87(September 2016), pp. 708–723. doi: 10.1016/j.bios.2016.09.032

Sharma, A. *et al.* (2014) Mycotoxins: Immunological techniques for detection and analysis. *Encyclopedia of Food Microbiology*. 2nd ed. Elsevier. doi: 10.1016/B978-0-12-384730-0.00233-0

Sharon A., Amsellem, Z. and Gressel, J. (1993) 'Quantification of infection by *Alternaria cassiae* using leaf immuno-autoradiography and radioimmunosorbent assays', *Journal of Phytopathology*, 138(3), pp. 233–243. doi: 10.1111/j.1439-0434.1993.tb01381.x

Shekhar, M.S., Kumar, D. and Gopikrishna, G. (2010) 'Development of immunodot blot assay for the detection of white spot syndrome virus infection in shrimps (*Penaeus monodon*)', *Aquaculture Research*, 41(11), pp. 1683–1690. doi: 10.1111/j.1365-2109.2010.02553.x

Singh, J. and Mehta, A. (2020) 'Rapid and sensitive detection of mycotoxins by advanced and emerging analytical methods: A review', *Food Science and Nutrition*, 8(5), pp. 2183–2204. doi: 10.1002/fsn3.1474

Sreenivasa, M.Y., Dass, R.S., Charith Raj, A.P. and Janardhana, G.R. (2008) PCR-based detection of genus *Fusarium* and Fumonisin-producing isolates from freshly harvested Sorghum grains grown in Karnataka, India. *Journal of Food Safety*, 28, pp. 236–237.

Sreenivasa, M.Y., Dass, R.S., Charith Raj, A.P. and Janardhana, G.R. (2011b) Mycological evaluation of maize grains produced in Karnataka (India) for the post-harvest fungal contamination. *World Applied Sciences Journal*, 13(4), pp. 688–692.

Sreenivasa, M.Y., Dass, R.S., Charith Raj, A.P., Prasad, M.N.N., Achar, P.N. and Janardhana, G.R. (2011a) Assessment of the growth inhibiting effect of some plant essential oils on different *Fusarium* species isolated from sorghum and maize grains. *Journal of Plant diseases and Protection*, 118(6), pp. 208–213.

Tola, M. and Kebede, B. (2016) 'Occurrence, importance and control of mycotoxins: A review', *Cogent Food and Agriculture*, 2(1). doi: 10.1080/23311932.2016.1191103

Wakeham, A.J. and White, J.G. (1996) 'Serological detection in soil of *Plasmodiophora brassicae* resting spores', *Physiological and Molecular Plant Pathology*, 48(5), pp. 289–303. doi: 10.1006/pmpp.1996.0024

Ward, E. *et al.* (2004) 'Plant pathogen diagnostics: Immunological and nucleic acid-based approaches', *Annals of Applied Biology*, 145(1), pp. 1–16. doi: 10.1111/j.1744-7348.2004.tb00354.x

Werres, S. and Steffens, C. (1994) 'Immunological techniques used with fungal plant pathogens: Aspects of antigens, antibodies and assays for diagnosis', *Annals of Applied Biology*, 125(3), pp. 615–643. doi: 10.1111/j.1744-7348.1994.tb04998.x

Yang, Y. and Ma, H. (2009) 'Western blotting and ELISA techniques', *Researcher*, 1(2), pp. 67–86.

Zheng, M.Z., Richard, J.L. and Binder, J. (2006) 'A review of rapid methods for the analysis of mycotoxins', *Mycopathologia*, 161(5), pp. 261–273. doi: 10.1007/s11046-006-0215-6

12 Detection of Fungal Phytopathogens by the Nanotechnology Method

Manjunatha Dadayya, Sowmya Hirakannavar Veeranna, and Thippeswamy Basaiah
Kuvempu University

Plant pathogens play essential roles as yield-limiting agents, having a significant negative impact on crop productivity worldwide. This phenomenon poses significant threats to food security and remains a global agricultural issue of extreme significance. Repeated application of chemical treatment can lead to resistance to pathogens. In addition to contributing to environmental pollution, excessive spraying can have a significant effect on soil microbes. To ensure agricultural sustainability and food security, the use of efficient diagnostic methods for the rapid identification of plant pathogens during the initial stages of infection is of the utmost importance. In order to achieve this aim, numerous molecular techniques have been developed to expedite the identification of plant pathogens. However, these techniques are time-consuming, expensive, dependent on expert operators and insufficient for in situ studies in general. Utilizing nanotechnology tools such as microneedle patches, nanopore sequencing, nano-barcoding, nanobiosensors, quantum dots, nano-diagnostic kit equipment, metal nanoparticles, miRNA-based nano-diagnosis and array-based nanosensors enables the achievement of plant protection through efficient plant pathogen diagnosis. As they are increasingly recognised as a promising technique for enhancing the sensitivity, accuracy and speed of plant pathogen identification, and for facilitating high-throughput analysis, they are growing in popularity.

DOI: 10.1201/9781003399704-12

GRAPHICAL ABSTRACT

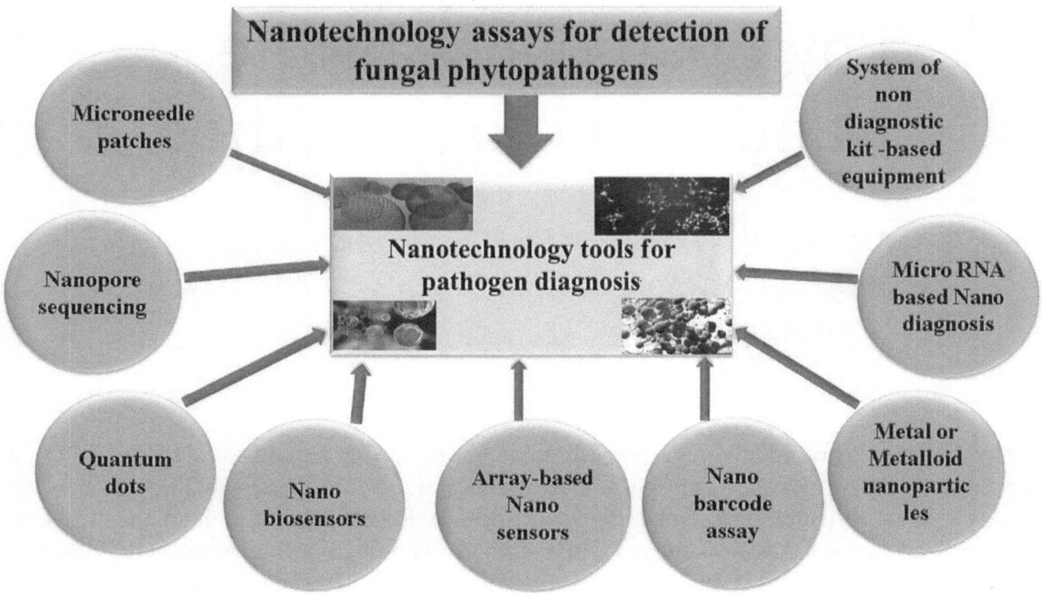

12.1 INTRODUCTION

More than a billion people worldwide are facing hunger, dealing with varying levels of food scarcity and inadequate nutrition. As our global population exceeds 7 billion, the demand for food is intensifying, putting increased pressure on agricultural methods to boost crop yields. Every year, food crops encounter threats from biotic stresses, leading to reduced yields and compromised quality. Among these challenges, plant diseases caused by pathogen infections and pest infestations have become major obstacles to agricultural productivity, presenting a significant global concern. Approximately 14% and 13% of crop losses globally are attributed to insect pests and plant diseases, respectively, amounting to an annual cost of around 2 trillion dollars (FAO, 2018). For instance, attacks by pathogens on crucial economic crops like potatoes, maize, peanuts and soybeans result in annual losses ranging from 10% to 25% of total production (Janardhana et al., 2009; Dass et al., 2007). Notably, the costs associated with pathogen detection and control remain relatively low, making up less than 3% of the overall expenses linked to crop production (Debnath, 2013; Kaminski and Christiaensen, 2014; Kumar and Kalita, 2017).

To improve crop productivity, foster agricultural sustainability and counteract the harmful impacts of plant diseases, it is crucial to employ diagnostic and evaluation methods known for their speed, accuracy and reliability. Currently, the identification of many plant diseases relies on visually observing plant features and leaf characteristics. Unfortunately, these diseases are often identified at an advanced stage, making it challenging to implement effective treatment strategies (Tran et al., 2020; Sreenivasa et al., 2008; Deepa et al., 2016a). In the past, advancements in controlling pathogens heavily leaned on biological identification approaches, involving the multiplication of pathogens within symptomatic plants and analysing their biological properties (Deepa and Sreenivasa, 2019b). While this approach offers satisfactory diagnostic accuracy, its implementation is impractical due to its time-intensive nature and limited suitability for widespread use. The identification of plant pathogenic organisms is typically conducted in laboratories using traditional molecular diagnostic techniques, such as polymerase chain reaction (PCR), enzyme-linked immunosorbent assay (ELISA), colony counting and fluorescence in situ hybridization (FISH) (Oluwaseun et al., 2018; Deepa et al., 2016b, 2016c; Deepa and Sreenivasa, 2019a). These established methods are renowned for providing precise and accurate results. However, challenges arise from the cost and limited shelf life of molecular biology reagents like enzymes and primers. Additionally, their utilization requires specialized equipment and laboratory infrastructure, restricting their applicability in economically challenged nations (Mark et al., 2014).

Nanotechnology, a fascinating interdisciplinary field merging principles from physics, chemistry and related disciplines focuses on manipulating individual atoms, molecules or molecular groups to create structures with distinct functions, as Joseph and Morrison (2006) elucidate. In agriculture, the application of nanotechnology has the potential to revolutionize research and enable advanced techniques for promptly identifying plant diseases and infections. Nanomaterials, characterized by their nanoscale dimensions typically ranging from 1 to 100 nm, prove exceptionally suitable for detecting plant pathogens. Within this size range, nanomaterials boast an increased surface area-to-volume ratio, resulting in unique chemical, optical and electrical characteristics not found in larger materials. They come in various shapes, including spherical, cubic, rod-like, wire-like, plate-like, prismatic core-shell and intricate three-dimensional structures. Notably, nanomaterials have the remarkable ability to change shape in response to external stimuli, thereby modifying their physical and chemical properties. This adaptability of nanomaterials serves as a significant sensing mechanism, holding the potential to enhance the speed and accuracy of plant pathogen detection compared to existing diagnostic methods. For a detailed comparison, please refer to Table 12.1.

12.2 MICRONEEDLE PATCHES

Micro-analytical devices, like microneedle (MN) patches designed to penetrate the skin, have gained popularity in healthcare applications. Their appeal lies in their non-invasive drug delivery, the ability to collect diagnostic analytes and real-time monitoring of essential clinical indicators. In the realm of plant disease diagnostics, there is a challenge in cost-effectively collecting pathogenic DNA from plant samples during the genomic DNA amplification process. This challenge has been effectively addressed by introducing a rapid DNA extraction technique that utilizes a reusable polymeric MN patch, providing a straightforward and economical solution. The MN patch roughly the size of a postcard stamp, is equipped with multiple needles, each measuring 0.8 millimetres in length and distributed across its surface. By applying microneedle patches to plant leaves (Figure 12.1) the extraction of amplification-ready DNA from various plant species can be accomplished in less than 1 minute. This eliminates the need for additional purification

TABLE 12.1
Comparison of Methods for the Detection of Plant Pathogens

Currently Available Methods	Detection Limit	Time Required for AN ASSAY	Time Before Results	Advantages
Conventional methods	10^7–10^8 cfuml^{-1}	72 h	72 h	Low cost of assay and high specificity
Nucleic acid-based methods	10^3–10^4 cfuml^{-1}	1–3 h	6–12 h	Mature and common technology, easy to handle, high sensitivity, high accuracy
Immunological methods	10^5–10^6 cfuml^{-1}	1–3 h	6–12 h	Low cost, portable, provide quantification and distinguish pathogens taxonomically
Nanotechnology based methods	1 fmol $^{-1}$	1 h	1 h	Rapid, easy to handle, cost effective, high specificity and sensitivity, excellent on-field portability, high throughput analysis

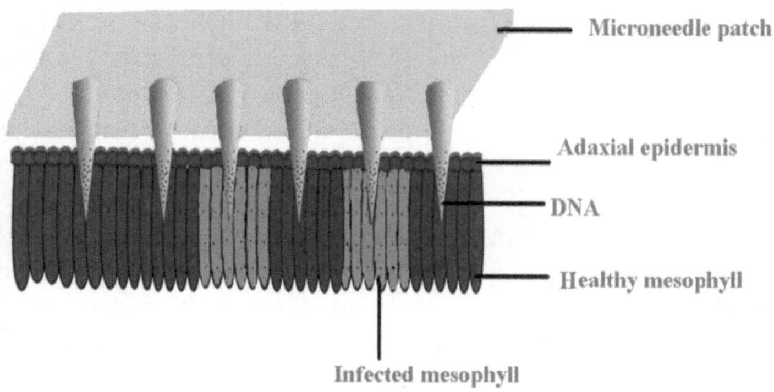

FIGURE 12.1 The extraction of plant pathogen DNA through the application of microneedle patches onto plant leaves. As depicted, the microneedles delicately puncture the leaf tissue, establishing channels through which DNA from both the plant and potential pathogens is absorbed onto the surface of the microneedle tips.

procedures. It's noteworthy that DNA from a plant plastid can be immediately amplified using PCR with DNA extracted via microneedle.

Significant strides have been made in the agricultural sector, particularly in the swift and accurate identification of *Phytophthora infestans* – an oomycete responsible for late blight diseases in potatoes and tomatoes. This progress is attributed to the innovative use of a polymer-based microneedle patch technique. Building upon these achievements, a comprehensive MN-smartphone amplification platform has been developed to facilitate on-site diagnostics for the rapid detection of plant diseases, achieving a remarkable turnaround time of just 30 minutes. The platform seamlessly integrates a thermally activated Loop-Mediated Isothermal Amplification (LAMP) cassette, a polymeric MN patch and a smartphone viewing system crafted using cutting-edge 3D printing technology. The MN employs a swift compression and retraction method to efficiently extract nucleic acids from plant leaves, completing the entire process within a minute. As elucidated by Paul et al. (2021), isothermal amplification experiments like LAMP can be conducted by placing extracted RNA or DNA samples and pre-dried assay chemicals onto an assay strip. The autofluorescence emitted from the LAMP cassette is then captured by a smartphone reader, enabling the detection of both DNA and RNA with a sensitivity as low as 1 picogram.

A comprehensive approach that analyses both DNA and RNA has proven to be highly accurate in identifying pathogens like the tomato spotted wilt virus (TSWV) and *Phytophthora infestans* in tomato leaves, whether sourced from field collections or laboratory infections. While current microneedle designs lack the required specificity for efficient isolation of nucleic acids and other intracellular analytes, the integration of adept microneedle patch techniques with compact DNA amplification assays provides a practical method for the rapid and portable detection of plant pathogens (Koo et al., 2013).

Advantages of Microneedle Patches:

- Non-Invasive: Microneedle patches offer a non-invasive method for various applications, including drug delivery and DNA extraction. This reduces the discomfort associated with traditional needles.
- Cost-Effective DNA Extraction: The described microneedle patch provides a cost-effective solution for DNA extraction, eliminating the need for additional purification steps. This is particularly advantageous in resource-limited settings.
- Rapid DNA Extraction: The microneedle patch allows for rapid DNA extraction from plant samples, completing the process in less than 1 minute. This quick turnaround time is crucial for on-site diagnostics and real-time monitoring.
- Ease of Use: Microneedle patches are relatively easy to use, making them suitable for applications in various fields, including healthcare and agriculture.

- Integration with Diagnostic Platforms: The integration of microneedle patches with diagnostic platforms, such as the MN-smartphone amplification platform, enhances their versatility and utility for on-site diagnostics.
- Reduced Need for Additional Purification: The microneedle patch's ability to extract amplification-ready DNA without the need for additional purification steps simplifies the overall diagnostic process.

Disadvantages of Microneedle Patches:

- Specificity Issues: Some microneedle designs may lack the required specificity for efficient isolation of nucleic acids and other analytes. This could lead to potential issues in the accuracy of diagnostics.
- Limited Depth of Penetration: Microneedle patches typically have a limited depth of penetration. This may affect their suitability for certain applications that require deeper tissue access.
- Potential for Cross-Contamination: The reuse of microneedle patches, as described in the context of DNA extraction, may raise concerns about potential cross-contamination between samples.
- Technological Challenges: The development of advanced microneedle technologies, such as those integrated with diagnostic platforms, may pose technological challenges and require specialized manufacturing processes.
- Adoption Challenges: The adoption of microneedle patches in certain applications may face resistance due to concerns related to their efficacy, safety and acceptance among end-users.

Microneedle patches offer several advantages, addressing specificity issues and ensuring their safe and effective use in various applications will be crucial for their widespread adoption. Continuous research and development are essential to overcome existing challenges and enhance the performance of microneedle patch technologies.

12.3 NANOPORE SEQUENCING PLATFORM

Nanopore sequencing faces two primary challenges. The first issue involves accurately identifying the nucleotides within a DNA strand as it moves through the nanopore. The second challenge is effectively controlling the speed at which the DNA strand traverses the nanopore. In this study, a method is employed using protein nanopores and specially engineered enzymes to manipulate individual DNA strands, allowing for direct electrical analysis as they move through the nanopore. The protein nanopore is positioned on the upper surface of a microwell embedded in a polymer bilayer membrane. Each microwell contains a sensor chip to measure the ionic current generated when a single molecule passes through the nanopore. However, achieving precise identification faces a significant obstacle due to the rapid movement of the DNA molecule across the nanopore (Clarke et al., 2009).

The creation of the Nano-SBS (Nanopore-based Single-Molecule Sequencing) DNA sequence involved incorporating four unique tags generated from nucleotides modified with a 5'-phosphate group. As outlined by Kumar et al. (2012), these tags play a crucial role in distinguishing the four DNA bases. Nano-scanning blotting is employed to sequence elongated RNA and DNA sequences. Additionally, Zhang et al. (2011) introduced the concept of a "DNA transistor," an advanced sequencing approach that records the nucleotide array by guiding the template through a nanopore sensor. IBM and Roche jointly contributed to the development of this innovative approach.

Recent research efforts employing nanopore sequencing technology have primarily focused on diagnosing plant diseases. In a study conducted by Bronzato Badial et al. (2018), whole transcriptome amplification and nanopore sequencing techniques were employed to rapidly identify two viral species, *Candidatus Liberibacter asiaticus* and plum pox virus, in peaches. Impressively, the study yielded results within a remarkably short timeframe of 24 hours. This success was made possible by using the MinION nanopore sequencing device, developed by Oxford Nanopore Technologies, allowing real-time sequencing of extensive DNA or RNA readings. During this process, the protein nanopore facilitates the passage of nucleic acids while simultaneously observing and analysing variations in electrical current. This unique mechanism enables the identification and interpretation of distinct DNA or RNA sequences based on the resulting signal. The adaptability of plant virus detection is showcased by the use of portable sequencers like MinION, combined with adaptive data analytics from Oxford Nanopore Technologies. Consequently, they find extensive use in various activities related to monitoring plant viruses, identifying novel viruses and other pertinent endeavours. The effectiveness of nanopore sequencing in diagnosing plant infections, such as tomato yellow leaf curl virus and watermelon chlorotic stunt virus, has been recently evaluated by sequencing DNA or RNA extracted from virus-infected plant samples. Notably, the assay is efficiently conducted within a time frame of less than two hours, producing results that align closely with outcomes from established diagnostic techniques like PCR and ELISA (Chalupowicz et al., 2019).

This study underscores the nanopore platform's capability to directly sequence DNA or RNA for virus detection. The research delves into a novel and comprehensive diagnostic method utilizing nanopore sequencing, particularly focusing on the detection of various diseases, notably *Xylella fastidiosa*. The method allows for the simultaneous assessment and identification of multiple diseases, demonstrating a sensitivity equivalent to traditional diagnostic assays. Moreover, the results consistently provided quantitative data in alignment with studies based on quantitative polymerase chain reaction (QPCR). A noteworthy feature of this approach is the incorporation of an internal control mechanism, enabling the cessation of sequencing upon obtaining sufficient data. This innovation significantly expedites the process of pathogen identification and genotyping, achieving an impressive timeframe of just 13 minutes (Marcolungo et al., 2022).

The incorporation of this feature, coupled with the use of portable devices, is poised to streamline the development of advanced diagnostic tests for effective monitoring of various plant diseases. However, it is crucial to note that nanopore sequencing techniques exhibit lower read fidelity compared to short-read sequencers. Additionally, there is currently a lack of readily available, fast, accurate, and user-friendly bioinformatics resources for data interpretation. Despite these challenges, nanopore sequencing methods have demonstrated significant effectiveness in diagnosing plant diseases. They hold the potential to provide precise, rapid, and on-site diagnostic solutions to address diverse plant protection needs in the foreseeable future.

Advantages of Nanopore Sequencing Platform:

- Real-Time Sequencing: One of the key advantages of nanopore sequencing is its ability to perform real-time sequencing, allowing for immediate data generation and analysis.
- Long Read Lengths: Nanopore sequencing can generate long read lengths, facilitating the sequencing of complex genomic regions, repetitive elements, and structural variations.
- Single-Molecule Sequencing: It enables single-molecule sequencing, eliminating the need for PCR amplification and reducing biases associated with amplification steps.
- Portability: Nanopore sequencing devices, such as MinION, are portable and compact, allowing for field applications and on-site diagnostics.
- Adaptability: The platform is adaptable to various applications, including whole genome sequencing, transcriptomics, and diagnostics for plant diseases.
- Cost-Effective: The initial costs of nanopore sequencing devices are relatively lower compared to some other sequencing technologies.

Disadvantages of Nanopore Sequencing Platform:

- Read Fidelity: Nanopore sequencing tends to have lower read fidelity compared to short-read sequencing platforms, leading to higher error rates in the generated sequences.

- Base Calling Accuracy: Accurate base calling can be challenging, especially in homopolymeric regions, resulting in errors in the identified nucleotides.
- Throughput: While nanopore sequencing has improved in throughput, it may still lag behind some high-throughput sequencing technologies, limiting its scalability for certain applications.
- Bioinformatics Challenges: Interpretation of nanopore sequencing data requires sophisticated bioinformatics tools. As of now, there is a need for more user-friendly and accurate bioinformatics resources.
- Sample Preparation: The preparation of samples for nanopore sequencing can be complex, and the platform may be sensitive to the quality and quantity of input DNA or RNA.
- Instrument Cost: Although the initial instrument cost is lower, the per-sample cost can be higher due to consumables and reagents.

Nanopore sequencing offers unique advantages such as real-time sequencing and portability; it also poses challenges related to accuracy and bioinformatics analysis. Ongoing technological advancements and research are likely to address some of these limitations, making nanopore sequencing an increasingly valuable tool in genomics and diagnostics, particularly for plant diseases.

12.4 QUANTUM DOTS

Quantum dots (QDs), semiconductor nanocrystals with luminescent properties, emit light at a wavelength determined by their size. According to Edmundson et al. (2014), the wavelength of the infrared light produced by QDs increases as their size grows. Quantum dots offer several advantages over organic dyes, including a broad range of excitation spectra. These substances exhibit molar extinction coefficients approximately 10–100 times higher, feature tunable fluorescence peaks, and possess prolonged fluorescence lifespans, effectively resisting photobleaching. The inherent characteristics of quantum dots enable the generation of multiple colours from a single source without signal interference, surpassing traditional fluorophores in terms of brightness (Zhao and Zeng, 2015). The widespread use of QD-FRET (Quantum Dot Forster Resonance Energy Transfer)-based nanosensors in agriculture and related sectors can be attributed to their exceptional characteristics. One prevalent application of these sensors is in detecting activities related to enzymes and nucleic acids (Stanisavljevic et al., 2015).

In the early stages of exploring semiconductor nanomaterial mycosynthesis, scientists observed the remarkable ability of unicellular yeast to generate cadmium sulphide (CdS) crystallites when exposed to cadmium salt stress. This intriguing phenomenon led researchers to investigate the biosynthetic capabilities of microorganisms, particularly fungi, which seemed to produce nanoparticles through a process called biosynthesis. The presence of enzymes in the cell membrane, capable of reducing metal ions, was identified as a likely factor contributing to this ability (Yadav et al., 2015).

For instance, *Fusarium oxysporum* was found to be proficient in producing highly luminescent CdTe quantum dots by interacting with a combination of $CdCl_2$ and $SeCl_4$ at normal ambient temperature (Kashyap et al., 2013). The potential applications of QDs extended to the field of nanosensors, with studies by Knudsen et al. (2013) demonstrating their effectiveness in detecting a wide range of enzyme activity.

Recent advancements in biosensors included the use of CdTe quantum dots coated with specific antibodies targeting the Polymyxa betae glutathione S-transferase (GST) protein. This innovative approach leveraged the comparable attraction between rhodamine and CdTe quantum dots, bringing them into close proximity to facilitate Forster resonance energy transfer (FRET). The resulting immunosensor exhibited exceptional specificity and sensitivity, enabling the rapid screening of plant samples on a large scale, with results obtainable in as little as 30 minutes.

In a notable study, Rad et al. (2012) enhanced the accuracy of a nanobiosensor using QDs to detect phytoplasma (*P. aurantifolia*) infection in lime plants, achieving an impressive specificity rate of 100%.

Another significant development involved the creation of a DNA biosensor utilizing optical FRET and artificial QD to detect a specific DNA sequence of Ganoderma boninense (MohdBakhori et al., 2013). This advancement included the conjugation of a single-stranded DNA probe (ssDNA) to enhanced quantum dots (5–8 nm) possessing carboxylic acid groups through amide coupling. The efficacy of the QD-FRET-based sensor in detecting Polymyxa betae, the exclusive carrier of beet necrotic yellow vein virus (BNYVV) causing infection in beet plants, was demonstrated (Joshi et al., 2019).

Furthermore, researchers employed CdSe QDs to enhance the precision of detecting the banana bunchy top virus (BBTV). The primary antibody utilized in this study was derived from the recombinant coat protein of BBTV. The incorporation of CdSe quantum dots significantly amplified the electrical signals in the test, making it particularly suitable for laboratory applications.

Despite the recognition of QD-based biosensors as a distinctive sensor type, offering novel possibilities for the detection of plant diseases, there is a notable shortage of dedicated research in this specific area. The unique photophysical properties of quantum dots as an interface component suggest the potential for a transformative impact on the identification of plant pathogens in the future (Majumder et al., 2020).

Advantages of Quantum Dots:

- Size-Tunable Emission: QDs exhibit size-dependent emission, allowing for precise tuning of the emitted light wavelength based on their size.
- Broad Excitation Spectra: QDs have a broad range of excitation spectra, enabling their use in various imaging and sensing applications with different excitation sources.
- High Extinction Coefficients: QDs possess high molar extinction coefficients, resulting in enhanced brightness and signal intensity compared to traditional fluorophores.
- Resistance to Photobleaching: QDs have prolonged fluorescence lifespans and are resistant to photobleaching, ensuring stable and long-lasting fluorescence signals.
- Multiplexing Capability: The inherent characteristics of QDs enable the generation of multiple colours from a single source without signal interference, allowing for multiplexing in biological applications.
- Exceptional Brightness: Quantum dots surpass traditional fluorophores in terms of brightness, making them suitable for high-sensitivity imaging and detection.
- Biosynthetic Potential: QDs can be biosynthesized by microorganisms, particularly fungi, offering a sustainable and environmentally friendly method for their production.
- Specificity and Sensitivity: Quantum dot-based biosensors exhibit exceptional specificity and sensitivity, enabling the rapid and accurate detection of target molecules, enzymes, or nucleic acids.

Disadvantages of Quantum Dots:

- Toxicity Concerns: Some quantum dots, especially those containing heavy metals like cadmium, may raise toxicity concerns, limiting their use in certain biological and medical applications.
- Biocompatibility Challenges: Achieving full biocompatibility of quantum dots for in vivo applications can be challenging, and surface modifications are often required.
- Cost: Quantum dots, particularly those made from rare or expensive materials, can be costlier compared to traditional fluorophores, impacting their widespread adoption.
- Complex Synthesis: The synthesis of quantum dots can be complex, requiring precise control over conditions to achieve the desired size, morphology and surface properties.
- Photostability: While quantum dots are generally resistant to photobleaching, their photostability may vary based on environmental conditions and the specific composition.
- Limited Research Focus: Despite their potential, there is a shortage of dedicated research in the specific application of quantum dot-based biosensors, limiting the exploration of their full capabilities.

- Environmental Impact: The potential release of heavy metals from quantum dots, if not managed properly, could have environmental implications.

Quantum dots offer unique advantages in terms of optical properties and biosynthetic potential, addressing concerns related to toxicity, biocompatibility and cost will be crucial for their widespread and safe application in various fields, including plant disease detection.

12.5 NANOBIOSENSORS

Nanobiosensors feature specialized bioreceptor probes immobilized to target molecules in samples, offering a range of advantages such as compact size, mobility, immediate response, accuracy, computational reliability, stability, repeatability and resilience. These characteristics make them highly effective in identifying both anticipated and complex plant health issues. Currently, these technologies play a crucial role in detecting and measuring minute amounts of harmful substances in agriculture and food systems, covering viruses, bacteria, fungi, poisons and other hazardous chemicals (Srinivasan and Tung, 2015).

The implications of nanosensors extend significantly to precision agricultural approaches (Rai and Ingle, 2012). They facilitate the study of disease patterns and the development of preventive strategies, enabling the early detection of plant pathogens through portable nanobiosensors. Placing these sensors strategically in various locations within fields, connected to a Global Positioning System (GPS), allows for real-time monitoring of disease occurrence, soil quality and crop health (Nezhad, 2014). The integration of biotechnological and nanotechnological techniques enhances biosensor sensitivity, enabling prompt detection and response to environmental changes and disease outbreaks. This proactive approach enables the identification of plant illnesses before visible symptoms appear, leading to more efficient disease management. Biosensors contribute to precision farming by providing accurate data and decision-making support, ultimately enhancing agricultural production and productivity (Rai and Ingle, 2012).

Advancements in nanotechnology and biotechnology have transformed nanostructures into advanced sensing platforms. These structures offer significant specific surface area, size-dependent electrical properties and the potential to reduce machinery size (Sertova, 2015). The use of nanomaterials, including carbon nanomaterials like nanotubes and graphene, nanofibers, nanotubes and nanostructured metal oxide nanoparticles, has greatly accelerated infection detection (Malhotra et al., 2014). Microfluidic systems incorporating nanostructures are exceptionally effective in swiftly and accurately detecting infections. A system developed by Bhattacharya et al. in 2007 showcases diverse nanostructured platforms to detect food deterioration caused by bacterial and fungal germs. This system employs thousands of nanomaterials that change colours upon interaction with food pathogens.

For the efficient immobilization of r-IgG and bovine serum albumin (BSA) to detect OTA generated by *Aspergillus ochraceus*, an immunoelectrode made of nanoSiO2 and chitosan proves to be the most effective method. The utilization of magnetic nanoparticles enables the development of an electrochemical immunosensor capable of detecting extremely low levels of AFM1 (Aflatoxin M1) produced by *A. flavus* in food, with sensitivity extending to as low as 0.01 parts per billion (ppb). A preliminary model using a microfluidic chip and an electrokinetic magnetic bead-based electrochemical immunoassay has been created for the quick, accurate and specific detection of zearalenone generated by *Fusarium* sp. Additionally, a highly sensitive immunosensor utilizing continuous-flow technology and multiwall carbon nanotubes has been developed for the precise and rapid detection of zearalenone in corn silage samples (Panini et al., 2010).

Traditional ELISA faces limitations, particularly related to diffusion constraints. To address these challenges, a highly sensitive magnetic nanoparticle immunoassay was developed for the detection of a wide range of mycotoxins. This innovative approach utilized magnetic nanoparticles as the solid phase, effectively immobilizing reactants and evenly dispersing them in the reaction media. This led to a significant increase in accessible surface area. The application of a magnetic field facilitated swift and efficient separation of substances, streamlining the cleansing procedures compared to conventional microplate-based ELISA. The method, referenced from Mak et al. (2010), successfully identified mycotoxins, including aflatoxin B1, zearalenone, and HT-2 mycotoxins.

Colorimetric biosensors, recognized for their ability to rapidly and visually confirm the presence of hazardous bacteria by detecting colour changes, stand out among optical biosensors. In solution-based colorimetric sensors, colloidal gold nanoparticle-mounted receptors interact with target pathogens, leading to nanoparticle aggregation and a visible colour change from red to purple. Colloidal gold-plated nanoparticles have been employed in horizontal flow immunoassays for various agricultural diseases, such as Potato Virus X in potatoes, *Fusarium* sp. in maize, and *Pantoea stewartii* bacterium in corn. However, a notable drawback of biosensors based on lateral flow assays is their limited sensitivity. The integration of biosensing technology with synthetic biology for plant pathogen diagnosis requires careful consideration of factors such as familiarity with farming systems, suitable in-field sampling techniques, and determining the optimal biological concentration of target pathogens. Despite the importance of research-based biosensors, further studies are needed in this field. Given the significance of plant pathogen diagnosis and the strides made in nanobiosensors, it is expected that ongoing efforts will inspire scientists to redirect their focus towards plant pathogen diagnosis and other aspects of pathology using nanobiosensors (Drygin et al., 2012, and Pohlmann et al., 2014).

Advantages of Nanobiosensors:

- Compact Size: Nanobiosensors are compact, making them suitable for portable and field applications.
- Mobility: Nanobiosensors offer mobility, allowing for on-site and real-time monitoring of environmental and biological parameters.
- Immediate Response: Nanobiosensors provide immediate response, enabling quick detection of target substances.
- Accuracy: These biosensors demonstrate high accuracy in detecting and measuring minute amounts of substances, including viruses, bacteria and chemicals.
- Computational Reliability: Nanobiosensors can integrate with computational systems, enhancing the reliability of data analysis and interpretation.
- Stability and Repeatability: Nanobiosensors exhibit stability and repeatability, ensuring consistent and reliable performance over multiple uses.
- Resilience: Nanobiosensors are resilient, with the potential to withstand challenging environmental conditions.
- Enhanced Sensitivity: Integration of nanotechnology enhances biosensor sensitivity, allowing for the detection of low concentrations of target molecules.
- Precision Agriculture: Nanobiosensors contribute to precision agriculture by providing accurate data for monitoring disease occurrence, soil quality and crop health.

Disadvantages of Nanobiosensors:

- Toxicity Concerns: Certain nanomaterials used in biosensors may raise toxicity concerns, especially in applications involving contact with living organisms.
- Biocompatibility Challenges: Achieving full biocompatibility of nanomaterials can be challenging, requiring careful surface modifications.
- Cost: Some nanomaterials can be expensive, impacting the cost-effectiveness of nanobiosensors.
- Complex Synthesis: The synthesis of nanomaterials for biosensors can be complex, requiring precise control over properties.
- Limited Research: Despite advancements, there is a shortage of dedicated research in specific areas of nanobiosensor applications, limiting their full exploration.

- Diffusion Constraints: Traditional biosensors, including nanobiosensors, may face diffusion constraints, impacting their performance.
- Sensitivity Limitations: Certain biosensors, like lateral flow assays, may have limited sensitivity compared to other detection methods.
- Integration Challenges: Integrating biosensing technology with synthetic biology for plant pathogen diagnosis requires careful consideration of various factors.
- Ongoing Research Needs: Further studies are needed to explore the full potential of nanobiosensors, particularly in plant pathogen diagnosis and other aspects of pathology.

Nano biosensors offer numerous advantages; addressing concerns related to toxicity, cost, and integration challenges is crucial for their widespread and safe application in various fields. Continued research and development are needed to unlock their full potential.

12.6 ARRAY-BASED NANOSENSORS

Sensor assemblies arranged in arrays, similar to the human olfactory system, utilize synthetic nanomaterials or chromophores to detect and distinguish diverse analytes simultaneously. One well-known variant is the electronic nose (e-nose), which replaces chemical sensors with electronic transducers. Arrays prove advantageous in discerning analyte combinations with similar traits due to cross-reactivity, and they can evaluate chemical fingerprints. Analysing the multidimensional output data from sensor arrays through chemometric methods, whether guided or unsupervised, can yield highly accurate and predicted outcomes (Li et al., 2018).

In contrast to e-nose signals, array-based chemical sensors can generate optical readouts, such as colorimetric or fluorescent signals, which are simpler to interpret. These chemical sensor arrays are often more cost-effective, easier to construct, and less susceptible to external factors like humidity and temperature fluctuations compared to most electronic nasal devices. Additionally, they are established based on precise chemical interactions, leading to increased chemical specificity. Creating these arrays is easily achievable by incorporating different dyes responsive to analytes, such as Lewis acid-base, Bronsted acidic or basic, solvatochromic or vapochromic, and redox indicator dyes, onto hydrophobic nanoporous substrates like modified silica sol-gels (Askim et al., 2013).

An effective indirect method for diagnosing plant illnesses involves detecting the volatile chemical signature released by diseased plants. Stressed plants emit or alter the composition of specific volatile organic compounds (VOCs), serving as markers for encountered biotic or abiotic stressors. E-noses, coated with conductive polymers or made of metal oxides, are gaining attention as gas sensors with multiple layers. They can detect VOC biomarkers in response to plant stress events such as pathogen infections, insect infestations, or physical traumas. Detection techniques include chemiresistive or colorimetric methods for identifying phytohormone ethylene. Innovative nanoplasmonic sensor arrays, like gold nanoparticles (Au-NPs) and molecularly imprinted sol-gels (AuNP-s @ MISG), aim for specific identification of terpenes. Smartphone-integrated VOC sensing systems, using plasmonic nanoparticles, distinguish tomato late blight from similar fungal diseases through unique colour patterns. These systems, applied onto paper strips and analysed with a smartphone microscope, prove suitable for detecting various abiotic challenges in plants, including physical damage, drought, nutritional deficits and widespread diseases (Shang et al., 2018).

Portable vapour analysers, incorporating gas chromatography (GC), tandem mass spectroscopy (MS) or ion mobility spectroscopy (IMS), serve as mobile devices for on-site plant health evaluation. These devices enable the detection and measurement of individual VOC species. Particularly noteworthy are handheld MEMS devices, paired with small GC-MS/IMS sensors, for their effectiveness in non-optical plant health monitoring. These devices provide convenience in sample collection and quick analysis, making them valuable tools in practical applications (Sharma et al., 2019).

Advantages of Array-Based Nanosensors:

- Simultaneous Detection: Array-based nanosensors can detect and distinguish diverse analytes simultaneously, similar to the human olfactory system.
- Cross-Reactivity Advantage: Arrays are advantageous in discerning analyte combinations with similar traits due to cross-reactivity, providing a broader understanding of chemical fingerprints.
- Chemometric Analysis: The multidimensional output data from sensor arrays can be analysed through chemometric methods, leading to highly accurate and predicted outcomes.
- Optical Readouts: Array-based chemical sensors can generate optical readouts, such as colorimetric or fluorescent signals, making interpretation simpler.
- Cost-Effectiveness: Chemical sensor arrays are often more cost-effective than electronic noses, easier to construct and less susceptible to external factors like humidity and temperature fluctuations.
- Chemical Specificity: Chemical sensor arrays are established based on precise chemical interactions, leading to increased chemical specificity in detection.
- Simplicity of Interpretation: Optical readouts, generated by array-based sensors, are simpler to interpret compared to electronic nose signals.
- Versatility: Different types of dyes responsive to analytes can be incorporated into array-based sensors, enhancing their versatility.
- Less Susceptibility to Environmental Factors: Chemical sensor arrays are less susceptible to external factors like humidity and temperature fluctuations compared to electronic noses.

Disadvantages of Array-Based Nanosensors:

- Limited Sensitivity: Array-based nanosensors may have limited sensitivity compared to some specialized sensors, affecting their ability to detect trace amounts of analytes.
- Complexity in Array Construction: Constructing sensor arrays with various dyes and substrates may involve complexity in the manufacturing process.
- Need for Chemometric Analysis: The chemometric analysis required for interpreting the data from sensor arrays may demand specialized expertise.
- Potential for False Positives: Cross-reactivity, while providing an advantage, can also lead to potential false positives in certain situations.
- Dependency on Volatile Organic Compounds (VOCs): Detection of plant illnesses through VOCs depends on the emission or alteration of specific volatile organic compounds by stressed plants.
- Less Specificity in Some Cases: In certain instances, the specificity of chemical sensor arrays may be lower compared to highly specialized sensors.
- Integration Challenges: Integrating array-based sensors with practical applications, such as smartphone-based systems, may pose technical challenges.
- Reliability in Field Conditions: The reliability of array-based nanosensors in field conditions, particularly in diverse environmental settings, needs careful consideration.

Array-based nanosensors offer several advantages; addressing limitations related to sensitivity, construction complexity and potential false positives is essential for their successful application in various fields. Ongoing research and development are crucial for improving their performance and expanding their practical applications.

12.7 NANO-BARCODE ASSAY

Advancements in nanotechnology have given rise to a novel technique known as bio-barcoding, revolutionizing the highly sensitive identification of proteins and DNAs without the reliance on enzymes. Protein barcode tests offer increased complexity, sensitivity and versatility compared to traditional ELISA-based assays, which hinge on target and sample density. The nanoparticle-based bio-barcode test, surpassing techniques like ELISA and qPCR, exhibits exceptional sensitivity in early-stage pathogen detection for plant diseases (Bao et al., 2006). The bio-barcoding technique employs a pair of probes:

1. Magnetic Microbeads (MMB): Specially engineered to transport antibodies or DNA, MMB acts as a biological probe binding to a specific target.
2. Au-NPs (Gold Nanoparticles): Comprising either polyclonal antibodies or oligonucleotides, Au-NPs serve as bio-barcodes. This highly sensitive technique identifies proteins and DNA without the need for enzymes or PCR amplification.

The field of bio-barcode technology has witnessed significant progress owing to nanotechnology breakthroughs, offering an efficient means of identifying protein complexes and nucleic acids without relying on enzymes or PCR amplification. Ongoing enhancements aim to bolster the efficacy of bio-barcode technology across diverse applications. For fungal identification, the utilization of DNA barcoding has been proposed, emphasizing the importance of a uniform and scalable DNA barcode, particularly in the protein-coding domain.

In the detection of *P. aeruginosa*, a fluorescent bio-barcode approach, in conjunction with gold nanoparticles, has been implemented. Two probes, one for target identification and the other as a signal identifier, create covalent bonds with their respective sequences. This assay, conducted via fluorescence spectroscopy, ensures rapid, straightforward, highly responsive and wide-ranging detection capability (5–200 ng/mL).

The hybrid structure formed by Oligo-AuNP sensors and oligonucleotide-functionalized magnetic microparticle (MMP) probes allows molecular targets to serve as connectors. After magnetic separation of the complexes, the oligonucleotides are liberated from the Oligo-AuNP tags. This scientometric technique enables quantitative evaluation of emitted bio-barcodes and exhibits significant potential, particularly in optimal conditions, for swiftly detecting protein complexes at extremely low concentrations (attomole) and nucleotides at extremely high concentrations (zeptomolar). Amini et al. (2017) propose the possibility of developing similar methods for promptly identifying plant illnesses, specifically viruses, with the aim of reducing crop losses.

Advantages of Nano-Barcode Assay:

- Increased Sensitivity: Nano-barcode assays, especially those utilizing gold nanoparticles and magnetic microbeads, offer exceptional sensitivity, allowing for the detection of proteins and DNA at very low concentrations.
- Versatility: Nano-barcode assays are versatile and can be applied to various targets, including proteins and nucleic acids, without the need for enzymes or PCR amplification.
- Efficient Identification: The bio-barcoding technique efficiently identifies protein complexes and nucleic acids, providing a reliable means of detection.
- Complexity and Specificity: Protein barcode tests offer increased complexity, sensitivity and versatility compared to traditional ELISA-based assays, providing more specific and accurate results.
- No Reliance on Enzymes: Nano-barcode assays do not rely on enzymes for detection, reducing the complexity of the assay and eliminating potential enzyme-related issues.
- Potential for Rapid Detection: The fluorescent bio-barcode approach, in conjunction with gold nanoparticles, allows for rapid and straightforward detection, making it suitable for early-stage pathogen detection.
- Wide Detection Range: The bio-barcoding technique exhibits a wide-ranging detection capability, making it applicable for a broad spectrum of concentrations (5–200 ng/mL in the case of *P. aeruginosa* detection).

- Quantitative Evaluation: Nano-barcode assays, such as the hybrid structure formed by Oligo-AuNP sensors and oligonucleotide-functionalized magnetic microparticle probes, enable quantitative evaluation of emitted bio-barcodes.
- Reduced Crop Losses: The potential application of nano-barcode assays in promptly identifying plant illnesses, especially viruses, holds promise for reducing crop losses.

Disadvantages of Nano-Barcode Assay:

- Technical Complexity: Despite advancements, nano-barcode assays may still involve technical complexities in their design and implementation.
- Optimization Challenges: Achieving optimal conditions for the efficient detection of protein complexes and nucleotides may pose challenges and require careful optimization.
- Cost Considerations: The cost of implementing nano-barcode assays, including the required nanoparticles and equipment, may be a consideration for widespread adoption.
- Limited Scope: While versatile, nano-barcode assays may have a somewhat limited scope compared to other detection methods, depending on the specific application.
- Need for Specialized Expertise: The design and interpretation of nano-barcode assays may require specialized expertise in nanotechnology and molecular biology.
- Compatibility Issues: The compatibility of nano-barcode assays with different sample types and conditions needs to be carefully considered and may vary based on the application.
- Limited Information on DNA Barcoding: The field of DNA barcoding for fungal identification is still evolving, and there may be challenges in developing a uniform and scalable DNA barcode, especially in the protein-coding domain.

Nano-barcode assays offer numerous advantages in terms of sensitivity, versatility and efficiency; addressing technical complexities, optimization challenges and cost considerations is essential for their successful implementation in various applications. Ongoing research and development will likely contribute to overcoming these challenges and expanding the practical use of nano-barcode assays.

12.8 METAL OR METALLOID NANOPARTICLES

Nanosensors, utilizing inorganic nanomaterials such as Silver (Ag), Gold (Au), Silicon (Si), and metal oxide nanoparticles, play a pivotal role in identifying and measuring target molecules. These nanosensors employ signal transducers and recognition ligands like antibodies or DNA oligos, serving as distinct sensing platforms for studying interactions, particularly relevant in plant systems and disease detection. The use of localized surface plasmon resonance (LSPR) in nanoparticles is a common approach, capturing specific sensor responses by exploiting alterations in surface-enhanced optical characteristics.

Extensive research has focused on self-assembled nanoparticles incorporating oligonucleotides or proteins to create nanobiosensors. These sensors, commonly using surface-enhanced Raman spectroscopy (SERS), often employ Ag nanorods as substrates to detect various plant diseases and toxins. For instance, nanosensors coated with IgG antibodies significantly improved accuracy in detecting *Bacillus thuringiensis* and *B. subtilis* in rhizosphere soil and carrot roots using mass spectrometry. Fluorescent silica nanoparticles, infused with a Ru (II) complex, proved effective in identifying plant pathogens like *X. campestris*, causing bacterial spot disease in nightshade plants. While these approaches are responsive, it's crucial to note that the use of heavy metal complexes raises environmental concerns. Modified Au nanoparticles, with specific single-stranded DNA, effectively identified genetic material from *R. solanacearum*, responsible for potato bacterial wilt, even at low concentrations (15 ng). Although the test exhibited selectivity, quantitative calibration data within the lower detectable range was lacking.

Another quick and accurate method for identifying *P. syringae* DNA utilized Au-NPs in an electrochemical biosensor. Rapid isothermal amplification (RPA) amplified specific pathogenic DNA sequences in under 60 minutes, demonstrating a significant accuracy enhancement (10,000-fold) compared to traditional PCR/gel electrophoresis methods. This technique could detect *P. syringae* infection even before visible signs of illness appeared.

Silver nanoparticles (AgNPs) are not only effective in combating bacteria and fungi in antimicrobial applications but also show promise in diagnosing plant pathogens. AgNPs have demonstrated efficacy against various fungal phytopathogens, including *Sclerotinia sclerotiorum*, *Rhizoctonia solani* and *S. minor*, with potential economic benefits. They inhibit fungal invasion and colonization in plant tissue cells, even at low concentrations.

Research on AgNPs against *Raffaelea* sp., responsible for oak wilt disease, showed their effectiveness in suppressing fungal hyphae development and conidia germination in a concentration-dependent manner. Moreover, AgNPs extended the vase life of *Gerbera jamesonii* cv. Ruikou flowers by suppressing bacterial growth. The utilization of nano-silver and silica compounds, exhibiting antibacterial properties, has shown effectiveness against various plant infections. These compounds resist pathogenic fungi and improve disease resistance by physically obstructing them. However, the optimal concentration for effective microbiological protection needs careful consideration. The current trend in using metal/metal oxide nanoparticles for the rapid and precise detection of plant diseases indicates their potential impact on mitigating crop diseases in economically important crops. Ongoing research suggests that metal/metal oxide nanoparticles may play a substantial role in the precise detection and management of crop diseases.

Advantages of Metal or Metalloid Nanoparticles:

- Antimicrobial Properties: Metal and metalloid nanoparticles, such as silver (Ag), gold (Au), and silicon (Si), exhibit antimicrobial properties, making them effective against bacteria and fungi. This property is advantageous for combating plant pathogens.
- Sensitive Detection: Metal nanoparticles, especially gold and silver, can be employed in sensitive detection methods, such as SERS and electrochemical biosensors, providing accurate and quick identification of target molecules, including pathogens.
- Localized Surface Plasmon Resonance: The use of LSPR in nanoparticles enhances their sensing capabilities, capturing specific sensor responses by exploiting alterations in surface-enhanced optical characteristics, leading to improved sensitivity in detection.
- Self-Assembled Nanobiosensors: Self-assembled nanoparticles incorporating ligands like antibodies or DNA oligos contribute to the development of nanobiosensors, allowing for the study of interactions in plant systems and disease detection.
- Rapid Isothermal Amplification: Metal nanoparticles, particularly gold, can be utilized in rapid isothermal amplification methods, enhancing the speed and accuracy of DNA detection compared to traditional PCR methods.
- Broad-Spectrum Efficacy: Silver nanoparticles (AgNPs) have demonstrated efficacy against various fungal phytopathogens, indicating a broad-spectrum effectiveness in combating plant diseases.
- Economic Benefits: Metal nanoparticles, particularly AgNPs, may offer economic benefits by inhibiting fungal invasion and colonization in plant tissue cells, potentially reducing crop losses.
- Versatile Applications: Metal and metalloid nanoparticles have versatile applications, ranging from antimicrobial coatings to diagnostic tools, showcasing their potential impact in various fields, including agriculture.

Disadvantages of Metal or Metalloid Nanoparticles:

- Environmental Concerns: The use of heavy metal nanoparticles, such as silver, raises environmental concerns due to their potential persistence and impact on ecosystems.

- Quantitative Calibration Challenges: Some detection methods utilizing metal nanoparticles may face challenges in quantitative calibration, especially in the lower detectable range, impacting the accuracy of concentration measurements.
- Specificity Issues: While certain nanobiosensors exhibit selectivity, ensuring specificity in the detection of target molecules, including pathogens, can be challenging.
- Optimal Concentration Requirements: The optimal concentration of metal or metalloid nanoparticles for effective microbiological protection needs careful consideration, as excessively high concentrations may have unintended consequences.
- Limited Data on Long-Term Effects: The long-term effects of the application of metal or metalloid nanoparticles in agriculture, especially in terms of plant and soil health, are still under investigation, and limited data may be available.
- Ethical and Regulatory Considerations: The use of nanoparticles in agriculture raises ethical and regulatory considerations related to their potential impact on human health, the environment, and food safety.

Metal and metalloid nanoparticles offer several advantages for plant disease detection and management, addressing environmental concerns, ensuring specificity, and understanding their long-term effects are critical aspects that require ongoing research and consideration. Balancing the benefits and potential risks is essential for the responsible and sustainable application of these nanoparticles in agriculture.

12.9 MICRORNA-BASED NANO-DIAGNOSIS

RNA molecules, with their remarkable diversity in structure and function, including ribozymes, aptamers, microRNAs, and tiny interfering RNAs, have garnered significant attention. In the realm of RNA nanotechnology, an innovative approach is employed for detecting and treating cancers and viral infections in plants. Noncoding miRNAs, crucial for gene expression regulation, particularly those consisting of 18–22 nucleotides, play a pivotal role in controlling translation. RNA nanotechnology facilitates the intricate process of editing disease-specific miRNAs for therapeutic purposes, offering possibilities to study, control and govern protein and enzyme functions. The modification of miRNAs can also contribute to gene conservation, safeguarding crops. Notably, plant miRNAs, akin to animal miRNAs, are often produced by RNA polymerase II, providing substantial adaptability (Chaudhary et al., 2018).

Given their crucial role in regulating gene products affected by diseases, miRNAs hold promise as biomarkers for early disease detection across various species. Previous approaches using real-time polymerase chain reaction and microarrays for identifying polluted miRNAs faced limitations due to the short sequences of miRNAs, requiring specific timings and complex coded primers. The integration of nanotechnology emerges as a potential solution to overcome these challenges, enhancing sensitivity and precision in miRNA identification.

Gold nanoparticles functionalized with thiols, amines and disulfides play a significant role in advancing highly sensitive and durable detection systems. The graphene field-effect transistor (FET) biosensor, exemplifying this advancement, is label-free, extremely sensitive, and highly selective. Coated with gold nanoparticles, this biosensor is specifically designed for detecting miRNA. By obviating the need for miRNA labelling, it offers enhanced precision and efficacy. Designed to identify and diagnose illnesses in various organisms, this biosensor utilizes gold nanoparticles attached to the surface of graphene, leading to a substantial improvement in sensitivity and specificity. It demonstrates the capability to identify specific miRNA molecules in serum samples, making it suitable for target identification and potential use as a point-of-care instrument for illness diagnostics. Despite the challenges in swiftly and precisely identifying plant miRNAs associated with illnesses, alternative methods, such as using nanoparticles as tags for miRNA tests, are being explored. Electrocatalytic nanoparticle tags, when connected to hybridized miRNA molecules, generate measurable electric currents, providing a higher detection threshold compared to optical miRNA tests. Anticipated advancements in miRNA detection techniques hold the potential to identify plant infections and address various crop protection concerns in the coming years (Cai et al., 2015).

Advantages of microRNA (miRNA)-Based Nano-Diagnosis:

- Early Disease Detection: miRNAs serve as promising biomarkers for early disease detection across various species, including plants. Nano-based diagnostics can enhance sensitivity and precision in identifying these miRNAs, enabling early detection of diseases.
- Specificity and Precision: Nanotechnology-based approaches, such as gold nanoparticles functionalized with thiols, amines and disulfides, offer high specificity and precision in detecting miRNAs. Functionalized nanoparticles enhance the selectivity of biosensors.
- Label-Free Detection: Certain nano-diagnostic platforms, like graphene FET biosensors coated with gold nanoparticles, enable label-free detection of miRNAs. This eliminates the need for complex labelling procedures, improving the overall efficiency of the diagnostic process.
- High Sensitivity: Nano-based biosensors, particularly those incorporating gold nanoparticles, can exhibit high sensitivity. This is crucial for detecting low concentrations of miRNAs, especially in complex biological samples like serum.
- Versatility: Nano-diagnostic techniques, utilizing diverse nanomaterials and platforms, offer versatility in detecting various miRNAs associated with different diseases. This adaptability is valuable for addressing a range of plant protection concerns.
- Point-of-Care Applications: Certain nano-diagnostic platforms, like the graphene FET biosensor, hold potential for point-of-care applications. This means that they can be used in on-site diagnostics, allowing for rapid and convenient disease identification.
- Non-Invasive Detection: miRNA-based nano-diagnosis can enable non-invasive detection of diseases. This is particularly important in plant systems where non-destructive testing methods are desirable for monitoring crop health.

Disadvantages of microRNA (miRNA)-Based Nano-Diagnosis:

- Challenges in Plant miRNA Identification: Swift and precise identification of plant miRNAs associated with illnesses poses challenges. The complexity of plant systems and the diversity of miRNAs require sophisticated detection methods.
- Detection Threshold: While electrocatalytic nanoparticle tags can enhance sensitivity, they may also introduce challenges related to the detection threshold. Optimization of detection conditions is crucial for reliable results.
- Complexity of Techniques: Certain nano-diagnostic techniques, especially those involving advanced platforms like graphene FET biosensors, may be complex to implement and require specialized knowledge for operation and interpretation.
- Limited Understanding of Plant miRNAs: The understanding of plant miRNAs and their roles in diseases is still evolving. Limited knowledge of specific miRNAs associated with certain plant diseases may hinder the development of targeted nano-diagnostic approaches.
- Cost Considerations: Some advanced nano-diagnostic platforms may involve higher costs, limiting their widespread adoption, especially in resource-limited agricultural settings.
- Ethical and Regulatory Concerns: The application of nano-diagnostics in agriculture raises ethical and regulatory considerations related to the potential environmental impact and safety of nanomaterials.

miRNA-based nano-diagnosis offers significant advantages for early disease detection and precision, addressing challenges related to plant miRNA identification, detecting threshold optimization, and ensuring the accessibility of techniques are critical for their successful implementation in plant health diagnostics.

12.10 SYSTEM OF NANO-DIAGNOSTIC KIT-BASED EQUIPMENT

The nano-diagnostic kit, often referred to as a lab in a box, is a compact container resembling a briefcase in size and weight. It contains advanced measuring devices, chemical substances, electrical sources and other standard laboratory elements. This technology serves as a valuable tool for specialists to assist farmers in efficiently identifying dangerous plant diseases right in the field. An exemplary illustration is the 4mycosensor, a tetraplex competitive antibody-based assay in dipstick format specifically designed for the immediate detection of multiple mycotoxins, including ZEA, T-2/HT-2, DON and FB1/FB2, in maize, oat, wheat and barley samples. Notably, it can detect these mycotoxins at or below the maximum residue limits (MRL) set by European regulations.

To enhance the effectiveness of molecular diagnostics and facilitate strain discrimination, it is imperative to identify and select crucial antigens, antibodies, and nucleotide receptors. Establishing universal standards for evaluating tests and detection thresholds becomes crucial for comparing detection limits across diverse research studies. While streamlining the purification process and isolating pivotal genes are essential steps in pinpointing precise disease-related genomic targets, it is noteworthy that the nano-diagnostic kit has not undergone thorough testing for the detection of plant infections in agricultural settings. Therefore, comprehensive further research is necessary in this domain to facilitate future progress (Khiyami et al., 2014; Shivashakarappaa et al., 2022).

Advantages of Nano-Diagnostic Kit-Based Equipment:

- Portability: Nano-diagnostic kits are designed to be compact and portable, resembling a briefcase. This portability allows for easy transportation to different locations, making it feasible for on-site testing in agricultural fields.
- Immediate Field Detection: The real-time nature of nano-diagnostic kits enables immediate detection of plant diseases right in the field. This rapid on-site testing can lead to prompt decision-making for disease management, preventing the spread of infections.
- Multiplex Detection: Some nano-diagnostic kits, like the 4mycosensor, are capable of multiplex detection, allowing simultaneous identification of multiple pathogens or mycotoxins. This efficiency is particularly advantageous for comprehensive disease screening.
- Compact Lab-in-a-Box Concept: The nano-diagnostic kit embodies a lab-in-a-box concept, providing a range of advanced measuring devices, chemicals and electrical sources in a compact form. This all-in-one design simplifies the testing process and makes it accessible to non-expert users.
- Specificity and Sensitivity: Nano-diagnostic kits often employ advanced technologies, such as antibodies and nucleotide receptors, ensuring high specificity and sensitivity in disease detection. This precision is crucial for accurate and reliable results.
- Compliance with Regulations: Kits like the 4mycosensor are designed to detect mycotoxins at or below the MRL set by regulations, ensuring compliance with safety standards.
- Facilitation of Molecular Diagnostics: Nano-diagnostic kits contribute to the effectiveness of molecular diagnostics by providing a platform for strain discrimination. The selection of crucial receptors enhances the accuracy of the diagnostic process.

Disadvantages of Nano-Diagnostic Kit-Based Equipment:

- Limited Testing for Plant Infections: Despite their potential, nano-diagnostic kits may not have undergone thorough testing for the detection of plant infections in agricultural settings. This lack of extensive testing may raise concerns about the reliability of the kits in diverse environmental conditions.
- Need for Universal Standards: The absence of universal standards for evaluating tests and detection thresholds can be a drawback. Standardization is crucial for ensuring consistency and comparability of results across different studies and applications.

- Requirement for Further Research: Comprehensive further research is necessary to validate the performance of nano-diagnostic kits in various agricultural scenarios. This includes understanding their efficacy under different environmental conditions and against a wide range of plant pathogens.
- Cost Considerations: The cost of nano-diagnostic kits may be a limiting factor, especially for small-scale farmers or in resource-limited agricultural settings. Affordability is a critical aspect for the widespread adoption of such technologies.
- Complexity of Genomic Target Identification: The process of pinpointing precise disease-related genomic targets involves streamlining the purification process and isolating crucial genes. This complexity may pose challenges in practical implementation.

Nano-diagnostic kit-based equipment offers significant advantages in terms of portability, rapid on-site detection and multiplex capabilities; addressing challenges related to testing validation, standardization and affordability is essential for their successful integration into plant health diagnostics in agriculture. Further research and development efforts are crucial to enhance the reliability and applicability of these technologies in diverse agricultural settings.

12.11 SUMMARY

The integration of nanotechnology for the swift and accurate detection of plant pathogens and diseases is gaining momentum, aiming to enhance agricultural productivity and ensure crop health. These advanced technologies offer significant advantages over traditional molecular-based identification methods, which can be time-consuming and require expensive equipment. Nanotechnology approaches provide accelerated and highly sensitive pathogen probes, serving as effective diagnostic tools for a wide range of plant diseases. Despite the promising developments, there are notable challenges that persist in current plant diagnostic technologies. Before deploying nanosensors in practical settings, addressing safety concerns is paramount, especially for certain nanoparticles like QDs that may pose potential toxicity risks. This necessitates a more rigorous safety evaluation and stricter regulatory measures. Additionally, there is a need to develop robust and enduring sensors capable of withstanding diverse environmental conditions. This involves in-depth research into cutting-edge sensor components, including materials that demonstrate resistance to environmental conditions and nanoparticles. In the long run, these cost-effective, rapid, highly responsive and precise nanotechnologies designed for real-time detection of plant pathogens are expected to be widely adopted in the future. Ongoing adjustments will be made to customize them for various environmental conditions, ensuring their effectiveness across different agricultural settings.

12.12 FUTURE PERSPECTIVE

Integrating nanoscale devices with distinct features into smart agricultural systems holds significant promise. These nanodevices could play a crucial role in detecting potential plant health issues before they become evident to farmers. They possess the ability to respond to specific conditions, identify problems and initiate appropriate actions for disease management, essentially serving as a proactive measure and an early warning system. Due to the remarkable progress in synthetic molecular recognition, nanoscale materials have become highly suitable for detecting plant illnesses. The integration of diagnostic devices with farmers' smartphones facilitates prompt decision-making regarding disease management measures. Modern technology enables precise field-based diagnosis with minimal sample manipulation and cost efficiency. Portable nanodevices exhibit high sensitivity to protein concentrations, detecting levels as low as a few nanograms/ml. Moreover, RNA can be utilized as a versatile detection tool to address complex issues related to post-harvest losses. Exploring nanophytopathology contributes to our understanding of the interactions between plants and pathogens, paving the way for the development of innovative and advanced methods to protect crops.

STUDY QUESTIONS

1. What are the advantages of nanotechnology tools over conventional methods for diagnosing plant pathogens?
2. How can safety concerns associated with nanoparticles like quantum dots (QDs) be addressed in the deployment of nanosensors in agriculture?
3. What role can resilient and long-lasting sensors play in improving plant pathogen identification in the field?
4. How might nanotechnologies be further modified to adapt to different environmental conditions in agriculture?
5. What are the potential benefits of integrating nanoscale devices into smart agricultural systems for disease management and crop protection?

CONFLICT OF INTEREST

The authors confirm that they have no conflicts of interest.

ACKNOWLEDGMENT

The authors expressed their gratitude to all individuals who provided support in the effective completion of this chapter.

FURTHER READING

Kutawa, A.B., Ahmad, K., Ali, A., Hussein, M.Z., Wahab, M.A.A., Adamu, A., Ismaila, A.A., Gunasena, M.T., Rahman, M.Z. and Hossain, M.I., 2021. Trends in nanotechnology and its potentialities to control plant pathogenic fungi: A review. *Biology*, 10, 881. https://doi.org/10.3390/biology10090881

REFERENCES

Amini, B., Kamali, M., Salouti, M., Yaghmaei, P., 2017. Fluorescence bio-barcode DNA assay based on gold and magnetic nanoparticles for detection of Exotoxin A gene sequence. *Biosensors and Bioelectronics* 92, 679–686. https://doi.org/10.1016/j.bios.2016.10.030

Askim, J.R., Mahmoudi, M., Suslick, K.S., 2013. Optical sensor arrays for chemical sensing: the optoelectronic nose. *Chemical Society Reviews* 42 (22), 8649–8682. https://doi.org/10.1039/C3CS60179J

Bao, Y.P., Wei, T.F., Lefebvre, P.A., An, H., He, L., Kunkel, G.T., Müller, U.R., 2006. Detection of protein analytes via nanoparticle-based bio bar code technology. *Analytical Chemistry* 78 (6), 2055–2059. https://doi.org/10.1021/ac051798d

Bhattacharya, S., Jang, J., Yang, L., Akin, D., Bashir, R., 2007. BioMEMS and nanotechnology-based approaches for rapid detection of biological entities. *Journal of Rapid Methods and Automation in Microbiology* 15 (1), 1–32. https://doi.org/10.1111/j.1745-4581.2007.00073.x

Bronzato Badial, A., Sherman, D., Stone, A., Gopakumar, A., Wilson, V., Schneider, W., King, J., 2018. Nanopore sequencing as a surveillance tool for plant pathogens in plant and insect tissues. *Plant Disease* 102 (8), 1648–1652. https://doi.org/10.1094/PDIS-04-17-0488-RE

Cai, B., Huang, L., Zhang, H., Sun, Z., Zhang, Z., Zhang, G.J., 2015. Gold nanoparticles decorated graphene field-effect transistor biosensor for femtomolar MicroRNA detection. *Biosensors and Bioelectronics* 74, 329–334. https://doi.org/10.1016/j.bios.2015.06.068

Chalupowicz, L., Dombrovsky, A., Gaba, V., Luria, N., Reuven, M., Beerman, A., Manulis-Sasson, S., 2019. Diagnosis of plant diseases using the Nanopore sequencing platform. *Plant Pathology* 68 (2), 229–238. https://doi.org/10.1111/ppa.12957

Chaudhary, V., Jangra, S., Yadav, N.R., 2018. Nanotechnology based approaches for detection and delivery of microRNA in healthcare and crop protection. *Journal of Nanbiotechnology* 16 (1), 1–18. https://doi.org/10.1186/s12951-018-0368-8

Clarke, J., Wu, H.C., Jayasinghe, L., Patel, A., Reid, S., Bayley, H., 2009. Continuous base identification for single-molecule nanopore DNA sequencing. *Nature Nanotechnology* 4 (4), 265–270. https://doi.org/10.1038/nnano.2009.12

Dass, R.S., Sreenivasa, M.Y., Janardhana, G.R. 2007. High incidence of *Fusarium verticillioides* in animal and poultry feed mixtures produced in Karnataka, India. *Plant Pathology Journal* 6 (2), 174–178.

Debnath, K., 2013. One billion hungry: Can we feed the world? 0–8014-7802–2. *Journal of Agriculture*, 738–740. https://doi.org/10.1111/1477-9552.12029

Deepa, N., Charith Raj, A.P., Sreenivasa, M.Y. 2016b. Nested PCR method for the early detection of fumonisin producing *Fusarium verticillioides* in pure cultures, cereal samples and plant parts. *Food Biotechnology* 30 (1), 18–29.

Deepa, N., Charith Raj, A.P., Sreenivasa, M.Y. 2016c. Multiplex PCR method for the early detection of fumonisin producing *Fusarium verticillioides*. *Food Bioscience* 13, 84–88.

Deepa, N., Nagaraja, H., Sreenivasa, M.Y. 2016a. Prevalence of fumonisin producing *Fusarium verticillioides* associated with cereals grown in Karnataka (India). *Food Science and Human Wellness* 5, 156–162. http://doi.org/10.1016/j.fshw.2016.07.001

Deepa, N., Sreenivasa, M.Y. 2019a. Molecular methods and key genes targeted for the detection of fumonisin producing *Fusarium verticillioides* – An updated review. *Food Bioscience* 32, 100473.

Deepa, N., Sreenivasa, M.Y. 2019b. Biocontrol strategies for effective management of phytopathogenic fungi associated with cereals. In J.S. Singh, D.P. Singh (Eds) *New and future developments in microbial biotechnology and bioengineering*, Elsevier, pp, 177–189, ISBN 9780444641915.

Drygin, Y.F., Blintsov, A.N., Grigorenko, V.G., Andreeva, I.P., Osipov, A.P., Varitzev, Y.A., 2012. Highly sensitive field test lateral flow immunodiagnostics of PVX infection. *Applied Microbiology and Biotechnology* 93, 179–189. https://doi.org/10.1007/s00253-011-3522-x

Edmundson, M.C., Capeness, M., Horsfall, L., 2014. Exploring the potential of metallic nanoparticles within synthetic biology. *New Biotechnology*, 31 (6), 572–578. https://doi.org/10.1016/j.nbt.2014.03.004

FAO. 2018. The future of food and agriculture: Alternative pathways to 2050. Summary version. Rome. 64 pp. http://www.fao.org/3/CA1553EN/ca1553en.pdf

Janardhana, G.R., Dass, R.S., Sreenivasa, M.Y. 2009. Fumonisins – A new class of Fusarial toxins. In K.A. Sridhar et al. (Eds.), *Frontiers in fungal ecology, diversity and metabolites*. I. K. International Publishing House Pvt. Ltd.

Joseph T., Morrison, M. 2006. *Nanotechnology in agriculture and food*. London: Institute of Nanotechnology. www.nanoforum.org

Joshi, S., de Britto, S., Jogaiah, S., Ito, S.I. 2019. Mycogenic selenium nanoparticles as potential new generation broad spectrum antifungal molecules. *Biomolecules* 9 (9), 419. https://doi.org/10.3390/biom9090419

Kaminski, J., Christiaensen, L., 2014. Post-harvest loss in sub-Saharan Africa—what do farmers say? *Global Food Security* 3 (3–4), 149–158. https://doi.org/10.1016/j.gfs.2014.10.002

Kashyap, P.L., Kumar, S., Srivastava, A.K., Sharma, A.K., 2013. Myconanotechnology in agriculture: A perspective. *World Journal of Microbiology and Biotechnology* 29 (2), 191–207. https://doi.org/10.1007/s11274-012-1171-6

Khiyami, M.A., Almoammar, H., Awad, Y.M., Alghuthaymi, M.A., Abd-Elsalam, K.A., 2014. Plant pathogen nanodiagnostic techniques: forthcoming changes. *Biotechnol and Equip* 28 (5), 775–785. https://doi.org/10.1080/13102818.2014.960739

Knudsen, B.R., Jepsen, M.L., Ho, Y.P., 2013. Quantum dot-based nanosensors for diagnosis via enzyme activity measurement. *Expert Review of Molecular Diagnostics* 13 (4), 367. https://doi.org/10.1586/erm.13.17

Koo, C., Malapi-Wight, M., Kim, H.S., Cifci, O.S., Vaughn-Diaz, V.L., Ma, B., Han, A., 2013. Development of a real-time microchip PCR system for portable plant disease diagnosis. *PLoS One* 8 (12), e82704. https://doi.org/10.1371/journal.pone.0082704

Kumar D., Kalita, P., 2017. Reducing postharvest losses during storage of grain crops to strengthen food security in developing countries. *Food* 6 (1), 8. https://doi.org/10.3390/foods6010008

Kumar S., Tao, C., Chien, M., Hellner, B., Balijepalli, A., Robertson, J.W., Ju, J., 2012. PEG-labeled nucleotides and nanopore detection for single molecule DNA sequencing by synthesis. *Scientific Reports* 2 (1), 1–8. https://doi.org/10.1038/srep00684

Li, Z., Askim, J.R., Suslick, K.S., 2018. The optoelectronic nose: colorimetric and fluorometric sensor arrays. *Chemical Reviews* 119 (1), 231–292. https://doi.org/10.1021/acs.chemrev.8b00226

Majumder, S., Bhattacharya, B., Singh, P.K., Johari, S., Singh, B., Rahman, R., 2020. Impedimetric detection of banana bunchy top virus using CdSe quantum dots for signal amplification. *SN Appl. Sci* 2 (4). https://doi.org/10.1007/s42452-020-2345-8

Mak, A.C., Osterfeld, S.J., Yu, H., Wang, S.X., Davis, R.W., Jejelowo, O.A., Pourmand, N., 2010. Sensitive giant magnetoresistive-based immunoassay for multiplex mycotoxin detection, *Biosensors and Bioelectronics* 25 (7), 1635–1639. https://doi.org/10.1016/j.bios.2009.11.028

Malhotra, B.D., Srivastava, S., Ali, M., Singh, C., 2014. Nanomaterial-based biosensors for food toxin detection. *Applied Biochemistry and Biotechnology* 174 (3), 880–896. https://doi.org/10.1007/s12010-014-0993-0

Marcolungo, L., Passera, A., Maestri, S., Segala, E., Alfano, M., Gaffuri, F., Marturano, G., Casati, P., Bianco, P.A., Delledonne, M., 2022. Real-time on-site diagnosis of quarantine pathogens in plant tissues by nanopore-based sequencing. *Pathogens* 11, 199. https://doi.org/10.3390/pathogens11020199

Mark, D., John, D., John, D., Tara, 2014. QPCR analysis apparatus. US2015/0165440A1. https://lens.org/009–919-572–680-277

MohdBakhori, N., Yusof, N.A., Abdullah, A.H., Hussein, M.Z., 2013. Development of a fluorescence resonance energy transfer (FRET)-based DNA biosensor for detection of synthetic oligonucleotide of Ganoderma boninense. *Biosensors* 3 (4), 419. https://doi.org/10.3390/bios3040419

Nezhad, A.S., 2014. Future of portable devices for plant pathogen diagnosis. *Lab on a Chip* 14 (16), 2887–2904. https://doi.org/10.1039/C4LC00487F

Oluwaseun, A.C., Phazang, P., Sarin, N.B., 2018. Biosensing technologies for the detection of pathogens - a prospective way for rapid analysis. *Biosensing Technol. Detect. Pathog. - A Prospect. Way Rapid Anal* https://doi.org/10.5772/intechopen.74668

Panini, N.V., Bertolino, F.A., Salinas, E., Messina, G.A., Raba, J., 2010. Zearalenone determination in corn silage samples using an immunosensor in a continuous-flow/stopped-flow systems. *Biochemical Engineering Journal* 51 (1–2), 7–13. https://doi.org/10.1016/j.bej.2010.04.005.

Paul, R., Ostermann, E., Chen, Y., Saville, A.C., Yang, Y., Gu, Z., Wei, Q., 2021. Integrated microneedle-smartphone nucleic acid amplification platform for in-field diagnosis of plant diseases. *Biosensors and Bioelectronics* 187, 113312. https://doi.org/10.1016/j.bios.2021.113312

Pohlmann, C., Dieser, I., Sprinzl, M., 2014. A lateral flow assay for identification of *Escherichia coli* by ribosomal RNA hybridisation. *Analyst* 139, 1063–1071. https://doi.org/10.1039/c3an02059b

Rad, F., Mohsenifar, A., Tabatabaei, M., Safarnejad, M.R., Shahryari, F., Safarpour, H., Fotokian, M., 2012. Detection of *Candidatus Phytoplasma aurantifolia* with a quantum dots fret-based biosensor. *Journal of Plant Pathology* 525–534. https://www.jstor.org/stable/45156279

Rai, M., Ingle, A., 2012. Role of nanotechnology in agriculture with special reference to management of insect pests. *Applied Microbiology and Biotechnology* 94 (2), 287–293. https://doi.org/10.1007/s00253-012-3969-4

Sertova, N.M., 2015. Application of nanotechnology in detection of mycotoxins and in agricultural sector. *Journal of Central European Agriculture* 16 (2), 117–130. https://doi.org/10.5513/jcea.v16i2.34724.

Shang, L., Liu, C., Chen, B., Hayashi, K., 2018. Plant biomarker recognition by molecular imprinting based LSPR sensor array: performance improvement by enhanced hotspot of Au nanostructure. *ACS Sensors*, 3, 1531–1538.

Sharma, R., Zhou, M., Hunter, M.D., Fan, X., 2019. Rapid in situ analysis of plant emission for disease diagnosis using a portable gas chromatography device. *Journal of Agricultural and Food Chemistry* 67 (26), 7530–7537. https://doi.org/10.1021/acs.jafc.9b02500

Shivashakarappaa, K., Reddyb, V., Tupakulaa, V.K., Farniana, A., Vuppulac, A., and Gunnaiah, R., 2022. Nanotechnology for the detection of plant pathogens. *Plant Nano Biology*, 2, 100018. https://doi.org/10.1016/j.plana.2022.100018

Sreenivasa, M.Y., Dass, R.S., Charith Raj, A.P., Janardhana, G.R. 2008. PCR-based detection of genus *Fusarium* and Fumonisin-producing isolates from freshly harvested Sorghum grains grown in Karnataka, India. *Journal of Food Safety* 28, 236–237.

Srinivasan, B., Tung, S., 2015. Development and applications of portable biosensors. *Journal of Laboratory Automation* 20 (4), 365–389. https://doi.org/10.1177/2211068215581349

Stanisavljevic, M., Krizkova, S., Vaculovicova, M., Kizek, R., Adam, V., 2015. Quantum dots-fluorescence resonance energy transfer-based nanosensors and their application. *Biosensors & Bioelectronics* 74: 562–574. https://doi.org/10.1016/j.bios.2015.06.076

Tran, T.T., Clark, K., Ma, W., Mulchandani, A., 2020. Detection of a secreted protein biomarker for citrus Huanglongbing using a single-walled carbon nanotubes-based chemiresistive biosensor. *Biosensors and Bioelectronics* 147, 111766. https://doi.org/10.1016/j.bios.2019.111766

Yadav, A., Kon, K., Kratosova, G., 2015. Fungi as an efficient mycosystem for the synthesis of metal nanoparticles: Progress and key aspects of research. *Biotechnology Letters* 37 (11), 2099–2120. https://doi.org/10.1007/s10529-015-1901-6.

Zhang, X., Li, H., Zhang, J., Zhang, C., Gong, P., Ziaf, K., Ye, Z., 2011. Expression of artificial microRNAs in tomato confers efficient and stable virus resistance in a cellautonomous manner. *Transgenic Research* 20 (3), 569–581. https://doi.org/10.1007/s11248-010-9440-3

Zhao, M.X., Zeng, E.Z., 2015. Application of functional quantum dot nanoparticles as fluorescence probes in cell labeling and tumor diagnostic imaging. *Nanoscale Research Letters* 10 (1), 1–9. https://doi.org/10.1186/s11671-015-0873-8.

13 Aptamers
A Promising Tool for the Detection of Phytopathogenic Fungal Toxins

R. Shylaja, R. Rakesh, and K. Joseph
Defence Food Research Laboratory, DRDO

Phytopathogenic fungi and their toxic metabolites pose a major threat to the agricultural industry incurring a high economic burden in terms of plant diseases. Moreover, contamination of food crops with prominent phytopathogenic genera results in the entry of highly toxic, carcinogenic, stable fungal toxins into the food chain leading to global safety concerns. Early detection and identification of pathogens is of utmost importance to reduce risks to health and economy. In light of these detrimental effects, maximum tolerance limits have been set by regulators in the majority of the countries on mycotoxins. Accurate, sensitive and rapid detection of fungal toxins is essential to facilitate effective mitigation of the deleterious effects. Chromatography, immunoassays, molecular and conventional methods are the most commonly employed and accepted methods. In addition, a few extremely sophisticated methods, such as high-resolution mass spectrometry, tandem MS/MS, surface plasmon resonance, micellar electrokinetic capillary chromatography, and nano-assays, have been employed. However, these analytical systems though highly sensitive, and reliable, require sophisticated instruments, making them unsuitable for on-site detection. This chapter focuses on the application of new-generation affinity ligand aptamers as the promising alternate for on-site detection of phytopathogenic fungi and their toxins. The selection of aptamers, SELEX technology and the development of various aptasensors will be discussed.

DOI: 10.1201/9781003399704-13

GRAPHICAL ABSTRACT

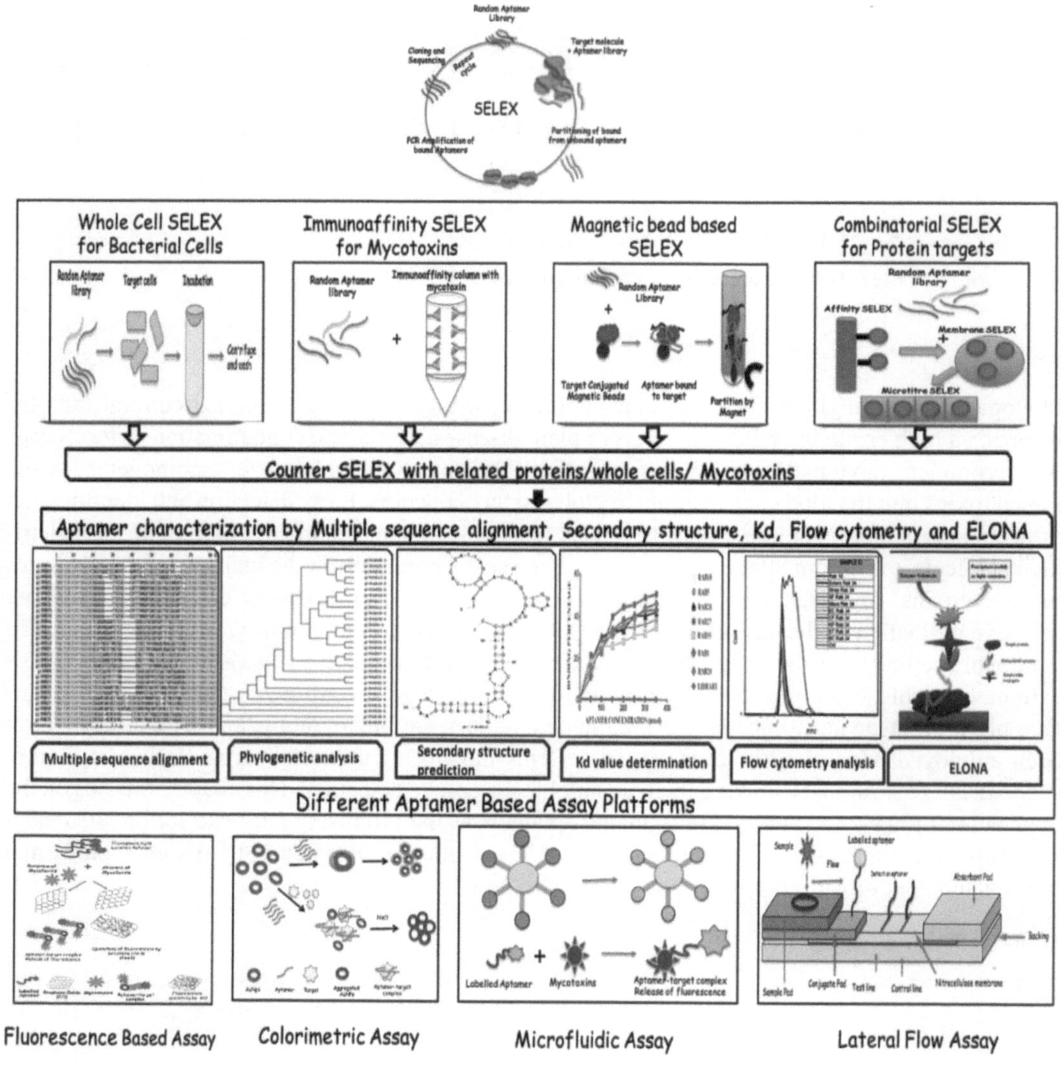

13.1 INTRODUCTION

Phytopathogenic fungi are a diverse group of microorganisms that have successfully occupied most natural habitats and are well known for developing a wide array of strategies to infect and colonize their plant hosts, leading to a negative impact on crop productivity (Peng et al. 2021; Singh et al. 2023; Sreenivasa et al. 2008). Colonization of the host plant species is achieved through the secretion of phytotoxins, which in turn manipulate the plant physiology in favour of fungal phytopathogens (Knogge 1996).

Phytotoxins are well known to act on the cell membrane, mitochondria and chloroplasts of host plants by causing disruption in plant physiology and metabolism, leading to the eventual demise of plants (Meena & Samal 2019; Zeilinger et al. 2016; Dass et al. 2007).

Fungal phytotoxins are natural secondary metabolites with distinct chemical structures and biological characteristics exhibiting a wide range of toxic effects on plants (Xu et al. 2021a; Janardhana et al. 2009). They are classified as host-specific toxins (HSTs) and non-host-specific toxins (NHSTs). HSTs are essential for pathogenicity and are specific for plants that serve as hosts for toxigenic fungi (Meena & Samal 2019). Conversely, NHSTs have a broader range of action, eliciting symptoms in other plant species in addition to the hosts of the pathogenic fungi (Strange 2007). However, these fungal toxins are not only pernicious for plants but also for vertebrates and invertebrates, especially when it comes to toxin-contaminated food crops and grains (Zain 2011).

Mycotoxins are another class of specific and non-specific toxins, with highly toxigenic characteristics and phytotoxic effects. The phytotoxic mycotoxins have the potential to directly or indirectly affect plant growth and metabolism (Abbas et al. 2002; Hasan 1999; Deepa & Sreenivasa 2017a, 2017b). Among the hundreds of known phytotoxic-mycotoxins, citrinin, aflatoxins, cytochalasins, patulin, penicillic acid, tenuazonic acid, ochratoxin A, deoxynivalenol, fumonisins, fusaric acid and zearalenone are known to frequently contaminate food grains and crops. In addition, these fungal toxins are found to inhibit seed germination, plant elongation and photosynthesis, and affect the normal growth and metabolism of plants. However, these toxins can nevertheless have a biological impact on the physiology of plants even in the absence of symptoms (Ismaiel & Papenbrock 2015; Sreenivasa et al. 2010).

13.2 EFFECT OF PHYTOTOXINS ON ECONOMY

The fungal phytotoxins are of significant concern in the fields of agriculture and forestry. Huge economic losses are incurred due to the possible propensity of fungal phytotoxins to induce serious diseases and infections in crop plants (Xu et al. 2021a; Deepa et al. 2016a). A highly specialized and complex relationship has emerged during the evolution history of plants and phytopathogenic fungi, resulting in mutual selection and co-evolution (Quintanilha-Peixoto et al. 2019; Nagaraja et al. 2016).

Among the thousands of secondary metabolites produced by the pathogenic fungal species, only a few of the fungal toxins are categorized to be important from a safety and economic perspective, which are described in Table 13.1. These fungal toxins are produced under specific environmental conditions and are ubiquitous in agricultural crops, including wheat, barley, maize, peanuts, coffee beans, nuts, spices and cotton, ultimately affecting crop yield and quality. As a consequence, this results in a huge economic loss to agricultural and industrial commodities (Yu & Pedroso 2023; Xu et al. 2021a; Sreenivasa et al. 2011a, 2011b).

13.3 REGULATORY LIMITS ON FUNGAL TOXINS IN FOOD AND FEED

Considering the deleterious effects of fungal phytotoxins, more than 100 countries around the globe have set stringent rules. Regulations on fungal toxins have been established in many countries to protect consumers from the harmful toxic effects by setting the permissible amounts of fungal toxins in food and feed (Henson & Caswell 1999; Egmond et al. 2007; Food Safety and Standards Authority of India- FSSAI, 2011), which in turn results in rendering commodities unacceptable to national or international trade, for lacking national standards. This is an overarching issue contributing to food waste and crop loss (Balendres et al. 2019). Maximum permissible limits set by regulatory agencies for a few of the important mycotoxins are mentioned in Table 13.2.

13.4 CHALLENGES IN MYCOTOXIN ANALYSIS

Mycotoxin analysis poses various challenges. The major challenges in mycotoxin analysis are the complex matrices and interfering substances that make it difficult for mycotoxin extraction and analysis. Sample size is another drawback as mycotoxins are unevenly distributed in samples. Regions known as "mycotoxin pockets" contain very high concentrations of toxin while the rest of the sample may be unaffected. Hence, a representative sampling is necessary for mycotoxin analysis. Sample preparation for different sample matrixes is another challenge. Mycotoxins can be

TABLE 13.1
Important Fungal Toxins with Phytotoxic Effects

Toxins	Fungal Species	Affected Plant	Chemical Group	Phytotoxic Effects	Ref.
T-2 toxin	*Fusarium*	Wheat, maize, Lemna minor	Sesquiterpenoid	Significant inhibition of root and shoot growth of germinating seedlings, inhibition of callus growth, decrease of leaf growth rate	Wakulinski, 1989; Van Asch et al. 1992; Vesonder et al. 1992
Cytochalasin	*Phoma*	Maize, apple, *Strigahermonthica*	Polyketide-amino acid hybrid	LD_{50} seedling viability, limpness of shoot	Masi et al. 2014
Aflatoxins	*Aspergillus*	Barley, wheat, sorgum, mustard, *Sesame*	Difuranocoumarin derivative	Maximal reduction in the growth rate, reduces seedling viability, and remarkable inhibitory effect on synthesis of chlorophyll, carotenoid	Crisan, 1973; McLean et al. 1994; Hasan, 1999
Citrinin	*Penicillium, Aspergillus*	Bean, cotton, sorghum	Benzopyran compound	Inhibit seed germination, seedling and induces wilting	Damodaran et al. 1975
Tenuazonic acid	*Alternaria*	Rice, *Daturainnoxia*, rye, wheat, lettuce and green gram	3-acetyltetramic acid	Exhibits non-specific phytotoxic effect on both monocotyledonous and dicotyledonous plantsLeaf browning, localized chlorotic spots, growth inhibition of germinating seeds and 80% root growth inhibition	Janardhanan and Husain et al. 1984; Lebrun et al. 1988;
Ochratoxin-A	*Aspergillus, Penicillium*	*Arabidopsis thaliana*	Benzopyran compound	Formation of lesions on leaves, blocking of root elongation	Peng et al. 2010; Wang et al. 2012
Patulin	*Penicillium, Aspergillus*	Maize, wheat	Benzopyran compound	Decreases in internodal elongation, floret number, seed weight and seed number	Ellis and McCalla et al. 1973; Ismaiel and Papenbrock et al. 2014
Penicillic acid	*Penicillium, Aspergillus*	Corn, lettuce, Picea glehnii	Isopropylidene Tetronic acid	Growth inhibition of lettuce seedlings, reduced growth rate, blocking of root elongation and inhibition of metabolic enzyme activity	Keromnes and Thouvenot, 1985; Yamaji et al. 2005
Deoxynivalenol (DON),	*Fusarium*	Wheat, corn and barley	Trichothecenes, sesquiterpenoid	Inhibition of root, shoot growth, complete inhibition of germinating, plantlet growth reduction	Shimada and Otani et al. 1990; Bandurska et al. 1994
Ergot alkaloids	*Claviceps*	Corns, barley, oats, triticale rye and other grass species	Alkaloids derived from amino acids and isoprenoids precursors	Seed loss, yield reduction, development of ergot sclerotium	Carlson, 2011

TABLE 13.2
Permissible Limits of Important Mycotoxins Set by Different Regulatory Agencies

Toxins	Regulatory Agency	Sources	Maximum Permissible Limit (ppb)
Aflatoxins (AFs)	FDA	Peanuts	20
		Corn	20
AFM1		Milk and dairy products	0.5
Afs	European Union	Nuts	10
		Maize and corn	4
		Rice	4
		Processed products intended for direct human consumption	4
AFB1		Groundnuts, nuts and dried fruits	2
AFM1		Milk and dairy products	0.05
AFB1	FSSAI	All foods intended for human consumption	30
		Cereal and cereal products	15
		Nuts	
		Maize	
		Spices	30
AFM1		Milk	0.5
Ochratoxin A	European Union	Maize	5
	Codex	Wheat, rye and barley	5
	FSSAI	Wheat, rye and barley	20
Fumonisins	FDA	Maize, maize products	2000–4000
	European Union	Maize, maize products	200–2000
	Codex	Maize, maize products	2000–4000
Patulin	FDA	Apple, and apple puree	25–50
	European Union	Solid fruits products	10–25
		Fruit juices	50
	Codex	Fruit juices	50

highly toxic even at low concentrations; hence, the analysis method must be sensitive and precise. The presence of multiple mycotoxins and masked mycotoxins with altered chemical structures is the major challenge and requires more sophisticated detection methods. There is a continuous search and constant need for sensitive and on-site analytical methods for monitoring and surveillance of phytotoxins to ensure food safety and security (Prasad et al. 2010).

Hitherto, conventional culture-based detection systems (Diba et al. 2007), different chromatographic techniques, viz., TLC, HPLC, LC-MS and GC-MS and molecular techniques (PCR, RT-PCR) have been reported for effective detection of phytotoxic mycotoxins. Thin-layer chromatography was first used in the year 1964 by De Iongh and his co-workers and has been regarded as the method of choice since 1990 by the Association of Official Analytical Chemists (AOAC) for the detection of mycotoxins. High-performance liquid chromatography (HPLC) is the widely approved, reliable and sensitive method for the separation and determination of fungal toxins (Li et al. 2011; Wacoo et al. 2014; Spadaro et al. 2020). Molecular methods like polymerase chain reaction (PCR) are employed for specific identification of many fungal species (Deepa et al. 2016b, 2016c). Understanding the molecular pathways in the biosynthesis of mycotoxins and genes responsible for the pathways has helped in the development of various PCR formats for the detection of mycotoxins (Deepa & Sreenivasa 2019a, 2019b). Though immuno methods such as ELISA can be used for screening of mycotoxins, the high cost and stability of antibodies restrict their use. Liquid chromatography-tandem mass spectrometry (LC-MS/MS) is the preferred method for qualitative and quantitative confirmation of single/multiple mycotoxins in complex matrices. MS/MS enhances the accuracy of detecting and identifying mycotoxins in a single analysis (Verma et al. 2004; Sulyok et al. 2006; Capriotti et al. 2012; Serrano et al. 2015; Laganà 2017; Guo et al. 2020a; Mehdizadeh et al. 2022). The other types of mass spectroscopic methods include time-of-flight (TOF) MS, the triple quadrupole MS–MS system and ion-trap MS for quantitative and qualitative analyses of mycotoxins from food and environmental samples (Ramos Catharino et al. 2005; Cavaliere et al. 2007; Lattanzio et al. 2007; Magnoli et al. 2018). Electrokinetic chromatography is an alternative chromatographic technique wherein HPLC is combined with other techniques. Micellar electrokinetic chromatography (MEKC), reversed flow micellar electrokinetic chromatography (RFMEKC), capillary electrophoresis-laser-induced fluorescence (CE-LIF), multiphoton excited fluorescence (MPE), and detection-based capillary electrokinetic chromatography (CEKC) are the other methods. However, these analytical methods require huge revenue for sophisticated instruments, skilled personnel, expensive reagents and complicated sample pre-treatments that have limited their utility in field analysis in developing countries (Gilbert & Vargas 2003; Shim et al. 2009).

Biosensors, electrochemical and photoelectrochemical methods; surface-enhanced Raman spectroscopy (SERS); and immunoassays employ antibodies as the recognition molecule. Features like detection of analytes with dynamic range, accuracy and specificity make antibody-based immunosensors a popular tool for mycotoxin detection. On the other hand, factors like the sensitive nature of antibodies, stringency of antibodies to certain chemicals and expensive instrumentations limit their applicability to laboratories and not at field-level analysis. In the search for a replacement for the recognition element to increase sensitivity, and stability to overcome the limitations of existing methods and to meet the challenges of diagnostics, a new class of oligonucleotides, the aptamers emerged as a promising alternative. Aptamers are next-generation recognition ligands gaining importance in the field of diagnostics. Aptamers can be effectively used to simultaneously detect thousands of targets in multiplex discovery platforms (Caglayan et al. 2022).

13.5 APTAMERS

Deoxyribonucleic acid and Ribonucleic acids (DNA and RNA) are the genetic blueprints of life with the entire information of an organism and its functions. These molecules change their conformation into 3D structures like ribozymes to carry out the biological reactions (expression of genes, communication of cellular responses and protein synthesis) of life (Dunn et al. 2017). Though the importance of folded RNA structures has long been recognized (Onoa & Tinoco 2004), the artificial *in vitro* selection of nucleic acids with non-biological functions was made possible by identifying and amplifying the members by PCR.

In the year 1990, two groups (Tuerk & Gold 1990; Ellington & Szostak 1990) independently worked on the generation of RNA molecules, which could explicitly bind to target specifically by *in vitro* selection renowned technique called Systematic Evolution of Ligands by Exponential Enrichment (SELEX). Affinity reagents based on either DNA or RNA are termed "Aptamers", the excellent exemplars of *in vitro* selection. The "Aptamer" word is a combination of the Latin word *aptus* – 'to fit' and the Greek word *meros* – 'part' (Ellington & Szostak 1992; Jayasena 1999). Aptamers recognize and bind to respective targets based on factors like hydrogen bonding, Van der Waals forces of attraction, electrostatic, hydrophobic interactions and structural compatibility. To date, several publications have reported the generation of high-affinity aptamers that bind to targets by attaining secondary structural confirmations (folding) upon binding to the target molecules (Hasegawa et al. 2016). Aptamers known as affinity reagents are often compared with antibodies in recognizing and binding to the targets. The uniqueness of aptamers makes them powerful tools in the arsenal of affinity reagents and as an alternative to antibodies (Jayasena 1999). Table 13.3 depicts the characteristics and advantages of aptamers in comparison with antibodies.

TABLE 13.3
Comparison between Aptamer and Antibody

Characteristics	Aptamers	Antibodies
Nature	Oligonucleotide & protein	Protein
Uniformity	Aptamers are chemically synthesized, hence, consistent throughout	Batch-to-batch consistency in antibody expression and purification is a challenge
Chemical Modification	Aptamers can be easily modified with different signalling molecules at both the 5` and 3` ends	Antibodies can be conjugated with one type of signalling molecule
Immunogenicity	Absent (No evidence)	Significant
Effect of temperature	Stable at high temperature. & can be regenerated after denaturation	Temperature sensitive Denaturation is irreversible
Method of production	Selection is *in-vitro* chemical process	Selection requires a biological system, comparatively difficult
Structure	Single/double-stranded DNA or RNA oligonucleotide or peptides	Monoclonal & Polyclonal
Target	Wide range of targets	Targets which induces immune response
Site targeting	Investigator determines	Immune system determines
Cost	Cost effective	Expensive
Cross reactivity	Separation of the structural analogues and cross reactive substances is possible	Unable to separate cross reactive substances
Stability	Stable at wide environmental conditions	Highly sensitive

13.5.1 Selection of Aptamers

High-affinity aptamers are selected against a given target by *in vitro* SELEX technique from complex 10^{12}–10^{15} combinatorial aptamer libraries. The SELEX process is mainly comprised of four steps (Stoltenburg et al. 2007). Library generation is the first step. The next step is partitioning, which is binding and isolation of the target-specific sequences from the vast non-specific sequences. Later, the target-bound specific aptamers are amplified by PCR. This amplified DNA is the new library for the immediate next round of SELEX. The fourth step is cloning and sequencing of the aptamer selected from the final round. The first three steps are repeated a number of times in order to enrich high-affinity aptamers to the target.

13.5.2 Aptamer Library

SELEX process initiates with chemically synthesized oligonucleotide (nt) library comprised of 10^{12}–10^{15} molecules of ssDNA/RNA random fragments. Each fragment with middle-random sequences of 20–30 nt flanked by specific primer sequences of 18–21 nt on either side to facilitate PCR amplification and to generate specific aptamers against the target (Stoltenburg et al. 2007). Important parameters to be considered while designing library are the number of nucleotides of the random site, inclusion of modified nucleotides to improve the binding affinity and to enhance the resistance degradation by endonucleases. Longer random sequences improve the diversity by forming complex three-dimensional (3 D) structures and binding loops, thus enhancing the probability for the selection of novel highly specific aptamers against difficult targets (Szeto et al. 2013). RNA libraries have also been successfully used for SELEX. However, protection from RNase and amplification by T7 RNA polymerase and reverse transcription before PCR are the requirements of the RNA library.

13.5.3 Partitioning

The partitioning step of the SELEX is the most critical part of aptamer selection and optimization of this step can reduce selection time remarkably. The step consists of binding of the molecule of interest with the oligonucleotide library, elution of bound sequences and removal of unbound library by washing. To improve partitioning efficiency, a number of methods have been evolved in the SELEX process (Gopinath 2007; Ozer et al. 2014).

13.5.4 Amplification

Conventional PCR amplification leads to generating more by-products rather than specific aptamer sequences (Stoltenburg et al. 2007). To overcome this limitation, emulsion PCR (ePCR), asymmetric PCR with high Tm, and amplification with modified primers by blocking the reverse primer by di-deoxy nucleotide, incorporation of inverted nucleotide and modified primer having different lengths separable from one another by denaturing gel electrophoresis are practised. Enzymes are employed to generate ssDNA. Generation of single-stranded aptamers can additionally be accomplished by using a 5′-biotinylated reverse primer in conjunction with streptavidin beads. These are some of the different strategies introduced to the SELEX for the amplification of single-stranded aptamer pool after each round of SELEX (Hünniger et al. 2014; Yufa et al. 2015; Tabarzad et al. 2014; Nehdi et al. 2020).

13.5.5 Sequencing

Specific aptamers from the last round of SELEX are cloned and sequenced. High-throughput sequencing technology is used in many laboratories to identify high-affinity binding aptamers (Cox et al. 2002).

13.5.6 Different SELEX Methods Followed

Since 1990, different SELEX technologies have been applied for different targets such as inorganic and organic molecules, antibiotics, amino acids, vitamins, proteins, peptides and carbohydrates besides complex whole cells, organisms or mixtures of analytes have been employed for aptamer selection. There is no universal SELEX method available for the selection of aptamers till today, depending on the chemistry of the analyte various SELEX protocols have been developed. A few of the SELEX reported are whole cell SELEX, capillary electrophoresis (CE), fluorescence-activated cell sorting (FACS) SELEX, automated SELEX, combinatorial SELEX, atomic force microscopy, magnetic bead SELEX, chimeric, indirect, microfluidic, Electrophoretic mobility shift (EMSA), genomic, toggle, next generation, photo, SELEX SAGE, Fungal spore SELEX, etc. The SELEX methods are continuously evolving based on the necessity and type of the analyte and also to reduce the number of rounds and to enhance the specificity (Darmostuk et al. 2015; Banu et al. 2022).

13.5.7 Structural Characteristics of Aptamers in Identifying Targets

Single-stranded oligonucleotides fold into multiple three-dimensional spatial structures. These structures are thermodynamically stable and play a vital role in target binding. Generally, ssDNA and RNA sequences recognize the respective targets by adjusting their structures into inducible forms of hairpin, bulge, and pseudoknot, G-quartet, etc. The aptamers integrate with the target specifically and selectively through van der Waals force, electrostatic, hydrogen or non-covalent bond or by hydrophobic stacking, etc. A study conducted by Pate and his co-worker described that aptamers used unique pocket-shaped structures or used some functional groups to interact with target (Xi-Hui et al. 2013). The binding characteristic of an aptamer is a critical parameter for the development of aptamer-based assays that are determined by measuring the dissociation constant, Kd. Lower the Kd value, strongest is the affinity between the aptamer and the molecule of interest. The binding properties and kinetics of the aptamers are assessed by a number of techniques like gel electrophoresis, HPLC, fluorescence polarization, surface plasmon resonance (SPR), nuclear magnetic resonance (NMR), circular dichroism (CD), FACS, EMSA, Foot-printing assays and quartz crystal microbalance (QCM), etc. (Tan et al.

2016). Software such as DNA/RNA fold, Mfold or many online DNA structure prediction tools are employed for the determination of secondary structure and QGRS mapper program for predicting the guanine quadruplex structures in the aptamer. The 3D aptamer structure stability is determined by the G-quadruplexes that facilitate electrostatic interactions with the positively charged binding side at the ligands. The other forces and interactions are also important.

13.6 APPLICATION OF APTAMERS

Aptamers have gained immense attention in scientific communities for their potential to bind to multiple targets inclusive of drugs, metal ions and whole cells apart from protein, peptides and amino acids (Dunn et al. 2017; Song et al. 2008). The exceptional specificity and high affinity make them valuable tools rendering wide applications in various fields such as analytical, therapeutic, targeted drug delivery, point-of-care aptasensors, imaging, environmental monitoring and management of various contaminants, including heavy metals, antibiotics, bisphenol herbicides, insecticides, pesticides, cancer and stem cell, have been developed. Aptamers are gaining more importance with the advancement of rapid, automated selection SELEX and design technologies creating tailor-made aptamers to specific targets, facilitating its application in vast areas. Integration of aptamers nanoparticles and nanomaterials can synergistically enhance the sensitivity and performance of assay platforms. The continued exploration and refinement of aptamer technology holds great promise for addressing complex challenges in healthcare, environmental protection and food safety (Schmitz et al. 2020).

13.7 SELECTION OF APTAMERS FOR MYCOTOXINS

Mycotoxins are a kind of trace exogenous harmful secondary metabolites produced by various fungal members. These are small molecular weight (>1kDa), stable compounds known to exhibit toxicity in both human and animal kingdoms. First, aptamer against mycotoxins was selected against ochratoxin (OTA) by Cruz-Aguado and Penner in the year 2008 (Darmostuk et al. 2015). The aptamer named 1.12 and two truncated minimers of 1.12 were selected and used as molecular probes for OTA. This was followed by a selection of aptamers against other mycotoxins like Aflatoxins (AFs), deoxynivalenol (DON), fumonisin B1 (FB1) and zearalenone (ZEA). In 2011, NeoVentures Biotechnology, Canada obtained a patent for aflatoxin B1 aptamer (Le et al. 2012).

13.8 APTAMERS IN PRE-TREATMENT AND PURIFICATION OF MYCOTOXINS

Generally, mycotoxins present in food matrices are purified prior to application to the detection system. The purification and enhancement of the samples is a prerequisite because of the sophisticated instruments and complexity of the contaminated matrices. Popularly used purification methods are liquid-liquid extraction, chromatographic column and immunoaffinity columns (IAC). These methods are very effective in the purification of mycotoxins, however, the requirement of large volumes of organic solvent that pose health and environmental threats, low amount of purified toxins and sensitive nature of antibodies limit their application. These shortcomings have paved the way for new purification techniques employing aptamers as they limit the use of large volumes of solvents, more stable, reusable and less expensive compared to other methods. For example, Chapuis-Hugon et al. (2011) recovered OTA from red wines by using aptamers immobilized onto the cyanogen bromide-activated sepharose (Chapuis-Hugon et al. 2011).

13.9 APTAMER-BASED DETECTION PLATFORMS

Applications of new-generation probe aptamers as receptors in biosensors are surging with time. The evolution of transducers for signalling bioreceptor-target binding events is constantly changing to meet the requirements of ultra-sensitive detection platforms. Since the advent of aptamer technology, several mycotoxin detection formats have been developed. These assays include fluorescence, colorimetric, chemiluminescence, PCR, lateral flow, and microfluidics-based methods apart from SERS and electrochemical systems. Various nanomaterials in diverse forms as nanowires, nanoparticles, nanorods and nanotubes have been explored with their novel property to adsorb biomaterials as immobilization platforms.

13.10 COLORIMETRY-BASED SYSTEMS

Colorimetric detection of mycotoxins employs gold and silver nanoparticles (AgNps and AuNPs) and DNAzymes mimicking horseradish peroxidase (HRP). Gold nanoparticles are the preferred choice for colorimetric assays due to their excellent SPR properties. Assays employing aptamer conjugated gold nanoparticles exploit the spectroscopic property of gold nanoparticles and conformational flexibility of aptamers for the detection of analyte by measuring the absorbance using surface plasmon resonance-related colour modification of nanoparticles in the presence and absence of the analyte. Functionalization of gold nanoparticles with single-stranded aptamer takes place by the interaction of aptamer bases with the negative surface of AuNPs by van der Waals forces of attraction. A colorimetric detection system was developed by Luan et al. (2015) using AuNPs and AFB1-specific aptamer with a limit of detection (LoD) of 0.025 ng mL^{-1}. Sensitive colorimetric detection of OTA, AFB1, AFM1, Fumonisin B1, and Patulin based on aptamer and gold nanoparticles are reported (Duan et al. 2016; Samokhvalov et al. 2019; Mirón-Mérida et al. 2021; Ebanks et al. 2023).

Another colorimetric component extensively studied and used in colorimetric biosensing assay platforms is the catalytic nucleic acids termed as DNAzymes. DNAzymes are Guanine-rich sequences of ssDNA that fold into four-stranded G-quadruplex (GQ) structures. These GQs bind to hemin and mimic the peroxidase activity. Hemin/GQs catalyse H_2O_2-mediated oxidation of chromogenic substrates such as Tetra methyl benzidine (TMB) and generate colorimetric signals. DNAzyme conjugated aptamers as Functional Nucleic Acids (FNAs) synergistically amplifies the sensing ability. Colorimetry is an optical technique that accomplishes the requirements of on-site detection system. An aptamer-DNAzyme hairpin for biosensing of OTA was developed by Yang et al. (2012) using aptamer-DNAzyme hairpin as a biorecognition element with an LOD of 2.5 nM. Colorimetric biosensor platform designed with aptamer magnetic separation and DNAzyme-based detection of Aflatoxin B1 (AFB1) with an LOD of 22.6 ppb is reported by Setlem et al. (2020).

13.11 FLUORESCENCE-BASED SYSTEMS

Integration of aptamers with various fluorophores has been explored extensively for the development of assay platforms. Chemically stable, highly fluorescent, water-soluble, less expensive fluoro labels are preferred. These fluorophores contain active chemical groups (carboxyl, amino, isosulfan cyano) for facilitating conjugation with biorecognition molecules (Sharma et al. 2014; Byrne et al. 2015). Quantum dots (QDs), semiconductor nanocrystals are at the forefront of nanoscience and as transducers in the biosensors owing to their excellent optical functionalities like high quantum yields, wider absorption spectra, narrow and symmetric size-tunable emission (Liu et al. 2009; Liao et al. 2014; Ren et al. 2014). Graphene oxide (GO) is a nanomaterial that has attracted tremendous attention due to its excellent fluorescent quenching properties (Morales-Narváez & Merkoçi 2012 ; Pérez-López & Merkoçi 2012). These unique features of GO have rendered it a good candidate for the development of aptamer-based fluorescence quenching systems. The fluorescence quenching by GO has been studied and widely applied for the sensitive detection of biomolecules and mycotoxins (Dong et al. 2010; Wu et al. 2012; Duan et al. 2014; Setlem et al. 2020). Fluorescent labels are conjugated to aptamer by hydroxyl, carboxyl, amino and thiol groups. The most commonly used labelling is via avidin-biotin binding.

Fluorescent-labelled aptamer with a quencher bonded at the opposite end is the probe. Target interactions with aptamer result in structural conformations of the aptamer to attain a secondary structure to bind with the target. This phenomenon increases the distance between the fluorophore and quencher and the subsequent release of fluorophores from the GO matrix, increasing the fluorescence intensity.

This change in fluorescence intensity can be measured and quantified using appropriate spectroscopic techniques. This allows for the determination of mycotoxin concentration in the sample based on the intensity of the fluorescence signal. Liu and co-workers developed a photoelectrochemical system using CdS QDs (Cadmium sulphide- quantum dots) and Fe_3O_4 magnetic nanocomposites (Liu et al. 2015). In another study, QDS and GO (graphene Oxide) system was used for the detection of AFB1 as low as 1.4 nM from peanut oil (Lu et al. 2015). Velu et al. (2015) reported an OTA detection method by using QDs and gold nanorods (AuNRs)-labelled aptamers and called it a linkage inversion assembled nano-apta-sensor (LIANA). Several other fluorophores like Cynine 5, Alexafluor, FAM (carboxyfluorescein) and TAMRA (tetramethyl-6-carboxyrhodamine) have been used for the development of fluorescence-based aptamer assays for the detection of mycotoxins like OTA, Fumonisin B_1 and Trichothecene T-2 toxin (Xi-Hui et al. 2013). The other fluorophores employed in the fluorescent mycotoxin aptasensors are rhodamine 6G (Rh6), tetraphenylethene salt, thioflavin T, 2,3-diaminophenazine *N*-methyl mesoporphyrin (DAP), PicoGreen, N,N`-bis (propylenetrimethylammonium)-3,4,9,10-perylenediimide. Time-Resolved Fluorescent Nanoparticles (TRFNPs) include $Zn_2GeO_4:Mn^{2+}$, ZGGO, ZGO, Mn, Cr, $Zn_{1.25}Ga_{1.5}Ge_{0.25}O_4$: Cr^{3+},Yb^{3+},Er^{3+}, and also a few silica photonic crystal microsphere (SPCM) are used currently (Yan et al. 2022).

13.12 CHEMILUMINESCENCE-BASED SYSTEMS

Chemiluminescence (CL) is another analytical technique that works on the phenomenon of excitation of intermediate chemical molecules to emit photons that aim to return to the ground state. Chemiluminescence resonance energy transfer (CRET) is used for the amplification of Chemiluminescence signals by energy transfer between a chemiluminescent donor and a fluorophore acceptor. Hosseini and co-workers (2015) developed an assay for the detection of AFB1 by CL and AuNPs labelled aptamers with LOD of 7 nM. The phenomenon of aggregation of AuNPs in the presence of AFB1 gave a weak CL signal, in contrary to the intense chemiluminescence in the absence of AFB1 (Hosseini et al. 2015).

13.13 APTA-POLYMERASE CHAIN REACTION (PCR)-BASED SYSTEMS

PCR is the most popular and reliable method when compared to many conventional microbiological assays The real-time quantitative PCR (RT-qPCR) involving fluorescence is being employed in aptamer technology to develop detection systems with utmost sensitivity and specificity. McKeague et al. (2014) selected OTA-specific aptamers and developed a Label-Free Fluorescence assay with SYBR® Green I-labelled aptamers. Upon binding with the target, the fluorescence signal was exponentially enhanced by real-time PCR and a limit of detection of 9 nM was obtained (McKeague et al. 2014). Aptamer-based RT-qPCR method was developed by Guo et al. (2016) for the detection of AFM1 in infant food and milk products. The aptamer used for target recognition was immobilized by biotin-streptavidin interaction and its complementary ssDNA was employed for quantification of AFM1 by RT-qPCR with LOD of 0.03 ng L^{-1} (Guo et al. 2016).

13.14 ELECTROCHEMICAL-BASED SYSTEMS

In electrochemical method, the aptamer is immobilized onto the electrode surface, and electrical current variations are studied based on the recognition event. Electrochemical platforms such as amperometric and impedimetric have been developed using the interaction of aptamer and analytes. In amperometric or voltammetric aptasensors, the ampere of current generated by the redox reaction of electroactive species at fixed or variable volt is measured. An ampere of the current generated correlates to the concentration of the target. Impedimetric method also known as Electrochemical Impedance Spectroscopy (EIS) allows sensitive monitoring of changes in conductivities and resistances of electrochemical interfaces. In 2011, Prabhakar and his co-workers published the first aptamer-based impedimetric platform for OTA detection with an LOD of 0.1 μg L^{-1} (Rhouati et al. 2016). Another group, Castillo et al. (2012) developed an electrochemical platform for the detection of OTA in coffee and wine samples by immobilization of thiol-modified aptamer on a Au electrode. The charge resistance increased with the concentration of the Ochratoxin A and a redox probe $[Fe(CN)_6]^{3-/4-}$ amplifies the signal. Various electrochemical aptasensors for mycotoxin analysis are reported. Electrochemical aptasensors based on cyclic voltammetry, EIS, different, differential pulse voltammetry (DPV) and square wave voltammetry (SWV) have gained attention in the diagnostics of mycotoxins owing to their advantages in terms of sensitivity, selectivity and user friendly (Castillo et al. 2012; Guo et al. 2020b).

13.15 SURFACE PLASMON RESONANCE-BASED ASSAYS

SPR is a label-free real-time non-destructive optical platform based on spectroscopy that provides information on the affinity and kinetics of biomolecular interaction by measuring the change in adsorption of molecules on the nanomaterial metal surface caused by SPR. Nano metals such as Au, Ag and Cu that enhance the sensitivity are utilized. SPR has extensive applications in the biomedical field, food safety, forensic and environmental monitoring (Nguyen et al. 2015). Detection of mycotoxin AFB1 by SPR employing aptamer was first reported by Li et al. (2017) with LOD of 0.4 fg mL^{-1} with recovery from food matrix in the range of 89–121%. Bianco et al. (2017) reported an aptamer-based SPR-polarization platform for the detection of OTA. A Quartz Crystal Microbalance-based piezoelectric biosensor was used for real-time monitoring of aptamer-OTA interactions with the detection of 0.005 ng mL^{-1}. Wu et al. (2018a) reported an SPR assay employing aptamer and biotin-streptavidin chemistry to detect AFB1. This biotinylated aptasensor exhibited LOD of 0.19 ng mL^{-1} in vinegar, achieving recoveries in the range 96.3–117.8%. SPR combined with AIR-SE (attenuated internal reflection spectroscopic ellipsometry), a highly sensitive analytical biosensing tool employed for the sensitive and selective detection of Zearalenone (ZEN), a metabolite of Fusarium species detection in cereal products with 0.08 ng/mL and 0.01–1000 ng mL^{-1} detection range (Caglayan & Üstündağ 2020). Optical fibre-based localized surface plasmon resonance (LSPR) aptasensor for ZEN detection is reported by Xu et al. (2021b).

13.16 SURFACE-ENHANCED RAMAN SCATTERING SYSTEM

SERS is a highly sensitive biosensor platform for fingerprinting of mycotoxins. The precise recognition ability of bioreceptor aptamers in SERS makes it an excellent method. Due to the advantage of narrow bandwidth, shortened detection times, reduced photobleaching and sensitivity, SERS can be used for multiple analyses of targets. SERS system for AFB1 and OTA in natural samples with LOD of 0.006 ng mL^{-1} for OTA and 0.03 ng mL^{-1} for AFB1 is reported by Zhao et al. (2015). The non-polar structural groups can be identified by Raman spectroscopy. Raman spectral fingerprinting of mycotoxins Aflatoxins B1, B2, G1, G2, Ochratoxin A, Zearalenone, Deoxynivalenol, Patulin, Fumonisin B1, Citrinin, and band assignments of characteristic peaks are reported (Wu et al. 2021).

13.17 MICROFLUIDIC PLATFORMS

Microfluidics are miniaturized analytical formats which can be integrated with multiple sensor arrays that enhance the throughput. Microfluidic platforms are gaining importance as an alternative technology for on-site detection of mycotoxins due to their cost effectiveness and mass producibility. Microfluidics are being used both as SELEX matrices for the generation of aptamers and detection platforms as well. Microfluidic paper-based analytical device (μPAD) is reported for the colorimetric detection of AFM1 employing aptamer AuNPs in milk samples (Kasoju et al. 2020; Banu et al. 2022).

13.18 APTAMER-BASED LATERAL FLOW ASSAY (LFA)

Lateral flow assays are on-site, point-of-care detection platforms requiring one-step sample addition by the user without any requirement of a sophisticated instrument. LFAs have gained importance in pharma, animal health, environmental testing, food, plant and crop health testing, etc. Aptamer-based LFAs are replacing antibody LFAs due to their infinite potential. Sandwich or competitive formats of LFAs are available. The biotin-streptavidin chemistry is employed for the immobilization of aptamer as a capturing ligand, where the biotinylated aptamer binds to a streptavidin-immobilized membrane (Kaiser et al. 2018). A one-step competitive lateral flow fluorescent aptasensor strip for the detection of OTA in corn and red wine has been reported (Zhang et al. 2018a; Liu et al. 2022). A fluorescent-labelled aptamer-based lateral flow detection strips are reported for Zearalenone and AFB1 (Wu et al. 2018b; Zhao et al. 2020). Different aptamer-based mycotoxin detection platforms are depicted in Table 13.4.

13.19 APTASENSORS FOR THE DETECTION OF PATHOGENIC FUNGI AND FUNGAL SPORES

More than 19,000 species of fungi cause diseases in crop plants. These fungi remain in dormant spore form till conditions are conducive (Jain et al. 2019). An electrochemical apta assay for the concurrent collection and detection of airborne *P. pachyrhizi* Uredinio rust spores from air is developed by Krivitsky et al. (2021). Aptamers selected against the fungal spores linked to an enzymatic reaction lead to the identification of spores of interest. The assay is sensitive enough to detect 100 spores of *P. pachyrhizi* airborne spores/cm2 in less than 2 minutes under field conditions. Highly specific aptamers are selected against the spores of toxigenic fungi *Aspergillus* species (*A. fumigatus*, *A flavus* and *A. niger*) for the first time by Seo et al. (2021). Aptamers are selected against whole cells of *A. fumigatus* and *C. albicans* (Milnes 2019) The selected aptamers conjugated with suitable transducers or sensors could be a potential alternative for the rapid detection of fungal toxins.

13.20 ADVANTAGES AND LIMITATIONS OF APTAMER-BASED ASSAYS

New innovative avenues are created in the diagnostic platforms by aptamers. As aptamers are more flexible and exhibit inherent advantages in terms of stability, sensitivity, reusability and applicability in designing multiple sensing platforms and multiplexed assays when compared to antibodies (details as shown in Table 13.3). Aptamers can be selected against low molecular weight molecules as small as 60 daltons, inorganic metal ions, non-immunogenic molecules. Moreover, aptamers can be chemically modified with ease. Chemically synthesized without any batch variations. Easily scalable. No animal host/cells are required unlike in antibody generation thus overcoming the animal ethical issues. Denaturation and reforming of aptamer is possible.

The main bottleneck preventing the real-time deployment of aptamers is their quick degradation by nucleases in biological media. However, multiple approaches have emerged to address this limitation. One of the most effective methods used to address this limitation is by modifying the nucleotides at 3' and 5' ends while performing SELEX to obtain nuclease-resistant aptamers. Incorporation of a functional group or polyethylene glycol (PEG) enhances the resistance of aptamers. The concept of Spiegelmers with L-ribose backbone makes the aptamer more rugged (Lakhin et al. 2013).

TABLE 13.4
Aptamer Based Mycotoxin Detection Platforms

	Analytical Method	Principle	Aptamer Sequence	Sample	LOD	Ref.
OTA	Fluorescence	PVP-protected graphene oxide as quencher	5`-FAM-GAT CGG GTG TGG GTG GCG TAA AGG GAG CAT CGG ACA-3`	Wine	21.8 nM	Sheng et al. 2011
OTA	Fluorescence	Aptamer-conjugated MBs and QDs	5`-GAT CGG GTG TGG GTG GCG TAA AGG GAG CAT CGG ACA–SH-3′	Peanut	5.4 pg mL–1	Wang et al. 2015
OTA	Fluorescence	Carboxy-modified fluorescent particles	5′-GAT CGG GTG TGG GTG GCG TAA AGG GAG CAT CGG ACA-3′	Beer	0.005 nM	Hayat et al. 2015
OTA	Fluorescence	Guanine as the quenching agent	5`-GAT CGG GTG TGG GTG GCG TAA AGG GAG CAT CGG ACA -3`	Herbal matrics	28.4 nM	Guo et al. 2023
OTA	Chemiluminescence	Split DNAzyme mediated	5`-GAT CGG GTG TGG GTG GCG TAA AGG GAG CATC-3`	Wine	0.10 nmol L^{-1}	Cheng et al. 2021
OTA	Chemiluminescence	Cl-ELASA	5′-GAT CGG GTG TGG GTG GCG TAA AGG GAG CAT CGG ACA-3′	Peanut and coffee	1.29 pg mL^{-1}	Mukherjee et al. 2021
OTA	Chemiluminescence	HRP-mimicking DNAzyme (HRPzyme)	5`-GCA TCT GAT CGG GTG TGG GTG GCG TAA AGG AAA-3`	Coffee bean	0.22 ng mL^{-1}	Jo et al. 2016
OTA	Colorimetric	Dual-functional Au/Fe3O4 nanohybrid	5′-GAT CGG GTG TGG GTG GCG TAA AGG GAG CAT CGG ACA-3′	Peanut and corn	1.15 ng mL^{-1}	Huang et al. 2019
OTA	Colorimetric	Charge effect-assisted silver enhancement	5`-GAT CGG GTG TGG GTG GCG TAA AGG GAG CAT CGG ACA-SH-3′	Peanut, corn, and soybean	28.18 pg mL^{-1}	Yang et al. 2023
OTA	Colorimetric	Co-ELASA	5′-GAT CGG GTG TGG GTG GCG TAA AGG GAG CAT CGG ACA-3′	Peanut and coffee	0.84 pg mL^{-1}	Mukherjee et al. 2021
OTA	q-PCR	Target-induced dissociation beads-based Apta-qPCR	5`-TGG TGG CTG TAG GTC AGC ATC TGA TCG GGT GTG GGT GGC GTA AAG GGA GCA TCG GAC AACG-3`	Grains and grain derivatives	0.009 ng mL^{-1}	Modh et al. 2017
AFB1	Fluorescence	PicoGreen dye-based strategy	5`-CAGCTTATTCAATTGTGCAGGGGGAGGGGGAGTGGTGGCT CGCGGTGC-3`	Areca nuts and yam	0.1 ug L^{-1}	Zhang et al. 2018b
AFB1	Fluorescence	Exo I and SYBR Gold-mediated quenching	5′- GTT GGG CAC GTG TTG TCT CTC TGT GTC TCG TGC CCT TCG CTA GGC CCA CA-3′	Sauces	1.82 ng mL^{-1}	Guo et al. 2020a

AFB1	Fluorescence	Graphene oxide and TPE-Z mediated quenching	5`-TRE-Z-GTT GGG CAC GTG TTG TCT CTC TGT GTC TCG TGC CCT TCG CTA GGC CCA CA-GO-3`	Food sample	0.25 ng mL^{-1}	Jia et al. 2019
AFB1	Fluorescence	Alexafluor-Graphene oxide mediated quenching	5`-TTC TTC TGC CTT GGT GGT TGG TGT GTC TGC TGA TTT GGT A -3`	Food sample	40 ppb	Setlem et al. 2022
AFM1	Electro chemical	CRISPR-Cas12a, rolling circle amplification (RCA) and catalytic activity of AuNPs	5`-TGT TGT CTC TCT GTG TC-3`	Fruit Juice and corn flour meal	0.05 ng L^{-1}	Abnous et al. 2021
AFB1	Electro chemical	Polycarboxylic macrocycle modified with neutral red	5′-GTT GGG CAC GTG TTG TCT CTC TGT GTC TCG TGC CCT TCG CTA GGC CCA CA-3`	Peanuts and cashew nuts	0.1 nM	Evtugyn et al. 2014
AFM1	Electro chemical	Redox marker-assisted	5′-ACT GCT AGA GAT TTT CCA CAT-3′	Milk sample	8.47 ng mL^{-1}	Karapetis et al. 2018
AFM1	Electro chemical	SPGE mediated	5′-ATC CGT CAA CCT GCT CTG ACG CTG GGG TCG ACC CGG AGA AAT GCA TTC CCT GTG GTG TTG GCT CCC GAT-TEG Biotin-3′	-	-	Pandey et al. 2017
AFB1	Electro chemical	apt-BIO/AFB1-BSA/PANI-PAA	5′- GTT GGG CAC GTG TTG TCT CTC TGT GTC TCG TGC CCT TCG CTA GGC CCA CA-3	Maize	0.086 ng mL^{-1}	Selvolini et al. 2019
AFB1	Colorimetric	Exonuclease III-assisted signal amplification	5′-FAM-TGC ACG TGT TGT CTC TCT GTG TCT CGT GCT TTT TT-3`	Cereals	0.0032 ng mL^{-1}	Chen et al. 2021
AFB1	Colorimetric	PCPD, and AgNPs nanoprobes	5′-GTT GGG CAC GTG TT G TCT CTC TGT GTC TCG TGC CCT TCG CTA GGC CCA CA-3`	Peanut, corn	0.09 ng mL^{-1}	Lerdsri et al. 2021
T-2 Toxin	Fluorescence	Tungsten disulfide (ws2) as quencher	5′- biotin-CAG CTC AGA AGC TTG ATC CTG TAT TAC AAG CAT CGC GTG TTTACA CAT GCG AGA GGT GAA GAC TCG AAG TCG TGC ATC TG-3′	Maize	0.33 pg	Niazi et al. 2019
T-2 Toxin	Fluorescence	UCNPs and MIL-101 based quenching	5`-CAG CTC AGA AGC TTG ATC CTG TAT ATC AAG CAT CGC GTG TTT ACA CAT GCG AGA GGT GAA GAC TCG AAG TCG TGC ATC TG-3`	Corn	0.087 ng mL^{-1}	Zhao et al. 2021
T-2 Toxin	Electrochemical	rGO-TEPAAu@Pt nanorods with a dual signal amplification strategy	–	Food sample	1.79 fg/mL	Zhong et al. 2019

(*Continued*)

TABLE 13.4 (Continued)

	Analytical Method	Principle	Aptamer Sequence	Sample	LOD	Ref.
ZEN	Fluorescence	Aptasensors based on upconverting nanoparticles	–	Corn	0.126 ug kg^{-1}	Wu et al. 2017
ZEN	Chemiluminescence	Peroxidase-like G-quadruplex DNAzyme mediated quenching	5`-GAT TCA TCT ATC TAT GGT ACA TTA CTA TCT GTA ATG TGA-3`	Cereals	8.95 nM	Guan et al. 2023
FB1	Electrochemical	Au NPs driven electrochemiluminescence (ECL)	5`-SH-(CH2)6-ATA CCA GCT TAT TCA ATT AAT CGC ATT ACC TTA TAC CAG CTT ATT CAA TTA CGT CTG CAC ATA CCA GCT TAT TCA ATT AGA TAG TAA GTG CAA TCT-3`	Wheat	0.2 ng mL^{-1}	Zhao et al. 2014
FB1	Microcantilever array aptasensor	SAMs functionalized sensing cantilevers	5`-SH-(CH)6-ATA CCA GCT TAT TCA ATT AAT CGC ATT ACC TTA TAC CAG CTT ATT CAA TTA CGT CTG CAC ATA CCA GCT TAT TCA ATT AGA TAG TAA GTG CAA TCT-3`	Food and agro product	33 ng mL^{-1}	Chen et al. 2015
PAT	Electrochemical	AuNPs-BPNS nanocomposites mediated dual-signalling strategy	5′-MB-CCCGGCCCGCCAACCCGCATCATCTACACTGATATTTTACCTTCCC-SH-3′	Food sample	0.043 ng mL^{-1}	Xu et al. 2023
PAT	Fluorescence	rGO-Fe3O4 based quenching	5`-GGC CCG CCA ACC CGC ATC ATC TAC ACT GAT ATT TTA CCT T -3′	Fruit juice	0.28 µg L–1	Ma et al. 2018
PAT	Fluorescence	Exonuclease III mediated amplification	5`-GGC CCG CCA ACC CGC ATC ATC TAC ACT GAT ATT TTA CCT T-3`	Fruit juice	4.7 ng L^{-1}	Deng et al. 2022
PAT	Impedimetric aptasensor	Redox label assisted	5′-GGC CCG CCA ACC CGC ATC ATC TAC ACT GAT ATT TTA CCT T-NH2-3′	Apple	2.8 ng L^{-1}	Khan et al. 2019
PAT	Electro analytical	DPV/Au@PANI/SeS2@Co MOF/SPE	-	Apple juice	0.00015–15.4 ng mL^{-1}	Selvam et al. 2021

The aptamer-based assay platforms still have other challenges such as screening of high-affinity aptamers against various structural analogues of mycotoxin variants.

13.21 SUMMARY

Timely detection of fungal pathogens, their toxins and spores is the need of the hour for the effective prevention and management of crop disease spread. Various promising aptamer-based assays are reported for detecting important mycotoxins OTA (Ochratoxin A) and AFB1 (Aflatoxin B1). However, compared to immunoassays, aptamer-based assays for fungal pathogens, and phytopathogenic fungal toxins are still in the development phase. Although aptamer assays are well-established in the literature for bacterial pathogens and few of the mycotoxins, viz., OTA and AFB1, ZEN, T2, FB1, for other phytotoxic fungal toxins are either recently initiated or yet to be developed. As advancements continue, aptamer-based assays hold considerable promise for revolutionizing the detection and management of fungal diseases in crops.

13.22 FUTURE PERSPECTIVES

Aptamer-based assays have the potential to be multiplexed, allowing the simultaneous detection of multiple fungal toxins in a single assay. This could provide a more comprehensive analysis of food and agricultural samples, saving time and resources. Aptamer-based assays could be employed in the field for monitoring fungal toxin levels in crops. Real-time data collection could enable timely decision-making for farmers, helping them manage and mitigate the impact of fungal phytotoxins on crop yields. Aptamer-based assays may become integral components of broader surveillance systems for monitoring fungal contamination in agricultural products. This could contribute to the early detection and prevention of mycotoxin-related issues in the food supply chain. The method of extraction of toxin from the complex matrix, identifying and differentiating between closely related mycotoxin analogues is also equally challenging. The development of multiple mycotoxin aptasensors in a single platform is the need of the hour.

STUDY QUESTIONS

1. In the field of therapeutics and diagnostics, can aptamers completely supersede antibodies?
2. Can aptamer chimaeras be an answer to detect multiple analogues of the mycotoxins or closely related toxins in a single platform?
3. As the aptamers offer numerous advantages, can they be a universal solution for several concerns or challenges?

ACKNOWLEDGEMENT

The authors are indebted to Director, DFRL, Mysore for providing support to write this book chapter.

FURTHER READING

Caglayan, M.O., Şahin, S. and Üstündağ, Z., 2022. Detection strategies of zearalenone for food safety: A review. *Critical Reviews in Analytical Chemistry*, 52(2), pp. 294–313.

Caglayan, M.O. and Üstündağ, Z., 2020. Detection of zearalenone in an aptamer assay using attenuated internal reflection ellipsometry and it's cereal sample applications. *Food and Chemical Toxicology*, 136, p. 111081.

REFERENCES

Abbas, H.K., Johnson, B.B., Shier, W.T., Tak, H., Jarvis, B.B. and Boyette, C.D., 2002. Phytotoxicity and mammalian cytotoxicity of macrocyclic trichothecene mycotoxins from Myrothecium verrucaria. *Phytochemistry*, 59(3), pp. 309–313.

Abnous, K., Danesh, N.M., Ramezani, M., Alibolandi, M., Nameghi, M.A., Zavvar, T.S. and Taghdisi, S.M., 2021. A novel colorimetric aptasensor for ultrasensitive detection of aflatoxin M1 based on the combination of CRISPR-Cas12a, rolling circle amplification and catalytic activity of gold nanoparticles. *Analytica Chimica Acta*, 1165, p. 338549.

Balendres, M.A.O., Karlovsky, P. and Cumagun, C.J.R., 2019. Mycotoxigenic fungi and mycotoxins in agricultural crop commodities in the Philippines: A review. Food, 8(7), p. 249.

Bandurska, H., Chelkowski, J. and Wisniewska, H., 1994. Free proline accumulation in wheat seedlings influenced by *Fusarium culmorum* infection and the pathogen metabolite deoxynivalenol. *Acta Physiologiae Plantarum*, 16(2).

Banu, K., Mondal, B., Rai, B., Monica, N. and Hanumegowda, R., 2022. Prospects for the application of aptamer based assay platforms in pathogen detection. *Biocybernetics and Biomedical Engineering*, 42(3), pp. 934–949.

Bianco, M., Sonato, A., De Girolamo, A., Pascale, M., Romanato, F., Rinaldi, R. and Arima, V., 2017. An aptamer-based SPR-polarization platform for high sensitive OTA detection. *Sensors and Actuators B: Chemical*, 241, pp. 314–320.

Byrne, B., Gilmartin, N., Lakshmanan, R.S. and O'Kennedy, R., 2015. Antibodies, enzymes, and nucleic acid sensors for high throughput screening of microbes and toxins in food. In *High throughput screening for food safety assessment* (pp. 25–80). Woodhead Publishing.

Caglayan, M.O., Şahin, S. and Üstündağ, Z., 2022. Detection strategies of zearalenone for food safety: A review. *Critical Reviews in Analytical Chemistry*, 52(2), pp. 294–313.

Caglayan, M.O. and Üstündağ, Z., 2020. Detection of zearalenone in an aptamer assay using attenuated internal reflection ellipsometry and it's cereal sample applications. *Food and Chemical Toxicology*, 136, p. 111081.

Capriotti, A.L., Caruso, G., Cavaliere, C., Foglia, P., Samperi, R. and Laganà, A., 2012. Multiclass mycotoxin analysis in food, environmental and biological matrices with chromatography/mass spectrometry. *Mass Spectrometry Reviews*, 31(4), pp. 466–503.

Carlson, M.P., 2011. Ergot of small grain cereals and grasses and its health effects on humans and livestock.

Castillo, G., Lamberti, I., Mosiello, L. and Hianik, T., 2012. Impedimetric DNA aptasensor for sensitive detection of ochratoxin A in food. *Electroanalysis*, 24(3), pp. 512–520.

Cavaliere, C., Foglia, P., Guarino, C., Nazzari, M., Samperi, R. and Laganà, A., 2007. Determination of aflatoxins in olive oil by liquid chromatography–tandem mass spectrometry. *Analytica Chimica Acta*, 596(1), pp. 141–148.

Chapuis-Hugon, F., du Boisbaudry, A., Madru, B. and Pichon, V., 2011. New extraction sorbent based on aptamers for the determination of ochratoxin A in red wine. *Analytical and Bioanalytical Chemistry*, 400, pp. 1199–1207.

Chen, X., Bai, X., Li, H. and Zhang, B., 2015. Aptamer-based microcantilever array biosensor for detection of fumonisin B-1. *RSC Advances*, 5(45), pp. 35448–35452.

Chen, Y., Zhang, F., Liu, R., Liu, M., Sang, Y., Wang, S. and Wang, X., 2021. A novel colorimetric nano aptasensor for ultrasensitive detection of aflatoxin b1 based on the exonuclease iii-assisted signal amplification approach. *Food*, 10(11), p. 2568.

Cheng, Y.A.N.G., Zhang, Y.Q., Lin, D.Q., Yang, L.I.U. and Bing-Bing, S.U.N., 2021. Chemiluminescence sensor based on composite functional nucleic acid for detection of ochratoxin A in wine. *Chinese Journal of Analytical Chemistry*, 49(4), pp. 496–503.

Cox, J., Rajendran, M., Riedel, T., Davidson, E.A., Sooter, L.J., Bayer, T.S., Schmitz-Brown, M. and Ellington, A.D., 2002. Automated acquisition of aptamer sequences. *Combinatorial Chemistry & High Throughput Screening*, 5(4), pp. 289–299.

Crisan, E.V., 1973. Effects of aflatoxin on seeding growth and ultrastructure in plants. *Applied Microbiology*, 26(6), pp. 991–1000.

Cruz-Aguado, J.A. and Penner, G., 2008. Determination of ochratoxin A with a DNA aptamer. *Journal of Agricultural and Food Chemistry*, 56(22), pp. 10456–10461.

Damodaran, C., Kathirvel-Pandian, S., Seeni, S., Selvam, R., Ganesan, M.G. and Shanmugasundaram, S., 1975. Citrinin, a phytotoxin? *Experientia*, 31(12), pp. 1415–1417.

Darmostuk, M., Rimpelova, S., Gbelcova, H. and Ruml, T., 2015. Current approaches in SELEX: An update to aptamer selection technology. *Biotechnology Advances*, 33(6), pp. 1141–1161.

Dass, R.S., Sreenivasa, M.Y., Janardhana, G.R. 2007. High incidence of *Fusarium verticillioides* in animal and poultry feed mixtures produced in Karnataka, India. *Plant Pathology Journal* 6(2), 174–178.

De Iongh, H., Vles, R.O. and De Vogel, P., 1964, March 18–19. The occurrence and detection of aflatoxin in food. In Proceedings of the Symposium on Mycotoxins in Foodstuffs, Cambridge, MA, USA.

Deepa, N. and Sreenivasa, M.Y. 2019a. Biocontrol strategies for effective management of phytopathogenic fungi associated with cereals. In J. S. Singh, D.P. Singh (Eds) *New and future developments in microbial biotechnology and bioengineering*, Elsevier, pp, 177–189, ISBN 9780444641915.

Deepa, N., Charith Raj, A. P., Sreenivasa, M.Y. 2016b. Nested PCR method for the early detection of fumonisin producing *Fusarium verticillioides* in pure cultures, cereal samples and plant parts. *Food Biotechnology*, 30(1), 18–29.

Deepa, N., Charith Raj, A.P., Sreenivasa, M.Y. 2016c. Multiplex PCR method for the early detection of fumonisin producing *Fusarium verticillioides*. *Food Bioscience*, 13, 84–88.

Deepa, N., Nagaraja, H., Sreenivasa, M.Y. 2016a. Prevalence of fumonisin producing *Fusarium verticillioides* associated with cereals grown in Karnataka (India). *Food Science and Human Wellness*, 5, 156–162. http://doi.org/10.1016/j.fshw.2016.07.001

Deepa, N., and Sreenivasa, M.Y. 2017a. Fumonisins: A review on its global occurrence, epidemiology, toxicity and detection. *Journal of Veterinary Medicine and Research*, 4(6), 1093.

Deepa, N., and Sreenivasa, M.Y. 2017b. *Fusarium verticillioides*, a globally important pathogen of agriculture and livestock: A review. *Journal of Veterinary Medicine and Research*. 4(4), 1084.

Deepa, N. and Sreenivasa, M. Y. 2019b. Molecular methods and key genes targeted for the detection of fumonisin producing *Fusarium verticillioides* - An updated review. *Food Bioscience*, 32, 100473.

Deng, L., Hu, J., Zhao, J., Zhang, Z., Wang, Q. and Wu, R., 2022. A novel ratiometric fluorescent aptasensor accurately detects patulin contamination in fruits and fruits products. *Arabian Journal of Chemistry*, 15(2), p. 103569.

Diba, K., Kordbacheh, P., Mirhendi, S.H., Rezaie, S. and Mahmoudi, M., 2007. Identification of *Aspergillus* species using morphological characteristics. *Pakistan Journal of Medical Sciences*, 23(6), p. 867.

Dong, H., Gao, W., Yan, F., Ji, H. and Ju, H., 2010. Fluorescence resonance energy transfer between quantum dots and graphene oxide for sensing biomolecules. *Analytical Chemistry*, 82(13), pp. 5511–5517.

Duan, N., Xu, B., Wu, S. and Wang, Z., 2016. Magnetic nanoparticles-based aptasensor using gold nanoparticles as colorimetric probes for the detection of *Salmonella typhimurium*. *Analytical Sciences*, 32(4), pp. 431–436.

Duan, Y.F., Ning, Y., Song, Y. and Deng, L., 2014. Fluorescent aptasensor for the determination of *Salmonella typhimurium* based on a graphene oxide platform. *Microchimica Acta*, 181, pp. 647–653.

Dunn, M.R., Jimenez, R.M. and Chaput, J.C., 2017. Analysis of aptamer discovery and technology. *Nature Reviews Chemistry*, 1(10), p. 0076.

Ebanks F., Nasrallah, H., Garant, T.M., McConnell, E.M. and DeRosa, M.C., 2023. Colorimetric detection of aflatoxins B1 and M1 using aptamers and gold and silver nanoparticles. *Advanced Agrochem*, 2(3), pp. 221–230.

Egmond, H.P., Schothorst, R.C. and Jonker, M.A., 2007. Regulations relating to mycotoxins in food. *Analytical and Bioanalytical Chemistry*, 1(389), pp. 147–157.

Ellington, A.D. and Szostak, J.W., 1990. In vitro selection of RNA molecules that bind specific ligands. *Nature*, 346(6287), pp. 818–822.

Ellington, A.D. and Szostak, J.W., 1992. Selection in vitro of single-stranded DNA molecules that fold into specific ligand-binding structures. *Nature*, 355(6363), pp. 850–852.

Ellis, J.R. and McCalla, T.M., 1973. Effects of patulin and method of application on growth stages of wheat. *Applied Microbiology*, 25(4), pp. 562–566.

Evtugyn, G., Porfireva, A., Stepanova, V., Sitdikov, R., Stoikov, I., Nikolelis, D. and Hianik, T., 2014. Electrochemical aptasensor based on polycarboxylic macrocycle modified with neutral red for aflatoxin B1 detection. *Electroanalysis*, 26(10), pp. 2100–2109.

Food Safety and Standards Authority of India- FSSAI. 2011. LEX-FAOC132615.

Gilbert, J. and Vargas, E.A., 2003. Advances in sampling and analysis for aflatoxins in food and animal feed. *Journal of Toxicology - Toxin Reviews*, 22(2–3), pp. 381–422.

Gopinath, S.C.B., 2007. Methods developed for SELEX. *Analytical and Bioanalytical Chemistry*, 387, pp. 171–182.

Guan, Y., Ma, J., Neng, J., Yang, B., Wang, Y. and Xing, F., 2023. A novel and label-free chemiluminescence detection of zearalenone based on a truncated aptamer conjugated with a G-quadruplex DNAzyme. *Biosensors*, 13(1), p. 118.

Guo, L., Li, Y., Gao, S. and Ren, L., 2023. Detection of ochratoxin A using a guo "turn-on" fluorescence assay based on guanine quenching of the aptamer. *Analytical Sciences*, 39(1), pp. 51–57.

Guo, X., Wen, F., Zheng, N., Li, S., Fauconnier, M.L. and Wang, J., 2016. A qPCR aptasensor for sensitive detection of aflatoxin M 1. *Analytical and Bioanalytical Chemistry*, 408, pp. 5577–5584.

Guo, X., Wen, F., Zheng, N., Saive, M., Fauconnier, M.L. and Wang, J., 2020a. Aptamer-based biosensor for detection of mycotoxins. *Frontiers in Chemistry*, 8, p. 195.

Guo, Z., Lv, L., Cui, C., Wang, Y., Ji, S., Fang, J., Yuan, M. and Yu, H., 2020b. Detection of aflatoxin B 1 with a new label-free fluorescent aptasensor based on exonuclease I and SYBR Gold. *Analytical Methods*, 12(22), pp. 2928–2933.

Hasan, H.A.H., 1999. Phytotoxicity of pathogenic fungi and their mycotoxins to cereal seedling viability. *Mycopathologia*, 148, pp. 149–155.

Hasegawa, H., Savory, N., Abe, K. and Ikebukuro, K., 2016. Methods for improving aptamer binding affinity. *Molecules*, 21(4), p. 421.

Hayat, A., Mishra, R.K., Catanante, G. and Marty, J.L., 2015. Development of an aptasensor based on a fluorescent particles-modified aptamer for ochratoxin A detection. *Analytical and Bioanalytical Chemistry*, 407, pp. 7815–7822.

Henson, S. and Caswell, J., 1999. Food safety regulation: An overview of contemporary issues. *Food Policy*, 24(6), pp. 589–603.

Hosseini, M., Khabbaz, H., Dadmehr, M., Ganjali, M.R. and Mohamadnejad, J., 2015. Aptamer-based colorimetric and chemiluminescence detection of aflatoxin B1 in foods samples. *Acta Chimica Slovenica*, 62(3), pp. 721–728.

Huang, R., Xiong, L.L., Chai, H.H., Fu, J.J., Lu, Z. and Yu, L., 2019. Sensitive colorimetric detection of ochratoxin A by a dual-functional Au/Fe_3O_4 nanohybrid-based aptasensor. *RSC Advances*, 9(66), pp. 38590–38596.

Hünniger, T., Wessels, H., Fischer, C., Paschke-Kratzin, A. and Fischer, M., 2014. Just in time-selection: A rapid semiautomated SELEX of DNA aptamers using magnetic separation and BEAMing. *Analytical Chemistry*, 86(21), pp. 10940–10947.

Ismaiel, A.A. and Papenbrock, J., 2014. The effects of patulin from Penicilliumvulpinum on seedling growth, root tip ultrastructure and glutathione content of maize. *European Journal of Plant Pathology*, 139, pp. 497–509

Ismaiel, A.A. and Papenbrock, J., 2015. Mycotoxins: Producing fungi and mechanisms of phytotoxicity. *Agriculture*, 5(3), pp. 492–537.

Jain, A., Sarsaiya, S., Wu, Q., Lu, Y. and Shi, J., 2019. A review of plant leaf fungal diseases and its environment speciation. *Bioengineered*, 10(1), pp. 409–424.

Janardhana, G. R., Dass, R.S., Sreenivasa, M.Y. 2009. Fumonisins – A new class of Fusarial toxins. In K. A. Sridhar et al. (Eds.), *Frontiers in Fungal Ecology, Diversity and Metabolites*. New Delhi: I. K. International Publishing House Pvt. Ltd.

Janardhanan, K.K. and Husain, A., 1984. Phytotoxic activity of tenuazonic acid isolated from *Alternaria alternata* (Fr.) Keissler causing leaf blight of *Datura innoxia* Mill. and its effect on host metabolism. *Journal of Phytopathology*, 111(3–4), pp. 305–311.

Jayasena, S.D., 1999. Aptamers: an emerging class of molecules that rival antibodies in diagnostics. *Clinical Chemistry*, 45(9), pp. 1628–1650.

Jia, Y., Wu, F., Liu, P., Zhou, G., Yu, B., Lou, X. and Xia, F., 2019. A label-free fluorescent aptasensor for the detection of Aflatoxin B1 in food samples using AIEgens and graphene oxide. *Talanta*, 198, pp. 71–77.

Jo, E.J., Mun, H., Kim, S.J., Shim, W.B. and Kim, M.G., 2016. Detection of ochratoxin A (OTA) in coffee using chemiluminescence resonance energy transfer (CRET) aptasensor. *Food Chemistry*, 194, pp. 1102–1107.

Kaiser, L., Weisser, J., Kohl, M. and Deigner, H.P., 2018. Small molecule detection with aptamer based lateral flow assays: Applying aptamer-C-reactive protein cross-recognition for ampicillin detection. *Scientific Reports*, 8(1), p. 5628.

Karapetis, S., Nikolelis, D. and Hianik, T., 2018. Label-free and redox markers-based electrochemical aptasensors for aflatoxin M1 detection. *Sensors*, 18(12), p. 4218.

Kasoju, A., Shahdeo, D., Khan, A.A., Shrikrishna, N.S., Mahari, S., Alanazi, A.M., Bhat, M.A., Giri, J. and Gandhi, S., 2020. Fabrication of microfluidic device for aflatoxin M1 detection in milk samples with specific aptamers. *Scientific Reports*, 10(1), p. 4627.

Keromnes, J. and Thouvenot, D., 1985. Role of penicillic acid in the phytotoxicity of *Penicillium cyclopium* and *Penicillium canescens* to the germination of corn seeds. *Applied and Environmental Microbiology*, 49(3), pp. 660–663.

Khan, R., Ben Aissa, S., Sherazi, T.A., Catanante, G., Hayat, A. and Marty, J.L., 2019. Development of an impedimetric aptasensor for label free detection of patulin in apple juice. *Molecules*, 24(6), p. 1017.

Knogge, W., 1996. Fungal infection of plants. *The Plant Cell*, 8(10), p. 1711.

Krivitsky, V., Granot, E., Avidor, Y., Borberg, E., Voegele, R.T. and Patolsky, F., 2021. Rapid collection and aptamer-based sensitive electrochemical detection of soybean rust fungi airborne urediniospores. *ACS Sensors*, 6(3), pp. 1187–1198.

Laganà, A., 2017. Introduction to the toxins special issue on LC-MS/MS methods for mycotoxin analysis. *Toxins*, 9(10), p. 325.

Lakhin, A.V., Tarantul, V.Z. and Gening, L., 2013. Aptamers: problems, solutions and prospects. Acta Naturae (англоязычная версия), 5(19), pp. 34–43.

Lattanzio, V.M.T., Solfrizzo, M., Powers, S. and Visconti, A., 2007. Simultaneous determination of aflatoxins, ochratoxin A and Fusarium toxins in maize by liquid chromatography/tandem mass spectrometry after multitoxin immunoaffinity cleanup. *Rapid Communications in Mass Spectrometry: An International Journal Devoted to the Rapid Dissemination of Up-to-the-Minute Research in Mass Spectrometry*, 21(20), pp. 3253–3261.

Le, L.C., Cruz-Aguado, J.A. and Penner, G.A., 2012. DNA ligands for aflatoxin and zearalenone. U.S. Patent Application 13/391,426.

Lebrun, M.H., Nicolas, L., Boutar, M., Gaudemer, F., Ranomenjanahary, S. and Gaudemer, A., 1988. Relationships between the structure and the phytotoxicity of the fungal toxin tenuazonic acid. *Phytochemistry*, 27(1), pp. 77–84.

Lerdsri, J., Thunkhamrak, C. and Jakmunee, J., 2021. Development of a colorimetric aptasensor for aflatoxin B1 detection based on silver nanoparticle aggregation induced by positively charged perylene diimide. *Food Control*, 130, p. 108323.

Li, P., Zhang, Q., Zhang, D., Guan, D., Liu, D.X., Fang, S., Wang, X. and Zhang, W., 2011. Aflatoxin measurement and analysis. In *Aflatoxins-Detection, Measurement and Control*. IntechOpen.

Li, Q., Lu, Z., Tan, X., Xiao, X., Wang, P., Wu, L., Shao, K., Yin, W. and Han, H., 2017. Ultrasensitive detection of aflatoxin B1 by SERS aptasensor based on exonuclease-assisted recycling amplification. *Biosensors and Bioelectronics*, 97, pp. 59–64.

Liao, Y., Zhou, X. and Xing, D., 2014. Quantum dots and graphene oxide fluorescent switch based multivariate testing strategy for reliable detection of Listeria monocytogenes. *ACS Applied Materials & Interfaces*, 6(13), pp. 9988–9996.

Liu, G., Mao, X., Phillips, J.A., Xu, H., Tan, W. and Zeng, L., 2009. Aptamer– nanoparticle strip biosensor for sensitive detection of cancer cells. *Analytical Chemistry*, 81(24), pp. 10013–10018.

Liu, Y., Liu, D., Cui, S., Li, C., Yun, Z., Zhang, J. and Sun, F., 2022. Design of a signal-amplified aptamer-based lateral flow test strip for the rapid detection of ochratoxin A in red wine. *Food*, 11(11), p. 1598.

Liu, Y., Yan, T., Li, Y., Cao, W., Pang, X., Wu, D. and Wei, Q., 2015. A simple label-free photoelectrochemical immunosensor for highly sensitive detection of aflatoxin B1 based on CDS–Fe_3O_4 magnetic nanocomposites. *RSC Advances*, 5(25), pp. 19581–19586.

Lu, Z., Chen, X., Wang, Y., Zheng, X. and Li, C.M., 2015. Aptamer based fluorescence recovery assay for aflatoxin B1 using a quencher system composed of quantum dots and graphene oxide. *Microchimica Acta*, 182, pp. 571–578.

Luan, Y., Chen, Z., Xie, G., Chen, J., Lu, A., Li, C., Fu, H., Ma, Z. and Wang, J., 2015. Rapid visual detection of aflatoxin B1 by label-free aptasensor using unmodified gold nanoparticles. *Journal of Nanoscience and Nanotechnology*, 15(2), pp. 1357–1361.

Ma, L., Guo, T., Pan, S. and Zhang, Y., 2018. A fluorometric aptasensor for patulin based on the use of magnetized graphene oxide and DNase I-assisted target recycling amplification. *Microchimica Acta*, 185, pp. 1–6.

Magnoli, A.P., Pereyra, M.L.G., Monge, M.P., Cavaglieri, L.R. and Chiacchiera, S.M., 2018. Validation of a liquid chromatography/tandem mass spectrometry method for the detection of aflatoxin B1 residues in broiler liver. *Revista Argentina de Microbiología*, 50(2), pp. 157–164.

Masi, M., Evidente, A., Meyer, S., Nicholson, J. and Muñoz, A., 2014. Effect of strain and cultural conditions on the production of cytochalasin B by the potential mycoherbicide Pyrenophora semeniperda (Pleosporaceae, Pleosporales). *Biocontrol Science and Technology*, 24(1), pp. 53–64.

McKeague, M., Velu, R., Hill, K., Bardóczy, V., Mészáros, T. and DeRosa, M.C., 2014. Selection and characterization of a novel DNA aptamer for label-free fluorescence biosensing of ochratoxin A. *Toxins*, 6(8), pp. 2435–2452.

McLean, M., Watt, M.P., Berjak, P. and Dutton, M.F., 1994. Effects of aflatoxin B 1 on in vitro cultures of Nicotianatabacum var. Samsun: 1: Callus growth and differentiation. *Mycopathologia*, 125, pp. 93–105.

Meena, M. and Samal, S., 2019. Alternaria host-specific (HSTs) toxins: An overview of chemical characterization, target sites, regulation and their toxic effects. *Toxicology Reports*, 6, pp. 745–758.

Mehdizadeh, M., Mushtaq, W., Siddiqui, S.A., Al-Taey, D.K.A. and Tampubolon, K., 2022. Assessment of phytotoxins using different technologies. *Current Applied Science and Technology*, 22(4), pp. 10–55003.

Milnes, B.E., 2019. Identification and validation of novel aptamers against fungal cells (Doctoral dissertation, University of Central Lancashire).

Mirón-Mérida, V.A., González-Espinosa, Y., Collado-González, M., Gong, Y.Y., Guo, Y. and Goycoolea, F.M., 2021. Aptamer–target–gold nanoparticle conjugates for the quantification of fumonisin B1. *Biosensors*, 11(1), p. 18.

Modh, H., Scheper, T. and Walter, J.G., 2017. Detection of ochratoxin A by aptamer-assisted real-time PCR-based assay (Apta-qPCR). *Engineering in Life Sciences*, 17(8), pp. 923–930.

Morales-Narváez, E. and Merkoçi, A., 2012. Graphene oxide as an optical biosensing platform. *Advanced Materials*, 24(25), pp. 3298–3308.

Mukherjee, M., Nandhini, C. and Bhatt, P., 2021. Colorimetric and chemiluminescence based enzyme linked apta-sorbent assay (ELASA) for ochratoxin A detection. *Spectrochimica Acta Part A: Molecular and Biomolecular Spectroscopy*, 244, p. 118875.

Nagaraja, H., Chennappa, G., Poorna Chandra Rao, K., Mahadevprasad, G., and Sreenivasa, M.Y. 2016. Diversity of toxic and phytopathogenic *Fusarium* species occurring on cereals grown in Karnataka state, India. *3Biotech*, 6, p. 57.

Nehdi, A., Samman, N., Aguilar-Sánchez, V., Farah, A., Yurdusev, E., Boudjelal, M. and Perreault, J., 2020. Novel strategies to optimize the amplification of single-stranded DNA. *Frontiers in Bioengineering and Biotechnology*, 8, p. 401.

Nguyen, H.H., Park, J., Kang, S. and Kim, M., 2015. Surface plasmon resonance: A versatile technique for biosensor applications. *Sensors*, 15(5), pp. 10481–10510.

Niazi, S., Khan, I.M., Yu, Y., Pasha, I., Shoaib, M., Mohsin, A., Mushtaq, B.S., Akhtar, W. and Wang, Z., 2019. A "turnon" aptasensor for simultaneous and time-resolved fluorometric determination of zearalenone, trichothecenes A and aflatoxin B1 using WS2 as a quencher. *Microchimica Acta*, 186, pp. 1–10.

Onoa, B. and Tinoco Jr, I., 2004. RNA folding and unfolding. *Current Opinion in Structural Biology*, 14(3), pp. 374–379.

Ozer, A., Pagano, J.M. and Lis, J.T., 2014. New technologies provide quantum changes in the scale, speed, and success of SELEX methods and aptamer characterization. *Molecular Therapy--Nucleic Acids*, 3, e183.

Pandey, A.K., Rajput, Y.S., Sharma, R. and Singh, D., 2017. Immobilized aptamer on gold electrode senses trace amount of aflatoxin M1. *Applied Nanoscience*, 7, pp. 893–903.

Peng, X.L., Xu, W.T., Wang, Y., Huang, K.L., Liang, Z.H., Zhao, W.W. and Luo, Y.B., 2010. Mycotoxin ochratoxin A-induced cell death and changes in oxidative metabolism of Arabidopsis thaliana. *Plant Cell Reports*, 29, pp. 153–161.

Peng, Y., Li, S.J., Yan, J., Tang, Y., Cheng, J.P., Gao, A.J., Yao, X., Ruan, J.J. and Xu, B.L., 2021. Research progress on phytopathogenic fungi and their role as biocontrol agents. *Frontiers in Microbiology*, 12, p. 670135.

Pérez-López, B. and Merkoçi, A., 2012. Carbon nanotubes and graphene in analytical sciences. *Microchimica Acta*, 179, pp. 1–16.

Prabhakar, N., Matharu, Z. and Malhotra, B.D., 2011. Polyaniline Langmuir–Blodgett film based aptasensor for ochratoxin A detection. *Biosensors and Bioelectronics*, 26(10), pp. 4006–4011.

Prasad M.N.N., Shankara Bhat, S., Sreenivasa, M.Y. 2010. Antifungal activity of essential oils against *Phomopsis azadirachtae* – the causative agent of die-back disease of neem. *Journal of Agricultural Technology*, 6 (1), pp. 127–133.

Quintanilha-Peixoto, G., Torres, R.O., Reis, I.M.A., de Oliveira, T.A.S., Bortolini, D.E., Duarte, E.A.A., Azevedo, V.A.D.C., Brenig, B., Aguiar, E.R.G.R., Soares, A.C.F. and Góes-Neto, A., 2019. Calm before the storm: A glimpse into the secondary metabolism of Aspergillus welwitschiae, the etiologic agent of the sisal bole rot. *Toxins*, 11(11), p. 631.

Ramos Catharino, R., de Azevedo Marques, L., Silva Santos, L., Baptista, A.S., Gloria, E.M., Calori-Domingues, M.A., Facco, E.M. and Eberlin, M.N., 2005. Aflatoxin screening by MALDI-TOF mass spectrometry. *Analytical Chemistry*, 77(24), pp. 8155–8157.

Ren, M., Xu, H., Huang, X., Kuang, M., Xiong, Y., Xu, H., Xu, Y., Chen, H. and Wang, A., 2014. Immunochromatographic assay for ultrasensitive detection of aflatoxin B1 in maize by highly luminescent quantum dot beads. *ACS Applied Materials & Interfaces*, 6(16), pp. 14215–14222.

Rhouati, A., Catanante, G., Nunes, G., Hayat, A. and Marty, J.L., 2016. Label-free aptasensors for the detection of mycotoxins. *Sensors*, 16(12), p. 2178.

Samokhvalov, A.V., Safenkova, I.V., Eremin, S.A., Zherdev, A.V. and Dzantiev, B.B., 2019. Application of Gold nanoparticles for high-sensitivity fluorescence polarization aptamer assay for ochratoxin A. *Nanotechnologies in Russia*, 14, pp. 397–404.

Schmitz, F.R.W., Valério, A., de Oliveira, D. and Hotza, D., 2020. An overview and future prospects on aptamers for food safety. *Applied Microbiology and Biotechnology*, 104, pp. 6929–6939.

Selvam, S.P., Kadam, A.N., Maiyelvaganan, K.R., Prakash, M. and Cho, S., 2021. Novel SeS2-loaded Co MOF with Au@ PANI comprised electroanalytical molecularly imprinted polymer-based disposable sensor for patulin mycotoxin. *Biosensors and Bioelectronics*, 187, p. 113302.

Selvolini, G., Lettieri, M., Tassoni, L., Gastaldello, S., Grillo, M., Maran, C. and Marrazza, G., 2019. Electrochemical enzyme-linked oligonucleotide array for aflatoxin B1 detection. *Talanta*, 203, pp. 49–57.

Seo, J.W., Kim, J.Y., Oh, J.J., Kim, Y.J. and Kim, G.H., 2021. Selection and characterization of toxic Aspergillus spore-specific DNA aptamer using spore-SELEX. *RSC Advances*, 11(5), pp. 2608–2615.

Serrano, A.B., Capriotti, A.L., Cavaliere, C., Piovesana, S., Samperi, R., Ventura, S. and Laganà, A., 2015. Development of a rapid LC-MS/MS method for the determination of emerging Fusarium mycotoxins enniatins and beauvericin in human biological fluids. *Toxins*, 7(9), pp. 3554–3571.

Setlem, K., Mondal, B., Shylaja, R. and Parida, M., 2020. Dual Aptamer-DNAzyme based colorimetric assay for the detection of AFB1 from food and environmental samples. *Analytical Biochemistry*, 608, p. 113874.

Setlem, S.K., Mondal, B. and Ramlal, S., 2022. A fluorescent aptasensor for the detection of Aflatoxin B1 by graphene oxide mediated quenching and release of fluorescence. *Journal of Microbiological Methods*, 193, p. 106414.

Sharma, A., Pillai, M.R.A., Gautam, S. and Hajare, S.N., 2014. Mycotoxinsl Immunological techniques for detection and analysis. pp. 869–879.

Sheng, L., Ren, J., Miao, Y., Wang, J. and Wang, E., 2011. PVP-coated graphene oxide for selective determination of ochratoxin A via quenching fluorescence of free aptamer. *Biosensors and Bioelectronics*, 26(8), pp. 3494–3499.

Shim, W.B., Dzantiev, B.B., Eremin, S.A. and Chung, D.H., 2009. One-step simultaneous immunochromatographic strip test for multianalysis of ochratoxin A and zearalenone. *Journal of Microbiology and Biotechnology*, 19(1), pp. 83–92.

Shimada, T. and Otani, M., 1990. Effects of Fusarium mycotoxins on the growth of shoots and roots at germination in some Japanese wheat cultivars. *Cereal Research Communications*, 18(3) pp. 229–232.

Singh, B.K., Delgado-Baquerizo, M., Egidi, E., Guirado, E., Leach, J.E., Liu, H. and Trivedi, P., 2023. Climate change impacts on plant pathogens, food security and paths forward. *Nature Reviews Microbiology*, 21(10), pp. 1–17.

Song, S., Wang, L., Li, J., Fan, C. and Zhao, J., 2008. Aptamer-based biosensors. *TrAC Trends in Analytical Chemistry*, 27(2), pp. 108–117.

Spadaro, D., Meloni, G.R., Siciliano, I., Prencipe, S. and Gullino, M.L., 2020. HPLC-MS/MS method for the detection of selected toxic metabolites produced by *Penicillium* spp. in nuts. *Toxins*, 12(5), p. 307.

Sreenivasa, M.Y., Dass, R.S., Charith Raj, A.P., Janardhana, G.R. 2008. PCR-based detection of genus *Fusarium* and Fumonisin-producing isolates from freshly harvested Sorghum grains grown in Karnataka, India. *Journal of Food Safety*, 28, pp. 236–237.

Sreenivasa, M.Y., Dass, R.S., Charith Raj, A.P., Janardhana, G.R. 2011b. Mycological evaluation of Maize grains produced in Karnataka (India) for the post-harvest fungal contamination. *World Applied Sciences Journal*, 13(4), pp. 688–692.

Sreenivasa, M.Y., Dass, R.S., Charith Raj, A.P., Prasad, M.N.N., Premila N. Achar, Janardhana, G.R. 2011a. Assessment of the growth inhibiting effect of some plant essential oils on different *Fusarium* species isolated from sorghum and maize grains. *Journal of Plant Diseases and Protection*, 118(6), pp. 208–213.

Sreenivasa, M.Y., Dass, R.S., Janardhana, G.R. 2010. Post harvest fungi associated with sorghum grains produced in Karnataka (India). *Journal of Plant Protection Research*, 50(3), pp. 335–339.

Stoltenburg, R., Reinemann, C. and Strehlitz, B., 2007. SELEX—A (r) evolutionary method to generate high-affinity nucleic acid ligands. *Biomolecular Engineering*, 24(4), pp. 381–403.

Strange, R.N., 2007. Phytotoxins produced by microbial plant pathogens. *Natural Product Reports*, 24(1), pp. 127–144.

Sulyok, M., Berthiller, F., Krska, R. and Schuhmacher, R., 2006. Development and validation of a liquid chromatography/tandem mass spectrometric method for the determination of 39 mycotoxins in wheat and maize. *Rapid Communications in Mass Spectrometry: An International Journal Devoted to the Rapid Dissemination of Up-to-the-Minute Research in Mass Spectrometry*, 20(18), pp. 2649–2659.

Szeto, K., Latulippe, D.R., Ozer, A., Pagano, J.M., White, B.S., Shalloway, D., Lis, J.T. and Craighead, H.G., 2013. Rapid-SELEX for RNA aptamers. *PLoS One*, 8(12), p. e82667.

Tabarzad, M., Kazemi, B., Vahidi, H., Aboofazeli, R., Shahhosseini, S. and Nafissi-Varcheh, N., 2014. Challenges to design and develop of DNA aptamers for protein targets. I. Optimization of asymmetric PCR for generation of a single stranded DNA library. *Iranian Journal of Pharmaceutical Research: IJPR*, 13(Suppl), p. 133.

Tan, S.Y., Acquah, C., Sidhu, A., Ongkudon, C.M., Yon, L.S. and Danquah, M.K., 2016. SELEX modifications and bioanalytical techniques for aptamer–target binding characterization. *Critical Reviews in Analytical Chemistry*, 46(6), pp. 521–537.

Tuerk, C. and Gold, L., 1990. Systematic evolution of ligands by exponential enrichment: RNA ligands to bacteriophage T4 DNA polymerase. *Science*, 249(4968), pp. 505–510.

Van Asch, M.A.J., Rijkenberg, F.H.J. and Coutinho, T.A., 1992. Phytotoxicity of fumonisin B1, moniliformin, and T-2 toxin to corn callus cultures. *Phytopathology*, 82(11), pp. 1330–1332.

Velu, R., Frost, N. and DeRosa, M.C., 2015. Linkage inversion assembled nano-aptasensors (LIANAs) for turn-on fluorescence detection. *Chemical Communications*, 51(76), pp. 14346–14349.

Verma, J., Johri, T.S., Swain, B.K. and Ameena, S., 2004. Effect of graded levels of aflatoxin, ochratoxin and their combinations on the performance and immune response of broilers. *British Poultry Science*, 45(4), pp. 512–518.

Vesonder, R.F., Labeda, D.P. and Peterson, R.E., 1992. Phytotoxic activity of selected water-soluble metabolites of Fusarium against Lemna minor L.(duckweed). *Mycopathologia*, 118, pp. 185–189.

Wacoo, A.P., Wendiro, D., Vuzi, P.C. and Hawumba, J.F., 2014. Methods for detection of aflatoxins in agricultural food crops. Journal of applied chemistry, 2014(1), p. 706291.

Wakulinski, W., 1989. Phytotoxicity of the secondary metabolites of fungi causing wheat head fusariosis (head blight). *Acta Physiologiae Plantarum*, 11(4), pp. 301–306.

Wang, C., Qian, J., Wang, K., Wang, K., Liu, Q., Dong, X., Wang, C. and Huang, X., 2015. Magnetic-fluorescent-targeting multifunctional aptasensor for highly sensitive and one-step rapid detection of ochratoxin A. *Biosensors and Bioelectronics*, 68, pp. 783–790.

Wang, Y., Peng, X., Xu, W., Luo, Y., Zhao, W., Hao, J., Liang, Z., Zhang, Y. and Huang, K., 2012. Transcript and protein profiling analysis of OTA-induced cell death reveals the regulation of the toxicity response process in Arabidopsis thaliana. *Journal of Experimental Botany*, 63(5), pp. 2171–2187.

Wu, S., Duan, N., Ma, X., Xia, Y., Wang, H., Wang, Z. and Zhang, Q., 2012. Multiplexed fluorescence resonance energy transfer aptasensor between upconversion nanoparticles and graphene oxide for the simultaneous determination of mycotoxins. *Analytical Chemistry*, 84(14), pp. 6263–6270.

Wu, S., Liu, L., Duan, N., Li, Q., Zhou, Y. and Wang, Z., 2018b. Aptamer-based lateral flow test strip for rapid detection of zearalenone in corn samples. *Journal of Agricultural and Food Chemistry*, 66(8), pp. 1949–1954.

Wu, W., Zhu, Z., Li, B., Liu, Z., Jia, L., Zuo, L., Chen, L., Zhu, Z., Shan, G. and Luo, S.Z., 2018a. A direct determination of AFBs in vinegar by aptamer-based surface plasmon resonance biosensor. *Toxicon*, 146, pp. 24–30.

Wu, Z., Pu, H. and Sun, D.W., 2021. Fingerprinting and tagging detection of mycotoxins in agri-food products by surface-enhanced Raman spectroscopy: Principles and recent applications. *Trends in Food Science & Technology*, 110, pp. 393–404.

Wu, Z., Xu, E., Chughtai, M.F., Jin, Z. and Irudayaraj, J., 2017. Highly sensitive fluorescence sensing of zearalenone using a novel aptasensor based on upconverting nanoparticles. *Food Chemistry*, 230, pp. 673–680.

Xi-Hui, Y.A.N.G., Wei-Jun, K.O.N.G., Mei-Hua, Y.A.N.G., Ming, Z.H.A.O., & Ouyang, Z. (2013). Application of aptamer identification technology in rapid analysis of mycotoxins. *Chinese Journal of Analytical Chemistry*, 41(2), pp. 297–306.

Xu, D., Xue, M., Shen, Z., Jia, X., Hou, X., Lai, D. and Zhou, L., 2021a. Phytotoxic secondary metabolites from fungi. *Toxins*, 13(4), p. 261.

Xu, J., Liu, J., Li, W., Wei, Y., Sheng, Q. and Shang, Y., 2023. A dual-signaling electrochemical aptasensor based on an in-plane gold nanoparticles–black phosphorus heterostructure for the sensitive detection of patulin. *Food*, 12(4), p. 846.

Xu, Y., Xiong, M. and Yan, H., 2021b. A portable optical fiber biosensor for the detection of zearalenone based on the localized surface plasmon resonance. *Sensors and Actuators B: Chemical*, 336, p. 129752.

Yamaji, K., Fukushi, Y., Hashidoko, Y. and Tahara, S., 2005. Penicillium frequentans isolated from Picea glehnii seedling roots as a possible biological control agent against damping-off. *Ecological Research*, 20, pp. 103–107.

Yan, X., Chen, H., Du, G., Guo, Q., Yuan, Y. and Yue, T., 2022. Recent trends in fluorescent aptasensors for mycotoxin detection in food: Principles, constituted elements, types, and applications. *Food Frontiers*, 3(3), pp. 428–452.

Yang, C., Lates, V., Prieto-Simón, B., Marty, J.L. and Yang, X., 2012. Aptamer-DNAzyme hairpins for biosensing of ochratoxin A. *Biosensors and Bioelectronics*, 32(1), pp. 208–212.

Yang, X., Huang, R., Xiong, L., Chen, F., Sun, W. and Yu, L., 2023. A colorimetric aptasensor for ochratoxin a detection based on tetramethylrhodamine charge effect-assisted silver enhancement. *Biosensors*, 13(4), p. 468.

Yu, J. and Pedroso, I.R., 2023. Mycotoxins in cereal-based products and their impacts on the health of humans, livestock animals and pets. *Toxins*, 15(8), p. 480.

Yufa, R., Krylova, S.M., Bruce, C., Bagg, E.A., Schofield, C.J. and Krylov, S.N., 2015. Emulsion PCR significantly improves nonequilibrium capillary electrophoresis of equilibrium mixtures-based aptamer selection: allowing for efficient and rapid selection of aptamer to unmodified ABH2 protein. *Analytical Chemistry*, 87(2), pp. 1411–1419.

Zain, M.E., 2011. Impact of mycotoxins on humans and animals. *Journal of Saudi Chemical Society*, 15(2), pp. 129–144.

Zeilinger, S., Gupta, V.K., Dahms, T.E., Silva, R.N., Singh, H.B., Upadhyay, R.S., Gomes, E.V., Tsui, C.K.M. and Nayak, S.C., 2016. Friends or foes? Emerging insights from fungal interactions with plants. *FEMS Microbiology Reviews*, 40(2), pp. 182–207.

Zhang, C., Dou, X., Zhang, L., Sun, M., Zhao, M., OuYang, Z., Kong, D., Antonio, F.L. and Yang, M., 2018b. A rapid label-free fluorescent aptasensor picoGreen-based strategy for aflatoxin B1 detection in traditional Chinese medicines. *Toxins*, 10(3), p. 101.

Zhang, G., Zhu, C., Huang, Y., Yan, J. and Chen, A., 2018a. A lateral flow strip based aptasensor for detection of ochratoxin A in corn samples. *Molecules*, 23(2), p. 291.

Zhao, X., Wang, Y., Li, J., Huo, B., Huang, H., Bai, J., Peng, Y., Li, S., Han, D., Ren, S. and Wang, J., 2021. A fluorescence aptasensor for the sensitive detection of T-2 toxin based on FRET by adjusting the surface electric potentials of UCNPs and MIL-101. *Analytica Chimica Acta*, 1160, p. 338450.

Zhao, Y., Luo, Y., Li, T. and Song, Q., 2014. Au NPs driven electrochemiluminescence aptasensors for sensitive detection of fumonisin B1. *RSC Advances*, 4(101), pp. 57709–57714.

Zhao, Y., Yang, Y., Luo, Y., Yang, X., Li, M. and Song, Q., 2015. Double detection of mycotoxins based on SERS labels embedded Ag@ Au core–shell nanoparticles. *ACS Applied Materials & Interfaces*, 7(39), pp. 21780–21786.

Zhao, Z., Wang, H., Zhai, W., Feng, X., Fan, X., Chen, A. and Wang, M., 2020. A lateral flow strip based on a truncated aptamer-complementary strand for detection of type-B aflatoxins in nuts and dried figs. *Toxins*, 12(2), p. 136.

Zhong, H., Yu, C., Gao, R., Chen, J., Yu, Y., Geng, Y., Wen, Y. and He, J., 2019. A novel sandwich aptasensor for detecting T-2 toxin based on rGO-TEPA-Au@ Pt nanorods with a dual signal amplification strategy. *Biosensors and Bioelectronics*, 144, p. 111635.

14 On-Site Diagnostic Techniques for Fungal Pathogens

Vignesh Dhandapani
Université de Sherbrooke

Madanraj Appiya Santharam
Université de Sherbrooke

Suriya Tamilselvan-Nattar-Amutha
Leibniz Institute of Plant Genetics and Crop Plant Research (IPK) OT Gatersleben

Charith Raj Adkar Purushothama
Université de Sherbrooke

Majority of the plant diseases are caused by fungi. Currently, over 19,000 plant pathogenic fungi are known worldwide. They are capable of remaining dormant on both living and dead plant tissues until conditions are conducive to their proliferation. As fungal pathogens devastate crop production and thus impact the economy, it is important to timely identify the pathogen for implementing an action plan. In this scenario, on-site detection kits are crucial as these kits provide results within a couple of hours. In this chapter, three on-site fungal pathogen diagnostic methods are discussed.

 DOI: 10.1201/9781003399704-14

GRAPHICAL ABSTRACT

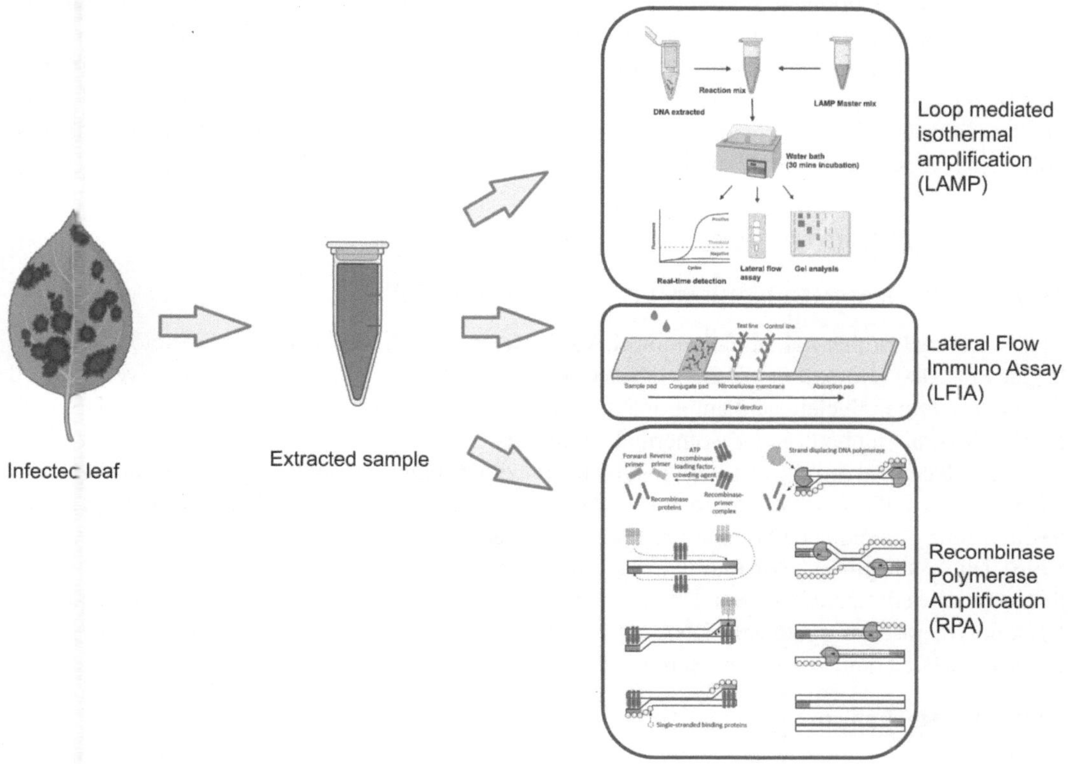

14.1 INTRODUCTION

Agricultural ecosystems have grown to cover 40% of the planet's surface to accommodate the needs of the increasing population in the past 300 years (Foley et al., 2011). One of the major constraints on food resources is the loss of 10–30% of the overall plant production due to the deterioration of plant products (Okawa, 2015). Plant diseases are primarily caused by microorganisms such as bacteria, fungi and viruses. These pathogens spread widely among the plantation through infected plants or the introduction of the disease-causing agent to the field (Sankaran et al., 2010). Food safety is a significant issue to address as pathogens and their associated toxins are one of the major reasons for the loss (Sreenivasa et al., 2008; Janardhana et al., 2009; Dass et al., 2007). Plant diseases contribute up to 15% of the major crop losses per year, leading to financial losses of millions of dollars, the majority of them through fungal pathogens (Chatterjee et al., 2016; Li et al., 2017). Visual inspection of symptoms such as growth inhibition, wilting and necrosis may not be enough to establish if a disease is caused by phytopathogenic fungus or not (Mehetre et al., 2021; Yin et al., 2016). However, these visual symptoms appear only when the pathogen is in high concentration, and by then the disease might be widespread in the field (Deepa and Sreenivasa, 2019a; 2019b).

Early detection of plant diseases and their severity on the crops is essential for putting forth field plant protection actions (Gautam and Kumar, 2020). Samples may have to be sent to a lab for processing to ascertain the presence of pathogens. Conventional lab-based gold standard methods such as polymerase chain reaction (PCR), enzyme-linked immunosorbent assays (ELISA), protein profiling through (SDS-PAGE), propagating a pathogen *in vitro* (e.g., growing fungus on potato dextrose agar) for further analyses are laborious and labour intensive (Deepa et al., 2016a; Byzova et al.,

2018; Fraaije et al., 2002; Gautam and Kumar, 2020; Hewison and Kuras, 2005; López-Soriano et al., 2017). Even if these methods produce accurate and reliable results, their major drawback stems from the requirement of a dedicated laboratory, expensive instruments to process the samples as well as the requirement of trained personnel. Due to the broad spectrum in the varieties in a variety of pathogenic fungi, rapid, affordable, accurate and on-site diagnosis is essential in their identification for specific targeting (Deepa et al., 2016b; Deepa et al., 2016c). This chapter focuses on the different on-site detection techniques, including loop-mediated detection, recombinase polymerase amplification and lateral flow assays.

14.2 ON-SITE DETECTION TECHNIQUES

14.2.1 Loop-mediated Isothermal Amplification (LAMP)

Loop-mediated isothermal amplification (LAMP) was proposed and developed by Notomi and colleagues in 2000 (Notomi et al., 2000). Since then, it has been one of the most extensively studied techniques among nucleic acid amplification techniques. In LAMP, millions of copies of DNA are amplified within an hour under isothermal conditions. The technology uses the polymerase and the strand displacement activity of a DNA polymerase enzyme obtained from *Bacillus stearothermophilus* (also known as, *Geobacillus stearothermophilus*; *Bs*t DNA polymerase) to amplify the target DNA through four primers specific to six regions of the target at a constant temperature (Aliotta et al., 1996). Due to the presence of auto-cycling strand displacement activity of the *Bs*t DNA polymerase, this technique eliminated the heat denaturation step of the classical PCR technique. Additionally, *Bs*t DNA polymerase can synthesize DNA between 60 °C and 65 °C, both primer annealing and primer extension can be done at the same temperature.

14.2.1.1 Mechanism

LAMP is performed in a reaction mix composed of nucleotides, primer sets, *Bst* DNA polymerase and magnesium ions (Nagamine et al., 2002; Notomi et al., 2000). LAMP can be performed at a constant temperature between 60 °C and 65 °C and can be used directly on tissue samples or crude DNA samples (Nagamine et al., 2002). The technique is primarily based on four primers *including* outer-forward (F3), inner-forward (FIP), outer-backward (B3) and inner-backward (BIP) primers recognizing six specific sequences on the target DNA in the presence of DNA polymerase having high strand displacement activity. An additional loop-primer pair enhances the process by hybridizing with the stem-loop structures and increasing the number of starting points for exponential target amplification. The whole process is accomplished in less than an hour using simple equipment such as a heating block or water bath. Stem-loop and cauliflower-like structures are generated by the LAMP. Different loops are formed as a result of auto-cycling, strand displacement DNA synthesis and the action of the primers (Iwamoto et al., 2003). For more details on LAMP amplification refer to Chapter 5 "Loop-mediated Isothermal Amplification (LAMP) for the Detection and Quantification of Fungal Phytopathogens".

Traditionally, the result of LAMP assays is visualized by the turbidity of the reaction mix. Often, this is not the best method to interpret the result because turbidity might go unnoticed in opaque reaction tubes or if a sample contains low amounts of target DNA. Hence, over the years many visualization techniques have been developed. Currently, the LAMP results can be obtained at the end of the assay (end point-based methods) or reading the result or as the assay progresses (real-time-based methods) (Panno et al., 2020).

Various visualization techniques are used to visualize the LAMP DNA amplicons including gel electrophoresis or by adding dyes such as SYBR Green, propidium iodide, Hydroxynapththol blue (HNB), calcein or Quant-iT PicoGreen (Duan et al., 2014; Zhou et al., 2014). Since the use of dyes can be performed in a closed-tube setup, contamination from external sources in an open-tube setup could be avoided. LAMP outweighs PCR because it is labour saving, has better sensitivity and is

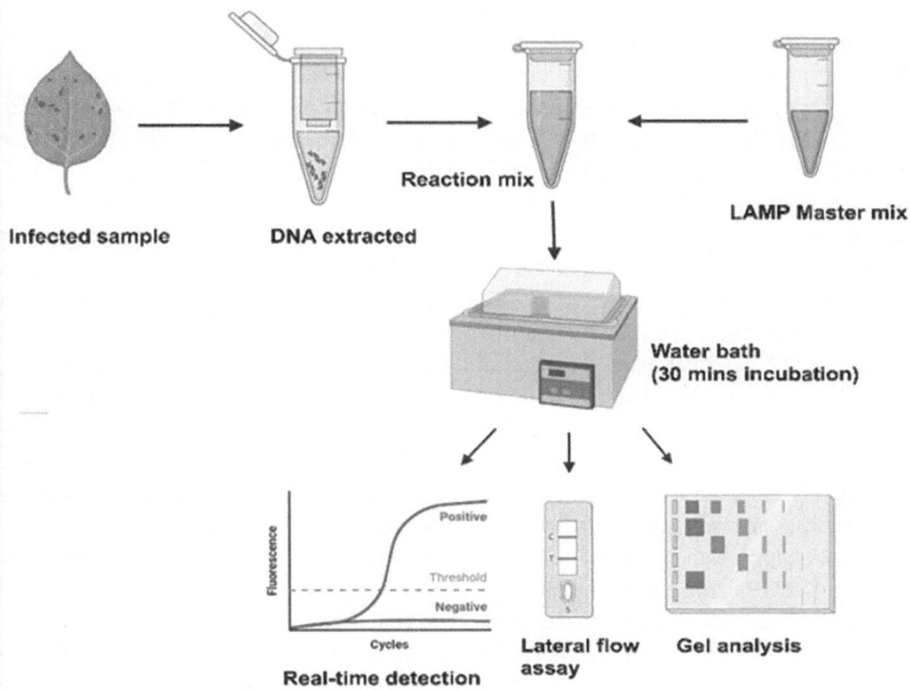

FIGURE 14.1 Detection of plant pathogens using loop-mediated isothermal amplification (LAMP) technique.

time-efficient compared to the other amplification methods (Panno et al., 2020). The key steps of the LAMP assay are shown in Figure 14.1.

Recent developments have led to real-time LAMP based on the turbidity of the solution or the use of fluorescent dyes (Gomez-Gutierrez and Goodwin, 2022). The use of fluorescence probes increases the sensitivity of the assay by several times when compared to the turbidity-based measurement (Quyen et al., 2019). Multiplex LAMP has been used to detect target-specific amplicons in a mixture of LAMP products by introducing an endonuclease recognition site in the sequence resulting in endonuclease-digested amplicons specific to the target sequence (Liang et al., 2012). Electric LAMP is a recent method where electronic stimulations are used to perform putative tests of LAMP primer sequence's compatibility with the target sequence (Wong et al., 2018). In-disc LAMP (iD-LAMP) combines the principle of LAMP and compact disc technology, where the genomic assays are performed using centrifugal devices and the analytical steps are controlled by the rotation rate (Santiago-Felipe et al., 2016). Biotin-labelled amplicons can be hybridized with fluorescence-labelled DNA probe coupled to a gold-labelled anti-fluorescein antibody to form a LAMP-Lateral Flow Dipstick (LFD) device (Ivanov et al., 2021; Rigano et al., 2014). The latest approaches include lyophilized LAMP reaction mix kit, which is activated by adding water and template DNA or RNA sample (Le and Vu, 2017; Panno et al., 2020). In most cases, a portable heating device is also supplied making it usable under field conditions.

14.2.1.2 Applications

Fungi account for 40–60% of the total agricultural produce loss due to diseases and pests as there are approximately 15,000 plant disease-causing fungi (Mahy and van Refenmortel, 2009). Therefore, the research community is focused on developing techniques for the earlier detection of fungal pathogens. LAMP assays have been used to detect *Fusarium oxysporum* f. sp. *Lycopersici*, associated with vascular wilts disease; *Bursaphelenchus xylophilus* (pine wilt disease) and *Pythium aphanidermatum* (tomato roots rot disease) (Almasi et al., 2013; Fukuta et al., 2014; Jędryczka et al., 2013; Kikuchi et al., 2009). LAMP is found to be cheaper and faster than real-time PCR in detecting *Leptosphaeria* populations causing stem canker disease in oilseed rape (Jędryczka et al., 2013). The enzyme GspSSD DNA polymerase enzyme replaced the conventional LAMP enzyme, that is, *Bst* polymerase allowing for enhanced amplification efficiency (Jędryczka et al., 2013). A colorimetric LAMP developed to detect *F. oxysporum* f. sp. *melonis* was faster and more sensitive (100-fold) than the conventional PCR (Almasi, 2019). The use of reagents such as SYBR Green I, HNB and GeneFinder™ has resulted in longer stability to colour change and lesser cross-contamination as these indicator dyes are added along with LAMP reagents at the beginning of the assay. LAMP including primers designed from the galactose oxidase precursor gene (gaoA gene) from *F. austroamericanum* isolate NRRL 2903 was used to detect *F. graminearum*, the causal agent of Fusarium Head Blight (FHB) (Niessen and Vogel, 2010).

Real-time LAMP has been used to detect and quantify wheat dwarf disease (Hao et al., 2021; Trzmiel and Hasiów-Jaroszewska, 2020), *F. graminearum* (Gupta et al., 2020), *Fusarium* species produced trichothecene (Denschlag et al., 2014), *Pyricularia oryzae* Triticum lineage (Yasuhara-Bell et al., 2018) and *Tilletia* species (Pieczul et al., 2018). A real-time fluorescence LAMP was designed to detect *F. oxysporum* f. sp. *cubense*, which causes Panama disease in bananas and the sensitivity of the assay was 0.4 pg/µl plasmid DNA mixed with solid DNA (Zhang et al., 2013). *Botrytis cinerea*, which causes rot root of vegetables, flowers and fruits, was successfully detected using real time-LAMP using an intercalating fluorescent dye (Tomlinson et al., 2010a). A spore trap system coupled to an assimilated-probe method was used to develop a field-detection kit for *Magnoporthe oryzae*. The specificity of LAMP in detecting spores of *Puccinia striiformis* f. sp. *Tritici*, which causes wheat stripe rust, was from 2 pg/µl and the sensitivity increased on DNA extraction from isolated wheat leaves (Huang et al., 2011). *Pythium spp.* causing root rot was detected using LAMP with a detection limit of 10fg and ten times more sensitivity than PCR (Fukuta et al., 2013). The sensitivity of LAMP to detect *Ustilago tritici*, which causes loose smut of wheat, was 100 fg/µl DNA, which was 100 times lower than qPCR (Yan et al., 2019).

14.2.1.3 Advantages and Disadvantages

The LAMP assay has several advantages over other molecular biology diagnostics assays such as PCR, dot blot and microarray. One of the foremost is that reaction can be carried out in a water bath or a thermos-block, thus eliminating the usage of expensive thermocycler as in the case of PCR (Siegieda et al., 2021). Usage of such simple equipment to achieve isothermal temperature makes this assay suitable for field application (Hodgetts et al., 2011; Tomlinson et al., 2007; Wilisiani et al., 2019). Visualization of results with the naked eye could be facilitated using colorimetric dyes such as HNB in the reaction mixture. As the *Bst* polymerase contains both reverse transcriptase, and DNA polymerase activity, a separate reverse transcription (RT) reaction is not required to detect RNA pathogens. Further, compared to PCR, LAMP is 10–100 times more sensitive (Bhat et al., 2013; Le et al., 2010; Li et al., 2007; Tomlinson et al., 2010b). Finally, LAMP enzyme shows less sensitivity to PCR inhibitors in the amplification reaction (Kaneko et al., 2007).

On the contrary, the main disadvantage of the technique includes design of the primers, which is a complex procedure. Hence, the development and optimization of the method are restricted to the specialists (Le and Vu, 2017). However, the assay could be performed by non-specialists with ease. Excess concentration of indicators such as colorimetric dyes (calcein) or reaction components (containing ionic forms of manganese) may inhibit the polymerase reaction, thus affecting the sensitivity and efficiency of the method (Tanner et al., 2015). Since LAMP is a very sensitive technique, there are chances of carryover contamination if the reaction is performed repeatedly in the same place without proper sanitation. As four to six primers are used in the LAMP assay, it is important to validate the method for each plant sample and its incubation time. Often, incubating longer time results in false positives (Francois et al., 2011). Therefore, development of field deployable and rapid detection approaches to detect pathogens is essential (Miles et al., 2015).

14.2.2 Lateral Flow Immunoassays

Lateral flow immunoassays (LFIA), biochemical assays based on the usage of polyclonal antibodies specific to an antigen, have been developed as a strategy to obtain point-of-care (POC) diagnosis (Van Emon, 2006). Ever since paraffin paper-based paper-chromatography was proposed by Rosalyn S. Yalow and Solomon A. Berson in 1960 as a simple, yet fast and efficient method to determine insulin in human blood plasma, LFIA has been successfully enhanced over 60 years to detect a range of analyte test systems (Andryukov, 2020; Yalow and Berson, 1960). Majorly used for detecting human conditions, this method of detection was initially developed for detecting and quantifying phytopathology through fungi in soil (Thornton et al., 2004). A murine hybridoma GD2 able to secrete IgM, against a range of surface antigens from AG (anastomosis group) 4 isolate (AB1) of *Rhizoctonia solani*, was generated for this purpose, injected into a BALB/c mouse to obtain an ascetic fluid after 3 weeks. Conjugating filtered ascites to 40nm-gold particles, it was applied to the test line, and a commercial anti-mouse Ig was applied to the control line of the immunochromatographic strip (Thornton et al., 2004). Use of LFIA in screening a huge number of plants for other fungal species *Phytophthora ramorum* and *P. kernoviae* was further demonstrated by Lane C.R. et al., in 2007 (Lane et al., 2007). The technique has now been expanded to screen for viruses (Byzova et al., 2018; De Boer and López, 2011; Safenkova et al., 2012) and bacteria (Boutal et al., 2022; Braun-Kiewnick et al., 2011; Hodgetts et al., 2015; Safenkova et al., 2014). Due to its user-friendly and non-technical characteristics, it has been preferred as a primary method for screening plant diseases and monitoring plant health.

14.2.2.1 Construction of LFIA and Mechanism

An LFIA strip is made of four parts mounted on a laminated card. They are (i) the sample pad, (ii) the conjugate pad, (iii) the membrane and (iv) the absorbent pad. The capture receptors form the test and control lines on the membrane. When the sample is dropped on the sample pad, the presence of the target analyte in the sample forms conjugates with the target-specific antibodies tagged with gold nanoparticles and flows towards the absorption pad. If there is no target analyte in the sample, then there will not be any conjugation between gold nanoparticle-tagged antibody with the target analyte. As the sample migrates, the presence of the target analyte in the sample results in the accumulation of gold nanoparticles on the test and control lines, and appears red colour. If there is no target analyte in the sample, then the gold nanoparticles accumulate only on the control line, giving a single red-coloured line output.

Modern LFIA strip comprises a nitrocellulose membrane between pads on either side held on a backing support, as shown in Figure 14.2. The support aids in the rigidity and assembly of various components (Bahadır and Sezgintürk, 2016). At the beginning of the strip, a sample pad (SP) is attached and at the end, an absorption pad (AP). Migration of a liquid sample from SP to the AP occurs through capillary action. Cellulose-based SP channels the fluid sample by absorption, allowing minimalization of matrix effect by gradual filtration and infusion of buffer salts to enhance sample interaction with the conjugate in polyester-based conjugate pad (CP) (Bahadır and Sezgintürk, 2016). Conjugate pad holds capture reagents specific to the target analyte, tagged to a fluorochrome particle – primarily gold nanoparticles or polystyrene microspheres; however, advancement in LFIA has led to the usage of additional types of nanomaterials (Lou et al., 2022; Quesada-González and Merkoçi, 2015). Subsequently, the conjugated sample migrates through the porous nitrocellulose membrane towards the detection zone, in which two or more immobilized antibody or antigen lines would be present. First, test lines contain exclusive biological components specific to the target conjugated antibody in the conjugate release pad and, second, a control line comprising of a biological probe, which binds to the conjugate independently of the presence of a target. Biological response at the test line indicates the presence of target, and a response at the control line indicates appropriate liquid flow across the strip. Capillary movement across the strip forces the fluid to flow to the absorption pad (AP) and excess fluid is absorbed (Boutal et al., 2022; Di Nardo et al., 2021; Koczula and Gallotta, 2016). Additional test lines could be added, either to test multiple analytes

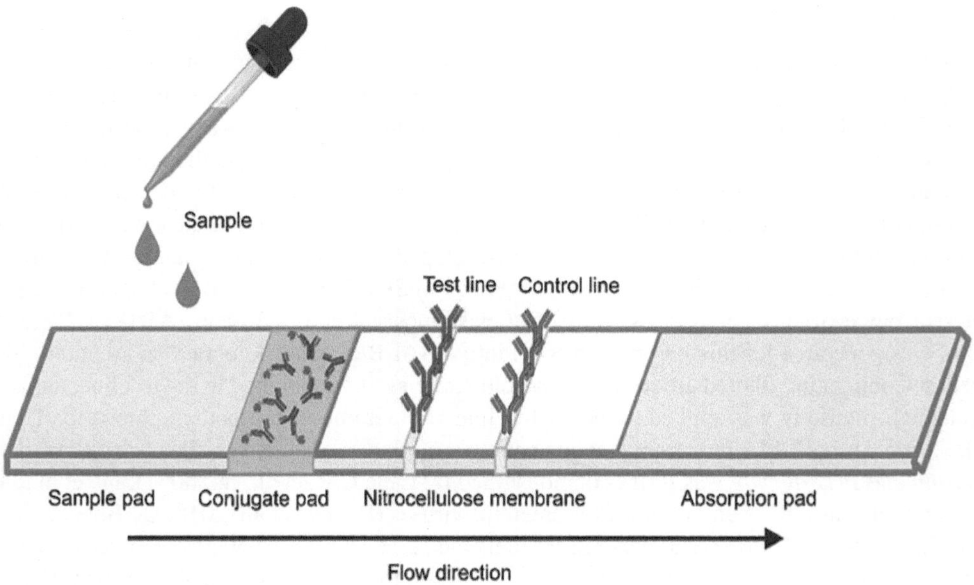

FIGURE 14.2 Design and principle of lateral flow immunoassay (LFIA).

simultaneously in the same LFIA strip using a mix of immobilized probes in a multiplex format or to have a semi-quantitative approach by having multiple lines of immobilized probes against the same target, and the number of lines determines the amount of target present in the sample known as a "ladder bars" LFIA (Fung et al., 2009; Koczula and Gallotta, 2016; Yen et al., 2015).

Test results can be obtained rapidly within 10–30 min after a sample has been added to the sample pad. Based on the size of the analyte, two formats of LFIA are used – direct and competitive assays. Direct LFIA is for larger antigens like *Xanthomonas arboricola pv. pruni*, a sandwich-based assay where the target is captured by one antibody on the conjugate pad and then it is tagged again by another antibody on the test line, giving a positive line (López-Soriano et al., 2017). The read-out is analysed by the development of colour, first, on the control line to validate the result obtained on the test line and the development of colour on the test line conveys the positivity of the sample towards the target (Thornton et al., 2004). The second type, competitive LFIA, is utilized when the analyte is smaller or has only a solitary antigenic determinant and hence does not have binding sites for two different antibodies. The analyte competes with the test line's blocking sites, thereby inhibiting the development of colour on the test line. In this case, once the result is validated with the development of a line on the control line, the absence of a colour on the test line indicates a positive result for the presence of the analyte (Boutal et al., 2022; Koczula and Gallotta, 2016). The steps involved in LFIA are shown in Figure 14.3.

The sample obtained is ground in an appropriate buffer to obtain the crude extract. One to two drops of the extract are dropped onto the sample pad, where the target analyte reacts with the gold-nanoparticle conjugated antibody and migrates towards the absorption pad (AP) on the nitrocellulose membrane. As the sample migrates, the target analyte that is bound to the target-specific gold-nanoparticle conjugated antibody will be captured in the test line, whereas the unbound target-specific gold-nanoparticle conjugated antibody will be detected on the control line. Hence, if there is a target (pathogen), both the test and the control lines turn red. If there is no target, then only the control line turns red. If only the test line turns red, then the test is invalid.

14.2.2.2 Applications

The idea behind generating a test that could be used outside the laboratories led to the creation of LFIA. Initially, LFIA was used majorly in the clinical setup for aiding clinicians such as nurses,

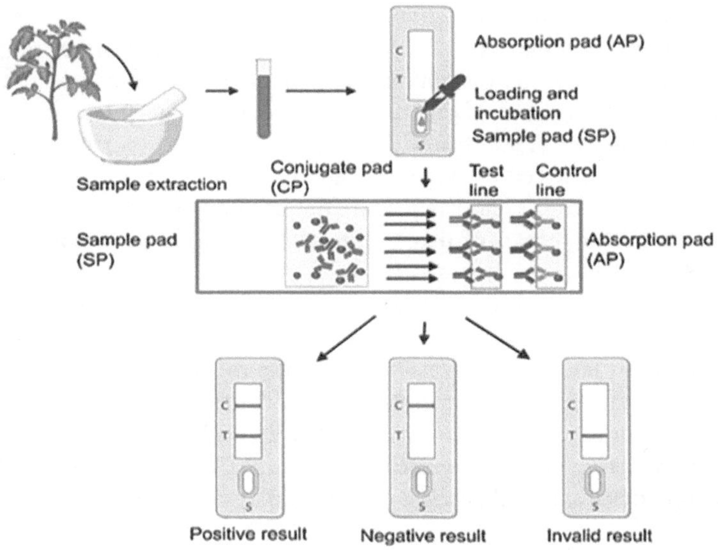

FIGURE 14.3 Steps involved in the detection of pathogens using Lateral Flow Immunoassay (LFIA).

doctors, health centre workers in quick diagnosis, surveillance, and prognosis, helping in decision-making to impact patient lives and more importantly limiting costs (Pai et al., 2012). Simplicity of the test and its analysis allowed LFIA to be used by normal people at home, especially by women for pregnancy tests, reducing congestion at the hospitals (Weihs et al., 2021). Gradually, LFIA application has been increased to screen food safety products, veterinary medicine and environmental pollutants (Di Nardo et al., 2021).

Since two decades ago, LFIA tests have been expanded to be used in the agriculture field to test for plant diseases and the condition of the soil (Lane et al., 2007; Thornton et al., 2004). The idea gained momentum in the early 2010s, and multiple groups started developing LFIA kits to test for viral, bacterial and fungal contamination in the field (Braun-Kiewnick et al., 2011; Byzova et al., 2018; De Boer and López, 2011; Safenkova et al., 2014, 2012). Mainly fungal contamination has garnered attraction in the field as, according to Food and Agriculture Organization (FAO), fungus-produced mycotoxins affect up to one-fourth of the world's crops (Reddy et al., 2010; Wu, 2007). Fungal infection produced by fungi genera *Aspergillus*, *Fusarium*, and *Penicillium* spreads easily, and it is essential to detect infection early on to manage the disease (Rahman et al., 2019; Xing et al., 2020). Several multiplex lateral flow immunoassays have been developed for qualitative determination of mycotoxins such as aflatoxin B1 (AB1), deoxynivalenol (DON), Ochratoxin A (OTA), zearalenone (ZEA) in maize samples (Jiang et al., 2023; Song et al., 2014), corn, peanut (Wang et al., 2019) and rice (Chen et al., 2016). To control the fungal spread, chemical fungicides are sprayed on the plants and their products; however, they are harmful for ingestion. LFIAs have been established to detect fungicide residue levels in vegetables and fruits (Liu et al., 2022; Qin et al., 2021; Xu et al., 2022). Due to its sensitivity, high throughput and rapidness, groups have preferred to develop LFIA over other methods in screening fungal presence in agriculture as well as levels of fungicide in agricultural produce.

14.2.2.3 Advantages and Disadvantages

Designed to be small, robust and portable, LFIAs have multiple advantages but regulatory requirements instigate the need to process potential positive samples through the traditional diagnostic tests performed in the laboratory (Koczula and Gallotta, 2016). However, this method can be used as the point-of-care/need technique to screen several hundred samples in the field. As LFIA does not require numerous steps of washing or adding reagents, it is a one-step assay and yields a result within 30 min (Posthuma-Trumpie et al., 2009). LFIA gives a qualitative or semi-qualitative result that can be performed and analysed by an untrained professional. Due to its ease of use and sensitivity, it relatively shortens the amount of time to develop a novel application, eventually making it affordable and high throughput. Additionally, the test requires a low volume of sample to start with and multiple test lines are analysed by adding several immobilized antibody lines that can detect multiple targets in the same test. Fluid samples do not need any pre-treatment and the LFIA test strips can be stored for long periods of time before use (Koczula and Gallotta, 2016; Posthuma-Trumpie et al., 2009).

Even though LFIA has several advantages over other diagnostic techniques, a few limitations of the technique include pre-conversion of solid sample into a fluid sample and requirement of an accurate volume of sample to have reliable results. Furthermore, sample viscosity and any residual matrix effect determine the amount of time required to obtain a result. Good quality and specific antibodies are essential to have a reliable test strip. Therefore, highest standard of quality control needs to be met during assay development (Boutal et al., 2022; Koczula and Gallotta, 2016). As LFIA contains nanoparticles and antibodies, it is important to store the assay reagents at an ambient temperature.

14.2.3 Recombinase Polymerase Amplification (RPA)

Recombinase polymerase amplification (RPA) is a high nucleic acid-based isothermal amplification technique, which operates at 37–42 °C. RPA is capable of amplifying as low as 1–10 DNA target copies in less than 20 min (Lobato and O'Sullivan, 2018). RPA is capable of amplifying both DNA and RNA targets, it is widely used to amplify diverse targets, from a wide variety of organisms and samples. Unlike LAMP, RPA uses only a pair of primers. This reduces non-specific amplification.

14.2.3.1 Mechanism

Recombinase, a protein, aids in the hybridization of oligonucleotide primers to the homologous target sequence. RPA was first invented in 2006, and it involves recombinase-dependent hybridization of primers to the target DNA instead of the temperature-dependent denaturation-annealing reactions for PCR (Piepenburg et al., 2006). Initially, the primers form a complex with the phage recombinase uvrX (Ivanov et al., 2021). During the amplification, the recombinase-primer complex scans the template DNA for homologous sequences. Once a suitable site is found, the recombinase enzyme facilitates the opening of the double strand to enable the primer hybridization to the target sequence. This reaction is facilitated by another recombinase factor urvY and ATP hydrolysis. The single-stranded DNA (ssDNA) binding proteins bind to the "melted" DNA stabilizing the DNA. Following the binding, the primer elongates replacing the old strand with a newly synthesized strand. *Bsu*I DNA polymerase enables DNA synthesis using the primers. Exponential amplification is achieved as the newly synthesized duplex acts as a template for the next cycle (Piepenburg et al., 2006). The *Bsu*I DNA polymerase is derived from *Bacillus subtilis* polymerase I (Lobato and O'Sullivan, 2018) (Figure 14.4).

14.2.3.2 Applications

Bipolaris sorakiniana, a causal agent for root rot and spot blotch in wheat and barley, was detected at a low detection limit of 10 pg and high sensitivity using RPA targeting the calmodulin gene (*cal* gene) (Zhao et al., 2021). *Leptosphaeria maculans* could cause phoma stem canker Brassica napus to result in massive crop loss. A dual RPA-CRISPR/Cas12a detection kit was developed to detect *L. maculans* and *L. biglobosa* (Lei et al., 2019) within 45 min and the detection limit was 4.7 genomic DNA copy (Lei et al., 2022). A diagnostic assay based on the RPA technology was used to detect *Plenodomus tracheiphilus*, fungus responsible for causing mal secco in lemon. The assay was more sensitive than PCR, and the detection threshold was 1 pg of gDNA (Rovetto et al., 2023). A rapid RPA assay enabled

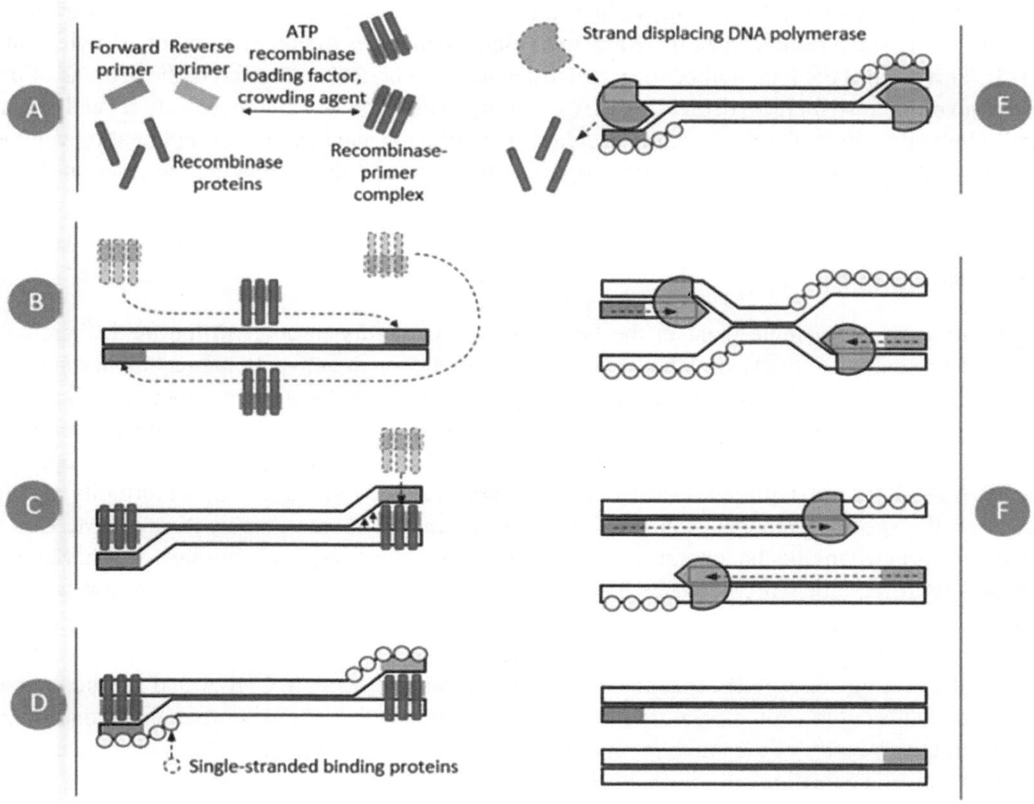

FIGURE 14.4 Mechanism of recombinase polymerase amplification.

the detection of Phytophthora fragariae, causal agent of red stele root rot of strawberry, at a very low detection limit of 10 fg genomic DNA by targeting the mitochondrial DNA *atp9-nad9* marker (Munawar et al., 2020). *Ustilaginoidea virens* is the causal agent of rice false smut. A sensitive RPA assay was able to detect *U.virens* in crude mycelial extracts from rice spikelets whereas PCR and LAMP were not able to detect (Banerjee et al., 2023). An assay system coupling RPA to Crispr/Cas systems with an integrated LFIA was able to effectively detect the presence of *Elsinoë fawcetti*, causal agent of citrus scab, in crude extracts of the citrus leaves (Shin et al., 2021).

14.2.3.3 Advantages and Disadvantages

RPA is significant in its operation at a constant low temperature without requiring an initial denaturation step or multiple primers. The two main advantages of RPA are the reaction temperature and time. The reaction conditions for RPA are tough, and detection in crude or raw samples could be problematic. The reaction operates at a temperature range of 30–42 °C, but the manufacturer recommends 39 °C. The recommended time is 20 min but more sensitive detection occurs at 10 min (Li et al., 2021; Yu et al., 2019). RPA could be multiplexed to detect several pathogens simultaneously (Larrea-Sarmiento et al., 2021). Considering the features of RPA such as selectivity, high sensitivity, simplicity, rapid amplification and multiplexing compatibility, it has often been favoured for designing user-friendly and affordable point-of-care and point-of-need kits and assays (Yan et al., 2014). For on-site testing, portable heating devices are often supplied with the kits. However, certain reactions have been performed using the body's physiological heat (Crannell et al., 2014; Larrea-Sarmiento et al., 2021; Liu et al., 2018).

Although RPA has been associated with several advantages, certain limitations still exist. RPA has a limit on the length of the amplified product of 1500 bp compared to 150 bp for lateral flow assays (Safenkova et al., 2020). The second limitation is RPA's tolerance to mismatch in primers, giving false positive results when there are up to nine mismatches in both the primers (Boyle et al., 2013). This shrinks the total number of target sequences that could be detected in the genome of the organism of interest, thereby, decreasing the possibility to discriminate among closely related organisms. High viscosity of the reaction mix decreases the diffusion and affects the amplification volume and event. Kit manufacturers recommend vigorous shaking for 4 min following the reaction (Lillis et al., 2016).

14.3 SUMMARY

Phytopathology of a single plant in the field can spread rapidly throughout the field; therefore, repeated screening of diseased plants becomes critical. Although conventional techniques such as PCR, ELISA and SDS-PAGE exist as gold standards to detect phytopathogens, there is still a need for an easier technique to be used in field conditions. LAMP proves to be a very rapid and sensitive technique where a set of primers with an enzyme containing strand displacement activity can amplify the DNA for detection. Different visualization techniques such as fluorescence and turbidity help in detecting the specific pathogen. Latex Microsphere Immunochromatographic Assay (LMIA) is a rapid detection technique that operates primarily based on the principle of capillary action. Sample interacts with the labelled conjugates and migrates to the detection zone. The detection is often observed when the labelled conjugate interacts with a biological component. Although LFIA is a rapid detection technique, capillary action requires the product to be in liquid form and hence cannot be used for testing solids. RPA uses the enzyme recombinase to amplify DNA and can be a rapid technique requiring lower temperature (30–42°C), making it feasible to operate in field conditions. However, the viscosity of the reaction mixture could cause an issue. Even if the gold standards for amplification and detection are still PCR, ELISA and SDS-PAGE, the development of on-site detection has enabled rapid and easier ways of detecting pathogens. Research directed towards the development of miniature devices could be an avenue for extensive exploration.

14.4 FUTURE PERSPECTIVE

Integrating microfluidic technology into the detection methods would aid in developing portable on-site assay kits. Miniaturized detection systems would thereby require reduced quantities of sample and still maintain sensitivity, specificity and rapidity (Panno et al., 2020). Reducing the cost by creating devices to detect a minimal volume of sample could further enhance the rapidity.

STUDY QUESTIONS

1 What are the different on-site detection methods for detecting pathogens? What are their advantages when compared to the conventional tests?
2 What are the advantages of loop-mediated isothermal amplification of nucleic acids technique?
3 Describe the principle of lateral flow immunoassays (LFIA).
4 Describe the working principle of recombinase polymerase amplification (RPA).
5 How is loop-mediated isothermal amplification of nucleic acids (LAMP) different from Recombinase polymerase amplification (RPA)?

FURTHER READING

Koczula, K.M., Gallotta, A., 2016. Lateral flow assays. *Essays in Biochemistry* 60, 111–120. https://doi.org/10.1042/EBC20150012

Lobato, I.M., O'Sullivan, C.K., 2018. Recombinase polymerase amplification: Basics, applications and recent advances. *Trends in Analytical Chemistry* 98, 19–35. https://doi.org/10.1016/j.trac.2017.10.015

Panno, S., Matić, S., Tiberini, A., Caruso, A.G., Bella, P., Torta, L., Stassi, R., Davino, S., 2020. Loop Mediated Isothermal Amplification: Principles and Applications in Plant Virology. *Plants* 9, 461. https://doi.org/10.3390/plants9040461

REFERENCES

Aliotta, J.M., Pelletier, J.J., Ware, J.L., Moran, L.S., Benner, J.S., Kong, H., 1996. Thermostable Bst DNA polymerase I lacks a 3'-->5' proofreading exonuclease activity. *Genetic Analysis* 12, 185–195.

Almasi, M.A., 2019. Development of a Colorimetric Loop-mediated Isothermal Amplification Assay for the Visual Detection of *Fusarium oxysporum* f.sp. *melonis*. *Horticultural Plant Journal* 5, 129–136. https://doi.org/10.1016/j.hpj.2019.01.004

Almasi, M.A., Dehabadi, S.M.H., Moradi, A., Eftekhari, Z., Ojaghk, M.A., Aghaei, S., 2013. Development and Application of Loop-Mediated Isothermal Amplification Assay for Rapid Detection of Fusarium Oxysporum f. *Sp. lycopersici*. *Journal of Plant Pathology & Microbiology* 4, 1–8. https://doi.org/10.4172/2157-7471.1000177

Andryukov, B.G., 2020. Six decades of lateral flow immunoassay: From determining metabolic markers to diagnosing covid-19. *AIMS Microbiology* 6, 280–304. https://doi.org/10.3934/microbiol.2020018

Bahadır, E.B., Sezgintürk, M.K., 2016. Lateral flow assays: Principles, designs and labels. *TrAC Trends in Analytical Chemistry* 82, 286–306. https://doi.org/10.1016/j.trac.2016.06.006

Banerjee, A., Bag, M.K., Chandra, A.K., Roy, S., Raghu, S., Mandal, N.P., 2023. Development and application of recombinase polymerase amplification for rapid detection of rice false smut pathogen (Ustilaginoidea virens). *Crop Protection* 167, 106204. https://doi.org/10.1016/j.cropro.2023.106204

Bhat, A.I., Siljo, A., Deeshma, K.P., 2013. Rapid detection of Piper yellow mottle virus and Cucumber mosaic virus infecting black pepper (Piper nigrum) by loop-mediated isothermal amplification (LAMP). *Journal of Virological Methods* 193, 190–196. https://doi.org/10.1016/j.jviromet.2013.06.012

Boutal, H., Moguet, C., Pommiès, L., Simon, S., Naas, T., Volland, H., 2022. The Revolution of Lateral Flow Assay in the Field of AMR Detection. *Diagnostics* 12, 1–26. https://doi.org/10.3390/diagnostics12071744

Boyle, D.S., Lehman, D.A., Lillis, L., Peterson, D., Singhal, M., Armes, N., Parker, M., Piepenburg, O., Overbaugh, J., 2013. Rapid Detection of HIV-1 Proviral DNA for Early Infant Diagnosis Using Recombinase Polymerase Amplification. *MBio* 4, e00135. https://doi.org/10.1128/mBio.00135-13

Braun-Kiewnick, A., Altenbach, D., Oberhänsli, T., Bitterlin, W., Duffy, B., 2011. A rapid lateral-flow immunoassay for phytosanitary detection of Erwinia amylovora and on-site fire blight diagnosis. *Journal of Microbiological Methods* 87, 1–9. https://doi.org/10.1016/j.mimet.2011.06.015

Byzova, N.A., Vinogradova, S. V., Porotikova, E. V., Terekhova, U.D., Zherdev, A. V., Dzantiev, B.B., 2018. Lateral flow immunoassay for rapid detection of grapevine leafroll-associated virus. *Biosensors* 8, 1–14. https://doi.org/10.3390/bios8040111

Chatterjee, S., Kuang, Y., Splivallo, R., Chatterjee, P., Karlovsky, P., 2016. Interactions among filamentous fungi Aspergillus Niger, Fusarium verticillioides and Clonostachys rosea: Fungal biomass, diversity of secreted metabolites and fumonisin production. *BMC Microbiology* 16, 1–13. https://doi.org/10.1186/s12866-016-0698-3

Chen, Y., Chen, Q., Han, M., Zhou, J., Gong, L., Niu, Y., Zhang, Y., He, L., Zhang, L., 2016. Development and optimization of a multiplex lateral flow immunoassay for the simultaneous determination of three mycotoxins in corn, rice and peanut. *Food Chemistry* 213, 478–484. https://doi.org/10.1016/j.foodchem.2016.06.116

Crannell, Z.A., Rohrman, B., Richards-Kortum, R., 2014. Equipment-Free Incubation of Recombinase Polymerase Amplification Reactions Using Body Heat. *PLoS One* 9, e112146. https://doi.org/10.1371/journal.pone.0112146

Dass, R.S., Sreenivasa, M.Y., Janardhana, G.R. 2007. High incidence of *Fusarium verticillioides* in animal and poultry feed mixtures produced in Karnataka, India. *Plant Pathology Journal* 6(2), 174–178.

De Boer, S.H., López, M.M., 2011. New Grower-Friendly Methods for Plant Pathogen Monitoring. *Annual Review of Phytopathology* 50, 197–218. https://doi.org/10.1146/annurev-phyto-081211-172942.

Deepa, N., Charith Raj, A. P., Sreenivasa, M.Y. 2016b. Nested PCR method for the early detection of fumonisin producing *Fusarium verticillioides* in pure cultures, cereal samples and plant parts. *Food Biotechnology* 30(1), 18–29.

Deepa, N., Charith Raj, A.P., Sreenivasa, M.Y. 2016c. Multiplex PCR method for the early detection of fumonisin producing *Fusarium verticillioides*. *Food Bioscience* 13, 84–88.

Deepa, N., Nagaraja, H., Sreenivasa, M.Y. 2016a. Prevalence of fumonisin producing Fusarium verticillioides associated with cereals grown in Karnataka (India). Food Science and Human Wellness 5, 156-162. https://doi.org/10.1016/j.fshw.2016.07.001.

Deepa N. and Sreenivasa M.Y. 2019a. Biocontrol Strategies for Effective Management of Phytopathogenic Fungi Associated with Cereals, In: Singh JS, Singh D.P.. (eds) *New and Future Developments in Microbial Biotechnology and Bioengineering*, Elsevier, pp, 177–189, ISBN 9780444641915.

Deepa, N. and Sreenivasa M. Y. 2019b. Molecular methods and key genes targeted for the detection of fumonisin producing *Fusarium verticillioides* - An updated review. *Food Bioscience* 32, 100473.

Denschlag, C., Rieder, J., Vogel, R.F., Niessen, L., 2014. Real-time loop-mediated isothermal amplification (LAMP) assay for group specific detection of important trichothecene producing Fusarium species in wheat. *International Journal of Food Microbiology* 177, 117–127. https://doi.org/10.1016/j.ijfoodmicro.2014.02.010

Di Nardo, F., Chiarello, M., Cavalera, S., Baggiani, C., Anfossi, L., 2021. Ten years of lateral flow immunoassay technique applications: Trends, challenges and future perspectives. *Sensors* 21. https://doi.org/10.3390/s21155185

Duan, Y.-B., Ge, C.-Y., Zhang, X.-K., Wang, J.-X., Zhou, M.-G., 2014. Development and Evaluation of a Novel and Rapid Detection Assay for Botrytis cinerea Based on Loop-Mediated Isothermal Amplification. *PLoS One* 9, e111094. https://doi.org/10.1371/journal.pone.0111094

Foley, J.A., Ramankutty, N., Brauman, K.A., Cassidy, E.S., Gerber, J.S., Johnston, M., Mueller, N.D., O'Connell, C., Ray, D.K., West, P.C., Balzer, C., Bennett, E.M., Carpenter, S.R., Hill, J., Monfreda, C., Polasky, S., Rockström, J., Sheehan, J., Siebert, S., Tilman, D., Zaks, D.P.M., 2011. Solutions for a cultivated planet. *Nature* 478, 337–342. https://doi.org/10.1038/nature10452

Fraaije, B.A., Lovell, D.J., Baldwin, S., 2002. Septoria epidemics on wheat: combined use of visual assessment and PCR-based diagnostics to identify mechanisms of disease escape. *Plant Protection Science* 38, 421–424. https://doi.org/10.17221/10512-PPS

Francois, P., Tangomo, M., Hibbs, J., Bonetti, E.-J., Boehme, C.C., Notomi, T., Perkins, M.D., Schrenzel, J., 2011. Robustness of a loop-mediated isothermal amplification reaction for diagnostic applications. *FEMS Immunology and Medical Microbiology* 62, 41–48. https://doi.org/10.1111/j.1574-695X.2011.00785.x

Fukuta, S., Takahashi, R., Kuroyanagi, S., Ishiguro, Y., Miyake, N., Nagai, H., Suzuki, H., Tsuji, T., Hashizume, F., Watanabe, H., Kageyama, K., 2014. Development of loop-mediated isothermal amplification assay for the detection of Pythium myriotylum. *Letters in Applied Microbiology* 59, 49–57. https://doi.org/10.1111/lam.12244

Fukuta, S., Takahashi, R., Kuroyanagi, S., Miyake, N., Nagai, H., Suzuki, H., Hashizume, F., Tsuji, T., Taguchi, H., Watanabe, H., Kageyama, K., 2013. Detection of Pythium aphanidermatum in tomato using loop-mediated isothermal amplification (LAMP) with species-specific primers. *European Journal of Plant Pathology* 136, 689–701. https://doi.org/10.1007/s10658-013-0198-3

Fung, K.-K., Chan, C.P.-Y., Renneberg, R., 2009. Development of enzyme-based bar code-style lateral-flow assay for hydrogen peroxide determination. *Analytica Chimica Acta* 634, 89–95. https://doi.org/10.1016/j.aca.2008.11.064

Gautam, A.K., Kumar, S., 2020. Chapter 12 - Techniques for the Detection, Identification, and Diagnosis of Agricultural Pathogens and Diseases, in: Egbuna, C., Sawicka, B. (Eds.), *Natural Remedies for Pest, Disease and Weed Control*. Academic Press, pp. 135–142. https://doi.org/10.1016/B978-0-12-819304-4.00012-9

Gomez-Gutierrez, S.V., Goodwin, S.B., 2022. Loop-Mediated Isothermal Amplification for Detection of Plant Pathogens in Wheat (Triticum aestivum). *Frontiers in Plant Science* 13, 857673. https://doi.org/10.3389/fpls.2022.857673

Gupta, S., Saharan, M.S., Gurjar, M.S., Singh, J., Bashyal, B.M., Aggarwal, R., 2020. Molecular detection of Fusarium graminearum causing head blight of wheat by loop mediated isothermal amplification (LAMP) assay. *Indian Phytopathology* 73, 667–672. https://doi.org/10.1007/s42360-020-00261-8

Hao, X., Wang, L., Zhang, X., Zhong, Q., Hajano, J.-U.-D., Xu, L., Wu, Y., 2021. A Real-Time Loop-Mediated Isothermal Amplification for Detection of the Wheat Dwarf Virus in Wheat and the Insect Vector *Psammotettix alienus*. *Plant Disease* 105, 4113–4120. https://doi.org/10.1094/PDIS-10-20-2279-RE

Hewison, D., Kuras, M., 2005. Book reviews. *The Journal of Analytical Psychology* 50, 395–403. https://doi.org/10.1111/j.0021-8774.2005.00541.x

Hodgetts, J., Karamura, G., Johnson, G., Hall, J., Perkins, K., Beed, F., Nakato, V., Grant, M., Studholme, D.J., Boonham, N., Smith, J., 2015. Development of a lateral flow device for in-field detection and evaluation of PCR-based diagnostic methods for Xanthomonas campestris pv. musacearum, the causal agent of banana xanthomonas wilt. *Plant Pathology* 64, 559–567. https://doi.org/10.1111/ppa.12289

Hodgetts, J., Tomlinson, J., Boonham, N., Gonzalez-Martin, I., Nikolić, P., Swarbrick, P., Yankey, E.N., Dickinson, M., 2011. Development of rapid in-field loop-mediated isothermal amplification (LAMP) assays for phytoplasmas. *Bulletin of insectology* 64, S41–S42.

Huang, C., Sun, Z., Yan, J., Luo, Y., Wang, H., Ma, Z., 2011. Rapid and Precise Detection of Latent Infections of Wheat Stripe Rust in Wheat Leaves using Loop-Mediated Isothermal Amplification. *Journal of Phytopathology* 159, 582–584. https://doi.org/10.1111/j.1439-0434.2011.01806.x

Ivanov, A.V., Safenkova, I.V., Zherdev, A.V., Dzantiev, B.B., 2021. The Potential Use of Isothermal Amplification Assays for In-Field Diagnostics of Plant Pathogens. *Plants* 10, 2424. https://doi.org/10.3390/plants10112424

Iwamoto, T., Sonobe, T., Hayashi, K., 2003. Loop-Mediated Isothermal Amplification for Direct Detection of *Mycobacterium tuberculosis* Complex, *M. avium*, and *M. intracellulare* in Sputum Samples. *Journal of Clinical Microbiology* 41, 2616–2622. https://doi.org/10.1128/JCM.41.6.2616-2622.2003

Janardhana, G. R., Dass, R.S., Sreenivasa, M.Y. 2009. Fumonisins – A new class of Fusarial toxins. In *Frontiers in Fungal Ecology, Diversity and Metabolites*. Ed: K A Sridhar et al. I. K. International Publishing House Pvt. Ltd. New Delhi.

Jędryczka, M., Burzyński, A., Brachaczek, A., Langwiński, W., Song, P., Kaczmarek, J., 2013. Loop-mediated isothermal amplification as a good tool to study changing Leptosphaeria populations in oilseed rape plants and air samples. *Acta Agrobotanica* 66, 93–100. https://doi.org/10.5586/aa.2013.055

Jiang, H., Su, H., Wu, K., Dong, Z., Li, X., Nie, L., Leng, Y., Xiong, Y., 2023. Multiplexed lateral flow immunoassay based on inner filter effect for mycotoxin detection in maize. *Sensors and Actuators B: Chemical* 374, 132793. https://doi.org/10.1016/j.snb.2022.132793

Kaneko, H., Kawana, T., Fukushima, E., Suzutani, T., 2007. Tolerance of loop-mediated isothermal amplification to a culture medium and biological substances. *Journal of Biochemical and Biophysical Methods* 70, 499–501. https://doi.org/10.1016/j.jbbm.2006.08.008

Kikuchi, T., Aikawa, T., Oeda, Y., Karim, N., Kanzaki, N., 2009. A Rapid and Precise Diagnostic Method for Detecting the Pinewood Nematode Bursaphelenchus xylophilus by Loop-Mediated Isothermal Amplification. *Phytopathology* 99, 1365–1369. https://doi.org/10.1094/PHYTO-99-12-1365

Koczula, K.M., Gallotta, A., 2016. Lateral flow assays. *Essays in Biochemistry* 60, 111–120. https://doi.org/10.1042/EBC20150012

Lane, C., Hobden, E., Walker, L., Barton, V., Inman, A., Hughes, K., Swan, H., Colyer, A., Barker, I., 2007. Evaluation of a rapid diagnostic field test kit for identification of Phytophthora species, including P. ramorum and P. kernoviae at the point of inspection. *Plant Pathology* 56, 828–835. https://doi.org/10.1111/j.1365-3059.2007.01615.x

Larrea-Sarmiento, A., Stack, J.P., Alvarez, A.M., Arif, M., 2021. Multiplex recombinase polymerase amplification assay developed using unique genomic regions for rapid on-site detection of genus Clavibacter and C. nebraskensis. *Scientific Reports* 11, 12017. https://doi.org/10.1038/s41598-021-91336-7

Le, D.T., Netsu, O., Uehara-Ichiki, T., Shimizu, T., Choi, I.-R., Omura, T., Sasaya, T., 2010. Molecular detection of nine rice viruses by a reverse-transcription loop-mediated isothermal amplification assay. *Journal of Virological Methods* 170, 90–93. https://doi.org/10.1016/j.jviromet.2010.09.004

Le, D.T., Vu, N.T., 2017. Progress of loop-mediated isothermal amplification technique in molecular diagnosis of plant diseases. *Applied Biological Chemistry* 60, 169–180. https://doi.org/10.1007/s13765-017-0267-y

Lei, R., Kong, J., Qiu, Y., Chen, N., Zhu, S., Wang, X., Wu, P., 2019. Rapid detection of the pathogenic fungi causing blackleg of Brassica napus using a portable real-time fluorescence detector. *Food Chemistry* 288, 57–67. https://doi.org/10.1016/j.foodchem.2019.02.089

Lei, R., Li, Y., Li, L., Wang, J., Cui, Z., Ju, R., Jiang, L., Liao, X., Wu, P., Wang, X., 2022. A CRISPR/Cas12a-based portable platform for rapid detection of Leptosphaeria maculans in Brassica crops. *Frontiers in Plant Science* 13, 976510. https://doi.org/10.3389/fpls.2022.976510

Li, C., Ju, Y., Shen, P., Wu, X., Cao, L., Zhou, B., Yan, X., Pan, Y., 2021. Development of Recombinase Polymerase Amplification Combined with Lateral Flow Detection Assay for Rapid and Visual Detection of *Ralstonia solanacearum* in Tobacco. *Plant Disease* 105, 3985–3989. https://doi.org/10.1094/PDIS-04-21-0688-RE

Li, J., Gu, F., Wu, R., Yang, J.K., Zhang, K.Q., 2017. Phylogenomic evolutionary surveys of subtilase superfamily genes in fungi. *Scientific Reports* 7, 1–15. https://doi.org/10.1038/srep45456

Li, W., Hartung, J.S., Levy, L., 2007. Evaluation of DNA Amplification Methods for Improved Detection of " *Candidatus* Liberibacter Species" Associated with Citrus Huanglongbing. *Plant Disease* 91, 51–58. https://doi.org/10.1094/PD-91-0051

Liang, C., Chu, Y., Cheng, S., Wu, H., Kajiyama, T., Kambara, H., Zhou, G., 2012. Multiplex Loop-Mediated Isothermal Amplification Detection by Sequence-Based Barcodes Coupled with Nicking Endonuclease-Mediated Pyrosequencing. *Analytical Chemistry* 84, 3758–3763. https://doi.org/10.1021/ac3003825

Lillis, L., Siverson, J., Lee, A., Cantera, J., Parker, M., Piepenburg, O., Lehman, D.A., Boyle, D.S., 2016. Factors influencing Recombinase polymerase amplification (RPA) assay outcomes at point of care. *Molecular and Cellular Probes* 30, 74–78. https://doi.org/10.1016/j.mcp.2016.01.009

Liu, L., Wang, J., Geng, Y., Wang, J., Li, R., Shi, R., Yuan, W., 2018. Equipment-free recombinase polymerase amplification assay using body heat for visual and rapid point-of-need detection of canine parvovirus 2. *Molecular and Cellular Probes* 39, 41–46. https://doi.org/10.1016/j.mcp.2018.04.004

Liu, Y., Xu, X., Liu, L., Xu, L., Kuang, H., Xu, C., 2022. Gold-based lateral-flow strip for the detection of penconazole in watermelon and cucumber samples. *Food Quality and Safety* 6, fyac007. https://doi.org/10.1093/fqsafe/fyac007

Lobato, I.M., O'Sullivan, C.K., 2018. Recombinase polymerase amplification: Basics, applications and recent advances. *Trends in Analytical Chemistry* 98, 19–35. https://doi.org/10.1016/j.trac.2017.10.015

López-Soriano, P., Noguera, P., Gorris, M.T., Puchades, R., Maquieira, Á., Marco-Noales, E., López, M.M., 2017. Lateral flow immunoassay for on-site detection of Xanthomonas arboricola pv. Pruni in symptomatic field samples. *PLoS One* 12, 1–13. https://doi.org/10.1371/journal.pone.0176201

Lou, D., Fan, L., Jiang, T., Zhang, Y., 2022. Advances in nanoparticle-based lateral flow immunoassay for point-of-care testing. *View* 3, 20200125. https://doi.org/10.1002/VIW.20200125

Mahy, B.W.J., van Refenmortel, M.H.V., 2009. *Desk encyclopedia of plant and fungal virology*. Academic Press, USA.

Mehetre, G.T., Leo, V.V., Singh, G., Sorokan, A., Maksimov, I., Yadav, M.K., Upadhyaya, K., Hashem, A., Alsaleh, A.N., Dawoud, T.M., Almaary, K.S., Singh, B.P., 2021. Current developments and challenges in plant viral diagnostics: A systematic review. *Viruses* 13, 1–31. https://doi.org/10.3390/v13030412

Miles, T.D., Martin, F.N., Coffey, M.D., 2015. Development of Rapid Isothermal Amplification Assays for Detection of *Phytophthora* spp. in Plant Tissue. *Phytopathology* 105, 265–278. https://doi.org/10.1094/PHYTO-05-14-0134-R

Munawar, M.A., Toljamo, A., Martin, F., Oksanen, E., Kokko, H., 2020. Development and evaluation of a recombinase polymerase amplification assay for rapid detection of strawberry red stele pathogen. *Phytopathology Research* 2, 26. https://doi.org/10.1186/s42483-020-00069-4

Nagamine, K., Hase, T., Notomi, T., 2002. Accelerated reaction by loop-mediated isothermal amplification using loop primers. *Molecular and Cellular Probes* 16, 223–229. https://doi.org/10.1006/mcpr.2002.0415

Niessen, L., Vogel, R.F., 2010. Detection of Fusarium graminearum DNA using a loop-mediated isothermal amplification (LAMP) assay. *International Journal of Food Microbiology* 140, 183–191. https://doi.org/10.1016/j.ijfoodmicro.2010.03.036

Notomi, T., Okayama, H., Masubuchi, H., Yonekawa, T., Watanabe, K., Amino, N., Hase, T., 2000. Loop-mediated isothermal amplification of DNA. *Nucleic Acids Research* 28, e63.

Okawa, K., 2015. Market and Trade Impacts of Food Loss and Waste Reduction. *OECD, Paris* https://doi.org/10.1787/5js4w29h0wr2-en

Pai, N.P., Vadnais, C., Denkinger, C., Engel, N., Pai, M., 2012. Point-of-Care Testing for Infectious Diseases: Diversity, Complexity, and Barriers in Low- And Middle-Income Countries. *PLoS Medicine* 9, e1001306.

Panno, S., Matić, S., Tiberini, A., Caruso, A.G., Bella, P., Torta, L., Stassi, R., Davino, S., 2020. Loop Mediated Isothermal Amplification: Principles and Applications in Plant Virology. *Plants* 9, 461. https://doi.org/10.3390/plants9040461

Pieczul, K., Perek, A., Kubiak, K., 2018. Detection of Tilletia caries, Tilletia laevis and Tilletia controversa wheat grain contamination using loop-mediated isothermal DNA amplification (LAMP). *Journal of Microbiological Methods* 154, 141–146. https://doi.org/10.1016/j.mimet.2018.10.018

Piepenburg, O., Williams, C.H., Stemple, D.L., Armes, N.A., 2006. DNA Detection Using Recombination Proteins. *PLoS Biology* 4, e204. https://doi.org/10.1371/journal.pbio.0040204

Posthuma-Trumpie, G.A., Korf, J., Van Amerongen, A., 2009. Lateral flow (immuno)assay: Its strengths, weaknesses, opportunities and threats. A literature survey. *Analytical and Bioanalytical Chemistry* 393, 569–582. https://doi.org/10.1007/s00216-008-2287-2

Qin, J., Lu, Q., Wang, C., Luo, J., Yang, M., 2021. Colloidal gold-based lateral flow immunoassay with inline cleanup for rapid on-site screening of carbendazim in functional foods. *Analytical and Bioanalytical Chemistry* 413, 3725–3735. https://doi.org/10.1007/s00216-021-03321-8

Quesada-González, D., Merkoçi, A., 2015. Nanoparticle-based lateral flow biosensors. *Biosensors and Bioelectronics* 73, 47–63. https://doi.org/10.1016/j.bios.2015.05.050

Quyen, T.L., Ngo, T.A., Bang, D.D., Madsen, M., Wolff, A., 2019. Classification of Multiple DNA Dyes Based on Inhibition Effects on Real-Time Loop-Mediated Isothermal Amplification (LAMP): Prospect for Point of Care Setting. *Frontiers in Microbiology* 10, 2234. https://doi.org/10.3389/fmicb.2019.02234

Rahman, H.U., Yue, X., Yu, Q., Xie, H., Zhang, W., Zhang, Q., Li, P., 2019. Specific antigen-based and emerging detection technologies of mycotoxins. *Journal of the Science of Food and Agriculture* 99, 4869–4877. https://doi.org/10.1002/jsfa.9686

Reddy, K.R.N., Salleh, B., Saad, B., Abbas, H.K., Abel, C.A., Shier, W.T., 2010. An overview of mycotoxin contamination in foods and its implications for human health. *Toxin Reviews* 29, 3–26. https://doi.org/10.3109/15569541003598553

Rigano, L.A., Malamud, F., Orce, I.G., Filippone, M.P., Marano, M.R., Do Amaral, A.M., Castagnaro, A.P., Vojnov, A.A., 2014. Rapid and sensitive detection of Candidatus Liberibacter asiaticus by loop mediated isothermal amplification combined with a lateral flow dipstick. *BMC Microbiology* 14, 86. https://doi.org/10.1186/1471-2180-14-86

Rovetto, E.I., Garbelotto, M., Moricca, S., Amato, M., Spada, F.L., Cacciola, S.O., 2023. Recombinase Polymerase Amplification Assay for the Field Detection of Mal Secco Disease by *Plenodomus tracheiphilus* (preprint). *Molecular Biology* https://doi.org/10.1101/2023.10.21.563392

Safenkova, I., Zaytsev, I., Pankratova, G., Varitsev, Y., Zherdev, A., Dzantiev, B., 2014. Lateral Flow Immunoassay for Rapid Detection of Potato Ring Rot Caused by Clavibacter michiganensis subsp sepedonicus. *Applied Biochemistry and Microbiology* 50, 675–682. https://doi.org/10.1134/S0003683814120011

Safenkova, I., Zherdev, A., Dzantiev, B., 2012. Factors influencing the detection limit of the lateral-flow sandwich immunoassay: a case study with potato virus X. *Analytical and Bioanalytical Chemistry* 403, 1595–1605. https://doi.org/10.1007/s00216-012-5985-8

Safenkova, I.V., Ivanov, A.V., Slutskaya, E.S., Samokhvalov, A.V., Zherdev, A.V., Dzantiev, B.B., 2020. Key significance of DNA-target size in lateral flow assay coupled with recombinase polymerase amplification. *Analytica Chimica Acta* 1102, 109–118. https://doi.org/10.1016/j.aca.2019.12.048

Sankaran, S., Mishra, A., Ehsani, R., Davis, C., 2010. A review of advanced techniques for detecting plant diseases. *Computers and Electronics in Agriculture* 72, 1–13. https://doi.org/10.1016/j.compag.2010.02.007

Santiago-Felipe, S., Tortajada-Genaro, L.A., Carrascosa, J., Puchades, R., Maquieira, Á., 2016. Real-time loop-mediated isothermal DNA amplification in compact disc micro-reactors. *Biosensors and Bioelectronics* 79, 300–306. https://doi.org/10.1016/j.bios.2015.12.045

Shin, K., Kwon, S.-H., Lee, S.-C., Moon, Y.-E., 2021. Sensitive and Rapid Detection of Citrus Scab Using an RPA-CRISPR/Cas12a System Combined with a Lateral Flow Assay. *Plants* 10, 2132. https://doi.org/10.3390/plants10102132

Siegieda, D.G., Panek, J., Frąc, M., 2021. "Shining a LAMP" (Loop-Mediated Isothermal Amplification) on the Molecular Detection of Phytopathogens Phytophthora spp. and Phytophthora cactorum in Strawberry Fields. *Pathogens* 10, 1453. https://doi.org/10.3390/pathogens10111453

Song, S., Liu, N., Zhao, Z., Njumbe Ediage, E., Wu, S., Sun, C., De Saeger, S., Wu, A., 2014. Multiplex Lateral Flow Immunoassay for Mycotoxin Determination. *Analytical Chemistry* 86, 4995–5001. https://doi.org/10.1021/ac500540z

Sreenivasa, M.Y., Dass, R.S., Charith Raj, A.P., Janardhana, G.R. 2008. PCR-based detection of genus *Fusarium* and Fumonisin-producing isolates from freshly harvested Sorghum grains grown in Karnataka, India. *Journal of Food Safety* 28, 236–237.

Tanner, N.A., Zhang, Y., Evans, T.C., 2015. Visual detection of isothermal nucleic acid amplification using pH-sensitive dyes. *BioTechniques* 58, 59–68. https://doi.org/10.2144/000114253

Thornton, C.R., Groenhof, A.C., Forrest, R., Lamotte, R., 2004. A one-step, immunochromatographic lateral flow device specific to Rhizoctonia solani and certain related species, and its use to detect and quantify R. solani in soil. *Phytopathology* 94, 280–288. https://doi.org/10.1094/PHYTO.2004.94.3.280

Tomlinson, J.A., Barker, I., Boonham, N., 2007. Faster, Simpler, More-Specific Methods for Improved Molecular Detection of Phytophthora ramorum in the Field. *Applied and Environmental Microbiology* 73, 4040–4047. https://doi.org/10.1128/AEM.00161-07

Tomlinson, J.A., Dickinson, M.J., Boonham, N., 2010a. Detection of Botrytis cinerea by loop-mediated isothermal amplification. *Letters in Applied Microbiology* 51, 650–657. https://doi.org/10.1111/j.1472-765X.2010.02949.x

Tomlinson, J. A., Dickinson, M.J., Boonham, N., 2010b. Rapid detection of Phytophthora ramorum and P. kernoviae by two-minute DNA extraction followed by isothermal amplification and amplicon detection by generic lateral flow device. *Phytopathology* 100, 143–149. https://doi.org/10.1094/PHYTO-100-2-0143

Trzmiel, K., Hasiów-Jaroszewska, B., 2020. Development of loop-mediated isothermal amplification assay for rapid detection of genetically different wheat dwarf virus isolates. *Molecular Biology Reports* 47, 8325–8329. https://doi.org/10.1007/s11033-020-05846-0

Van Emon, J.M., 2006. *Immunoassay and Other Bioanalytical Techniques*, 1st ed. CRC Press, Boca Raton.

Wang, D., Zhu, J., Zhang, Z., Zhang, Q., Zhang, W., Yu, L., Jiang, J., Chen, X., Wang, X., Li, P., 2019. Simultaneous lateral flow immunoassay for multi-class chemical contaminants in maize and peanut with one-stop sample preparation. *Toxins* 11, 1–11. https://doi.org/10.3390/toxins11010056

Weihs, F., Anderson, A., Trowell, S., Caron, K., 2021. Resonance Energy Transfer-Based Biosensors for Point-of-Need Diagnosis—Progress and Perspectives. *Sensors* https://doi.org/10.3390/s21020660

Wilisiani, F., Tomiyama, A., Katoh, H., Hartono, S., Neriya, Y., Nishigawa, H., Natsuaki, T., 2019. Development of a LAMP assay with a portable device for real-time detection of begomoviruses under field conditions. *Journal of Virological Methods* 265, 71–76. https://doi.org/10.1016/j.jviromet.2018.10.005

Wong, Y.-P., Othman, S., Lau, Y.-L., Radu, S., Chee, H.-Y., 2018. Loop-mediated isothermal amplification (LAMP): a versatile technique for detection of micro-organisms. *Journal of Applied Microbiology* 124, 626–643. https://doi.org/10.1111/jam.13647

Wu, F., 2007. Measuring the economic impacts of Fusarium toxins in animal feeds. *Animal Feed Science and Technology* 137, 363–374. https://doi.org/10.1016/j.anifeedsci.2007.06.010

Xing, K.-Y., Shan, S., Liu, D.-F., Lai, W.-H., 2020. Recent advances of lateral flow immunoassay for mycotoxins detection. *TrAC Trends in Analytical Chemistry* 133, 116087. https://doi.org/10.1016/j.trac.2020.116087

Xu, Z.-K., Meng, J.-N., Lei, Y., Yang, X.-X., Yan, Y.-Y., Liu, H.-H., Lei, H.-T., Wang, T.-C., Shen, X., Xu, Z.-L., 2022. Highly selective monoclonal antibody-based lateral flow immunoassay for visual and sensitive determination of conazole fungicides propiconazole in vegetables. *Food Additives & Contaminants: Part A* 39, 92–104. https://doi.org/10.1080/19440049.2021.1976423

Yalow, R.S., Berson, S.A., 1960. Immunoassay of endogenous plasma insulin in man. *Journal of Clinical Investigation* 7, 1157–1175. https://doi.org/10.1002/j.1550-8528.1996.tb00274.x

Yan, H., Zhang, J., Ma, D., Yin, J., 2019. qPCR and loop mediated isothermal amplification for rapid detection of *Ustilago tritici. PeerJ* 7, e7766. https://doi.org/10.7717/peerj.7766

Yan, L., Zhou, J., Zheng, Y., Gamson, A.S., Roembke, B.T., Nakayama, S., Sintim, H.O., 2014. Isothermal amplified detection of DNA and RNA. Molecular BioSystems 10, 970. https://doi.org/10.1039/c3mb70304e

Yasuhara-Bell, J., Pedley, K.F., Farman, M., Valent, B., Stack, J.P., 2018. Specific Detection of the Wheat Blast Pathogen (*Magnaporthe oryzae* Triticum) by Loop-Mediated Isothermal Amplification. *Plant Disease* 102, 2550–2559. https://doi.org/10.1094/PDIS-03-18-0512-RE

Yen, C.-W., de Puig, H., Tam, J.O., Gómez-Márquez, J., Bosch, I., Hamad-Schifferli, K., Gehrke, L., 2015. Multicolored silver nanoparticles for multiplexed disease diagnostics: distinguishing dengue, yellow fever, and Ebola viruses. *Lab on a Chip* 15, 1638–1641. https://doi.org/10.1039/C5LC00055F

Yin, Z., Ke, X., Kang, Z., Huang, L., 2016. Apple resistance responses against Valsa mali revealed by transcriptomics analyses. *Physiological and Molecular Plant Pathology* 93, 85–92. https://doi.org/10.1016/j.pmpp.2016.01.004

Yu, J., Shen, D., Dai, T., Lu, X., Xu, H., Dou, D., 2019. Rapid and equipment-free detection of *Phytophthora capsici* using lateral flow strip-based recombinase polymerase amplification assay. *Letters in Applied Microbiology*.13166. https://doi.org/10.1111/lam.13166

Zhang, X., Zhang, H., Pu, J., Qi, Y., Yu, Q., Xie, Y., Peng, J., 2013. Development of a Real-Time Fluorescence Loop-Mediated Isothermal Amplification Assay for Rapid and Quantitative Detection of Fusarium oxysporum f. sp. cubense Tropical Race 4 In Soil. *PLoS One* 8, e82841. https://doi.org/10.1371/journal.pone.0082841

Zhao, W., Chi, Y., Ye, M., Wang, T., Xu, A., Qi, R., 2021. Development and application of Recombinase Polymerase Amplification assay for detection of Bipolaris sorokiniana. *Crop Protection* 145, 105619. https://doi.org/10.1016/j.cropro.2021.105619

Zhou, D., Guo, J., Xu, L., Gao, S., Lin, Q., Wu, Q., Wu, L., Que, Y., 2014. Establishment and application of a loop-mediated isothermal amplification (LAMP) system for detection of cry1Ac transgenic sugarcane. *Scientific Reports* 4, 4912. https://doi.org/10.1038/srep04912

15 Policies, Regulations and Quality Assurance Requirements for Sustainable Agriculture

S. Sundaramoorthy
Horticultural College and Research Institute (HC&RI), Tamil Nadu Agricultural University (TNAU)

*V. Jaiganesh**
Citrus Research Station (CRS), Tamil Nadu Agricultural University (TNAU)

GRAPHICAL ABSTRACT

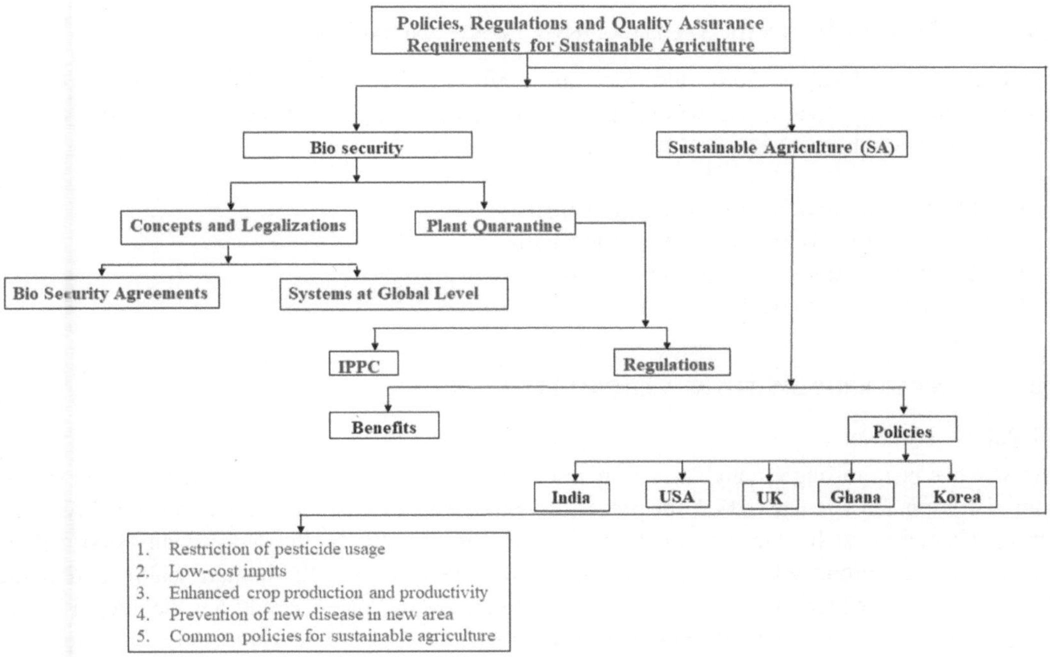

* SSM wrote the manuscript and collected the data on Plant Biosecurity and Policies, VJ contributed for collection of other relevant data. Both authors reviewed the final manuscript.

DOI: 10.1201/9781003399704-15

15.1 INTRODUCTION

Increasing volume of travel, trade and communications have changed the world, allowing enormous growth in the global movement of people, commodities and conveyances over the past century. Natural, state, and country borders that once were efficient barriers to the dissemination and introduction of harmful disease-causing organisms are now under pressure from the volume of international traffic. As a consequence, the global community has three important intergovernmental mechanisms setting the standards by which human, animal and plant health are protected, particularly from the undesirable impact of global movements of people and traded goods. These are OIE (Office International des Epizooties), CA (Codex Alimentarius) and IPPC (International Plant Protection Convention), respectively. Globally, the annual losses due to plant pests have been estimated to be 36.5% average of total losses, 14.1% are caused by plant diseases, 10.2% are caused by Insects and 12.2% are caused by weeds (Agrios, 2005).

15.2 BIOSECURITY

It is a planned and combined approach consisting of policies and regulation frameworks to analyse and manage risks in the life and health of humans, animals and plants. (FAO, 2007).

15.2.1 Basic Concepts of Plant Biosecurity

- Sustainability of Agriculture
- Broad-ranging features of health in human beings
- Environmental protection, including diversity of various biological sources (FAO, 2007)

15.2.2 Relevant International Legal Instruments and Agreements

In addition to the legal standards and also agreement-related policies formed by the CAC (*Codex Alimentarius Commission*), OIE including many other global legal instruments, rules, policy making are relevant to biosecurity. They are

- SPS (Sanitary and Phytosanitary) Agreement
- TBT (Technical Barriers to Trade) Agreement
- CBD (Convention on Biological Diversity)
- CP (Cartagena Protocol on Biosafety).

15.2.3 Biosecurity Systems at a Global Level

15.2.3.1 Norway

Norway has been through many fazes in relation to a reform process of the country's food safety administration. On 1 January 2004, the process concluded with the formation of a four-pillar restructuring, characterized by the development of a new group, that is, the Norwegian Food Safety Authority. This authority recognized scientific provisions related to the human health food chain through the development of an autonomous risk valuation body, a New Food Law, the integration of 13 individual Acts and a new explanation of the applicable Government ministry's constitutional responsibilities (FAO, 2007).

15.2.3.2 New Zealand

The government released New Zealand's first Biosecurity strategy in August 2003 published by the Biosecurity Council with the bringing together of Biosecurity New Zealand with the border services functions of MAF in 2007. MAF Biosecurity New Zealand was created. The strategy raises public awareness and understanding of Biosecurity. A number of other strategies such as surveillance strategy, science strategy, response foundations and border systems have been developed to support the Biosecurity system. New Zealand has been a role model in formulating a Biosecurity Act in 1993, well ahead of the global need. (FAO, 2007).

15.2.3.3 Australia

Biosecurity Australia is the unit within the Biosecurity Service Group, in the Agriculture, Fisheries and Forestry Department, which is responsible for recommendations for the development of Australia's Biosecurity Policy. Biosecurity in Australia provides science-based quarantine assessment and legal advice that defends the pest and disease status in Australia and improves the country's access to international plant and animal-related markets. (FAO, 2007).

15.2.3.4 USA

In the USA, agricultural biosecurity is looked after by the APHIS (Animal and Plant Health Inspection Service) headed by an Administrator under the USDA (US Department of Agriculture). The APHIS constitutes various divisions including animal care, Plant Protection and Quarantine, Biotechnology regulatory services, and wildlife services supported by the Office of Emergency Management and Homeland Security as well. Homeland Security of the Department was created in 2002 with as many as 22 agencies, including the US Department of Agriculture's APHIS Service. The Advisory System of Homeland Security is designed to give guidance on Protective actions when exact detail of a specific sector or geographic condition is received. (Roger, et al. 2009).

15.3 INTERNATIONAL PLANT PROTECTION CONVENTION (1951–1952)

It is an international treaty relating to plant health. In 2023, 185 governments were contracting parties. The IPPC convention has been placed with the DG, FAO of the United Nations (UN) since it's a primary adoption by the FAO conference in its 6th Session of 1951. The IPPC came into force in 1952, superseding all previous international plant protection agreements. The main principle of IPPC is to protected common and specific action to secure the dissemination and introduction of new pests of agricultural crop and their products, and to promote appropriate measures for their management. The IPPC develops, accepts and encourages the application of International Phytosanitary Measures (ISPMs) as an important tool for the protection of global food security, enabling safe trade and environmental protection. Its implementation involves the collaboration of National Plant Protection Organizations (NPPOs), which are the official services established by governments to discharge the functions specified by the IPPC, and regional Plant Protection Organizations (RPPOs), which may be working as coordinating agencies on a regional level for participation in the actions to accomplish the IPPC objectives (Guide to the IPPC, 2005).

15.4 PLANT QUARANTINE (PQ) REGULATIONS

The regulations initially started as a prohibition of agricultural crop and their plant materials from specific Countries due to past experience with Invasive Alien Species. The first effort towards the International Agreement on Plant Protection was completed in 1914 under the auspicious of the Agricultural Institute at the International level in Rome. It was trailed by an international convention on Plant Protection by more than 50 member countries of the Institute in 1919, where certain agreements regarding the Issue and acceptance of Phytosanitary Certificate were finalized. Appreciating the need for a convention to strengthen such international efforts, FAO sponsored an IPPC in 1951. Indian country became a member of the IPPC group in 1956 and agreed to accept legal measures quantified in the convention for the determination of protecting common and specific action to avoid the introduction and spread of new pests and diseases of agricultural crops and their crop materials and to improve measures for their control. Plant Quarantine regulations are entrusted with public

TABLE 15.1
Milestones of Plant Quarantine Regulations in the World

Sl. No	Year	Plant Quarantine Regulations
1	1877	Indonesia enacted a law to prohibit importation of coffee plants and beans from Sri Lanka.
2	1891	California State of USA initiated seaport inspection at San Pedro, probably first in the world.
3	1909	In Australia, the first set of regulation governing PQ came in the force.
4	1912	The Federal Plant Quarantine Act was enacted in USA.
5	1913	The German Government started to put ban on plants and plant products from USA.
6	1913	The very first PQ regulations were enacted by French Government.
7	1914	In India, the first regulation in the name of Destructive Insects and Pests (DIP) Act came in to force.
8	1923	The first PQ regulations of French Government came into effect.

authority to legally establish barriers against the dissemination of injurious pests. This has remained unchanged till today. The need for Plant Quarantine to be entrusted to public authority is considered one of the fundamental principles of Plant Quarantine. Now, almost all the countries have PQ regulations in force (Table 15.1).

15.5 IMPACT OF PLANT QUARANTINE REGULATIONS

15.5.1 The Great Famine/Irish Potato Famine (1845–1852)

Potato was introduced into Ireland as a garden crop. In the late 17th century, it was consumed as a supplementary food. By the early 18th century, it assumed the status of staple food for the poor during the winter season and as well as considered an agrarian economy as it fetched more money and grew in more acreage (<60%). In 1843–1844, a new disease started to destroy the crops and was identified as the late blight of potato, *Phytophthora infestans* (de Bary). This pathogen infection completely destroyed potato crops in 1845, which led to the starvation and death of 1 million people and the migration of another 1 million people from Ireland between 1845 and 1852. This unfortunate famine occurred due to the incursion of *Phytophthora infestans* from USA, which caused the loss of around £3,500,000 worth of potatoes.

15.5.2 Toppling of Grapevine Industry in Europe (1840–1885)

The introduction of powdery mildew of Grapevine (*Oidium tuckeri*) into Europe with American Grapevine rootstocks and its pathogenicity on European Grapevine rootstocks was unknown at that time, and the disease spread like wildfire on European Grapevines (1850). To control powdery mildew, rootstocks of resistant varieties were imported from America (1854). However, these Grapevine rootstocks carried *Phylloxera vastatrix*, a root inhabiting aphid of grape vines. To combat this pest, more American vines resistant to *Phylloxera* were introduced, but this additional introduction brought with them the Downy mildew (*Plasmopara viticola*), and black rot (*Guignardia bidwellii*). France, where the vine industry was thriving, had to face the burden of these pest incursions, and many businessmen abandoned vine production and emigrated to Algeria and other countries. Further, these incidents led to the discovery of Bordeaux Mixture (First Universal Fungicide) in 1885 for its management.

15.5.3 The DIP Act, 1914

In the direction to safeguard the agricultural and forest plant materials of Indian country, the GoI (Government of India) took many legal steps in 1914 and agreed to an Act mentioned, "The Destructive Insects and Pests Act" (Act II of 1914) on 3 February 1914 for regulation of the import of crop, plant and plant materials. Initially, the various rules and regulations of the DIP Act were enforced by the Sea Customs Act under the Customs Department. The outcome was that various new insects and pathogens entered India due to unscientific methods of crop inspection and clearance. The great Bengal famine occurred in undivided India (Bangladesh and West Bengal) due to the outbreak of Brown spot of Rice caused by *Bipolaris oryzae* (*Helminthosporium oryzae/Drechslera oryzae*), which wiped out 3 million people of India, and almost complete destruction of rice crops (97%) in 1942–1943 acted as a precursor for the establishment in Faridabad of Directorate of Plant Protection, Quarantine and Storage (DPPQ&S) in 1946 based on the Woodhead commission recommendation. It was decided to form PQ institutes at various land frontiers, seaports, and international airports and check posts at the domestic level for effective implementation of PQ regulations (https://www.indiacode.nic.in/handle/123456789/2354?sam_handle=123456789/1362).

15.5.4 PQ (Plant Quarantine – Regulation of Import into India) Order, 2003

This order came into force on 1 January 2004. The PQ Order, 2003, replaces all the preceding Orders/Notifications of Plant Quarantine Regulation. It was formulated on the scientific basis of Pest Risk Analysis (PRA). The commodities are categorized into XII Schedules based on associated pest risk either through pathways or from countries where the pest was known to be reported. This order clearly spells out notified points of entry, a list of prohibited, restricted, and regulated commodities; quarantine weeds of concern to India; Inspection Fees; list of notified authorities to issue Import Permits (PQ Officials), list of Designated Inspection Authorities (DIA) to certify Post Entry Quarantine Facilities (PEQ) and the deposition of samples to Gene Bank of NBPGR (National Bureau of Plant Genic Resources) (Table 15.2).

15.6 SUSTAINABLE AGRICULTURE

According to the U.S. Code Title 7, Section 3103, "sustainable agriculture" means an integrated system of plant and animal production practices having a site-specific application that will over the long term (https://www.nifa.usda.gov/).

That includes

- To satisfy human food and fibre needs
- To enhance the environmental quality
- To improve the natural resource based upon which the agriculture economy should be practised over the long term.

TABLE 15.2
Important Plant Quarantine Regulations in India

Sl. No	Year	Plant Quarantine Regulations
1	1914	The Destructive Insects and Pests Act
2	1936	Import Regulations
3	1972	Regulations to import cotton
4	1984	PFS Order
5	1988	National Policy on seed Development
6	1989	PFS Order (Revised)
7	1992	Amendment to DIP Act for levying inspection fee
8	2003	Plant Quarantine (Regulation of Import into India) Order

- To make the most efficient use of non-renewable resources and on-farm resources
- To sustain the economic viability of agricultural operations.
- To enhance the quality of farmers and societal life.

15.6.1 Benefits of Sustainable Agricultural Farming

- To maintain the environmental health
- To minimize the inputs which disturb the environment
- To reduce the pollution in all ways
- To optimize the resource, usage and environmental conservation
- To maximize crop production and productivity with low-cost inputs
- To improve healthy, quality food production

15.6.2 Sustainable Agriculture for Crop Production and Productivity

- Crop diversification
- Crop biodiversity
- Cover crops, intercrops along with perennial crops
- Reducing tillage process
- Integrated Pest Management (IPM), Integrated Disease Management (IDM) and Integrated Plant Health Management (IPHM),
- Integrated Farming System (IFS) along with livestock and aquaculture
- To adopt agroforestry and silviculture
- Landscape farming with medicinal, human aesthetic values

15.6.3 Sustainable Agriculture for Animal husbandry

- To maintain a healthy animal growth
- To adopt the macro and micro nutrition mix in animal health
- To adopt healthy medicines with low residues
- To adopt a single vaccine for multiple diseases
- To reduce the antibiotics in animal health

15.6.4 Sustainable Agriculture for Aquaculture

- To improve aquaculture industries
- To adopt the limitation of fish harvesting time
- To adopt the limitation of fish harvesting areas
- To reduce environmental pollution in seashore areas
- To enforce traceability standards
- To improve fish cultivation and processing technologies

15.6.5 Sustainable Agroforestry

- To adopt the reforestation
- To maintain afforestation
- To adopt sustainable pest and disease management in forests
- To adopt the pruning, specific logging, and thinning practices
- To adopt the elimination of mature trees and old diseased trees
- To manage the artificial and natural burning
- To adopt regular training and monitoring for forestry officials

15.7 SUSTAINABLE AGRICULTURE POLICIES IN INDIA

Farmers doubling income by increasing crop production and productivity increases the high resource use with low-cost investment in agriculture and also by increasing cropping intensity by intercrop, mixed, cover crop and second crop on fallow land cultivation and crop rotation methods etc.

15.7.1 National Policy for Farmers (2007)

Government of India has already implemented the National Policy for Farmers (NPF), 2007, which aims to improve the economic sustainability of crop farming and improve the farmer's profit. The policies are asset improvements with respect to agricultural land, irrigation, animal husbandry, aquaculture and biodiversity. Also, supply of certified healthy seeds and pest, disease-free seeds, issue of soil health cards to the farmers and integrated plant health management system are included in this policy. In addition to the above, agricultural machinery, women's involvement in agriculture timely, farmer-friendly insurance, use of Information and Communication Technology (ICT), initiation of farm schools, reorganize the agricultural extension, national social security scheme, minimum support price (MSP) implementation, foodgrain banks initiation, agricultural market infrastructure development, agricultural universities curriculum design change, organic farming, contract farming and rural energy were included (PIB, 2007).

15.7.2 National Agricultural Policy (2000)

The National Agricultural Policy (NAP) was constituted by the Government of India on 28 July 2000. The main objectives of NAP include

- Realizing the huge unused Indian agricultural potential sources
- To maintain a growth rate > 4% / year in agricultural sector
- To maintain sustainable agricultural growth
- To improve rural infrastructure development
- To discourage the migration from rural to urban areas
- Sustainable agriculture
- To maintain a food and nutritional security status
- To the successful establishment of ToT (Transfer of Technology)

Also, this policy includes agricultural inputs management, agricultural incentives, agricultural institutional structure, price protection to the farmers, crop risk management and crop management reforms. The policies were formulated under the provisions of the World Trade Organization (WTO). For the above policies Report on Policies and Action Plan for a Secure and Sustainable Agriculture (2019) mentioned that the "National Agricultural and Farmers' Welfare Policy (NAFWP)" must be formulated to achieve Sustainable Development Goals (SDGs) through secure and sustainable agriculture.

15.7.3 National Food Security Mission

National Food Security Mission (NFSM) was launched by the Government of India in 2007–2008 to improve the cereals and pulses production, and productivity by the end of the XI Plan. NFSM includes increasing soil fertility, and economic and employment opportunities in the farm sector. NFSM will encourage Farmer Producing Organizations (FPOs) to undertake procurement operations under the Minimum Support Prize on behalf of State and Central Procurement.

Promoting the use of microbial products in agriculture such as biofertilizers, bioinoculants, bioprotectants, and bioinsecticides is an important component in sustainable agriculture. The Central

Insecticide Board & Registration Committee (CIBRC) has formulated simplified guidelines for the registration of microbial pesticides as compared to agrochemicals.

Also, GoI (Government of India) through sustainable organic farming schemes of Paramparagat Krishi Vikas Yojana (PKVY), Mission Organic Value Chain Development for North Eastern Region (MOVCDNER) and Capital Investment Subsidy Scheme (CISS) aims for sustainable agricultural productivity with environmental progression in tune with nature, promoting organic inputs and pesticide-free agricultural product for increasing the Indian people health condition.

Under, PKVY, the assistance of approx. Rs. 20,000 (INR) per acre for 3 years is provided. Among them, 62% of the amount is given to the farmers directly through direct benefit transfer, for organic inputs (biofertilizers, bio-pesticides, vermicompost, botanical extracts, natural products, etc.) production/ procurement, post-harvest pest and disease management, etc. Under MOVCDNER, the organic farmers are given an assistance of Rs. 10,000 (INR) per acre for 3 years for both on-farm and off-farm organic agricultural inputs, and seeds.

Under, Capital Investment Subsidy Scheme, GoI promotes the production of microbial pesticides and fertilizers by providing 100% assistance to State level Government Agencies up to a maximum INR limit of one crore and sixty lakhs per unit for the production of art liquid/ carrier-based microbial product units of 200 tonnes per year. In the case of, individuals or private industries assistance up to 25% of cost as capital investment is given through NABARD (National Bank for Agriculture and Rural Development).

15.7.4 National Mission for Sustainable Agriculture

The mission was made operational from the year 2014–2015, which aims at making productive agriculture, sustainable, composite farming systems, soil and moisture conservation, soil health management, efficient irrigation management practices and mainstreaming rainfed technologies.

On Farm Water Management (OFWM) was implemented as one of the important of the National Mission for Sustainable Agriculture (NMSA) during 2014–2015 with the objective of enhancing WUE (Water Use Efficiency) by promoting irrigation technological interventions like microirrigation (drip and sprinkler irrigation) technologies. Hereafter, the above objectives are under the scheme of "Per Drop More Crop (PDMC)" component of Pradhan Mantri Krishi Sinchayee Yojana (PMKSY) during 2015–2016.

Rainfed Area Development Programme has been implemented as a component under the National Mission for Sustainable Agriculture since 2014–2015 in the country. The programme focuses on IFS (Integrated Farming System) for enhancing crop productivity and reducing the risks associated with climatic variability. IFS components, including agri crops, are integrated with horticulture, livestock, aquaculture, vermicompost, animal manure, green manuring, and apiculture, to enable farmers profitable farm returns for sustained livelihood.

Soil Health Management (SHM) is one of the components under NMSA. SHM aims at promoting Integrated Nutrient Management (INM) through the judicious use of chemical fertilizers with organic manures, and biofertilizers for improving soil health and its productivity for improving the soil fertility status. The NMSA components are as below:

15.7.4.1 Rainfed Area Development

Rainfed Area Development (RAD) will adopt natural resource conservation along with rainfed farming systems. This scheme will introduce specific farming systems with multiple components in agriculture such as crops, horticulture, livestock, fishery and forestry with agro-based income-generating activities.

15.7.4.2 Sub-Mission on Agroforestry (SMAF)

The scheme was launched in 2016–2017 to encourage tree planting on farmland "Har Medh Par Ped", along with a cropping system. That will give additional income and employment opportunities for farmers, and increase tree cover, which leads to climatic change adaptation and mitigation and also enriching soil organic matter.

15.7.4.3 National Bamboo Mission (NBM)

Bamboo is capable of providing ecological, economic and livelihood security to the people. India is a high bamboo importer. Also, it is especially bamboo-based potential livelihoods and employment for rejuvenating the rural economy and particularly for doubling farmers' income.

15.7.4.4 Soil Health Management

The scheme will aim at promoting specific sustainable crop and soil health management including agrochemical residue management and organic farming practices. This component will also provide support to the reclamation of acidic/alkaline/saline problematic soils. It will be implemented by the State Govt., NCOF (National Centre of Organic Farming), CFQC&TI (Central Fertilizer Quality Control & Training Institute) and SLUSI (Soil and Land Use Survey of India).

15.7.5 Climate Change and Sustainable Agriculture: Monitoring, Modelling and Networking (CCSAMMN)

This scheme will provide the creation of climate change adaptation and mitigation research projects in the domain of climate-smart sustainable management practices and IFS suitable to local agro-climatic conditions.

The scheme will also be supported under SAUs (State Agricultural Universities), ICAR (Indian Council of Agricultural Research) Institutes, CGIAR Institutes, KVKs, Public/Private R&D Organizations etc. for research studies, Documentation, Publication, Training, Seminars/Workshops/Conferences.

APEDA (Agricultural and Processed Food Products Export Development Authority) signed a MoU (Memorandum of Understanding) with the National Research Development Corporation in April 2022 to implement the Agri-Export Policy and strengthen agricultural commodities' export value chains through the dissemination of new ideas to promote climate-resilient sustainable agriculture (OECD, 2023). In May 2022, the Government of India introduced new technical support for the application of drones in agricultural operations such as spraying including chemical and organic inputs, pest and disease monitoring and digitising land records (OECD, 2023).

15.8 SUSTAINABLE AGRICULTURE IN UNITED STATES OF AMERICA (USA)

The development of sustainable agriculture programs in the USA such as the United States Department of Agriculture/Sustainable Agriculture Research Education program (USDA/SARE), the United States Department of Agriculture/National Organics Program (USDA/NOP), the Leonardo Academy/ANSI project and the Keystone Centre initiative provide fertile research grounds for investigation of sustainable agriculture.

USA Food Security Act (1985) policies included provisions to support the development of sustainable agriculture. Low-Input Sustainable Agriculture Program (USDA/LISA) was created by USA Govt (1988). The main aim of the LISA was to promote sustainable agricultural systems with low inputs to meet their people's needs and improve the quality of rural people's life (Madden, 1998).

The focused areas in the SARE (Sustainable Agriculture Research Education Program) are

- Environmentally agricultural ecosystems
- Low input or limited resource farmers
- Economical societal development
- Organic farming systems
- Women in sustainable agriculture (SARE, 2009)

The California Organic Foods Act of 1990 established a legal definition of organic food products – certified organic labels based on third-party certification schemes (Guthman, 2004). National Organic Program (NOP) standards would have an organic effect on the entire USA country.

15.8.1 US Farm Bill, 2023

This is to compete against climate change to support America's working lands, natural resources, and communities:

- To ensure America's agricultural system is equitable, resilient, and prosperous
- Foster an equitable and competitive marketplace for all agricultural producers
- To make safe, nutritious food available to all Americans
- To expand opportunities for economic development and improve the quality of life in rural and tribal communities.
- To attract, inspire and retain an engaged and motivated workforce that's proud to represent USDA (USDA, 2023).

15.9 SUSTAINABLE AGRICULTURAL POLICIES IN UK

UK Path to Sustainable Farming's main goals include (UK, 2021)

- Introducing the Environmental Land Management
- Investing in improving animal health and welfare
- New grants and scheme to the farmers
- To launch a Farming Investment Fund
- Simplifying existing schemes and their processes

The new and improved Sustainable Farming Incentive (SFI) 2023 (SFI, 2023) will start as the government has confirmed, offering farmers additional agricultural activities. Farmers will get paid for taking agricultural activities that improve food production, productivity and resilience, while also maintaining the sustainable ecosystem. The SFI concepts include soil health and moorland, as well as new actions on hedgerows, integrated pest management, nutrient management, farmland wildlife, buffer strips, and low-input grassland (SFI, 2023).

15.10 SUSTAINABLE AGRICULTURE LAWS IN GHANA

Environmental Protection Agency (EPA) Act, 1994 (Act 940) and the Environmental Assessment Regulations, 1999 (LI 1652), are the major environmental rules in Ghana that require farmers to reduce environmental pollution. The Plants and Fertiliser Act, 2010 (Act 803) and the Plant Protection Regulations, 2012 (L.I 2193) make provision for approved sustainable plant health practices for food production and productivity. The Fisheries Act, 2002 (Act 625) and regulations make provisions against unhealthy aquaculture practices in Ghana while the Forestry Commission Act, 1999 (Act 571) provides sustainable forestry practices. International laws such as the Rules, Laws and Regulations Relating to Traceability and Quality Management, Forest Law Enforcement Governance and Trade (FLEGT), the European Union Green Agricultural Policies, and International Fisheries Law maintain a sustainable agribusiness in Ghana.

15.11 SUSTAINABLE AGRICULTURE IN KOREA

Almost all agricultural crops in Korea were grown organically until the 1960s. Ever since, the Korean government's agricultural policy has focused on improving food production, the capitalistic

Saemaul Movement has accelerated the erosion of the traditional sense of rural community, putting rural communities out of place. Various sustainable agriculture movements have arisen in Korea since the 1990s, raising concerns for social integrity as well as the quality of food products and the natural ecosystem. The increasing cultivated area of organic farming over the past two decades in Korea is owing to various local organic cultivation movements. The Back-to-the-Land Migration Movement in particular has offered a holistic vision of rebuilding the rural economy. Agricultural farming community migration has yet to repopulate the rural society but has clearly alleviated or lessened a lack of labour force in rural Korea (Suh, 2018).

15.12 SUMMARY

Social, economic, eco-friendly and sustainability are closely entangled and essential components for sustainable agriculture. It works for elimination of poverty in rural people, adopt the low agri inputs, improve the soil fertility status, reduce the ecological hazards, etc. Only by creating policies in many countries that integrate many factors such as social, environmental, and economic interests can promote more sustainable agricultural farming systems for societal welfare.

15.13 FUTURE PERSPECTIVE

There is a need to create policies worldwide according to their resource availability to take care of the environmental and economic infrastructure of individual countries that promote sustainable agriculture.

STUDY QUESTIONS

1. Sustainable Agricultural Policies in India
2. Impact of plant quarantine regulations
3. Define sustainable agriculture

ACKNOWLEDGEMENT

I acknowledge the knowledge and support gained from the DPPQ&S, Faridabad, and the National Institute of Plant Health Management (NIPHM), Hyderabad. Also, I would like to thank Dr. S. Gnanasambandan, Joint Director (Plant Quarantine), NPPO-India for his advice and guidance.

FURTHER READING

https://pqms.cgg.gov.in/6653b4a7-de0a-489e-a95f-aee2be2063ae

REFERENCES

Agrios, G. N., 2005. *Plant Pathology*, 5th ed. Academic Press, USA. 922 p.

FAO. 2007. *FAO Biosecurity Toolkit.* Food and Agriculture Organization of the United Nations; Rome, Italy: 2007. [(accessed on 15 December 2021)]. Biosecurity Principles and Components. Part. 1; pp. 1–20. Available online: https://www.fao.org/3/a1140e/a1140e.pdf [Google Scholar].

Guide to the IPPC. (2005). *An official Publication on Guide to International Plant Protection Convention.* FAO Knowledge Repository.

Guthman, J. (2004). *Agrarian Dreams: The Paradox of Organic Farming in California.* University of California Press: Berkeley, CA, USA.

Madden, J.P. (1998). The Early Years of LISA, SARE, and ACE Programs: Reflections of the Founding Director—Conception; Western Region SARE, Utah State University: Logan, UT, USA; Available online: http://wsare.usu.eud/about/index/cfm?sub=hist_concept (accessed on 13 November 2006).

OECD, APME (Agricultural Policy Monitoring and Evaluation) (2023). Adapting Agriculture to Climate Change- Annual report, 2023.

PIB (National Policy for farmers). (2007). National Policy for farmers. https://pib.gov.in/newsite/PrintRelease.aspx?relid=117468

Report on Policies and Action Plan for a Secure and Sustainable Agriculture. (2019). Policy submitted to The Principal Scientific Adviser, Government of India. 224p.

Roger, D. M., Manuel, C.-G., and Daniel, A. F. (2009). Plant Biosecurity in the United States: Roles, Responsibilities and Information Needs, *BioScience*, 59(10), 875–885.

SARE. (2009). *Southern Region Sustainable Agriculture Research and Education: One Program, Six Grant Opportunities*; Southern SARE (SSARE): Griffin, GA, USA, 2009;

SFI. (2023). Sustainable Farming initiative. https://www.gov.uk/government/news/sustainable-farming-incentive-2023-expanded-offer-to-roll-out-fromugust#:~:text=SFI%202023%3A%20What's%20on%20offer%3A&text=A%20management%20payment%20of%20%C2%A3,in%20upland%20and%20lowland%20areas (accessed on 14th January, 2024).

Suh, J. (2018). Sustainable Agriculture in the Republic of Korea. E. Lichtfouse (ed.), *Sustainable Agriculture Reviews* 27, 193–211. https://doi.org/10.1007/978-3-319-75190-0_7

UK. (2021). Government unveils path to sustainable farming from 2021. https://www.gov.uk/government/news/government-unveils-path-to-sustainable-farming-from-2021. (accessed on 14th January, 2024).

USDA. (2023). *Budget summary – FY 2023*. 115p.

16 Policies, Regulations and Quality Assurance Requirements for the Agri-Food Industry

G. Balaji
Lactalis India

S. S. Hari Sankar
Food Safety and Standards Authority of India

Subhaprada Nishtala
International Training Centre Food Safety & Applied Nutrition

*Anuja Bejoy and Sanu Jacob**
Food Safety and Standards Authority of India, National Food Laboratory

Global agri-food industries are trying to mitigate the challenges encountered in the supply chain through various environmental agents such as pests, pathogenic diseases, locusts and their carryover residues throughout the supply chain. Fungal phytopathogens among food crops pose a severe threat of toxicosis and migration of fungal toxins into the food supply chain. Therefore, the agri-food supply chain is subject to stringent policies, regulations and quality assurance requirements to mitigate the impact of fungal phytopathogens on crop production and ensure food products' safety. Regulatory bodies worldwide have followed stringent and magnified measures to address the challenge posed by these fungal pathogens in crops, including the usage of pesticides. Countries such as the USA, India, Canada, EU and Japan have formulated measures proper for the farm with various plant health protection measures. Joint FAO/WHO Meeting on Pesticide Residues (JMPR) convenes annually to harmonize the requirement and the risk assessment on pesticide residues, thereby recommending the acceptable level of pesticide residues. Further, the Codex Alimentarius Commission (CAC), managed by FAO and WHO, fixes the MRLs based on the recommendations of JMPR through the assessment of risk assessment data. However, every country has standards and safety measures against food pathogens, contaminants or adulterants. The quality control of foods generally happens through active inspections (surveillance and enforcement) and subsequent laboratory testing as laid down by the act and regulations.

* The authors confirm contribution to the paper as follows: study conception and design by Dr. Sanu Jacob, data collection by Mr. Balaji G.; analysis and interpretation of results by Dr. Harishankar; draft manuscript preparation by Ms. Subhaprada Nishtala and Ms. Anuja Bejoy. All authors have reviewed the results and have approved the final version of the manuscript.

DOI: 10.1201/9781003399704-16

GRAPHICAL ABSTRACT

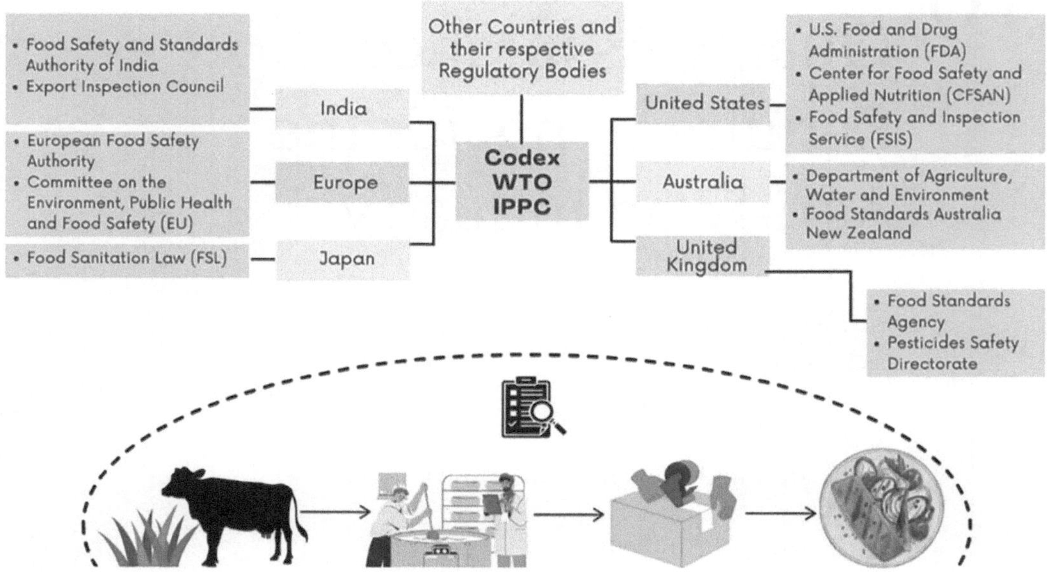

16.1 INTRODUCTION

The agri-food industry is the cornerstone of global sustenance, grappling with historic and emerging challenges by environmental agents such as pests, pathogenic diseases and the highly pervasive fungal phytopathogens. The global community is actively engaged in diverse mitigation approaches to counter the impact of phytopathogens. By combining traditional wisdom with cutting-edge technologies and fostering international cooperation, these strategies aim to fortify agricultural systems against the ever-evolving challenges posed by phytopathogens, ensuring a resilient and sustainable future for global food production. Global approaches include integrated pest management (IPM), growing resistant crop varieties, using biocontrol agents and precision agriculture (Dass et al., 2007; Janardhana et al., 2009; Sreenivasa et al., 2008). In the face of these challenges, there is a pressing need for harmonizing policies, regulations and quality assurance requirements to become unmistakable. These threats necessitate a meticulous approach to developing policies, regulations and quality assurance requirements to mitigate the impact of fungal phytopathogens.

Regulatory bodies worldwide have amplified their efforts to address the challenge of fungal pathogens in crops, employing measures including judicious use of pesticides. Countries like the USA, India, Canada, the EU and Japan have crafted tailored approaches incorporating diverse plant health protection measures. The Joint FAO/WHO Meeting on Pesticide Residues (JMPR) is a pivotal forum that harmonizes requirements and conducts risk assessments on pesticide residues to establish acceptable levels. The Joint FAO/WHO Expert Committee on Food Additives (JECFA) does the safety evaluation of chemicals (additives, residues of veterinary drugs in animal products, contaminants and natural toxins) in food toxicology and other relevant sciences. The Codex Alimentarius Commission (CAC), under the auspices of the Food and Agriculture Organization (FAO) and World Health Organization (WHO), further refines Maximum Residue Limits (MRLs) based on JMPR and JECFA recommendations, facilitating a globally aligned response (Deepa and Sreenivasa, 2019b). This chapter explores the significance of striking a nuanced balance: a balance pivotal in assuring consumer safety and protecting against environmental damage while fostering enduring and sustainable productivity for farmers.

16.2 AGRI-FOOD SUPPLY CHAIN

For a long time, it was believed that almost 25% of agricultural products were contaminated with aflatoxins, as per data published by FAO. However, it was reported to be underestimated, and recent analysis projects contamination levels of 60–80% (Eskola et al., 2019). Fungal producers of Aflatoxins are broadly classified into three groups or sections: Section Flavi, Nidulante and Ochraceorosei. Section Flavi comprises the most notorious species: *A. flavus* and *A. parasiticus* (Jallow et al., 2021). *A. flavus* is non-specific, infecting most crops, unlike *A. parasiticus*, which majorly infects groundnut crops (Kumar et al., 2017). They are majorly soil-borne fungi thriving on other organic matter. However, in some instances, they attack crops and produce mycotoxins as secondary metabolites. Predominant mycotoxins that contaminate common agricultural products are listed in Table 16.1.

Among cereals, maize crops are more prone to aflatoxin contamination. Kamika and Tekere (2016) observed that 32% of contamination at the preharvest level accumulates to 100% at the retail level in samples collected from Congo, thereby confirming the progressive accumulation of Aflatoxin along the value chain. The Aflatoxin levels translated to a 300x higher limit than what is prescribed by Codex Alimentarius. In a similar study conducted in Costa Rica, 38.6% of maize and 27.8% of maize products (Granados-Chinchilla et al., 2017; Deepa and Sreenivasa, 2019a) were contaminated, while in China, the contamination levels in maize ranged from 39.96 to 91.84% (Sun et al., 2017).

Rice is grown in very congenial conditions for fungal proliferation and is prone to aflatoxin contamination (Millan & Martinez, 2003). Therefore, postharvest drying to moisture levels below critical limits (14%) within the first day of harvest and further sanitary storage is crucial in preventing Aflatoxin in the most consumed cereal grain in the World (Masood et al., 2018; Deepa et al., 2016a). Lai et al. (2015) confirmed the prevalence of aflatoxins in rice from China, with 63.5% of samples testing positive, while this was 35% in samples collected from Pakistan (Asghar et al., 2016; Deepa et al., 2016b). The latter study also confirmed that brown rice has a higher mean of total aflatoxins.

Sorghum is a predominant cereal in the African continent, owing to its minimal requirements for growth. Hence, sorghum contamination with Aflatoxin is a serious issue as a malnourished population mostly consumes it. Several studies reaffirm the widespread contamination of Aflatoxin in Sorghum (Sirma et al., 2016; Kange et al., 2015; Ssepuuya et al., 2018). However, the concern is not so alarming in India, as only 35 of the 1606 sorghum samples tested positive beyond the Indian

TABLE 16.1
Common Mycotoxins, Commodity Affected, Sources and Health Effects

Mycotoxins	Commodities	Fungal source	Effects of ingestion
Aflatoxin B1, B2, G1, G2	Corn, peanuts	*Aspergillus flavusAspergillus parasiticusAspergillus nomius*	Aflatoxin B1 was identified as a potent human carcinogen by IARC[1]. Risk of human toxicosis.
Deoxynivalenol Nivalenol Vomitoxin	Wheat, corn and Barley	*Fusarium graminearumFusarium crookwellenseFusarium culmorum*	Human toxicoses in India, China, Japan.
Zearalenone	Corn, wheat	*Fusarium graminearumFusarium culmorumFusarium crookwellense*	Identified by IARC as a potent carcinogen. Affects the reproductive system.
Ochratoxin A	Barley, wheat	*Aspergillus ochraceusPenicillium verrucosum*	Suspected by IARC as a human carcinogen.
Fumonisin B1	Corn	*Fusarium moniliforme*	Suspected by IARC as a human carcinogen.
Patulin	Apple juice	*Penicillium expansum*	Liver and kidney damages.

regulatory limit (Ratnavathi et al., 2016; Deepa et al., 2016c). Another important crop, wheat, is not spared from the deadly fungi (Daou et al., 2020; Turksoy and Kabak 2020; Hathout et al., 2020).

Although all oilseed crops are susceptible to aflatoxin, groundnuts are more prone to it. In West Africa, groundnuts are a major source of aflatoxin poisoning (N'dede et al., 2012). Agroecological conditions where groundnuts are grown primarily favour the rapid multiplication of fungi and subsequent toxin accumulation (Bankole & Adebanjo, 2003). A study spread over 10 years revealed that 42% of groundnut samples in Gambia are contaminated with levels higher than what is prescribed by Codex Alimentarius (Jallow et al., 2021). Like in rice, although the initial attack occurs in the field stage, the establishment is more pronounced during postharvest storage (Waliyar et al., 2015). Aflatoxin is carried over in groundnut oil as well. The concentration is progressively accumulated along the process chain. 15.4 to 49.9 µg/kg Aflatoxin B1 was detected in groundnut oil sampled in China (Mahoney & Molyneux, 2010). Similar carry forward has also been reported from peanut butter (Schwartzbord & Brown, 2015; Njoroge et al., 2016).

Spices, the culinary backbone of many food cultures, are also not spared the wrath of aflatoxins. Contrary to the popular belief that, due to the presence of aromatic compounds, spices may be resistant to aflatoxins and other fungal metabolites, they are reported as sources of human aflatoxin poisoning (Kabak & Dobson, 2017). In pepper samples tested in India, the average amount of Aflatoxin detected was around 320 ppm, a 78.1% incidence (Jeswal & Kumar, 2015). The postharvest processing, handling and storage expose chilly to aflatoxin contamination non-complying levels as per different regulatory standards (Ezekiel et al., 2019; Golge et al., 2013; Klieber, 2001). Other minor spices like cinnamon, cassia, coriander, cloves, saffron, nutmeg, cumin, thyme and curry powder are also not spared under the infestation network of aflatoxins (Kabak and Dobson, 2017).

16.3 AFLATOXIN CARRY-OVER IN FOOD

As discussed earlier, the fungi secreting aflatoxins are predominantly soil-borne. Due to biotic and abiotic stress, they tend to attack weak and susceptible food crops. Once the infection is established, like any fungi, the hyphal ramifications spread across, utilizing the energy reserves within plant tissues and simultaneously secreting secondary metabolites, including Aflatoxin. These remnants are carried forward in storage due to improper drying and unhygienic storage conditions. Ensuring optimum plant health during crop growth stages are critical steps to ensure minimum aflatoxin contamination as it causes drying of crops leading to critical moisture content in crops. However, in the major food-growing regions of the world, hot and humid climates, supplemented with a lack of proper storage and drying facilities, facilitate faster establishment and accumulation of aflatoxin-producing fungi.

Aflatoxins enter the human food chain by directly consuming contaminated cereals, pulses, or products made with contaminated raw materials. Another major carryover source is biomagnification through animal products like milk and meat (M1 and M2 aflatoxins). Animals fed with aflatoxin-contaminated feed tend to accumulate these toxins and enter the human food chain at concentrated levels through milk, meat, eggs and animal fat (Fink-Gremmels & Van Der Merwe, 2019). An alarming 14% of human breast milk samples in Nigeria were contaminated with aflatoxins (Oluwafemi et al., 2014). It takes only 12 hours for milch animals to secrete Aflatoxin contaminated in food through their milk (Applebaum et al., 1982). The carryover is proportional to the relative milk yield of cows, with 1–2% in low milkers and up to 6% in high milkers (Britzi et al., 2013). Overall, studies reflect that the rate of carryover of Aflatoxin B1 varies among different milch animals (Prandini et al., 2009). Heat processing of milk failed to reduce aflatoxin contamination (Daou et al., 2020). Aflatoxin was detected in raw, pasteurized and ultra heat treatment-sterilized milk at 58.8%, 90.9%, and 66%, respectively.

Hussain et al. (2010) documented that broiler chickens fed with contaminated feed accumulated aflatoxin B1 in the liver and muscle tissues. The residual level was directly proportional to the quantity of Aflatoxin in feed and inversely proportional to the bird's age. They also observed that the Aflatoxin continued to persist in body tissues, even after the birds were withdrawn from feeding contaminated feed. Trucksess et al. (1983) identified the presence of Aflatoxin in the eggs, kidneys, muscles, blood and ova of chickens fed with contaminated feed. The muscle component of beef was less likely to be contaminated with Aflatoxin than offal, mainly kidney (Anthony et al., 2012).

16.4 MITIGATION STRATEGIES (GAP/GHP)

Aflatoxins are specific mycotoxins produced as secondary metabolites in food crops predominantly by fungi of the species *Aspergillus flavus* (Shephard, 2009). Although numerous biotic and abiotic factors contribute to the occurrence and proliferation of these toxins in agricultural products, stress-inducing factors like drought, pests and nutrient deficiency can aggravate their occurrence (Bhat & Vasanthi, 2003). Njoroge (2018) of ICRISAT has extensively summarized different levels of aflatoxin mitigation strategies and broadly classified them into pre- and postharvest management strategies. Preharvest techniques should be aimed at reducing plant stress that indirectly makes it susceptible to attack. The use of resistant varieties is a cheap cultural practice. Several crop breeding programs are already in place to develop crop varieties resistant/tolerant to *Aspergillus sp.* (Waliyar et al., 2015). However, cultural and biological control should be combined due to the non-availability of such highly resistant varieties (Nigam et al., 2009). Crop plants can be planted in synchrony with rains so that the preharvest period escapes flooded conditions, favouring Aspergillus infection. It should also be noted that optimum irrigation and critical growth stages shall be provided to prevent drought stress-induced susceptibility. Optimum nutrient management ensures a critical nutrient supply and makes plants healthy enough to resist fungal attacks. Biological control involves the use of non-toxin-producing strains of Aspergillus to compete with native strains for establishing infection in plants; thereby, they either displace or competitively exclude the aflatoxin-producing ones (Bandyopadhyay and Cotty, 2013). Such competitive exclusion can result in up to 80–90% reduction in aflatoxin contamination at harvest and during storage. Corn harvested at 20% moisture and dried immediately to 14% inhibits aflatoxin accumulation (Cassel et al., 2001).

Postharvest mitigation strategies involve predominantly drying the produce to moisture below critical limits. These limits vary with the crop; for rice and maize, they are 14%, and for groundnuts, they are 8%. Damp, moist conditions of storing harvested cereals in deep silage also contributed to Aspergillus contamination. Hence, periodical sterilization of silos is recommended (Cassel et al., 2001). Widespread use of moisture meters at large agricultural markets can screen potential aflatoxin risks (Emmott, 2013). Once the grains are dried, storing different batches separately is always recommended, which further helps avoid cross-contamination risks.

16.5 REGULATORY LANDSCAPE FOR MITIGATING FUNGAL PHYTOPATHOGENS AND ITS ASSOCIATED RAMIFICATIONS

Fungal phytopathogens are often characterized as trivial organisms that, upon magnification on each stage/cycle of the food chain, cause various human and animal health ailments. These Mycotoxin-contaminated food products can lead to mycotoxin exposure in humans and animals, resulting in various health issues. WHO technical report on toxicological evaluation of various mycotoxins found in food outlines the toxicological evaluation of various mycotoxins found in food, including aflatoxins, ochratoxin A, patulin and fumonisins.

Mitigating phytopathogens from the farm stage along the food supply chain has become a challenge, and various governments are making the utmost efforts to mitigate their effects on human and animal health. These measures are elaborated herein:

16.5.1 International Plant Protection Convention (IPPC)

The IPPC is a critical global framework that plays a pivotal role in controlling fungal phytopathogens and safeguarding global agriculture. Fungal phytopathogens, which cause devastating diseases in crops, significantly threaten food security, economic stability and environmental sustainability.

The IPPC, established under the Food and Agriculture Organization (FAO) of the United Nations, serves as the primary international standard-setting body for plant health. Its primary aim is to prevent the spread and introduction of plant pests, including fungal phytopathogens, that can cause severe damage to crop, thereby posing a significant threat to food security, economic stability, and environmental sustainability. It is a critical global framework for controlling fungal phytopathogens and safeguarding global agriculture. IPPC's role in controlling fungal phytopathogens is indispensable for global food security, economic stability and environmental protection. By establishing international standards, facilitating cooperation and promoting the adoption of best practices, the IPPC serves as a linchpin in the global effort to effectively manage fungal phytopathogens and other plant pests. It is a testament to the power of international collaboration in addressing the challenges that fungal diseases pose to agriculture worldwide. For more details, refer to: https://www.ippc.int/en/strategic-objectives/ippc-strategic-framework/.

16.5.2 Codex Alimentarius Commission

The Codex Alimentarius, often called Codex, was established by the FAO and the WHO in 1962. It develops a collection of internationally recognized food standards, guidelines and codes of practice. The CAC plays a central role in the global food system. Codex food safety standards are the "international standard" used as guidance for national food safety regulation. They are often used as a reference to litigate World Trade Organization (WTO) trade disputes. The JECFA is a scientific committee that assesses contaminants, residues, veterinary drugs, and food safety and gives necessary advice to the Codex Alimentarius, FAO, and WHO.

Diverse food standards across countries reflect national regulatory traditions and food culture. However, they are also highly contentious because they imply trade costs and market access difficulties for countries with differing regulatory systems. The CAC attempts to bridge these differences across countries by agreeing on internationally accepted food standards; these include provisions on food hygiene, food additives, pesticide residues, veterinary drug residues, contaminants, labelling and presentation, methods of analysis and sampling, and import and export inspection and certification. In practical terms, this means that standard by standard, food item by food item, for each potential contaminant, residue level, additive or hygiene issue, a standard must be negotiated in an often-time-consuming procedure. According to the Food and Agriculture Organization and World Health Organization (FAO/WHO, 2018), over its 56-year history, Codex has produced more than 4100 maximum levels (MLs) for food additives; over 5200 MRLs for over 300 pesticides; over 600 MRLs covering 63 veterinary drugs; more than 200 commodity standards; and over 100 MLs covering 18 contaminants in food (Figure 16.1).

Codex establishes MRLs for various pesticides and chemical substances used in agriculture, including those aimed at controlling fungal phytopathogens. These MRLs are essential to ensure that the levels of residues in agricultural products resulting from the use of pesticides remain within safe limits for human consumption. By defining these limits, Codex indirectly controls fungal pathogens, as many fungicides and pesticides target these pathogens. Codex develops guidelines and standards for the proper use of pesticides and fungicides in agriculture. These guidelines include information on the appropriate application of these chemicals to mitigate fungal pathogen infections while minimizing risks to human health and the environment.

Codex establishes standards and guidelines for mycotoxins, toxic secondary metabolites produced by some fungal phytopathogens. These standards set limits for mycotoxin contamination in food and feed products. Codex encourages the use of best practices in agriculture and food production. These practices encompass correctly handling and storing agricultural products to prevent fungal contamination. Codex's guidelines on good agricultural practices (GAP) and good manufacturing practices (GMP) indirectly support the control of fungal phytopathogens.

Thus, Codex Alimentarius plays a significant role in controlling fungal phytopathogens by providing internationally recognized standards, guidelines and best practices related to pesticides,

Development of international standards and guidelines

- The IPPC develops international standards and guidelines, including the International Standards for Phytosanitary Measures (ISPMs), which are widely recognized as essential tools for plant protection. These documents provide recommendations and best practices for managing fungal phytopathogens and other plant pests.

Implementation of SPS measures

- The IPPC assists countries in establishing and implementing phytosanitary measures to prevent the introduction and spread of fungal phytopathogens. Measures may include quarantine procedures, inspections, and certification of plant material, all of which are instrumental in reducing the risk of pathogen transmission.

Risk Assesment

- Through the IPPC, countries conduct risk assessments to evaluate the potential threat of fungal phytopathogens and develop risk management strategies. These assessments inform decisions about trade regulations, pest management, and plant health policies

International Cooperation

- IPPC promotes international cooperation and collaboration among countries, facilitating the exchange of information, research, and best practices for controlling fungal phytopathogens. This collaborative approach ensures that nations work together to mitigate the global impact of plant diseases.

FIGURE 16.1 International plant protection convention standards.

fungicides, mycotoxins and agricultural processes. These standards promote food safety and quality while indirectly contributing to the prevention and management of fungal pathogen-related issues in the agri-food industry. For more details, refer to: https://www.fao.org/fao-who-codexalimentarius/home/en/.

16.6 INDIA

In India, the control of fungal mycotoxins in crops is primarily regulated by various government bodies and agencies. The regulations and measures aim to ensure food safety and protect public health. Here are the key regulations and agencies involved.

16.6.1 Food Safety and Standards Authority of India (FSSAI)

FSSAI is an autonomous body established under the Ministry of Health & Family Welfare, Government of India. FSSAI is responsible for protecting and promoting public health through regulating and supervising food safety. The mandate of FSSAI is enlisted as below:

- Framing of regulations and laying down science-based food safety standards.
- Laying down guidelines for accreditation of laboratories for food testing.
- Providing scientific advice and technical support to the Central Government.
- Contributing to the development of international technical standards in food.
- Collect and collate data regarding food consumption, contamination, emerging risks, etc.
- Disseminating information and promoting awareness about food safety and nutrition in India.

FSSAI is the primary regulatory authority responsible for ensuring food safety in India. It has established standards and regulations for mycotoxins in food products, including maximum permissible limits for specific mycotoxins. FSSAI does periodic risk assessments by means of surveillance of various food groups and contaminants such as aflatoxins and heavy metals. FSSAI's Scientific Panel on Contaminants, Antibiotic Residues, and Pesticide Residues recommends the MRLs for heavy metals, aflatoxins, pesticide residues, and antibiotic residues. For more details, please refer to: https://www.fssai.gov.in/index.php?page=about-us.php.

16.6.2 Ministry of Agriculture and Farmers Welfare (MoAFW)

Department of Agriculture and Indian Council of Agricultural Research (ICAR), under the aegis of MoAFW, determines the use of fungicides to control fungal phytopathogens in food crops. ICAR conducts research and provides guidance to prevent and manage mycotoxin contamination in crops. It promotes developing and adopting resistant crop varieties and integrated pest management strategies. For more details, refer to https://ppqs.gov.in/divisions/integrated-pest-management/overview-and-activities.

It is important to note that regulations related to mycotoxins may vary based on the specific crop and product. Hence, growers, processors and exporters need to stay informed about the relevant standards and regulations applicable to their commodities. Regular monitoring and testing for mycotoxins in crops and food products are crucial to ensure compliance and protect public health.

16.7 EUROPE

16.7.1 European Food Safety Authority (EFSA)

The EFSA plays a significant role in ensuring the safety and quality of food and feed in the European Union (EU) (EU Plant Health Regulation (EU) 2016/2031. For more details, refer to: https://www.efsa.europa.eu/en.

16.7.2 EU Plant Health Regulation (EU) 2016/2031

The regulation aims to protect the European Union's plants, plant products and plant health. It establishes a harmonized framework for plant health, promoting preventive measures to reduce the risk of introducing and spreading harmful plant pests. This regulation specifies phytosanitary measures, including plant passporting, inspections and diagnostic testing, to ensure that the movement of plants and plant products within the EU is safe and traceable. The unique nature of this regulation is that it introduces the concept of plant passports, which are official labels providing information about the health status of plants and plant products. These passports facilitate traceability and control of movements within the EU. This regulation outlines the procedures for official controls at various supply chain stages, from importation to production, processing and distribution within the EU.

The EU Plant Health Regulation lays down rules to prevent the introduction and spreading of harmful organisms, including fungal phytopathogens, in the EU. It sets out requirements for plant health controls, inspections and the issuance of plant passports for certain plants and plant products.

16.7.3 Maximum Residual Limits

The EU sets MRLs for pesticide residues, including those from fungicides, in food and feed products. These MRLs ensure that pesticide residues in agricultural products are within safe limits for human consumption. EFSA conducts comprehensive risk assessments on aflatoxins, considering the latest scientific evidence and available data. This assessment evaluates the potential health risks associated with exposure to aflatoxins through consuming contaminated food. For more details, refer to: https://ec.europa.eu/food/plant/pesticides/eu-pesticides-database/start/screen/mrls.

16.7.4 Integrated Pest Management

The EU promotes using IPM practices to manage pests and diseases, including fungal phytopathogens. IPM emphasizes the reduction of pesticide use through the integration of various control methods. For more details, please refer to: https://food.ec.europa.eu/plants/pesticides/sustainable-use-pesticides/integrated-pest-management-ipm_en#principles.

EFSA provides scientific advice and risk assessments related to food safety issues, including aflatoxins. Aflatoxins are mycotoxins produced by certain fungi that can contaminate food crops, posing potential health risks. While EFSA does not implement mitigation measures directly, it plays a critical role in assessing risks and providing recommendations to inform risk management decisions by regulatory authorities. EFSA conducts comprehensive risk assessments on aflatoxins because of fungal phytopathogens, considering the latest scientific evidence and available data. This assessment evaluates the potential health risks associated with exposure to aflatoxins through consuming contaminated food. This would be done by collecting, collating and analysing aflatoxin occurrence data in food and feed across the European Union. EFSA monitors data submitted by member states, scientific literature and other relevant sources.

Based on the risk assessments, EFSA sets scientific criteria, including tolerable daily intake (TDI) values for specific aflatoxins. TDIs represent the amount of a substance that can be ingested daily over a lifetime without appreciable health risks. EFSA communicates its scientific opinions and recommendations to risk managers, including the European Commission and member states. These recommendations include mitigating the risks of aflatoxin contamination in food and feed.

16.8 JAPAN

The primary regulatory authority responsible for ensuring food safety and its associated matters is the Ministry of Health, Labour and Welfare (MHLW), which oversees food safety regulations. In Japan, the regulation and management of fungal phytopathogens in foods fall under various laws and regulations to ensure food safety and protect public health. For more details, refer to: https://www.japaneselawtranslation.go.jp/en/laws/view/3687/en.

16.8.1 Food Sanitation Law (FSL)

The FSL is the central piece of legislation governing food safety in Japan. It is a comprehensive law that addresses various aspects of food safety, hygiene and quality enforced by Japan's MHLW. It addresses the safety of food products, including contaminants such as fungal phytopathogens. The key aspects of food sanitation law are enumerated in (Figure 16.2).

Further, FSL sets maximum allowable levels for various contaminants, including mycotoxins produced by specific fungal pathogens. Mycotoxins, such as aflatoxins and ochratoxins, are closely monitored to ensure they do not exceed established limits in food products. Japan conducts regular monitoring and surveillance programs to assess the presence of mycotoxins and fungal pathogens in food products. These programs are designed to identify and prevent contaminated products from reaching consumers. Japan has specific import and export requirements for food safety and fungal phytopathogens. Imported food products are subject to inspection and compliance with Japanese food safety standards. The Food Safety Commission of Japan (FSCJ) is an independent scientific body assessing food safety risks, including fungal phytopathogens and mycotoxins. Its recommendations influence food safety policies and regulations.

FIGURE 16.2 Schematic representation of food sanitation law.

16.9 UNITED STATES

In the USA, the responsibility of ensuring safe food across the agri-food supply chain is being enforced by the US Food and Drug Administration (FDA) and the US Department of Agriculture (USDA). These agencies ensure food products' safety and quality, monitoring and regulating potential contaminants such as fungal phytopathogens.

16.9.1 Food Safety Modernization Act (FSMA)

The FSMA, signed into law in 2011, is a significant piece of legislation modernizing food safety practices in the USA. It places a strong emphasis on preventive measures. The primary goal of the FSMA is to shift the focus of the US food safety system from responding to contamination and preventing it. Simultaneously, the FSMA covers various aspects of food safety, its relevance to controlling contaminants is particularly notable. The primary goal of the FSMA is to shift the focus of the US food safety system from responding to contamination to preventing it. The act gives the US FDA-enhanced authority to regulate and enforce safety standards throughout the food supply chain. While the FSMA covers various aspects of food safety, its relevance to controlling contaminants is particularly notable. For more details, refer to: https://www.fda.gov/food/guidance-regulation-food-and-dietary-supplements/food-safety-modernization-act-fsma. Key components of the FSMA contaminants are explained in Figure 16.3.

The FSMA has significantly modernized the US approach to food safety, emphasizing prevention, science-based standards and collaboration across the supply chain. Its relevance to controlling contaminants is evident in the emphasis on preventive controls, risk analysis and swift response to potential threats to food safety. Implementing the FSMA continues to profoundly impact how food safety is managed and regulated in the USA.

16.9.2 Maximum Residual Limits

The Environmental Protection Agency (EPA) establishes and enforces pesticide tolerances, including those used to control fungal phytopathogens. MRLs are the maximum allowable pesticide

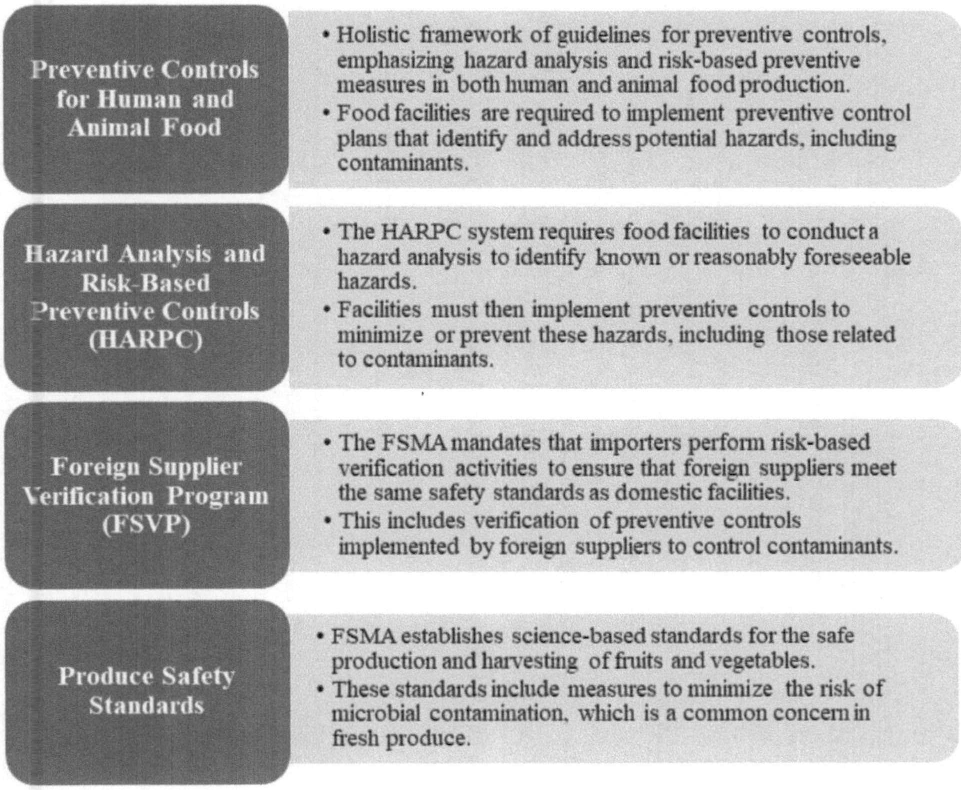

FIGURE 16.3 Key components of the FSMA contaminants.

residues in or on food products. For more details, refer to: https://www.nal.usda.gov/research-tools/food-safety-research-projects; https://fas.usda.gov/maximum-residue-limits-mrl-database; https://www.epa.gov/pesticide-tolerances

16.10 AUSTRALIA

Australia employs stringent biosecurity measures to control and manage the introduction and spread of fungal phytopathogens that can affect crops and food production. The country's biosecurity framework is designed to protect its agriculture and natural environment from the entry of harmful pests and diseases. The Department of Agriculture, Water, and the Environment (DAWE) is Australia's key agency responsible for biosecurity. It has strict regulations governing the importation of plant and plant products. These regulations aim to prevent the entry of fungal pathogens and diseases through imported goods. The import of certain plant products may be subject to specific conditions, inspections and treatments to mitigate the risk of introducing fungal phytopathogens. On arrival, consignments of plant products go through stringent quarantine and inspection procedures at border control points. Inspections are designed to detect and intercept any signs of fungal infections or other plant diseases that may threaten Australian agriculture. Certain imported plant products may be required to undergo treatment, such as fumigation or heat treatment, to eliminate or reduce the risk of fungal contamination. Treatment protocols are based on scientific assessments of the effectiveness of controlling specific pathogens. Australia's biosecurity measures are dynamic and responsive to emerging threats. The goal is to maintain the country's pest and disease-free status, safeguarding its agriculture, environment and food production systems from the impact of harmful fungal phytopathogens. For more details, refer to: https://www.agriculture.gov.au/

TABLE 16.2
List of Food Regulatory Authorities Ensuring a Safe Food Ecosystem

Continent	Country	Food Regulatory Authority	Weblink
Africa	African Union	Africa Food Safety Forum	https://afsf.icaf-ci.com/
	African Union	African Food Safety Network	https://www.africanfoodsafetynetwork.org/
	Gambia	The Food Safety and Quality Authority of The Gambia (FSQA)	https://fsqa.gm/
	Tanzania	Tanzania Bureau of Standards (TBS)	https://www.tbs.go.tz/
	Ethiopia	Ethiopian Food and Drug Authority (EFDA)	https://www.moh.gov.et/site/Ethiopian_Food_and_Drug_Authority
	Morocco	Moroccan National Office of Food Safety (ONSSA)	https://www.onssa.gov.ma/transversal-regulation/food-safety/?lang=en
	South Africa	South African National Regulator for Compulsory Specifications (NRCS)	https://www.sabs.co.za/
	Egypt	Egyptian Food Safety Authority	https://www.nfsa.gov.eg/
	Ghana	Ghana Food and Drugs Authority	https://www.moh.gov.gh/foods-and-drug-authority/
	Nigeria	National Agency for Food and Drug Administration and Control	https://www.nafdac.gov.ng/
		Kaduna State Livestock Regulatory Authority	https://kdsg.gov.ng/2021/03/06/kaduna-state-livestock-regulatory-authority/
America	Argentina	National Food Safety and Quality Service (SENASA) *Servicio Nacional de Sanidad y Calidad Agroalimentaria*	https://www.argentina.gob.ar/senasa
	Canada	Agriculture and Agri-Food Canada (AAFC)	https://agriculture.canada.ca/en
		Canadian Food Inspection Agency (CFIA)	https://inspection.canada.ca/eng/1297964599443/1297965645317
		Health Canada	https://www.canada.ca/en/health-canada.html
		Health Products and Food Branch	https://www.canada.ca/en/health-canada/corporate/about-health-canada/branches-agencies/health-products-food-branch.html
	United States of America	U.S. Food and Drug Administration (FDA)	https://www.fda.gov/
		Center for Food Safety and Applied Nutrition (CFSAN)	https://www.fda.gov/food
		United States Department of Agriculture (USDA)	https://www.usda.gov/
		Food Safety and Inspection Service (FSIS)	https://www.fsis.usda.gov/
Asia	Saudi Arabia	Saudi Food and Drug Authority	https://www.sfda.gov.sa/en
	Bangladesh	Bangladesh Food Safety Authority (BFSA)	https://www.bfsa.gov.bd/
	Myanmar (Burma)	Food and Drug Administration	https://www.fda.gov.mm/
	China	General Administration of Quality Supervision, Inspection and Quarantine	https://web.archive.org/web/20090209025401/https://english.aqsiq.gov.cn/
		National Medical Products Administration (NMPA) – founded under State Food and Drug Administration	https://www.nmpa.gov.cn/

	Hong Kong SAR	Centre for Food Safety	https://www.cfs.gov.hk/english/index.html
	India	Food Safety and Standards Authority of India (FSSAI)	https://www.fssai.gov.in/
		Export Inspection Council (EIC)	https://www.eicindia.gov.in/WebApp1/
	Indonesia	National Agency of Drug and Food Control of Indonesia	https://www.pom.go.id/
	Nepal	Department of Food Technology and Quality Control	https://dftqc.gov.np/?page_id=55
	Philippines	Food and Drug Administration	https://www.fda.gov.ph/
	South Korea	Ministry of Food and Drug Safety (MFDS)	https://www.mfds.go.kr/eng/index.do
		Ministry for Health, Welfare and Family Affairs	https://www.mohw.go.kr/
	Taiwan	Ministry of Health and Welfare (MHW)	https://www.mohw.gov.tw/mp-2.html
		Food and Drug Administration (TFDA)	https://www.fda.gov.tw/eng/
	Malaysia	Ministry of Health (MOH)	https://www.moh.gov.my/
	Kazakhstan	Islamic Organization for Food Security	https://www.iofs.org.kz/
Europe	European Union	Food Safety Promotion Board	https://www.gov.ie/en/organisation/16211907-food-safety-promotion-board-safefood/
		Committee on the Environment, Public Health and Food Safety (EU)	https://www.europarl.europa.eu/committees/en/envi/home/highlights
		European Food Safety Authority	https://www.efsa.europa.eu/en
		Safe Foods	https://www.safefoods.nl/en/safefoods.htm
	Belgium	Federal Agency for the Safety of the Food Chain	https://www.fasfc.be/
	France	French Agency for Food, Environmental and Occupational Health & Safety (ANSES)	https://www.anses.fr/en
	Greece	Hellenic Food Authority	https://www.efet.gr/index.php/en/
	Germany	Federal Ministry of Food, Agriculture	https://www.bmel.de/EN/Home/home_node.html
		Federal Institute for Risk Assessment (BfR) *Bundesinstitut für Risikobewertung*	https://www.bfr.bund.de/en/home.html
		Federal Office of Consumer Protection and Food Safety *Bundesamt für Verbraucherschutz und Lebensmittelsicherheit*	https://www.bvl.bund.de/EN/Home/home_node.html
		Bundesanstalt für Landwirtschaft und Ernährung	https://www.ble.de/DE/Startseite/startseite_node.html
	Netherlands	Ministry of Economic Affairs and Climate Policy	https://www.government.nl/ministries/ministry-of-economic-affairs-and-climate-policy
		nieuwe Voedsel en Waren Autoriteit (nVWA)	https://www.nvwa.nl/
		Rijks Kwaliteitsinstituut voor Land- en Tuinbouwproducten (RiKILT)	https://www.wur.nl/en/research-results/research-institutes/food-safety-research.htm
		National Institute for Public Health and the Environment (Rijksinstituut voor Volksgezondheid en Milieu, RiVM)	https://www.rivm.nl/en

(*Continued*)

TABLE 16.2
(Continued)

Continent	Country	Food Regulatory Authority	Weblink
		The Netherlands Nutrition Centre (Stichting Voedingscentrum Nederland)	https://www.voedingscentrum.nl/nl/service/english.aspx
	Norway	Ministry of Agriculture and Food (Landbruks-og matdepartementet)	https://www.regjeringen.no/no/dep/lmd/id627/
		Norwegian Food Safety Authority Mattilsynet	https://www.mattilsynet.no/
	Portugal	Ministry of Economy	https://www.portugal.gov.pt/en/gc21/ministries/economy/about
		Economic and Food Safety Authority (Autoridade de Segurança Alimentar e Económica, ASAE)	https://www.asae.gov.pt/
	Spain	Ministry of Health, Government of Spain *Ministerio de Sanidad, Consumo y Bienestar Social, Gobierno de España*	https://www.sanidad.gob.es/
		Spanish Agency for Food Safety and Nutrition*Agencia Española de Seguridad Alimentaria y Nutrición* (AESAN)	https://www.aesan.gob.es/AECOSAN/web/home/aecosan_inicio.htm
	United Kingdom	Department for Environment, Food and Rural Affairs	https://www.gov.uk/government/organisations/department-for-environment-food-rural-affairs
		Animal Health	https://www.gov.uk/government/organisations/animal-health
		Pesticides Safety Directorate	https://webarchive.nationalarchives.gov.uk/ukgwa/20060716031123/https://www.pesticides.gov.uk/
		Veterinary Medicines Directorate	https://www.gov.uk/government/organisations/veterinary-medicines-directorate
		Food Standards Agency	https://www.food.gov.uk/
		British Retail Consortium (BRC) Global Standard (Baines, R. (2010) Quality and Safety Standards in Food Supply Chains, Royal Agricultural University, UK)	https://www.brcgs.com/our-standards/food-safety/
		Advisory committee on the microbiological safety of food	https://acmsf.food.gov.uk/
	Scotland	Food Standards Scotland	https://www.foodstandards.gov.scot/
Oceania	Multinational	Food Standards Australia New Zealand (FSANZ)	https://www.foodstandards.gov.au/Pages/default.aspx
	Australia	Department of Agriculture, Water and the Environment	https://www.agriculture.gov.au/
		Government of New South Wales	https://www.nsw.gov.au/
		Minister for Primary Industries (NSW)	https://www.dpi.nsw.gov.au/
		New South Wales Food Authority	https://www.foodauthority.nsw.gov.au/
	New Zealand	Ministry for Food Safety	https://www.foodsafety.govt.nz/index.htm%20
		New Zealand Food Safety Authority (NZFSA)	https://www.mpi.govt.nz/food-business/

(The links were last accessed on 03 December 2023)

TABLE 16.3
Limit of the Fungal Mycotoxin Contaminant in Foods by Various Regulatory Bodies

Regulatory Limits for Aflatoxins as per European Commission (EC) No 1881/2006 (Referenced at # 9)

Foodstuffs	**Maximum Levels (μg/kg)**	
Aflatoxins	B_1	Sum of B_1, B_2, G_1 and G_2
Groundnuts to be subjected to sorting, or other physical treatment, before human consumption or use as an ingredient in foodstuff	8 [(1)]	15 [(1)]
Nuts to be subjected to sorting, or other physical treatment, before human consumption or use as an ingredient in foodstuffs	5 [(1)]	10 [(1)]
Groundnuts and nuts and processed products, thereof, intended for direct human consumption or use as an ingredient in foodstuffs	2 [(1)]	4 [(1)]
Dried fruit to be subjected to sorting, or other physical treatment, before human consumption or use as an ingredient in foodstuffs	5	10
Dried fruit and processed products, thereof, intended for direct human consumption or use as an ingredient in foodstuffs	2	4
All cereals and all products derived from cereals, including processed cereal products	2	4
Maize to be subjected to sorting or other physical treatment before human consumption or use as an ingredient in foodstuffs	5	10
Raw milk, heat-treated milk and milk for the manufacture of milk-based products	-	-
Following species of spices: Capsicum spp. (dried fruits thereof, whole or ground, including chillies, chilli powder, cayenne and paprika); Piper spp. (fruits thereof, including white and black pepper); Myristica fragrans (nutmeg); Zingiber officinale (ginger); Curcuma longa (turmeric)	5	10
Processed cereal-based foods and baby foods for infants and young children	0.1	-
Infant formulae and follow-on formulae, including infant milk and follow-on milk	-	-
Dietary foods for special medical purposes intended specifically for infants	0.1	-

[(1)] The maximum levels refer to the edible part of groundnuts and nuts. If groundnuts and nuts 'in shell' are analysed, it is assumed when calculating the aflatoxin content that all the contamination is on the edible part.

Regulatory Limits for Aflatoxins as per FSSAI (Food Safety and Standards (Contaminants, Toxins and Residues) Regulations, 2011) (Referenced at #15)

Name of Contaminants	**Article of the Food**	**Maximum Limit (μg/kg)**
Total Aflatoxins	Cereal and cereal products	15
	Dried figs	10
	Areca nut or betelnut	15
	Nuts: Nuts for further processing Ready to eat	1515
	Oilseeds or oil:Oilseeds for further processingReady to eat	1515
	Pulses	15
	Spices/Spice mix	30
	Food product containing any of the above-mentioned food articles	20
Aflatoxin B_1	Areca nut or betelnut	10
	Cereal and cereal products	10

(*Continued*)

TABLE 16.3
(Continued)

Regulatory Limits for Aflatoxins as per FSSAI (Food Safety and Standards (Contaminants, Toxins and Residues) Regulations, 2011) (Referenced at #15)

Name of Contaminants	Article of the Food	Maximum Limit (µg/kg)
	Dried figs	10
	Nuts: Nuts for further processing, ready to eat	10
	Oilseeds or oil: Oilseeds for further processing, ready to eat	10
	Pulses	10
	Spices/Spice mix	15
	Food product containing any of the above-mentioned food articles	10
Aflatoxin M_1	Milk (liquid)	0.5
	Skimmed milk powder	6
	Whole milk powder	4
Ochratoxin A	Wheat, rye, barley	20
Patulin	Apple juice	50
	Apple juice used as an ingredient in other beverages	50
Deoxynivalenol	Wheat	1000

Regulatory Limits for Aflatoxins as per *Codex Alimentarius*, International Food Standards, Food and Agriculture Organization of United Nations (Referenced at # 16)

Aflatoxins, Total

Commodity/ Product Name	Maximum Level (µg/kg)	Portion of the Commodity/ Product to which the Maximum Level applies	Notes/Remarks (ML refers to Maximum Level)
Almonds	10	Whole commodity after removal of shell	The ML applies to almonds "ready-to-eat" (**).
Almonds	15	Whole commodity after removal of shell	The ML applies to almonds intended for further processing (*).
Brazil nuts	10	Whole commodity	The ML applies to shelled Brazil nuts ready-to-eat (**).
Brazil nuts	15	Whole commodity	The ML applies to shelled Brazil nuts intended for further processing (*).
Hazelnuts	10	Whole commodity after removal of shell	The ML applies to hazelnuts, also known as filberts, "ready to eat" (**).
Hazelnuts	15	Whole commodity after removal of shell	The ML applies to hazelnuts, also known as filberts, intended for further processing (*).
Peanuts	15	Unless specified, seed or kernels, after removal of shell or husk	The ML applies for peanuts, also known as groundnuts, intended for further processing (*).
Pistachios	10	Whole commodity after removal of shell	The ML applies to pistachios "ready to eat" (**).
Pistachios	15	Whole commodity after removal of shell	The ML applies to pistachios intended for further processing (*).
Dried figs	10	Whole commodity	The ML applies to dried figs "ready-to-eat" (**).

(*Continued*)

TABLE 16.3 (Continued)

Regulatory Limits for Aflatoxins as per *Codex Alimentarius*, International Food Standards, Food and Agriculture Organization of United Nations (Referenced at # 16)

Commodity/ Product Name	Maximum Level (μg/kg)	Portion of the Commodity/ Product to which the Maximum Level applies	Notes/Remarks (ML refers to Maximum Level)
		Aflatoxins, Total	
Maize grain, destined for further processing	15	Whole commodity	"Destined for further processing" means intended to undergo an additional processing/ treatment that has proven to reduce levels of AFs before being used as an ingredient in foodstuffs, otherwise processed or offered for human consumption. Codex members may define the processes that have been shown to reduce levels. The ML does not apply to maize destined to animal feed or wet milling.
Flour, meal, semolina and flakes derived from maize	10	Whole commodity	
Husked rice	20	Whole commodity	
Polished rice	5	Whole commodity	
Sorghum grain, destined for further processing	10	Whole commodity	"Destined for further processing" means intended to undergo an additional processing/treatment that has proven to reduce levels of AFs before being used as an ingredient in foodstuffs, otherwise processed or offered for human consumption. Codex members may define the processes that have been shown to reduce levels.
Cereal-based foods for infants and young children	5	Whole commodity as sold; not reconstituted or otherwise prepared for consumption.	Relevant Codex commodity standard is CXS 74-1981. The Maximum Level applies to all cereal-based foods intended for infants (up to 12 months) and young children (12 to 36 months).
Cereal-based foods for infants and young children	10	Whole commodity as sold; not reconstituted or otherwise prepared for consumption.	Relevant Codex commodity standard is CXS 74-1981. The ML applies to cereal-based foods destined for food aid programs intended for infants (6 to 12 months) and young children (12 to 36 months).
		Aflatoxin M 1	
Milk	0.5	Whole commodity	Milk is the normal mammary secretion of milking animals obtained from one or more milkings without either addition to it or extraction from it, intended for consumption as liquid milk or for further processing. A concentration factor applies to partially or wholly dehydrated milks.

Regulatory Limits for Aflatoxins as per US Food and Drug Administration (Referenced at # 19)

Commodity	Action Level (ppb)
Animal Feeds	
a) Corn and peanut products intended for finishing (i.e., feedlot) beef cattle	300
b) Cottonseed meal intended for beef, cattle, swine or poultry (regardless of age or breeding status	300

(*Continued*)

TABLE 16.3
(Continued)

c) Corn and peanut products intended for finishing swine of 100 pounds or greater	200
d) Corn and peanut products intended for breeding beef cattle, breeding swine or mature poultry	100
e) Corn, peanut products and other animal feeds and feed ingredients but excluding cottonseed meal, intended for immature animals	20
f) Corn, peanut products, cottonseed meal and other animal feed ingredients intended for dairy animals, for animal species or uses not specified above, or when the intended use is not known	20
Brazil nuts	20
Foods	20
Milk	0.5(aflatoxin M_1)
Peanuts and peanut products	20
Pistachio nuts	20

(*) "destined for further processing" means intended to undergo an additional processing/treatment that has proven to reduce levels of aflatoxins before being used as an ingredient in foodstuffs, otherwise processed or offered for human consumption. Processes that have proven to reduce levels of aflatoxins are shelling, blanching followed by colour sorting, and sorting by specific gravity and colour (damage). There is some evidence that roasting reduces aflatoxins in pistachios, but for other nuts the evidence is still to be supplied.

(**) "ready-to-eat" means "not intended to undergo an additional processing/treatment that has proven to reduce levels of aflatoxins before being used as ingredient in foodstuffs, otherwise processed or offered for human consumption."

Certain other regulatory bodies and organizations that contribute towards food safety are listed in Table 16.2. Various regulatory bodies worldwide have set the permissible limit of mycotoxin contaminants in foods, including Codex, and they have been listed in Table 16.3 for reference.

16.11 SUMMARY

Fungal phytopathogen infection in crops has become a serious concern over recent years due to its potential implications in the carryover of its metabolites, such as aflatoxins, across the food chain. This carryover in the food chain can cause serious food safety concerns. These fungal phytopathogens directly and indirectly damage plants and crops, affecting humans. As the direct damage poses a serious threat to food security as it damages crops, many countries are implementing stringent measures to control the fungal phytopathogen by adopting good farming practices. Thus, this chapter provides a nutshell of the role of some countries and global regulatory bodies in mitigating the threat of fungal phytopathogens.

16.11.1 Future Perspectives

Within the agri-food industry, there is a growing emphasis on adopting best practices, such as GAP, to control fungal phytopathogens. GAP includes using resistant crop varieties, biocontrol agents, and IPM strategies. The management in the agri-food supply chain has been dynamic, with regulations and practices evolving to address the emerging challenges.

STUDY QUESTIONS

1 Explain Food Regulatory Authorities in Africa.
2 Mention FSSAI regulatory limits for Aflatoxin.

3 List regulatory authorities in Europe.
4 Briefly describe HARPC working line.

FURTHER READING

https://www.ippc.int/en/strategic-objectives/ippc-strategic-framework/
https://www.fao.org/fao-who-codexalimentarius/home/en/
https://ppqs.gov.in/divisions/integrated-pest-management/overview-and-activities.
https://www.efsa.europa.eu/en.
https://ec.europa.eu/food/plant/pesticides/eu-pesticides-database/start/screen/mrls
https://food.ec.europa.eu/plants/pesticides/sustainable-use-pesticides/integrated-pest-management-ipm_en#principles.
https://www.japaneselawtranslation.go.jp/en/laws/view/3687/en.
https://www.fda.gov/food/guidance-regulation-food-and-dietary-supplements/food-safety-modernization-act-fsma.
https://www.nal.usda.gov/research-tools/food-safety-research-projects
https://fas.usda.gov/maximum-residue-limits-mrl-database
https://www.epa.gov/pesticide-tolerances
https://www.agriculture.gov.au/

REFERENCES

Anthony, M.H., Francis, D.M., Berka, N.P., Ayinla, G.T., Haruna, O.G., 2012. Aflatoxin contamination in foods and feeds: A special focus on Africa. *Trends Vital Food Control Engineering*, pp.188–234.

Applebaum, R. S., Brackett, R. E., Wiseman, D. W., Marth, E. H., 1982. Responses of dairy cows to dietary aflatoxin: Feed intake and yield, toxin content, and quality of milk of cows treated with pure and impure aflatoxin. *Journal of Dairy Science*, *65*(8), 1503–1508. https://doi.org/10.3168/jds.S0022-0302(82)82374-6

Asghar M. A., Iqbal, J., Ahmed, A., Shamsuddin, Z., Khan, M. A., 2016. Incidence of aflatoxins in export quality basmati rice collected from different areas of Pakistan. *Science Technology and Development (Pakistan)*, *32*(2), 110–119.

Bandyopadhyay, R., Cotty, P. J., 2013. Biological controls for aflatoxin reduction. In: *Aflatoxins: Finding solutions for improved food safety*. D. Grace and L. Unnevehr, eds. International Food Policy Research Institute, Washington, DC.

Bankole, S., Adebanjo, A., 2003. Mycotoxins in food in West Africa: Current situation and possibilities of controlling it. *African Journal of Biotechnology*, 2(9), 254–263. https://doi.org/10.1016/j.jspr.2017.02.001

Bhat, R. V., Vasanthi, S., 2003. *Mycotoxin food safety risk in developing countries*. https://ideas.repec.org/p/fpr/2020br/1003

Britzi, M., Friedman, S., Miron, J., Solomon, R., Cuneah, O., Shimshoni, J. A., Soback, S., Ashkenazi, R., Armer, S., Shlosberg, A., 2013. Carry-over of aflatoxin B1 to aflatoxin M1 in high yielding Israeli cows in mid-and late-lactation. *Toxins*, *5*(1), 173–183. https://doi.org/10.3390/toxins5010173

Cassel E.K., Campbell B., Draper M., Epperson, B., 2001. Aflatoxins hazards in grain/Aflatoxicosis and livestock. South Dakota State University (SDSU).

Daou, R., Afif, C., Joubrane, K., Khabbaz, L. R., Maroun, R., Ismail, A., El Khoury, A., 2020. Occurrence of aflatoxin M1 in raw, pasteurized, UHT cows' milk, and dairy products in Lebanon. *Food Control*, *111*, 107055. https://doi.org/10.1016/j.foodcont.2019.107055

Dass, R.S., Sreenivasa, M.Y., Janardhana, G.R. 2007. High incidence of *Fusarium verticillioides* in animal and poultry feed mixtures produced in Karnataka, India. *Plant Pathology Journal*, *6*(2), 174–178.

Deepa, N., Charith Raj, A. P., Sreenivasa, M.Y. 2016b. Nested PCR method for the early detection of fumonisin producing *Fusarium verticillioides* in pure cultures, cereal samples and plant parts. *Food Biotechnology*, *30*(1), 18–29.

Deepa N., Charith Raj, A.P., Sreenivasa, M.Y. 2016c. Multiplex PCR method for the early detection of fumonisin producing *Fusarium verticillioides*. *Food Bioscience*, *13*, 84–88.

Deepa N., Nagaraja, H., Sreenivasa, M.Y. 2016a. Prevalence of fumonisin producing Fusarium verticillioides associated with cereals grown in Karnataka (India). *Food Science Human Wellness*, *5*, 156–162. https://doi.org/10.1016/j.fshw.2016.07.001.

Deepa, N and Sreenivasa M. Y. 2019a. Molecular methods and key genes targeted for the detection of fumonisin producing *Fusarium verticillioides* - An updated review. *Food Bioscience, 32*, 100473.

Deepa, N, Sreenivasa, M.Y. 2019b. Biocontrol Strategies for Effective Management of Phytopathogenic Fungi Associated with Cereals. In: Singh JS, Singh DP. (eds) *New and Future Developments in Microbial Biotechnology and Bioengineering*, Elsevier, pp, 177–189, ISBN 9780444641915.

Emmott, A., 2013. *Market-led aflatoxin interventions: Smallholder groundnut value chains in Malawi* (No. 20 (8)). International Food Policy Research Institute (IFPRI).

Eskola, M., Kos, G., Elliott, C. T., Hajšlova, J., Mayar, S., Krska, R., 2019. Worldwide contamination of food-crops with mycotoxins Validity of the widely cited 'FAO estimate' of 25%. *Critical Review in Food Science and Nutrition, 60*, 1–17. https://doi.org/10.1080/10408398.2019.1658570

Ezekiel, C. N., Ortega-Beltran, A., Oyedeji, E. O., Atehnkeng, J., Kössler, P., Tairu, F., Hoeschle-Zeledon, I., Karlovsky, P., Cotty, P., & Bandyopadhyay, R., 2019. Aflatoxin in chili peppers in Nigeria: Extent of contamination and control using atoxigenic *Aspergillus flavus* genotypes as biocontrol agents. *Toxins, 11*(7), 429. https://doi.org/10.3390/toxins11070429

FAO 2018. FAO, Ethiopia: report on feed inventory and feed balance. Food and Agriculture organization of the United Nations, Rome, Italy.

Fink-Gremmels, J., van der Merwe, D., 2019. Mycotoxins in the food chain: Contamination of foods of animal origin. *Chemical Hazards in Foods of Animal Origin*, 1190–1198. https://doi.org/10.3920/978-90-8686-877-3_10

Food Safety and Standards (Contaminants, Toxins and Residues) Regulations. 2011 https://www.fssai.gov.in/upload/uploadfiles/files/Compendium_Contaminants_Regulations_28_01_2022.pdf

Golge, O., Hepsag, F., Kabak, B., 2013. Incidence and level of aflatoxin contamination in chilli commercialised in Turkey. *Food Control, 33*(2), 514–520. https://doi.org/10.1016/j.foodcont.2013.03.048

Granados-Chinchilla, F. Alvarado, A. M., and Zamora-Sanabria, R., 2017. A focus on aflatoxins in feedstuffs: Levels of contamination, prevalence, control strategies, and impacts on animal health. *Aflatoxin-Control, Analysis, Detection and Health Risks*, 116–152. https://doi.org/10.1136/bmjgast-1630

Hathout, A. S., Abel-Fattah, S. M., Abou-Sree, Y. H., Fouzy, A. S., 2020. Incidence and exposure assessment of aflatoxins and ochratoxin A in Egyptian wheat. *Toxicology Reports, 7*, 867–873. https://doi.org/10.1016/j.toxrep.2020.07.003

Hussain, Z., Khan, M.Z., Khan, A., Javed, I., Saleemi, M.K., Mahmood, S. Asi, M.R., 2010. Residues of aflatoxin B1 in broiler meat: Effect of age and dietary aflatoxin B1 levels. *Food and Chemical Toxicology, 48*(12), pp.3304–3307.

Jallow A, Xie H, Tang X, Qi Z, Li, P., 2021. Worldwide aflatoxin contamination of agricultural products and foods: From occurrence to control. *Comprehensive Reviews in Food Science and Food Safety 20*, 2332–2381. https://doi.org/10.1111/1541-4337.12734

Janardhana, G. R., Dass, R.S., Sreenivasa, M.Y. 2009. Fumonisins – A new class of Fusarial toxins. In *Frontiers in Fungal Ecology, Diversity and Metabolites*. Ed: K. A. Sridhar et al. I. K. International Publishing House Pvt. Ltd. New Delhi.

Jeswal, P., Kumar, D., 2015. Natural occurrence of toxigenic mycoflora and ochratoxin A & aflatoxins in commonly used spices from Bihar state (India). *Journal of Environmental Science, Toxicology and Food Technology, 9*(2), 50–55. https://doi.org/10.1155/2015/242486

Kabak, B., Dobson, A. D., 2017. Mycotoxins in spices and herbs: An update. *Critical Reviews in Food Science and Nutrition, 57*(1), 18–34. https://doi.org/10.1080/10408398.2013.772891

Kamika, I., Tekere, M., 2016. Occurrence of aflatoxin contamination in maize throughout the supply chain in the Democratic Republic of Congo. *Food Control, 69*, 292–296. https://doi.org/10.1016/j.foodcont.2016.05.014

Kange, A. M., Cheruiyot, E. K., Ogendo, J. O., Arama, P. F. 2015. Effect of sorghum (Sorghum bicolor L. Moench) grain conditions on occurrence of mycotoxin-producing fungi. *Agriculture & Food Security*, 4(1), 15. https://doi.org/10.1186/s40066-015-0034-4

Klieber, A., 2001. Aflatoxin contamination and its management in chilli and paprika products in Australia. *Food Australia, 53*(3), 90–92.

Kumar, P., Mahato, D. K., Kamle, M., Mohanta, T. K., Kang, S. G., 2017. Aflatoxins: A global concern for food safety, human health and their management. *Frontiers in Microbiology, 7*, 2170. https//doi.org/10.3389/fmicb.2016.02170

Lai, X., Liu, R., Ruan, C., Zhang, H., Liu, C., 2015. Occurrence of aflatoxins and ochratoxin A in rice samples from six provinces in China. *Food Control, 50*, 401–404.

Mahoney, N., Molyneux, R. J., 2010. Rapid analytical method for the determination of aflatoxins in plant-derived dietary supplement and cosmetic oils. *Journal of Agricultural and Food Chemistry, 58*(7), 4065–4070

Masood, S., Un Nisa, A., Hina, S., Ahmad, I. 2018. Effects of drying pattern on aflatoxin in stored paddy rice. *Bangladesh Journal of Scientific and Industrial Research*, *53*(4), 253–258. https://doi.org/10.3329/bjsir.v53i4.39188

Millan, T. F., Martinez, Y.A., 2003. Efficacy and stability of ammoniation process as aflatoxin B1 decontamination technology in rice. *Archivos Latinoamericanos de Nutricion*, *53*(3), 287–292.

N'dede, C. B., Jolly, C., Vodouhe, S. D., Jolly, P., 2012. Economic risks of aflatoxin contamination in marketing of peanut in Benin. *Economics Research International*, *2012*(2012), 1–12. https://doi.org/10.1155/2012/230638

Nigam, S. N., Waliyar, F., Aruna, R., Reddy, S. V., Kumar, P. L., Craufurd, P. Q., Diallo, A. T., Ntare, B. R., Upadhyaya, H. D., 2009. Breeding for resistance to aflatoxin contamination at ICRISAT. *Peanut Science* *35*, 42–49.

Njoroge, S. M., 2018. A critical review of Aflatoxin contamination of peanuts in Malawi and Zambia: The past, present and Future. *Plant Disease*. *102*(12), 2394–2406.

Njoroge, S. M., Matumba, L., Kanenga, K., Siambi, M., Waliyar, F., Maruwo, J., Monyo, E. S. (2016). A case for regular aflatoxin monitoring in peanut butter in sub-Saharan Africa: lessons from a 3-year survey in Zambia. *Journal of Food Protection*, *79*(5), 795–800.

Oluwafemi, F., Badmos, A.O., Kareem, S.O., Ademuyiwa, O., Kolapo, A.L., 2014. Survey of aflatoxin M 1 in cows' milk from free-grazing cows in Abeokuta, Nigeria. *Mycotoxin Research*, *30*, pp.207–211.

Prandini, A., Tansini, G., Sigolo, S., Filippi, L., Laporta, M., Piva, G., 2009. On the occurrence of aflatoxin M_1 in milk and dairy products. *Food and Chemical Toxicology* *47*, 984–991

Ratnavathi, C., Komala, V., Chavan, U., 2016. Mycotoxin contamination in Sorghum. In *Sorghum Biochemistry* (An Industrial Perspective, 1st, 107–180). Elsevier.

Schwartzbord, J. R., Brown, D. L., 2015. Aflatoxin contamination in Haitian peanut products and maize and the safety of oil processed from contaminated peanuts. *Food Control*, *56*, 114–118.

Shephard, G. S., 2009. Aflatoxin analysis at the beginning of the twenty-first century. *Analytical and Bioanalytical Chemistry*, *395*(5), 1215–1224.

Sirma, A., Senerwa, D., Grace, D., Makita, K., Mtimet, N., Kang'ethe, E. K., Lindahl, J. F., 2016. Aflatoxin B1 occurrence in millet, sorghum and maize from four agro-ecological zones in Kenya. *African Journal of Food, Agriculture, Nutrition and Development*, *16*(3), 10991–11003.

Sreenivasa, M.Y., Dass, R.S., Charith Raj, A.P., Janardhana, G.R. 2008. PCR-based detection of genus *Fusarium* and Fumonisin-producing isolates from freshly harvested Sorghum grains grown in Karnataka, India. *Journal of Food Safety*. *28*, 236–237.

Ssepuuya, G., Van Poucke, C., Ediage, E. N., Mulholland, C., Tritscher, A., Verger, P., Kenny, M., Bessy, C., De Saeger, S., 2018. Mycotoxin contamination of sorghum and its contribution to human dietary exposure in four sub-Saharan countries. *Food Additives & Contaminants: Part A*, *35*(7), 1384–1393. https://doi.org/10.1080/19440049.2018.1461253

Sun, X. D., Su, P., Shan, H., 2017. Mycotoxin contamination of maize in China. *Comprehensive Reviews in Food Science and Food Safety*, *16*(5), 835–849. https://doi.org/10.1111/1541-2913

Trucksess, M.W., Stoloff, L., Young, K., Wyatt, R.D., Miller, B.L., 1983. Aflatoxicol and aflatoxins B1 and M1 in eggs and tissues of laying hens consuming aflatoxin-contaminated feed. *Poultry Science*, *62*(11), 2176–2182.

Turksoy, S., Kabak, B., 2020. Determination of aflatoxins and ochratoxin A in wheat from different regions of Turkey by HPLC with fluorescence detection. *Acta Alimentaria*, *49*(1), 118–124.

Waliyar, F., Osiru, M., Ntare, B., Kumar, K. V. K., Sudini, H., Traore, A., Diarra, B., 2015. Post-harvest management of aflatoxin contamination in groundnut. *World Mycotoxin Journal*, *8*(2), 245–252.

17 Further Directions in Fungal Phytopathogen Diagnostics

K. R. Sridhar
Mangalore University

S. Mahadevakumar
Andaman and Nicobar Regional Centre

The education of plant pathology is a significant area of agricultural science that focuses on the aetiology, epidemiology, diagnosis and management of plant diseases. It was intended to provide an updated description of fundamental crop disease diagnosis in a manual that will act as a guide for all stakeholders, given that plant disease identification techniques are constantly evolving. Though several landmark research outbreaks have been recorded, from disease reporting to outbreaks and reemergence to deciphering the molecular signatures, NGS sequencing, whole genome sequencing, metabolomics and transcriptomic (OMICS) platforms have completely transformed the understanding of plant pathogen biology and offered several avenues to understand and manage them. From establishing pure cultures to understanding biology through advanced molecular sequences, recent scientific findings have played a pivotal role in the successful combating of plant pathogens. Therefore, the successful integration of recent scientific advancements into the field of diagnosis is very essential. In this chapter, an outline of research development in plant disease diagnosis, their management, the future of diagnostics and the role to be played to achieve food security is provided.

 DOI: 10.1201/9781003399704-17

GRAPHICAL ABSTRACT

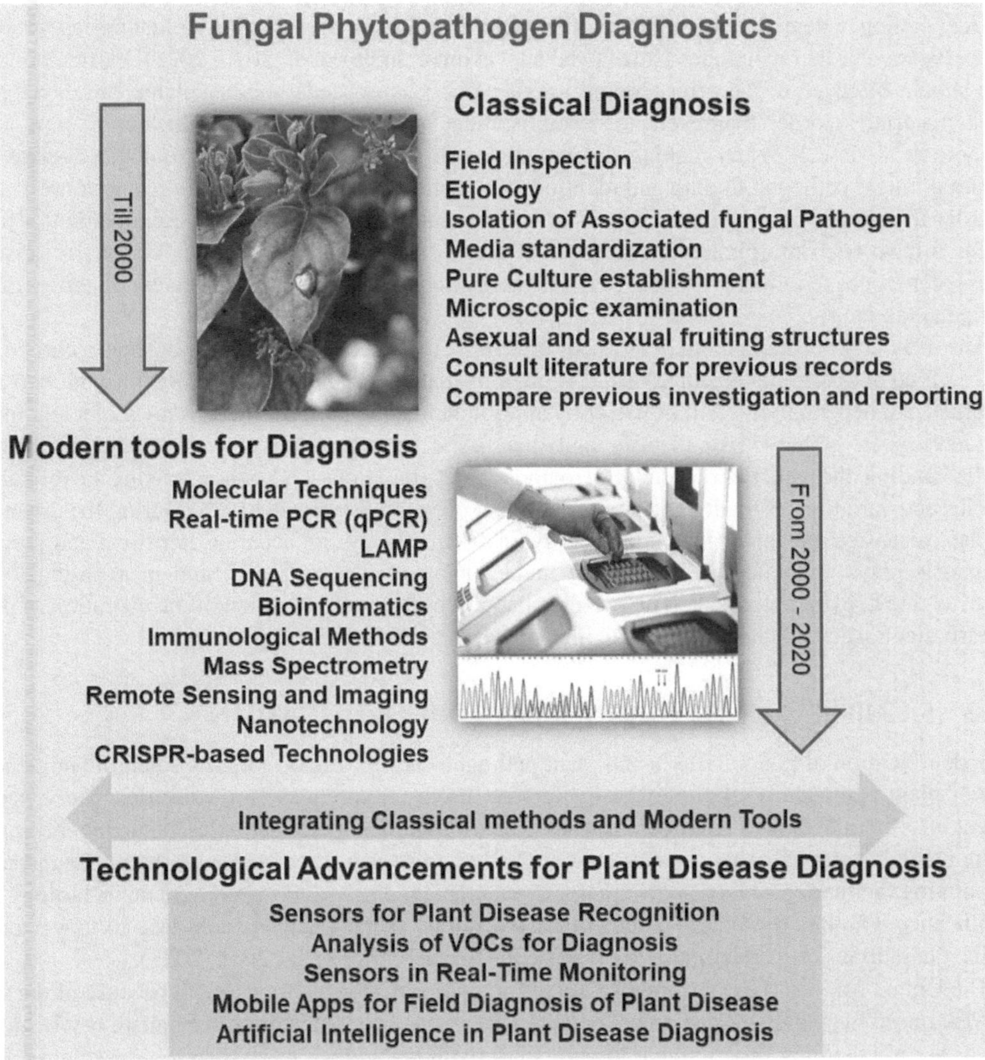

17.1 INTRODUCTION

Plant diseases are attributed to a yield loss of ≤12% of the world's total crop production (Reeleder et al., 2003). Both the incidence of newly emerging diseases and the severity of disease epidemics produced by virulent fungal plant infections have been constantly growing over the past 20 years (Fisher et al. 2012; Santini et al. 2013; Almeida 2018). The significance of precisely diagnosing plant diseases and the entities that cause them has led to an increase in global cooperation and the creation of multiple web-based databases to enhance the naming and documentation of plant infective fungi. The execution of plant disease management techniques depends on all of these initiatives.

Pathogenic fungi are the root cause of many serious plant diseases and economically important crops that inflict enormous damage on people around the globe. They can infect an array of plant species, or they can be confined to a handful of host plant species. Almost all of them are saprophytes, which can live in soil, water and the atmosphere without the assistance of a host, but a small number are obligatory parasites that must have a living host in order to thrive and multiply. Physical characteristics, host variety (*formae speciales*), virulence (pathotypes or races) and the capability to combine genetically different strains to create stable vegetative heterokaryons are some of the means by which pathogenic fungal isolates can be differentiated from one another (Capote et al. 2012). Correct pathogen detection and identification is the initial stage in managing a plant disease or starting preventive or therapeutic measures (Mahadevakumar and Sridhar, 2020; 2021). Particular attention should be given to the prior recognition of pathogens in seeds, mother plants and developing plant material in order to prevent the establishment and transmission of illnesses to new areas (Capote et al. 2012). When making judgements about disease management, quick, effective and accurate fungal pathogen finding and identification techniques are becoming more and more crucial in order to achieve these needs. Historically, the common techniques for the identification of plant diseases have been morphological methods based on cultural and morphological characteristics. These techniques require a great deal of time, energy and a thorough comprehension of conventional pathogen taxonomy.

The difficulty of establishing *in vitro* cultures for some infections is another barrier to early diagnosis (Goud and Termorshuizen 2003). It is obvious that overcoming these limitations and boosting accuracy and dependability will be easier by applying molecular approaches. Many molecular methods have been employed to identify and quantify a large number of plant pathogenic fungi. Understanding the genetic variations of pathogenic populations and characterizing novel fungal species and variations is made much easier with the use of molecular methods. During the diagnosis of plant diseases, molecular techniques are typically quicker, more accurate, sensitive and precise. Diagnosis plays an important role in the management of plant diseases and in avoiding losses incurred due to plant pathogens. The present chapter will furnish an outline of the evolution of plant disease diagnostic techniques, emphasizing the present and future perspectives.

17.2 SIGNIFICANCE OF DIAGNOSIS AND IDENTIFICATION

The identification of plant diseases and plant pathogens is very important. The scientific nomenclature of plant pathogenic fungi plays a major role in our comprehension of diseases. Since names reflect information about a pathogen's biology, distribution and potential harm, choosing the appropriate name is essential when discussing them and the diseases associated with them. For instance, pinpointing the fungal genus responsible for a certain plant disease may shed light on its biology and significance. On the other hand, outdated and inaccurate species names could lead to unwarranted limits, quarantines or inadvertent plant disease invasions (Crous et al., 2015, 2021).

The United Nations (UN) declared 2020 the International Year of Plant Health because of the serious risk raised by pathogens and newly emerging plant diseases. Plant diseases that are newly identified or are re-emerging pose a threat to particularly vulnerable regions of the global economy, food security, health and ecosystems. Numerous groups of plant pathogens, which consist of bacteria, fungi, phytoplasmas and viruses can give rise to new and developing plant diseases that can occasionally cause large crop losses with global economic ramifications. New developments and research, like the use of molecular biology have led to better technology for faster and more precise pathogen identification. Evolution, ecology and molecular factors are now viewed as novel issues for plant pathology research in traditional epidemiology, which has changed throughout time. Integrated disease control styles are becoming popular among farmers in response to consumer demands for healthier food options and the need for sustainable food production. However, contemporary agriculture is always on guard due to the appearance of new diseases or the resurgence of existing ones brought on by changes in the global environment and the increased movement of people and goods.

From visual examination of disease signs and symptoms to molecular-level pathogen detection, the knowledge of plant disease diagnostics has advanced due to technological developments. Plant disease management and avoiding the spread of pathogens to previously uninvaded areas depend on the sensitive and precise detection of plant pathogens. The fields of molecular biology, biotechnology, bioinformatics and plant pathology have all advanced, creating new opportunities for the construction of precise and delicate diagnostic techniques. Assays for plating, like the blotter method and the towel paper test, can reveal evidence of the presence and viability of pathogens, but they are labour-intensive, time-consuming and have limited sensitivity and specificity. Conventional approaches may be supplemented by molecular techniques such as quantitative polymerase chain reaction (PCR), flow cytometry and serological techniques. These techniques take less time and offer great sensitivity and specificity. However, in order to achieve standards of sensitivity and specificity, these procedures must also be standardized. Furthermore, for the procedure to be accepted globally as a standard procedure for quarantine reasons, standardized protocols are required.

17.3 RECENT DEVELOPMENTS IN DIAGNOSTICS

Phylogenetic studies have frequently enhanced our knowledge of fungal evolution and substantially changed fungal systematics. The ITS-rDNA (SSU) phylogeny was used in 2000 to classify fungi as a monophyletic group that was again classified as nine subgroups (Tehler et al., 2000). In 2007, the fungal kingdom's phylogeny was reconstructed using multi-locus DNA sequences. At the outset of molecular phylogenetic investigations of fungi, the tree was constructed solely from single-gene sequences (Liu et al., 1999; Tehler et al., 2000). In the following years, multi-locus phylogenies, which usually include less than six loci, were first documented in the literature (Blackwell et al., 2006; Spatafora et al., 2006; Zhang et al., 2006). With respect to single-gene trees, the latter had a somewhat higher resolving power. In the Assembling the Fungal Tree of Life (AFTOL) project, the large subunits (LSU) and small subunits (SSU) of rDNA, along with several genes coding proteins with RPB1, RPB2, TEF1, mtSSU and ATP6, were selected to explore the fungal phylogeny (http://aftol.org/about.php), which provided the basis for the classification of the fungi (James et al., 2006; Hibbett et al., 2007). Numerous recently released fungal monographs have also relied on multi-locus phylogenies, such as those in the families of Sordariomycetes, *Cladosporium*, *Phoma*, *Alternaria* and Massarineae (Maharachchikumbura et al., 2016).

The evolutionary connections among fungal taxa at all taxonomical levels have been investigated through a multitude of molecular investigations based on one or a few genes (Schoch et al., 2009). As the ribosomal RNA genes (rDNA) are in multiple copies that are isolated and cloned easily owing to different sections of rDNA evolving at different rates, making them valuable for taxonomic resolution of lineages at diverse taxonomic levels, rDNA is the most studied and frequently used locus in fungal phylogeny (Bruns et al., 1991; Hibbett, 1992). Since the beginning of molecular phylogenetics, wide research has been performed on the internal transcribed spacers (ITS), intergenic spacers, LSU and SSU of the rDNA of fungi (White et al., 1990). The SSU and LSU rDNA genes were commonly used to delineate higher taxonomic levels, whereas ITS was employed to resolve specific species or genera. The ITS of rDNA was designated as a universal DNA barcode for fungi due to its high-level discrimination rates for species as well as the success rate of PCR (Schoch et al., 2012).

In response to this issue, a group of mycologists and curators of the NCBI's fungal RefSeq have begun curating a set of common fungal ITS sequences fetched from the type resources (Schoch et al., 2014). Other than rDNA, single-copy as well as housekeeping protein genes make up the vast bulk of the cellular genome, are vital to biological processes and serve as useful markers for tracking the evolution of organisms. Actin (ACT), calmodulin (CAL), DNA replication licensing factor (MCM7), glycerol-3-phosphate dehydrogenase (GAPDH), histone (H3, H4), RNA polymerase II largest subunit (RPB1), RNA polymerase II 2nd largest subunit (RPB2) and translation elongation factor 1 (TEF1) are among the genes that are significant in fungal systematics (Templeton et al., 1992; Glass and Donaldson, 1995; Liu et al., 1999; Matheny et al., 2002; Guerber et al., 2003; Helgason et al., 2003; Hong et al., 2005; Rehner and Buckley, 2005; Schmitt et al., 2009). Additionally, mitochondrial genes like ATPase subunit 6 (ATP6), cytochrome c oxidase I (COI) and mitochondrial LSU and SSU of rDNA have been employed in fungal phylogenetics (Kretzer and Bruns, 1999; Zoller et al., 1999; Seif et al., 2005; Seifert et al., 2007).

17.4 CLASSICAL IDENTIFICATION

Identification of pathogens can be performed either by classical methods or by advanced techniques. Traditional methods of disease diagnosis include culturing pathogens on semi-defined media, observation of colony characteristics, sporulating structures and fruit bodies. Owing to advances in instrumentation and molecular biology, the schemes facilitate the diagnosis of plant pathogens more rapidly. Advanced methods often employed in finding plant pathogens are DNA/RNA probe technology, PCR and immunological methods by amplification of nucleic acid sequences. Such techniques have potential rewards over conventional identification methods, especially in rapidity, accuracy, quantification and could be followed by investigators who are not experts in taxonomy (Ward et al. 2004). More precisely, these techniques also assist in the diagnosis of non-culturable pathogens. Furthermore, molecular techniques are appropriate for revealing new diseases with indefinite aetiologies. Relative analysis of genomic sequences permits the phylogenetic reconstruction of pathogen relations at various taxonomic levels.

Molecular diagnosis of pathogens has gone through major advances over four decades, which have helped in the growth and application of these methods for various purposes. The discovery of antibody-based detection, specifically monoclonal antibodies (MA) and enzyme-linked immune sorbent assays (ELISA), was an important breakthrough. The application of these approaches was the turning point, especially in virology and bacteriology, owing to the rapidness in diagnosis. Thereafter, DNA-based technologies, particularly PCR, revolutionized by augmenting to magnify the original target DNA several million fold.

Although various selective media are obtainable for the isolation and identification of *Pythium* spp., they are time-consuming and demand high expertise. Whenever selective media are employed for the isolation of *Pythium*, there will be many colonies representing a greater number of *Pythium* spp., which need immediate transfer to fresh media for growth to ascertain morphological and cultural features. Further, identification of *Pythium* spp. to the species is difficult based exclusively on morphological characteristics. Other factors that complicate morphological identification include (i) the high number of non-typical isolates; (ii) the heterothallism in some species; (iii) sexually sterile isolates; (iv) variations in the size and shape of structures within species; (v) similar morphology among different groups of species (Hendrix and Papa, 1974; Ali-Shtayeh et al., 1985; Martin and Kistler 1990; Klassen et al. 1996; Barr et al., 1987). To circumscribe such difficulties, simple, rapid and accurate methods of detection are essential. Although serological methods are popular in the detection of *Pythium*, the inherent problem is the production of species-specific antibodies.

As novel information about the characteristics and phylogeny of each genus and species becomes available, the scientific names of fungi will continue to evolve. Since the days of characterizing fungi only by their shape, traits like spore colour and septation have frequently been inconclusive in classifying species or even defining them. Formerly widely defined genera have been split into many genera, frequently with species that were previously included in those genera being relocated to other genera. Hypothetically, each clearly defined genus should only contain closely linked species. Furthermore, a lot of species that were earlier believed to be solitary species have now been revealed to be complexes; as a result, each species is now better defined to reflect its biology and the scientific name is updated to reflect this newfound understanding (Lucking et al., 2020; Crous et al., 2021).

Growing international collaborations among mycologists over the past 10 to 15 years have been a welcome and notable trend in fungal systematics. The mycologists from top international research organizations and universities are playing significant roles in these kinds of multi-authored publications and the participants are becoming more international. This kind of cooperation also makes it easier to obtain cultures and specimens that span larger geographic and taxonomic ranges than in the past, guaranteeing that studies can become more thorough, representative and trustworthy. There has been a noticeable increase in collaboration in species-level mycological research, especially when it comes to genera with industrial, medical or plant pathogenic significance. Currently, fungal systematics is a more active and dynamic global field than it has ever been. This bodes well for the construction of an increasingly solid and reliable foundation for the naming of all fungal taxa, including those crucial to plant pathology, biological control and crop storage (Crous et al., 2015).

17.5 FUTURE OF PLANT DISEASE DIAGNOSIS

There are several uses for DNA barcoding in the realm of disease diagnosis. It can assist in the quick identification of phytopathogens, which can facilitate the prompt application of management or control measures. Crop production systems can become more environment-friendly by using fewer agrochemicals when phytopathogens are quickly recognized at the field level. DNA barcoding experts and taxonomists have to collaborate in the context of next-generation sequencing (NGS) and environmental sampling. The entire process of metabarcoding and environmental sampling, from sample collection to the identification of genetic areas for analysis, must be carefully organized.

17.6 MODERN TOOLS FOR THE IDENTIFICATION OF PLANT PATHOGENS

Plant pathogen identification has greatly benefited from advancements in modern tools and technologies. These tools enable more accurate, rapid and cost-effective identification of plant pathogens. Some of the key modern tools for plant pathogen identification are as follows.

- **Molecular Techniques**: PCR is a widely used technique to multiply specific DNA sequences, allowing for the detection and identification of pathogens through the amplification of their genetic material.
- **Real-time PCR (qPCR)**: This technique allows for the quantification of DNA during the PCR process, providing both qualitative and quantitative information about the presence of plant pathogens.
- **Loop-Mediated Isothermal Amplification (LAMP)**: It is referred to as an isothermal nucleic acid amplification technique that is highly specific and sensitive, making it suitable for onsite detection of plant pathogens.
- **DNA Sequencing**: NGS: NGS technologies, like Illumina and PacBio, have revolutionized DNA sequencing. They allow for the quick and high-throughput sequencing of entire genomes, enabling the identification and characterization of plant pathogens at the genomic level.
- **Bioinformatics**: Various databases and bioinformatics tools allow researchers to analyse and compare genomic data (genome database). This aids in the identification of pathogenic traits, virulence factors and genetic variations in plant pathogens.
- **Immunological Methods**: ELISA is a widely used immunological method for identifying precise antigens or antibodies associated with plant pathogens. It is particularly useful for large-scale screening.
- **Mass Spectrometry**: Matrix-Assisted Laser Desorption/Ionization Time-of-Flight Mass Spectrometry (MALDI-TOF MS) technique is used for the quick and accurate identification of plant pathogens based on their protein profiles. It is especially useful for bacterial and fungal identification.
- **Remote Sensing and Imaging**: Remote sensing technologies will be helpful in monitoring plant health and identifying potential pathogen outbreaks by analysing changes in vegetation patterns and spectral reflectance.
- **Nanotechnology**: Nanoparticles can be engineered to detect specific plant pathogens. They offer rapid detection and high sensitivity and can be integrated into portable diagnostic devices.
- **CRISPR-based Technologies**: CRISPR-based technologies enable precise genome editing and have applications in developing resistant plant varieties. They can also be used for the detection of specific pathogenic DNA sequences. These modern tools collectively contribute to a more comprehensive and efficient approach to plant pathogen identification, helping researchers and farmers better understand, monitor and manage plant diseases.

17.7 CHALLENGES IN PLANT PATHOGEN IDENTIFICATION

Scientific progress in the previous decade has led to advancements in the field of plant disease diagnosis. Shifting from conventional morphology, culture- and aetiology-based identification to molecular tools and barcoding phytopathogens to deciphering the genome of pathogen and host and their interaction unravelled various strata of information, which will be explored to develop better management strategies. However, emerging new diseases and re-emerging diseases are posing a severe threat to food security. The development of plant disease as an outcome of constant interaction with the host and prevailing environment leads to successfully evading several management protocols. Therefore, understanding the pathogens and scanning them with modern tools is necessary to successfully manage them. Yet, the plant disease diagnosis has faced several challenges.

There are several challenges to assessing the fungal kingdom.

- **Genetic Diversity**: Plant pathogens often exhibit significant genetic diversity within a species, making it challenging to design molecular probes or markers that can accurately identify all variants. Many of the fungal pathogens are evolving and several populations and variations within the population make plant disease more challenging.
- **Horizontal Gene Transfer**: The transfer of genetic material amongst different organisms can lead to the acquisition of new virulence factors, complicating the identification process.
- **Emerging Pathogens**: The emergence of new strains or variants of known pathogens can go undetected initially, especially if diagnostic tools are not updated to account for these variations.
- **Novel Pathogens**: Identifying completely new or unknown pathogens can be particularly challenging, as there may be no prior knowledge or reference sequences available.
- **Co-infections and multiple pathogens**: Plants can be infected with multiple pathogens simultaneously, and distinguishing between the different pathogens in a mixed infection can be difficult.
- **Synergistic Interactions**: Interactions between different pathogens in a co-infected plant may lead to enhanced virulence, making it challenging to predict disease outcomes.
- **Diagnostic Sensitivity and Specificity**: Diagnostic tools may produce false positives or false negatives, impacting the accuracy of plant pathogen identification. This is especially crucial in the context of quarantine and disease management.
- **Sample Quality and Complexity**: Obtaining high-quality samples from plants, soil, or water can be challenging and the occurrence of inhibitors in these samples can influence the performance of diagnostic assays.
- **Complex Microbial Communities**: Plant environments often contain diverse microbial communities, making it challenging to isolate and identify specific pathogens accurately.
- **Technological Limitations**: Some regions, particularly in developing countries, may lack the necessary resources and infrastructure to adopt advanced diagnostic technologies.
- **Access to Sequencing Facilities**: Next-generation sequencing technologies require specialized equipment and expertise, limiting access in certain areas.
- **Rapid Mutation and Adaptation**: Plant pathogens can rapidly evolve, leading to changes in their genetic makeup and virulence. This can outpace the development of diagnostic tools and strategies.
- **Interactions with Host Plants**: Host plants can develop resistance mechanisms or immune responses that may affect with the recognition of pathogens, especially during the initial stages of infection.
- **Data Integration and Standardization**: Integrating and managing large datasets from various sources, such as genomics, proteomics and environmental data, can be challenging.
- **Standardization**: Lack of standardized protocols and methods across laboratories can hinder the comparability of results and the development of universal diagnostic tools. Addressing these challenges requires ongoing research, collaboration and the development

of innovative solutions in the discipline of plant pathology. Advances in technology, bioinformatics and international cooperation can contribute to overcoming these obstacles and improving the precision and efficacy of plant pathogen identification.

17.8 TECHNOLOGICAL ADVANCEMENTS IN PLANT DISEASE DIAGNOSIS

17.8.1 Sensors for Plant Disease Recognition

The advancement of various sensors has been fuelled by the necessity for rapid, affordable and user-friendly technologies that can detect target pathogens with high sensitivity and specificity without the requirement for labelling. This method has the potential to avoid the limits of conventional diagnosis procedures and eliminate the necessity for specialized scientists. Many sensor-based diagnostic techniques have been established and are currently being employed in the field, particularly for the identity of viral infections. As an illustration, consider the application of a quartz crystal microbalance in combination with an immunosensor to distinguish the maize chlorotic mottle virus (Huang et al., 2014), which is characterized by a shift in the crystal's resonance frequency brought on by a mass change. This sensor's sensitivity is comparable to the ELISA test, but it also has other benefits such as easy use, affordability, speed, high sensitivity and real-time application potential (Montagut et al., 2011). Surface plasmon resonance (SPR) was used by Lin et al. (2014) as a label-free detection method to track differences in the refractive index on the sensor surface brought about by the interaction of an immobilized ligand and an analyte in solution. Because surface-enhanced Raman spectroscopy (SERS) may identify chemical fingerprints, it is another pertinent detection method. SERS demonstrated accuracy, high sensitivity, speed and a low LOD in contrast to conventional high-performance liquid chromatography (HPLC).

17.8.2 Analysis of VOCs for Diagnosis

Since plants create and release volatile organic compounds (VOCs) as a defence mechanism against pathogen attack, analysing the VOCs produced by different phytopathogens can serve as an indirect way of detecting plant pathogens (Scala et al., 2013). Plant VOCs are characterized by low molecular weight, low boiling point and high vapour pressure biomolecules and metabolites. Many VOCs that are essential to plant growth, defence, survival and communication are released by plants (Baldwin et al., 2006). Besides, pathogen infections in plants may cause the discharge of particular VOCs that are indicative of the physiological health status of the plants and can be employed for non-invasive plant disease monitoring. Because of this, VOC profiling is becoming a useful, quick and non-destructive method for detecting plant pathogens with high sensitivity and no use for chemical reagents. Gas chromatography-mass spectrometry (GC-MS)-based methods have historically been used to find out VOCs (Cellini et al., 2016; De Lacy et al., 2001; Ewen et al., 2004; Spadafora et al., 2016). These techniques are cumbersome, costly, time-consuming and require extensive training to operate correctly (Sharma et al., 2019).

17.8.3 Sensors in Real-Time Monitoring

Abiotic stressors, such as the combination of salt and heat, dryness and cold or heat and drought, are less widely documented in scientific literature than biotic stresses. The uniqueness of (Körner and Challa, 2003; Mortensen, 2000) means that their responses to various stresses applied separately cannot be readily extrapolated (Suzuki et al., 2014), which diminishes the validity of procedures that do not rely on diagnostic techniques. Recent advancements in smart sensor technology have been effectively utilized to track and manage significant environmental, biotic or physiological challenges by using wearable sensors to communicate plant health in great detail. These have a number of benefits, including supporting farm management within the context of smart agriculture and assisting botanists in comprehending growth requirements.

17.8.4 Mobile Apps for Field Diagnosis of Plant Disease

Numerous apps, or mobile applications, have been advanced recently to aid in the rapid diagnosis of plant diseases, thanks to developments in software technology and smartphones in particular. A few examples of these are "AgroAI-Plant Diseases Diagnosis", "Cropalyser", "Leaf Doctor", and "Plantix-Your Crop Doctor" (Pethybridge and Nelson, 2015; Siddiqua et al., 2022). Using colour photos, plant tissues with and without illness were identified and an interactive, reiterative smartphone application was employed to detect the disease severity. Using their finger, the user can select from up to eight different hues that denote healthy tissues on the application's display screen. Modern concepts in agricultural innovations include "wireless sensing" and "real-time communication". The term "smart farming" refers to the use of contemporary information and communication technologies to enable farmers to monitor field conditions remotely or with high-tech assistance on the field (Buja et al., 2021). Though these technological advancements are supportive in assisting plant disease diagnosis, they are not conclusive and there are several demerits inherent with them, which makes them unreliable.

17.9 ARTIFICIAL INTELLIGENCE IN PLANT DISEASE DIAGNOSIS

Fast and accurate disease detection and diagnosis are possible with artificial intelligence (AI). AI, particularly deep learning approaches, is showing enormous promise for quick, precise and automated diagnosis and identification of plant diseases from photos. This holds the potential to enable targeted real-time therapies and earlier interventions. In the upcoming years, it is anticipated that continued advancements in AI will strengthen phytopathology's capacity for disease diagnosis. Several studies show that AI is effective at correctly detecting complicated diseases. A deep convolutional neural network created by Ramcharan et al. (2017) can identify four common cassava diseases from leaf photos with an accuracy rate of more than 90%. Similar to this, Inoue et al. (1979) used machine learning to identify the severe wheat blast fungus from lesion photos with 95% accuracy. Fuentes et al. (2017) applied AI to the illnesses of tomato plants. The authors described a deep learning-based method that employs photos clicked by camera devices at different resolutions to identify pests and illnesses in tomato plants. To improve accuracy and lower false positives during training, they suggested a technique for both local and global class annotation and data augmentation. The suggested technique can handle intricate situations from a plant's surroundings and is capable of accurately identifying nine distinct kinds of illnesses and pests. Sladojevic et al. (2016) used deep neural networks and hyperspectral data to distinguish virus-infected grapevines from healthy vines with 97% performance.

The capacity of AI in the identification and diagnosis of plant ailments has been more advanced by recent studies. Feng et al. (2023) used a dataset of potato leaf disease photos with single and complex backdrops to create a convolutional neural network model for a deep learning-based potato late blight detection technique with good accuracy and quick inference speed. Dropping the number of convolutions, decreasing the depth of the network and adding an attention module all helped to advance the model's performance and minimize its numerous parameters. Bracino et al. (2023) used deep learning algorithms like EfficientNet-b0, MobileNet-v2 and Places365-GoogLeNet to conduct a study focused on the non-destructive cataloguing of paddy rice leaf diseases. Their objective is to determine if the leaves of rice paddy fields are free of any diseases, such as downy mildew, bacterial blight, leaf streaks, or rice tungro disease. Based on its average accuracy of 97.74%, the study suggests that EfficientNet-b0 is the finest model for this classification assignment. Using hyperspectral pictures, Jouini et al. (2023) created a deep convolutional neural network model that reached 94% testing accuracy in identifying wheat leaf rust. In resource-constrained contexts where timely disease identification as well as management are crucial for sustainable agriculture, the method proved the practicality of real-time wheat disease detection (Jouini et al., 2023).

Plant disease forecasting and monitoring are taking novel directions as a result of collaborative work and creativity in AI and cloud-based technologies. A cooperative AI and cloud-based platform for the identification, tracking and forecasting of plant diseases has been developed by Lavanya and Krishna (2022). This novel method provides real-time disease diagnosis and disease density mapping by fusing AI algorithms with a mobile application. Similar to the initiatives by Otero et al. (2022) and Zen and Fransisca (2022) in developing predictive models for vineyards and AI-based mobile applications for plant disease detection, respectively (Otero et al., 2022; Lavanya and Krishna, 2022; Zen and Fransisca (2022)), this collaborative and technology-driven approach reflects a shift towards more integrated and interconnected systems for plant disease management (González-Rodríguez et al., 2024). These advancements signify a new era in plant disease diagnosis and represent a significant turning point at the intersection of agronomy and cutting-edge technology. The combination of cloud-based collaborative platforms and machine learning techniques is changing how scientists and farmers tackle the problem of plant disease. These developments not only improve disease diagnosis and management precision, but they make it easier to respond quickly and effectively, which is essential for global sustainability and food security.

17.10 CHALLENGES OF AI IN PLANT DISEASE DIAGNOSIS

Though AI can detect recognizable diseases, it cannot distinguish any new diseases, species and races. It is a fact that AI uses image resources and then it can only detect a particular disease that is already in its domain. However, it's very hard to find solutions through AI for the diagnosis of new and emerging diseases. In addition, there are some drawbacks to AI in phytopathogen diagnosis. Though revolutionary advances in phytopathology are promised by artificial intelligence, careful measures must be taken before incorporating it into agricultural systems. There are further technical hurdles in the way of creating solid, trustworthy AI solutions for plant disease situations in the actual world (Jordan and Mitchell, 2015). The restricted emphasis of many machine learning models now in use on certain crops, diseases and controlled environments is a major drawback (Benos et al., 2021). It is common for algorithms that have been trained on small datasets to not generalize to different agricultural environments. Developing AI tools that are flexible for in situ use is a problem due to the wide range of differences in crop cultivars, growth phases, climates, soil conditions and disease strains (Singh et al., 2016; Singh et al., 2018). A closer look is necessary at the serious ethical issues that the use of AI in agriculture brings up, including data privacy, accountability, labour implications and environmental sustainability. As a consequence, inclusive stakeholder participation is required for the construction of phytopathology AI systems, with an emphasis on human needs and values (González-Rodríguez et al., 2024).

17.11 SUMMARY

Food losses caused by pathogens are currently to blame for a decrease in yield quantity and quality as well as a drop in yield value and financial returns. Thus, in plant pathology, "early detection" coupled with "fast, accurate and affordable" diagnostics has also become a necessity, particularly for emerging diseases or difficult-to-control pathogens that spread through asymptomatic individuals who display subtle early symptoms but become difficult to combat later on. Plant diagnosis has evolved as science and technology have advanced with the invention of molecular tools and technological advancements in software as well.

17.12 FUTURE PERSPECTIVE

For instance, though PCR and barcoding of the ITS region were treated as landmark events, the development of different sequencing platforms, ease of work and cost feasibility made molecular detection methods accessible. However, further innovative molecular tools such as NGS and onsite detection tools are not cost-effective. The challenge still persists in the era of advanced molecular biology, as it poses new challenges for accurate diagnosis.

STUDY QUESTIONS

1. What is AI tool?
2. List of modern methods for fungal identification.
3. What are the challenging methods in plant pathogen identification?

FURTHER READING

Dong, X., Wang, Q., Huang, Q., Ge, Q., Zhao, K., Wu, X., et al., 2023. PDDD-PreTrain: A Series of Commonly Used Pre-Trained Models Support Image-Based Plant Disease Diagnosis. *Plant Phenomics*, 5:0054. https://doi.org/10.34133/plantphenomics.0054

Crous, P. W., Hawksworth, D. L. and Wingfield, M. J., 2015. Identifying and Naming Plant-Pathogenic Fungi: Past, Present and Future. *Ann. Rev. Phytopathol.* 53:247–267. https://doi.org/10.1146/annurev-phyto-080614-120245

Lucking, R., Aime, M. C., Robbertse, B., Miller, A. N., Ariyawansa, H. A., et al., 2020. Unambiguous identification of fungi: Where do we stand and how accurate and precise is fungal barcoding? *IMA Fungus* 11:14. https://doi.org/10.1186/s43008-020-00033-z.

Crous, P. W., Rossman, A. Y., Aime, M. C., Allen, W. C., Burgess, T., Groenewald, J. Z. and Castlebury, L. A., 2021. Names of Phytopathogenic Fungi: A Practical Guide. *Phytopathol.*, 111, 1500-1508. https://doi.org/10.1094/PHYTO-11-20-0512-PER

Mancini, V., Murolo, S. and Romanazzi, G., 2016. Diagnostic methods for detecting fungal pathogens on vegetable seeds. *Plant Pathology*, 65(5), 691 – 703. https://doi.org/10.1111/ppa.12515

REFERENCES

Ali-Shtayeh, M. S., Jamous, R. M. and Yaghmour, R. M.-R., 1985. *Mycology Manual*. An-Najah University, Nablus.

Almeida, R. P. P., 2018. Emerging plant disease epidemics: Biological research is key but not enough. *PLoS Biol* 16(8): e2007020. https://doi.org/10.1371/journal.pbio.2007020

Baldwin, I., Halitschke, R., Paschold, A., Dahl, C. and Preston, C., 2006. Volatile Signaling in Plant-Plant Interactions: "Talking Trees" in the Genomics Era. *Science*, 311, 812–815. https://doi.org/10.1126/science.1118446

Barr, D. J. S., Warwick, S. I. and Desaulniers, N. L., 1987. Isozyme variation, morphology and growth response to temperature in *Pythium irregular*. *Can. J Bot.*, 75(12), 2073–2081. https://doi.org/10.1139/b97-918

Benos, L., Tagarakis, A.C., Dolias, G., Berruto, R., Kateris, D. and Bochtis, D., 2021. Machine Learning in Agriculture: A Comprehensive Updated Review. *Sensors (Basel)*, 21. https://doi.org/10.3390/S21113758

Blackwell, M., Hibbett, D. S., Taylor, J. W. and Spatafora, J. W., 2006. Research Coordination Networks: A phylogeny for kingdom Fungi (Deep Hypha). *Mycologia*, 98, 829–837. https://doi.org/10.1080/15572536.2006.11832613

Bracino, A. A., Evangelista, D. G. D., Concepcion, R. S., Dadios, E. P. and Vicerra, R. R. P., 2023. Non-Destructive Classification of Paddy Rice Leaf Disease Infected by Bacterial and Fungal Species Using Vision-Based Deep Learning. *J. Adv. Comput. Intell. Intell. Inform.*, 27(3), 333–339. https://doi.org/10.20965/jaciii.2023.p0333

Bruns, T. D., White, T. J. and Taylor, J. W., 1991. Fungal Molecular Systematics. *Annu. Rev. Ecol. Syst.* 22, 525–564.

Buja, I., Sabella, E., Monteduro, A. G., Chiriacò, M. S., De Bellis, L., et al., 2021. Advances in Plant Disease Detection and Monitoring: From Traditional Assays to In-Field Diagnostics. *Sensors*, 21, 2129 . https://doi.org/10.3390/s21062129

Capote, N., Pastrana, A. M., Aguado, A. and Sánchez-Torres, P., 2012. Molecular tools for detection of plant pathogenic fungi and fungicide resistance. In: *Plant Pathology*, Cumagun, C.J. (ed.)., In Tech. https://doi.org/978-953-51-0489-6

Cellini, A., Biondi, E., Blasioli, S., Rocchi, L., Farneti, B., et al., 2016. Early detection of bacterial diseases in apple plants by analysis of volatile organic compounds profiles and use of electronic nose. *Ann. Appl. Biol.*, 168. https://doi.org/10.1111/aab.12272

Crous, P. W., Hawksworth, D. L. and Wingfield, M. J., 2015. Identifying and naming plant-pathogenic fungi: Past, present and future. *Annu. Rev. Phytopathol.*, 53, 247–267. https://doi.org/10.1146/annurev-phyto-080614-120245

Crous, P. W., Rossman, A. Y., Aime, M. C., Allen, W. C., Burgess, T., Groenewald, J. Z. and Castlebury, L. A., 2021. Names of Phytopathogenic Fungi: A Practical Guide. *Phytopathology*, 111, 1500-1508. https://doi.org/10.1094/PHYTO-11-20-0512-PER

De Lacy, B. C., Evans, P., Ewen, R., Gunson, H., Jones, P., Ratcliffe, N. and Spencer-Phillips, P., 2001. Gas chromatography–mass spectrometry analyses of volatile organic compounds from potato tubers inoculated with Phytophthora infestans or *Fusarium coeruleum*. *Plant Pathol.*, 50, 489–496. https://doi.org/10.1046/j.1365-3059.2001.00594.x

Ewen, R.J., Jones, P.R.H., Ratcliffe, N.M., Spencer-Phillips, P.T.N., 2004. Identification by gas chromatography-mass spectrometry of the volatile organic compounds emitted from the wood-rotting fungi Serpula lacrymans and Coniophora puteana and from Pinus sylvestris timber. *Mycol. Res.*, 108, 806–814.

Feng, J., Hou, B., Yu, C., Yang, H., Wang, C., et al., 2023. Research and Validation of Potato Late Blight Detection Method Based on Deep Learning. *Agronomy*, 13, 1659. https://doi.org/10.3390/AGRONOMY13061659

Fisher, M. C., Henk, D. A., Briggs, C. J., Brownstein, J. S., Madoff, L. C., McCraw, S. L. and Gurr, S. J., 2012. Emerging fungal threats to animal, plant and ecosystem health. *Nature*, 484, 186–194. https://doi.org/10.1038/nature10947

Fuentes, A., Yoon, S., Kim, S. C. and Park, D. S., 2017. A Robust Deep-Learning-Based Detector for Real-Time Tomato Plant Diseases and Pests Recognition. *Sensors*, 17, 2022. https://doi.org/10.3390/S17092022

Glass, N. L. and Donaldson, G. C., 1995. Development of primer sets designed for use with the PCR to amplify conserved genes from filamentous Ascomycetes. *Appl. Environ. Microbiol.*, 61(4), 1323–1330. https://doi.org/10.1128/aem.61.4.1323-1330.1995

González-Rodríguez, V. E., Izquierdo-Bueno, I., Cantoral, J. M., Carbú, M. and Garrido, C., 2024. *Artificial Intelligence: A Promising Tool for Application in Phytopathology*. Preprints.org https://doi.org/10.20944/preprints202401.1882.v1

Goud, J. and Termorshuizen, A., 2003. Quality of Methods to Quantify Microsclerotia of *Verticillium dahliae* in Soil. *Eur. J. Plant Pathol.*, 109, 523–534. https://doi.org/10.1023/A:1024745006876

Guerber, J. C., Liu, B., Correll, J. C. and Johnston, P. R., 2003. Characterization of Diversity in Colletotrichum acutatum sensu lato by Sequence Analysis of Two Gene Introns, mtDNA and Intron RFLPs and Mating Compatibility. *Mycologia*, 95(5), 872-895. https://doi.org/10.2307/3762016

Helgason, T., Watson, I. J. and Young, J. P. W., 2003. Phylogeny of the Glomerales and Diversisporales (Fungi: Glomeromycota) from actin and elongation factor 1-alpha sequences. *FEMS Microbiol Lett.*, 229, 127-132. https://doi.org/10.1016/S0378-1097(03)00802-4

Hendrix, F. F. Jr. and Papa, K. E., 1974. Taxonomy and genetics of Pythium. *Proc. Am. Phytopathol. Soc.* 1, 200–207.

Hibbett, D.S., 1992. Ribosomal RNA and fungal systematics. *Trans. Mycol. Soc. Japan*. 33, 533–556.

Hibbett, D. S., Binder, M., Bischoff, J. F., Blackwell, M., Cannon, P. F. et al., 2007. A higher-level phylogenetic classification of the Fungi. *Mycol. Res.*, 111, 509–547. https://doi.org/10.1016/j.mycres.2007.03.004

Hong, P., Liu, X. S., Zhou, Q., Lu, X., Liu, J. S. and Wong, W. H., 2005. A boosting approach for motif modeling using ChIP-chip data. *Bioinformatics* 21(11), 2636-2643. https://doi.org/10.1093/bioinformatics/bti402

Huang, X., Xu, J., Ji, H.-F., Li, G. and Chen, H., 2014. Quartz crystal microbalance based biosensor for rapid and sensitive detection of maize chlorotic mottle virus. *Anal. Methods*, 6, 4530–4536. https://doi.org/10.1039/C4AY00292J

Inoue, Y., Vy, T.T.P., Yoshida, K., Asano, H., Mitsuoka, C., Asuke, S., Anh, V.L., Cumagun, C.J.R., Chuma, I., Terauchi, R., et al. 2017. Evolution of the Wheat Blast Fungus through Functional Losses in a Host Specificity Determinant. *Science* (1979):357, 80–83, https://doi.org/10.1126/SCIENCE.AAM9654

James, T. Y., Kauff, F., Schoch, C., Matheny, P. B., Hofstetter, V., Cox, C. J., et al., 2006. Reconstructing the early evolution of the fungi using a six–gene phylogeny. *Nature*, 443:818–822. https://doi.org/10.1038/nature05110

Jordan, M. I. and Mitchell, T. M., 2015. Machine Learning: Trends, Perspectives and Prospects. *Science*, 349, 255–260. https://doi.org/10.1126/SCIENCE.AAA8415

Jouini, O., Sethom, K. and Bouallegue, R., 2023. Wheat Leaf Disease Detection Using CNN in Smart Agriculture. In *Proceedings of the 2023 International Wireless Communications and Mobile Computing*, IWCMC 2023; Institute of Electrical and Electronics Engineers Inc., pp. 1660–1665. https://doi.org/10.1109/IWCMC58020.2023.10183348

Klassen, G. R., Balcerzak, M. and de Cock, A. W. A. M., 1996. 5S Ribosomal RNA Gene Spacers as Species-Specific Probes for Eight Species of Pythium. *Phytopathology*, 86, 581-587. https://doi.org/10.1094/Phyto-86-581

Körner, O. and Challa, H., 2003. Process-based humidity control regime for greenhouse crops. *Comput. Electron. Agric.*, 39, 173–192. https://doi.org/10.1016/S0168-1699(03)00079-6

Kretzer, A.M. and Bruns, T. D., 1999. Use of atp6 in fungal phylogenetics: an example from the boletales. *Mol Phylogenet Evol.* 13(3):483-492. https://doi.org/10.1006/mpev.1999.0680

Lavanya, A. and Krishna, T. M., 2022. An AI and Cloud Based Collaborative Platform for Plant Disease Identification, Tracking and Forecasting for Farmers. *Int. J. Eng. Technol. Manage. Sci.*, 6, 527–537, https://doi.org/10.46647/ijetms.2022.v06i06.091

Lin, H.Y., Huang, C.H., Lu, S.H., Kuo, I.T., Chau, L.K., 2014. Direct detection of orchid viruses using nanorod-based fiber optic particle plasmon resonance immunosensor. *Biosens. Bioelectron.* 51, 371–378. https://doi.org/10.1016/j.bios.2013.08.009

Liu, Y. J., Whelen, S. and Hall, B. D., 1999. Phylogenetic relationships among ascomycetes: evidence from an RNA polymerase II subunit. *Mol. Biol. Evol.* 16(12), 1799-1808. https://doi.org/10.1093/oxfordjournals.molbev.a026092

Lucking, R., Aime, M. C., Robbertse, B., Miller, A. N., Ariyawansa, H. A., et al., 2020. Unambiguous identification of fungi: Where do we stand and how accurate and precise is fungal barcoding? *IMA Fungus* 11, 14. https://doi.org/10.1186/s43008-020-00033-z

Mahadevakumar, S., Sridhar, K.R., 2020. Diagnosis of *Pythium* by Classical and Molecular Approaches. In: *The genus Pythium – Diagnosis, Diseases and Management.* Rai, M., Abd-Esalam, K.A., Ingle, A.P. (eds.). CRC Press, pp. 200–224. https://doi.org/10.1201/9780429296406

Mahadevakumar, S., Sridhar, K.R., 2021. Diversity of Plant Pathogenic Fungi in Agricultural Crops. In: *Plant, Soil and Microbes in Tropical Ecosystems.* Dubey, S.K., Verma, S.K. (eds.). Springer Nature, pp. 101-149. https://doi.org/10.1007/978-981-16-3364-5_6

Maharachchikumbura, S.S.N., Hyde, K.D., Jones, E.B.G., McKenzie, E.H.C., Bhat, J.D. et al. 2016. Families of Sordariomycetes. *Fungal Diversity*, 79, 1–317. https://doi.org/10.1007/s13225-016-0369-6

Martin, F. N. and Kistler, H. C., 1990. Species-specific banding patterns of restriction endonuclease-digested mitochondrial DNA from the genus Pythium. *Exp. Mycol.* 14, 33–46. https://doi.org/10.1016/0147-5975(90)90084-7

Matheny, P. B., Liu, Y. J., Ammirati, J. F. and Hall, B. D., 2002. Using RPB1 sequences to improve phylogenetic inference among mushrooms (Inocybe, Agaricales). *Am. J. Bot.* 89(4), 688–698. https://doi.org/10.3732/ajb.89.4.688

Montagut, Y., Narbon, J. G., Jimenez, Y., March, C., Montoya, A., Arnau, A., 2011. QCM Technology in Biosensors. *Biosens. Emerg. Mater. Appl.*, 153–178. https://doi.org/10.5772/17991

Mortensen, L., 2000. Effects of air humidity on growth, flowering, keeping quality and water relations of four short-day greenhouse species. *Sci. Hortic.*, 86, 299–310. https://doi.org/10.1016/S0304-4238(00)00155-2

Otero, M., Velasquez, L. F., Basile, B., Onrubia, J. R., Pujol, A. J. and Pijuan, J., 2022. Data Driven Predictive Models Based on Artificial Intelligence to Anticipate the Presence of Plasmopara Viticola and Uncinula Necator in Southern European Winegrowing Regions. *Front. Artif. Intell. Appl.*, 356, 164–167, https://doi.org/10.3233/FAIA220333

Pethybridge, S. J. and Nelson, S. C., 2015. Leaf Doctor: A New Portable Application for Quantifying Plant Disease Severity. *Plant Disease*, 99(10), 1310-1316. https://doi.org/10.1094/PDIS-03-15-0319-RE

Ramcharan, A., Baranowski, K., McCloskey, P., Ahmed, B., Legg, J. and Hughes, D.P., 2017. Deep Learning for Image-Based Cassava Disease Detection. *Front Plant Sci*, 8. https://doi.org/10.3389/FPLS.2017.01852/BIBTEX

Reeleder, R. D., Capell, B. B., Tomlinson, L. D. and Hickey, W. J., 2003. The extraction of fungal DNA from multiplelargesoilsamples. *Can. J. PlantPathol.*,25(2),182–191.https://doi.org/10.1080/07060660309507067

Rehner, S. A. and Buckley, A., 2005. A Beauveria phylogeny inferred from nuclear ITS and EF1-a sequences: evidence for cryptic diversification and links to Cordyceps teleomorphs. *Mycologia*, 97(1), 84–98. https://doi.org/10.3852/mycologia.97.1.84

Santini, A., Ghelardini, L., De Pace, C., Desprez-Loustau, M. L., Capretti, P., et al., 2013. Biogeographical patterns and determinants of invasion by forest pathogens in Europe. *New Phytologists*, 197(1), 238-250. https://doi.org/10.1111/j.1469-8137.2012.04364.x

Scala, A., Allmann, S., Mirabella, R., Haring, M. and Schuurink, R., 2013. Green Leaf Volatiles: A Plant's Multifunctional Weapon against Herbivores and Pathogens. *Int. J. Mol. Sci.*, 14, 17781–17811. https://doi.org/10.3390/ijms140917781

Schmitt, I., Crespo, A., Divakar, P. K., Fankhauser, J. D., Herman-Sackett, E., et al., 2009. New primers for promising single-copy genes in fungal phylogenetics and systematics. *Persoonia*, 23, 35–40. https://doi.org/10.3767/003158509X470602

Schoch, C. L., Robbertse, B., Robert, V., Vu, D., Cardinali, G., et al. 2014. Finding needles in haystacks: linking scientific names, reference specimens and molecular data for Fungi. *Database*. https://doi.org/10.1093/database/bau061

Schoch, C. L., Seifert, K. A., Huhndorf, S., Robert, V., Spouge, J. L. et al., 2012. Nuclear ribosomal internal transcribed spacer (ITS) region as a universal DNA barcode marker for Fungi. *Proc. Natl. Acad. Sci. U.S.A.*, 109(16), 6241–6246. https://doi.org/10.1073/pnas.1117018109

Schoch, C.L., Sung, G.H., Lopez-Giraldez, F., Townsend, J. P. and Miadlikowska, M., 2009. The Ascomycota Tree of Life: A Phylum-wide Phylogeny Clarifies the Origin and Evolution of Fundamental Reproductive and Ecological Traits. *Syst. Biol.*, 58(2):224-239. https://doi.org/10.1093/sysbio/syp020

Seif, E. Leight, J., Liu, Y., Roewer, I., Forget, L. and Lang, B. F., 2005. Comparative mitochondrial genomics in zygomycetes: bacteria-like RNase P RNAs, mobile elements and a close source of the group I intron invasion in angiosperms. *Nucl. Acid Res.* 33, 734–744. https://doi.org/10.1093/nar/gki199

Seifert, K. A., Samson, R. A., Dewaard, J. R., Houbraken, J., Lévesque, C. A., Moncalvo, J., et al., 2007. Prospects for fungus identification using CO1 DNA barcodes, with Penicillium as a test case. *Proc. Natl. Acad. Sci. U.S.A.*, 104, 3901–3906. https://doi.org/10.1073/pnas.0611691104

Sharma, R., Zhou, M., Hunter, M. D. and Fan, X., 2019. Rapid In Situ Analysis of Plant Emission for Disease Diagnosis Using a Portable Gas Chromatography Device. *J. Agric. Food Chem.*, 67, 7530–7537. https://doi.org/10.1021/acs.jafc.9b02500

Siddiqua, A., Kabir, M.A., Ferdous, T., Ali, I.B., Weston, L.A., 2022. Evaluating Plant Disease Detection Mobile Applications: Quality and Limitations. *Agronomy*, 12, 1869. https://doi.org/10.3390/agronomy12081869

Singh, A., Ganapathysubramanian, B., Singh, A.K. and Sarkar, S., 2016. Machine Learning for High-Throughput Stress Phenotyping in Plants. *Trends Plant Sci*, 21, 110–124. https://doi.org/10.1016/J.TPLANTS.2015.10.015

Singh, A.K., Ganapathysubramanian, B., Sarkar, S. and Singh, A., 2018. Deep Learning for Plant Stress Phenotyping: Trends and Future Perspectives. *Trends Plant Sci.*, 23, 883–898. https://doi.org/10.1016/J.TPLANTS.2018.07.004

Sladojevic, S., Arsenovic, M. A., Culibrk, D. and Stefanovic, D., 2016. Deep Neural Networks Based Recognition of Plant Diseases by Leaf Image Classification. *Comput. Intell. Neurosci.* 2016, 3289801. https://doi.org/10.1155/2016/3289801

Spadafora, N., Paramithiotis, S., Drosinos, E., Cammarisano, L., Rogers, H., Muller, C.T., 2016. Detection of Listeria monocytogenes in cut melon fruit using analysis of volatile organic compounds. *Food Microbiol.*, 54, 52–59. https://doi.org/10.1016/j.fm.2015.10.017

Spatafora, J.W., Sung, G.H., Johnson, D., Hesse, C., O'Rourke, B. et al., 2006. A five-gene phylogeny of Pezizomycotina. *Mycologia*, 98, 1018–1028. https://doi.org/10.1080/15572536.2006.11832630

Suzuki, N., Rivero, R. M., Shulaev, V., Blumwald, E. and Mittler, R., 2014. Abiotic and biotic stress combinations. *New Phytol.*, 203, 32–43. https://doi.org/10.1111/nph.12797

Tehler, A., Farris, J. S., Lipscomb, D. L. and Kallersjo, M., 2000. Phylogenetic analyses of fungi based on large rDNA data sets. *Mycologia*, 92, 459–474. https://doi.org/10.1080/00275514.2000.12061182

Templeton, A. R., Crandall, K. A. and Sing, C. F., 1992. A cladistic analysis of phenotypic associations with haplotypes inferred from restriction endonuclease mapping and DNA sequence data. III. Cladogram estimation. *Genetics*, 132(2):619-633. https://doi.org/10.1093/genetics/132.2.619

Ward, E., Foster, S., Fraaije B. and McCartney, H.A., 2004. Plant pathogen diagnostics: immunological and nucleic acid-based approaches. *Ann. Appl. Biol.* 145, 1–16. https://doi.org/10.1111/J.1744-7348.2004.TB00354.X

White, T.J., Bruns, T., Lee, S., Taylor, W., 1990. Amplification and direct sequencing of fungal ribosomal RNA genes for phylogenetics. In *PCR Protocols: A Guide to Methods and Applications*, ed. M Innis, D Gelfand, J Sninsky, T White, pp. 315–322. San Diego, CA: Academic Press. https://doi.org/10.1016/B978-0-12-372180-8.50042-1

Zen, B.P. and Fransisca, D.C., 2022. Applications for Detecting Plant Diseases Based on Artificial Intelligence. *SinkrOn*, 7, 2537–2546, https://doi.org/10.33395/SINKRON.V7I4.11833

Zhang, N., Castlebury, L.A., Miller, A.N., Huhndorf, S.M., Schoch, C.L., et al. 2006. An overview of the systematics of the Sordariomycetes based on a four-gene phylogeny. *Mycologia*, 98, 1076–1087. https://doi.org/10.3852/mycologia.98.6.1076

Zoller S., Christoph Scheidegger, C. and Sperisen, C., 1999. PCR primers for the amplification of mitochondrial small subunit ribosomal DNA of lichen-forming ascomycetes. *Lichenologist*, 31(5), 511–516. https://doi.org/10.1006/lich.1999.0220

Index

Pages in *italics* refer to figures and pages in **bold** refer to tables.

C

D

E

F

Q

R

S

T

U